UCLA Symposia on Molecular and Cellular Biology, New Series

Series Editor
C. Fred Fox

UCLA Symposia Published Previously

Publishers

(1) Alan R. Liss, Inc.
     150 Fifth Avenue
     New York, NY 10011

(2) Academic Press, Inc.
     111 Fifth Avenue
     New York, NY 10003

(3) W.A. Benjamin, Inc.
     2725 Sand Hill Road
     Menlo Park, CA 94025

(4) Plenum Publishing Corp.
     227 W. 17th Street
     New York, NY 10011

Symposia Board

C. Fred Fox, Director
Molecular Biology Institute
UCLA

Members

Ronald Cape, Ph.D., MBA
Chairman
Cetus Corporation

Pedro Cuatrecasas, M.D.
Vice President for Research
Burroughs Wellcome Company

Luis Glaser, Ph.D.
Professor and Chairman
of Biochemistry
Washington University School
of Medicine

Donald Steiner, M.D.
Professor of Biochemistry
University of Chicago

Ernest Jaworski, Ph.D.
Director of Molecular Biology
Monsanto

Paul Marks, M.D.
President
Sloan-Kettering Institute

William Rutter, Ph.D.
Professor and Chairman
of Biochemistry
University of California Medical
Center

Sidney Udenfriend, Ph.D.
Director
Roche Institute

The members of the board advise the director in identification of topics for future symposia.

MECHANISMS OF DNA REPLICATION AND RECOMBINATION

MECHANISMS OF DNA REPLICATION AND RECOMBINATION

Proceedings of a UCLA Symposium
held in Keystone, Colorado,
April 3–9, 1983

Editor

NICHOLAS R. COZZARELLI
Department of Molecular Biology
University of California
Berkeley

Alan R. Liss, Inc. • New York

Address all Inquiries to the Publisher
Alan R. Liss, Inc., 150 Fifth Avenue, New York, NY 10011

Library of Congress Cataloging in Publication Data
Main entry under title:
Mechanisms of DNA replication and recombination.

Includes bibliographical references and index.
1. DNA replication–Congresses.   2. Recombinant DNA—
Congresses.   3. Genetic recombination—Congresses.
I. Cozzarelli, Nicholas R.   II. Title: Mechanisms of
D.N.A. replication and recombination.   [DNLM: 1. DNA,
Recombinant—Congresses.   2. DNA replication—Congresses.
W3 U17N new ser. v. 10 / QU 58 U17m 1983]
QH450.M43 1983   574.87'3282   83-18710
ISBN 0-8451-2609-1

Contents

Contributors

P. Abarzua, Department of Developmental Biology and Cancer, Albert Einstein College of Medicine, Bronx, NY [125]

M. Abdel-Monem, Max-Planck-Institut für Medizinische Forschung, Abteilung Molekulare Biologie, Heidelberg, Federal Republic of Germany [65]

K. Abremski, Basic Research Program–LBI, Frederick Cancer Research Facility, Frederick, MD [671]

C.N. Ahlem, Department of Medicine, University of California at San Diego, La Jolla, CA [511]

W.F. Anderson, MRC Group on Protein Structure and Function, University of Alberta, Edmonton, Alberta, Canada [77]

Akira Aoyama, Department of Biology, University of California at San Diego, La Jolla, CA [115]

Josef Arendes, Laboratory of Genetics, National Institute of Environmental Health Sciences, Research Triangle Park, NC; present address Johannes Gutenberg Universität, Physiologisch–Chemisches Institut, Mainz, Federal Republic of Germany [527]

David W. Banner, European Molecular Biology Laboratory, Heidelberg, Federal Republic of Germany [327]

Benjamin B. Beauchamp, Department of Biological Chemistry, Harvard Medical School, Boston, MA [135]

Howard W. Benjamin, Department of Molecular Biology, University of California, Berkeley, CA [637]

K.I. Berns, Department of Immunology and Medical Microbiology, University of Florida, College of Medicine, Gainesville, FL [353]

LeRoy L. Bertsch, Department of Biochemistry, Stanford University School of Medicine, Stanford, CA [93, 275]

Ashok Bhagwat, Cold Spring Harbor Laboratory, Cold Spring Harbor, NY [797]

Marco E. Bianchi, Departments of Human Genetics and Molecular Biophysics and Biochemistry, Yale University School of Medicine, New Haven CT [697]

S.B. Biswas, Department of Biochemistry, Stanford University School of Medicine, Stanford, CA [93]

Luis Blanco, Centro de Biología Molecular (CSIC-UAM), Universidad Autónoma, Canto Blanco, Madrid, Spain [203]

V. Bonnewell, Biology Division, Oak Ridge National Laboratory, and University of Tennessee–Oak Ridge Graduate School of Biomedical Sciences, Oak Ridge, TN [849]

The boldface number in brackets indicates the opening page of that contributor's article.

xv

J.R. Broach, Department of Microbiology, State University of New York, Stony Brook, NY **[685]**

M. Brougham, Department of Immunology and Medical Microbiology, University of Florida College of Medicine, Gainesville, FL **[753]**

Steven S. Broyles, Department of Biochemistry, Biophysics, and Genetics, University of Colorado Health Sciences Center, Denver, CO **[19]**

Douglas L. Brutlag, Department of Biochemistry, Stanford University School of Medicine, Stanford, CA **[55]**

P.M.J. Burgers, Department of Biochemistry, Stanford University School of Medicine, Stanford, CA; *present address* Department of Biological Chemistry, Washington University School of Medicine, St. Louis, MO **[93]**

W. Burhans, Department of Medicine, University of California at San Diego, La Jolla, CA **[367]**

John Capobianco, Abbott Laboratories, North Chicago, IL **[661]**

C. Carton, Department of Medicine, University of California at San Diego, La Jolla, CA **[367]**

Gianni Cesareni, European Molecular Biology Laboratory, Heidelberg, Federal Republic of Germany **[327]**

L.E. Chalifour, Department of Biological Chemistry, Harvard Medical School, Boston, MA **[423]**

M.F. Charette, Department of Biological Chemistry, Harvard Medical School, Boston, MA **[423]**

Grace L. Chen, Department of Physiological Chemistry, Johns Hopkins Medical School, Baltimore, MD **[43]**

David A. Clayton, Department of Pathology, Stanford University School of Medicine, Stanford, CA **[581]**

Ronald C. Conaway, Department of Biochemistry, Stanford University School of Medicine, Stanford, CA **[495]**

Michael M. Cox, Department of Biochemistry, Stanford University School of Medicine, Stanford, CA; *present address* Department of Biochemistry, University of Wisconsin, Madison WI **[709]**

Nicholas R. Cozzarelli, Department of Molecular Biology, University of California, Berkeley, CA **[xxv, 637]**

Nancy L. Craig, Laboratory of Neurochemistry, National Institute of Mental Health, Bethesda, MD **[617]**

Donald J. Cummings, University of Colorado Health Sciences Center, Denver, CO **[595]**

M.E. Cusick, Department of Biological Chemistry, Harvard Medical School, Boston, MA; *present address* Biology Department, California Institute of Technology, Pasadena, CA **[423]**

Ginger M. Dani, Division of Genetics, Hutchinson Cancer Research Center, Seattle, WA **[553]**

M.L. DePamphilis, Department of Biological Chemistry, Harvard Medical School, Boston, MA **[423]**

E. Di Capua, Institute for Cell Biology, Swiss Federal Institute of Technology, ETH-Hönggerberg, Zürich, Switzerland **[723]**

Stephen DiNardo, Department of Biochemistry, State University of New York, Stony Brook, NY **[29]**

Nicholas E. Dixon, Department of Biochemistry, Stanford University School of Medicine, Stanford, CA **[93, 275]**

Mary K. Dolejsi, Institute of Molecular Biology and Department of Chemistry, University of Oregon, Eugene, OR **[153]**

Jan M. Dungan, Department of Molecular Biology, University of California, Berkeley, CA **[637]**

Randolph C. Elble, Program in Molecular, Cellular, and Developmental Biology, Department of Biology, Indiana University, Bloomington, IN [303]

Michael J. Engler, Department of Biological Chemistry, Harvard Medical School, Boston, MA; *present address* Department of Biochemistry and Molecular Biology, University of Texas Medical School at Houston, Houston, TX [135]

Cristina Escarmís, Centro de Biología Molecular (CSIC-UAM), Universidad Autónoma, Canto Blanco, Madrid, Spain [203]

Frederic R. Fairfield, Institute of Molecular Biology and Department of Chemistry, University of Oregon, Eugene, OR [153]

S.C. Falco, Department of Biophysics and Theoretical Biology, The University of Chicago, Chicago, IL [245]

Michael A. Fennewald, Department of Microbiology, University of Notre Dame, Notre Dame, IN [661]

Jeffrey Field, Department of Developmental Biology and Cancer, Albert Einstein College of Medicine, Bronx, NY [395]

James E. Flynn, Jr., Department of Biochemistry, Stanford University School of Medicine, Stanford, CA [93, 275]

R.F. Fowler, Biology Division, Oak Ridge National Laboratory, and University of Tennessee–Oak Ridge Graduate School of Biomedical Sciences, Oak Ridge, TN [849]

Carl W. Fuller, Department of Biological Chemistry, Harvard Medical School, Boston, MA [135]

Robert S. Fuller, Department of Biochemistry, Stanford University School of Medicine, Stanford, CA [93, 275]

Juan A. García, Centro de Biología Molecular (CSIC-UAM), Universidad Autónoma, Canto Blanco, Madrid, Spain [203]

Costa Georgopoulos, Department of Cellular, Viral, and Molecular Biology, University of Utah Medical Center, Salt Lake City, UT [317]

M. Goulian, Department of Medicine, University of California at San Diego, La Jolla, CA [367]

J. Greenbaum, Department of Developmental Biology and Cancer, Albert Einstein College of Medicine, Bronx, NY [125]

Jack Griffith, Department of Microbiology and Immunology, Cancer Research Center, University of North Carolina, Chapel Hill, NC [731]

Richard M. Gronostajski, Department of Developmental Biology and Cancer, Albert Einstein College of Medicine, Bronx, NY [395]

Ronald A. Guggenheimer, Department of Developmental Biology and Cancer, Albert Einstein College of Medicine, Bronx, NY [395]

D.R. Guinta, Departments of Biophysics and Theoretical Biology, The University of Chicago, Chicago, IL [245]

Brian D. Halligan, Department of Physiological Chemistry, Johns Hopkins Medical School, Baltimore, MD [43]

Robert K. Hamatake, Department of Chemistry, University of California at San Diego, La Jolla, CA [115]

Joyce L. Hamlin, Department of Biochemistry, University of Virginia, Charlottesville, VA [605]

Richard Harland, Laboratory of Molecular Biology, MRC Centre, Cambridge, England [563]

R.T. Hay, Department of Biological Chemistry, Harvard Medical School, Boston, MA; *present address* MRC Virology Unit, Glasgow, Scotland **[423]**

Masaki Hayashi, Department of Biology, University of California at San Diego, La Jolla, CA **[115]**

Fred Heffron, Cold Spring Harbor Laboratory, Cold Spring Harbor, NY; *present address* Scripps Clinic and Research Foundation, La Jolla, CA **[785, 797]**

Nicholas Heintz, Department of Biochemistry, University of Virginia, Charlottesville, VA **[605]**

E.A. Hendrickson, Department of Biological Chemistry, Harvard Medical School, Boston, MA **[423]**

José M. Hermoso, Centro de Biología Molecular (CSIC-UAM), Universidad Autónoma, Canto Blanco, Madrid, Spain **[203]**

James B. Hicks, Cold Spring Harbor Laboratory, Cold Spring Harbor, NY **[785]**

N. Patrick Higgins, Department of Biochemistry, University of Wyoming, Laramie, WY; *present address* Department of Biology, University of Alabama, Birmingham, AL **[187]**

Yukinori Hirota, National Institute of Genetics, Mishima, Japan **[257]**

Joel W. Hockensmith, Institute of Molecular Biology and Department of Chemistry, University of Oregon, Eugene, OR **[153]**

R. Hoess, Basic Research Program—LBI, Frederick Cancer Research Facility, Frederick, MD **[671]**

H. Hoffmann-Berling, Max-Planck-Institut für Medizinische Forschung, Abteilung Molekulare Biologie, Heidelberg, Federal Republic of Germany **[65]**

W. Holloman, Department of Immunology and Medical Microbiology, University of Florida College of Medicine, Gainesville, FL **[753]**

Marshall S. Horwitz, Department of Microbiology and Immunology and Cell Biology, Albert Einstein College of Medicine, Bronx, NY **[395]**

Paul Howard-Flanders, Departments of Therapeutic Radiology, and Molecular Biophysics and Biochemistry, Yale University, New Haven, CT **[739]**

Martha M. Howe, Department of Bacteriology, University of Wisconsin, Madison, WI **[187]**

Ulrich Hübscher, Institute for Pharmacology and Biochemistry, School of Veterinary Medicine, University of Zürich, Zürich, Switzerland **[517]**

Jerard Hurwitz, Department of Developmental Biology and Cancer, Albert Einstein College of Medicine, Bronx, NY **[395]**

Junetsu Ito, Department of Molecular and Medical Microbiology, University of Arizona, College of Medicine, Tucson, AZ **[225]**

Anthony A. James, Dana-Farber Cancer Institute and Department of Biological Chemistry, Harvard Medical School, Boston, MA **[761]**

M. Jayaram, Department of Microbiology, State University of New York, Stony Brook, NY **[685]**

Elaine V. Jones, Laboratory of Biology of Viruses, National Institute of Allergy and Infectious Diseases, Bethesda, MD **[449]**

Jon M. Kaguni, Department of Biochemistry, Stanford University School of Medicine, Stanford, CA **[93, 275]**

Laurie S. Kaguni, Department of Biochemistry, Stanford University School of Medicine, Stanford, CA **[495]**

Kwang C. Kim, Laboratory of Genetics, National Institute of Environmental Health Sciences, Research Triangle Park, NC [527]

Amar J.S. Klar, Cold Spring Harbor Laboratory, Cold Spring Harbor, NY [785]

E. Kmiec, Department of Immunology and Medical Microbiology, University of Florida College of Medicine, Gainesville, FL [753]

Ichizo Kobayashi, Institute of Molecular Biology, University of Oregon, Eugene, OR [773]

M. Kodaira, Department of Biochemistry, Stanford University School of Medicine, Stanford, CA; present address Department of Genetics, Osaka University Medical School, Osaka, Japan [93]

Tokio Kogoma, Department of Biology and Department of Cell Biology, Cancer Research and Treatment Center, University of New Mexico, Albuquerque, NM [337]

R. Kollek, Department of Medicine, University of California at San Diego, La Jolla, CA; present address Heinrich-Pette Institut, Universität Hamburg, Hamburg, Federal Republic of Germany [367]

Th. Koller, Institute for Cell Biology, Swiss Federal Institute of Technology, ETH-Hönggerberg, Zürich, Switzerland [723]

Richard Kolodner, Dana-Farber Cancer Institute and Department of Biological Chemistry, Harvard Medical School, Boston, MA [761]

Arthur Kornberg, Department of Biochemistry, Stanford University School of Medicine, Stanford, CA [3, 93, 275]

Richard Kostriken, Cold Spring Harbor Laboratory, Cold Spring Harbor, NY [785]

Mark A. Krasnow, Department of Molecular Biology, University of California, Berkeley, CA, and Department of Biochemistry, University of Chicago, Chicago, IL [637]

P. Kroeger, Department of Immunology and Medical Microbiology, University of Florida College of Medicine, Gainesville, FL [753]

Rosa M. Lacatena, European Molecular Biology Laboratory, Heidelberg, Federal Republic of Germany [327]

Ronald Laskey, Laboratory of Molecular Biology, MRC Centre, Cambridge, England [563]

José M. Lázaro, Centro de Biología Molecular (CSIC-UAM), Universidad Autónoma, Canto Blanco, Madrid, Spain [203]

Jonathan H. LeBowitz, Department of Biochemistry, The Johns Hopkins University, Baltimore, MD [819]

Robert L. Lechner, Department of Biological Chemistry, Harvard Medical School, Boston, MA [135]

Chao-Hung Lee, Cold Spring Harbor Laboratory, Cold Spring Harbor, NY [797]

I.R. Lehman, Department of Biochemistry, Stanford University School of Medicine, Stanford, CA [495, 709]

Margaret Levin, Vanderbilt University, Department of Molecular Biology, Nashville, TN [173]

Y.-Y. Li, Department of Microbiology, State University of New York, Stony Brook, NY [685]

Gene Lin, Vanderbilt University, Department of Molecular Biology, Nashville, TN [173]

Jeff Lindenbaum, Department of Developmental Biology and Cancer, Albert Einstein College of Medicine, Bronx, NY [395]

Leroy F. Liu, Department of Physiological Chemistry, Johns Hopkins Medical School, Baltimore, MD **[43]**

Zvi Livneh, Department of Biochemistry, Stanford University School of Medicine, Stanford, CA **[709]**

Christine Loehrlein, Department of Biochemistry, The Johns Hopkins University, Baltimore, MD **[819]**

Heinz Lother, Max-Planck-Institut für Molekulare Genetik, Berlin (Dahlem), Federal Republic of Germany **[289]**

Robert L. Low, Department of Biochemistry, Stanford, CA; *present address* Department of Pathology, Washington University School of Medicine, St. Louis, MO **[275]**

Rudi Lurz, Max-Planck-Institut für Molekulare Genetik, Berlin (Dahlem), Federal Republic of Germany **[289]**

Paul Macdonald, Vanderbilt University, Department of Molecular Biology, Nashville, TN **[173]**

Joanne Bednarz Mallory, Department of Biochemistry, The Johns Hopkins University, Baltimore, MD **[819]**

Purita Manlapaz-Ramos, Department of Biology, University of Utah, Salt Lake City, UT **[187]**

K.J. Marians, Department of Developmental Biology and Cancer, Albert Einstein College of Medicine, Bronx, NY **[125]**

Steven W. Matson, Department of Biological Chemistry, Harvard Medical School, Boston MA; *present address* Department of Biology, University of North Carolina at Chapel Hill, Chapel Hill, NC **[135]**

B.W. Matthews, Institute of Molecular Biology and Department of Physics, University of Oregon, Eugene, OR **[77]**

Martin M. Matzuk, Department of Molecular Biology, University of California, Berkeley, CA; *present address* Washington University School of Medicine, Washington University, St. Louis, MO **[637]**

Mary McCormick, Cold Spring Harbor Laboratory, Cold Spring Harbor, NY; *present address* National Institutes of Health, Bethesda MD **[797]**

M. McLeod, Department of Microbiology, State University of New York, Stony Brook, NY **[685]**

Roger McMacken, Department of Biochemistry, The Johns Hopkins University, Baltimore, MD **[819]**

Marcel Méchali, Laboratory of Molecular Biology, MRC Centre, Cambridge, England **[563]**

Rafael P. Mellado, Centro de Biología Molecular (CSIC-UAM), Universidad Autónoma, Canto Blanco, Madrid, Spain **[203]**

Walter Messer, Max-Planck-Institut für Molekulare Genetik, Berlin (Dahlem), Federal Republic of Germany **[289]**

Jeffrey Milbrandt, Department of Biochemistry, University of Virginia, Charlottesville, VA **[605]**

Dino Moncecchi, Department of Biochemistry, University of Wyoming, Laramie, WY **[187]**

Carolyn Moomaw, Cold Spring Harbor Laboratory, Cold Spring Harbor, NY **[785]**

Gisela Mosig, Vanderbilt University, Department of Molecular Biology, Nashville, TN **[173]**

Bernard Moss, Laboratory of Biology of Viruses, National Institute of Allergy and Infectious Diseases, Bethesda, MD **[449]**

Mark A. Muesing, Program in Molecular, Cellular, and Developmental Biology, Department of Biology, Indiana University, Bloomington, IN **[303]**

N. Muzyczka, Department of Immunology and Medical Microbiology, University of Florida, College of Medicine, Gainesville, FL [353]

Kyosuke Nagata, Department of Developmental Biology and Cancer, Albert Einstein College of Medicine, Bronx, NY [395]

Howard A. Nash, Laboratory of Neurochemistry, National Institute of Mental Health, Bethesda, MD [617]

Eric M. Nelson, Department of Physiological Chemistry, Johns Hopkins Medical School, Baltimore, MD [43]

John W. Newport, Institute of Molecular Biology and Department of Chemistry, University of Oregon, Eugene, OR; *present address* Department of Biochemistry and Biophysics, University of California Medical Center, San Francisco, CA [153]

Tohru Ogawa, Department of Biochemistry, Stanford University School of Medicine, Stanford, CA [275]

D.H. Ohlendorf, Institute of Moelcular Biology and Department of Physics, University of Oregon, Eugene, OR [77]

Eiichi Ohtsubo, Cold Spring Harbor Laboratory, Cold Spring Harbor, NY; *present address* University of Tokyo, Tokyo, Japan [797]

Atsuhiro Oka, Institute for Chemical Research, Kyoto University, Uji, Kyoto, Japan [257]

Baldomero M. Olivera, Department of Biology, University of Utah, Salt Lake City, UT [187]

Elisha Orr, Department of Genetics, University of Leicester, Leicester, England [289]

Neil Osheroff, Department of Biochemistry, Stanford University School of Medicine, Stanford, CA; *present address* Department of Biochemistry, Vanderbilt University School of Medicine, Nashville, TN [55]

Hans-Peter Ottiger, Institute for Pharmacology and Biochemistry, School of Veterinary Medicine, University of Zürich, Zürich, Switzerland [517]

Engin Özkaynak, Department of Biology, Massachusetts Institute of Technology, Cambridge, MA [463]

Richard Pan, Department of Biology, Massachusetts Institute of Technology, Cambridge, MA [463]

Ricardo Pastrana, Centro de Biología Molecular (CSIC-UAM), Universidad Autónoma, Canto Blanco, Madrid, Spain [203]

Leland S. Paul, Institute of Molecular Biology and Department of Chemistry, University of Oregon, Eugene, OR [153]

Miguel A. Peñalva, Centro de Biología Molecular (CSIC-UAM), Universidad Autónoma, Canto Blanco, Madrid, Spain [203]

David E. Pettijohn, Department of Biochemistry, Biophysics, and Genetics, University of Colorado Health Sciences Center, Denver, CO [19]

Ann F. Pluta, Division of Genetics, Hutchinson Cancer Research Center, Seattle, WA [553]

Barry Polisky, Program in Molecular, Cellular, and Developmental Biology, Department of Biology, Indiana University, Bloomington, IN [303]

Ignacio Prieto, Centro de Biología Molecular (CSIC-UAM), Universidad Autónoma, Canto Blanco, Madrid, Spain [203]

Arthur E. Pritchard, University of Colorado Health Sciences Center, Denver, CO [595]

C.G. Pritchard, Department of Biological Chemistry, Harvard Medical School, Boston, MA; *present address* Syva Company, Palo Alto, CA [423]

Charles M. Radding, Departments of Human Genetics and Molecular Biophysics and Biochemistry, Yale University School of Medicine, New Haven, CT **[697]**

James C. Register III, Department of Microbiology and Immunology, Cancer Research Center, University of North Carolina, Chapel Hill, NC **[731]**

D. Revie, Department of Medicine, University of California at San Diego, La Jolla, CA **[367]**

Charles C. Richardson, Department of Biological Chemistry, Harvard Medical School, Boston, MA **[135]**

J.K. Rist, Department of Biochemistry, The University of Chicago, Chicago, IL **[245]**

John D. Roberts, Department of Biochemistry, The Johns Hopkins University, Baltimore, MD **[819]**

Jean-Michel Rossignol, Department of Biochemistry, Stanford University School of Medicine, Stanford, CA **[495]**

L.B. Rothman-Denes, Department of Biophysics and Theoretical Biology, The University of Chicago, Chicago, IL **[245]**

Thomas C. Rowe, Department of Physiological Chemistry, Johns Hopkins Medical School, Baltimore, MD **[43]**

Akira Sakai, Laboratory of Genetics, National Institutes of Health, Research Triangle Park, NC; *present address* Department of Molecular and Population Genetics, University of Georgia, Athens, GA **[527]**

Margarita Salas, Centro de Biología Molecular (CSIC-UAM), Universidad Autónoma, Canto Blanco, Madrid, Spain **[203]**

R.J. Samulski, Department of Immunology and Medical Microbiology, University of Florida, College of Medicine, Gainesville, FL; *present address* Department of Microbiology, State University of New York, Stony Brook, NY **[353]**

Hitoshi Sasaki, Institute for Chemical Research, Kyoto University, Uji, Kyoto, Japan **[257]**

Jane C. Schneider, Program in Molecular, Cellular, and Developmental Biology, Department of Biology, Indiana University, Bloomington, IN **[303]**

Ruth Seaby, Vanderbilt University, Department of Molecular Biology, Nashville, TN **[173]**

Meng-Fu Shih, Department of Molecular and Medical Microbiology, University of Arizona, College of Medicine, Tucson, AZ **[225]**

Richard R. Sinden, Department of Biochemistry, Biophysics, and Genetics, University of Colorado Health Sciences Center, Denver, CO **[19]**

D.M. Skinner, Biology Division, Oak Ridge National Laboratory, and University of Tennessee–Oak Ridge Graduate School of Biomedical Sciences, Oak Ridge, TN **[849]**

Robert Snapka, Department of Biology, Massachusetts Institute of Technology, Cambridge, MA **[463]**

W. Soeller, Department of Developmental Biology and Cancer, Albert Einstein College of Medicine, Bronx, NY **[125]**

Mark Solomon, Department of Biology, Massachusetts Institute of Technology, Cambridge, MA **[463]**

Daniel A. Soltis, Department of Biochemistry, Stanford University School of Medicine, Stanford, CA **[709]**

Joan M. Sperrazza, Department of Microbiology and Immunology, Cancer Research Center, University of North Carolina, Chapel Hill, NC **[731]**

A. Srivastava, Department of Immunology and Medical Microbiology, University of Florida, College of Medicine, Gainesville, FL; *present address* Department of Medicine, University of Arkansas Medical School, Little Rock, AR **[353]**

Franklin W. Stahl, Institute of Molecular Biology, University of Oregon, Eugene, OR [773]

Mary M. Stahl, Institute of Molecular Biology, University of Oregon, Eugene, OR [773]

J. Stambouly, Department of Biophysics and Theoretical Biology, The University of Chicago, Chicago, IL [245]

A. Stasiak, Institute for Cell Biology, Swiss Federal Institute of Technology, ETH-Hönggerberg, Zürich, Switzerland [723]

M.M. Stayton, Department of Biochemistry, Stanford University School of Medicine, Stanford, CA; present address Advanced Genetic Sciences, Inc., Manhattan, KS [93]

N. Sternberg, Basic Research Program–LBI, Frederick Cancer Research Facility, Frederick, MD [671]

Rolf Sternglanz, Department of Biochemistry, State University of New York, Stony Brook, NY [29]

Bruce W. Stillman, Cold Spring Harbor Laboratory, Cold Spring Harbor, NY [381]

Jeffrey Strathern, Cold Spring Harbor Laboratory, Cold Spring Harbor, NY [785]

Nelda L. Subia, Department of Biology, University of New Mexico, Albuquerque, NM [337]

Kazunori Sugimoto, Institute for Chemical Research, Kyoto University, Uji, Kyoto, Japan [257]

Akio Sugino, Laboratory of Genetics, National Institute of Environmental Health Sciences, National Institutes of Health, Research Triangle Park, NC; present address Department of Molecular and Population Genetics, University of Georgia, Athens, GA [245,527]

Olof Sundin, Department of Biology, Massachusetts Institute of Technology, Cambridge, MA; present address Cold Spring Harbor Laboratory, Cold Spring Harbor, NY [463]

Satoshi Tabata, Institute for Chemical Research, Kyoto University, Uji, Kyoto, Japan [257]

Stanley Tabor, Department of Biological Chemistry, Harvard Medical School, Boston, MA [135]

L.C. Tack, Department of Biological Chemistry, Harvard Medical School, Boston, MA; present address Department of Biochemistry, Scripps Clinic and Research Foundation, La Jolla, CA [423]

Mituru Takanami, Institute for Chemical Research, Kyoto University, Uji, Kyoto, Japan [257]

Y. Takeda, Chemistry Department, University of Maryland, Baltimore County, Catonsville, MD [77]

Joe Tamm, Program in Molecular, Cellular, and Developmental Biology, Department of Biology, Indiana University, Bloomington, IN [303]

Douglas P. Tapper, Department of Pathology, Stanford University School of Medicine, Stanford, CA [581]

G. Taucher-Scholz, Max-Planck-Institut für Medizinische Forschung, Abteilung Molekulare Biologie, Heidelberg, Federal Republic of Germany [65]

Kathleen M. Tewey, Department of Physiological Chemistry, Johns Hopkins Medical School, Baltimore, MD [43]

Catherine Thrash, Department of Biochemistry, State University of New York, Stony Brook, NY [29]

Ben. Y. Tseng, Department of Medicine, University of California at San Diego, La Jolla, CA [367,511]

Alexander Varshavsky, Department of Biology, Massachusetts Institute of Technology, Cambridge, MA **[463]**

Karen A. Voelkel, Department of Biochemistry, State University of New York, Stony Brook, NY **[29]**

Peter H. von Hippel, Institute of Molecular Biology and Department of Chemistry, University of Oregon, Eugene, OR **[153]**

P.M. Wassarman, Department of Biological Chemistry, Harvard Medical School, Boston, MA **[423]**

Kounosuke Watabe, Department of Molecular and Medical Microbiology, University of Arizona, College of Medicine, Tucson, AZ **[225]**

D.T. Weaver, Department of Biological Chemistry, Harvard Medical School, Boston, MA **[423]**

Stephen C. West, Departments of Therapeutic Radiology, and Molecular Biophysics and Biochemistry, Yale University, New Haven, CT **[739]**

John H. White, Department of Biological Chemistry, Harvard Medical School, Boston, MA **[135]**

Jo Anne K. Wilkinson, Department of Biochemistry, The Johns Hopkins University, Baltimore, MD **[819]**

Francis Wilson-Coleman, Laboratory of Genetics, National Institute of Environmental Health Sciences, Research Triangle Park, NC **[527]**

Elaine Winters, Laboratory of Biology of Viruses, National Institute of Allergy and Infectious Diseases, Bethesda, MD **[449]**

D.O. Wirak, Department of Biological Chemistry, Harvard Medical School, Boston, MA **[423]**

Marc S. Wold, Department of Biochemistry, The Johns Hopkins University, Baltimore, MD **[819]**

M. Yarnall, Department of Immunology and Medical Microbiology, University of Florida College of Medicine, Gainesville, FL **[753]**

Seiichi Yasuda, National Institute of Genetics, Mishima, Japan **[257]**

Virginia A. Zakian, Division of Genetics, Hutchinson Cancer Research Center, Seattle, WA **[553]**

Maceij Zylicz, Department of Cellular, Viral and Molecular Biology, University of Utah Medical Center, Salt Lake City, UT **[317]**

Preface

This volume contains contributions from a conference entitled "Mechanisms of DNA Replication and Recombination" held in Keystone, Colorado in April 1983. The recombination portion was an option for those attending the succeeding conference, "Cellular Responses to DNA Damage." The unity in DNA replication, recombination, and repair is evident in the common enzymes and mechanisms. A half dozen genes important for DNA replication were first identified on the basis of the repair deficiency of mutants. DNA damage stimulates both recombination and repair. Many types of recombination require DNA synthesis and/or repair and, conversely, repair needs DNA synthesis and sometimes recombinational proteins. Origins of phage DNA replication may be generated by recombination. Despite the synergism in considering the 3 R's of DNA metabolism together, each field has matured to the point where it demands its own conference.

The conference showed that it is a fruitful time to study the molecular mechanisms of replication and recombination. The variety in DNA secondary, tertiary, and quaternary structure is now manifest and can be analyzed with precision. The complete replication of several coliphages, site-specific recombination, and critical parts of general recombination can be carried out at physiological rate and fidelity with homogeneous proteins. Thus the individual components and their interaction can be analyzed. Cloning has facilitated production of large amounts of pure protein and model DNA substrates making both amenable to physical chemical analysis. The 3-dimensional structure of several proteins of DNA metabolism is already known and co-crystals with DNA synthesized chemically have been prepared. The mechanisms proposed and their tests have become highly specific and sophisticated.

Perhaps the major difference between this meeting and the one held at Keystone on the same topic just three years ago is the greatly increased importance of studies of DNA replication in eukaryotes. Two plenary sessions were devoted exclusively to the topic and additional contributions were in other sessions. The nucleotide sequence of several replicons from higher organisms is now known and origin sequences are being studied. Many of the requisite enzymes have been purified and characterized; the properties of the primase and α-DNA polymerase are, in particular, the subject of intensive investigation. For the first time, the three powerful approaches of reconstitution, genetics, and specific inhibition can all be applied to eukaryotic DNA replication. Replication can be carried out faithfully *in vitro* for several systems and, in the case of adenovirus, almost entirely with purified components. In terms of conservation of mechanism between prokaryotes and eukaryotes, DNA replication seems to fall between the extremes represented by transcription and translation. The prokaryotic paradigms are very useful but new concepts are emerging from the work with higher organisms.

We gratefully acknowledge the financial support provided by the following companies: Abbott Diagnostics Division, American Cyanimid Company, Merck Sharp & Dohme Research Laboratories, New England Biolabs, Inc., and Shering Corporation.

Nicholas R. Cozzarelli

I. KEYNOTE ADDRESS

Mechanisms of DNA Replication and Recombination, pages 3–15
© 1983 Alan R. Liss, Inc., 150 Fifth Avenue, New York, NY 10011

ASPECTS OF DNA REPLICATION: 1983[1]

Arthur Kornberg

Department of Biochemistry, Stanford University
School of Medicine, Stanford, California 94025

Replication, recombination and repair are subjects of central importance that still remain essentially free of discrimination based on cellular size or origin. Attention to the basic mechanisms and control of DNA replication, recombination and repair in prokaryotic systems persists despite crusading forces for cells with a nucleus. At this Symposium, many of us will speak openly and fondly about the little beasties who teach us so much with so little.

This year, Nature magazine is convening two symposia to celebrate the 30th anniversary of the publication of a notable paper (1) in one of its 1953 issues. In this justly famous Watson-Crick paper, DNA is described as a helix of two strands linked by hydrogen-bonded base pairing of adenine to thymine and of guanine to cytosine. A component of this paper, at least as important as the description of its physical form, was the attractive and correct proposal that the base-pairing complementarity of the structure provided a facile means for DNA replication.

With the spotlight already so strongly focused on DNA, convening a symposium in England this spring and another in the United States this fall seems excessive. More significant, the programs of these two symposia show a relative lack of interest in the enzymology of DNA. Perhaps not unrelated to this 1983 attitude is the only flaw I find in the 1953 paper. It is the biochemically naive proposal that nucleotides, base-paired along each of the parental template strands, would zipper up spontaneously. I confess that for

[1]This work was supported by grants from the National Institutes of Health and the National Science Foundation.

some years I thought that DNA polymerase I, a single polypeptide, might by itself manage DNA replication, but for any
cellular process of this refinement to be managed without
enzymes was, even in 1953, unprecedented and unreasonable.

In this introductory talk, I will dwell on DNA replication. Recombination and repair depend considerably on
replication and my connections with recombination and repair
are not as intimate or sensitive. I attempted an overview
of DNA replication five years ago at Cold Spring Harbor (2).
Whatever merit that paper had was <u>not</u> in its prophecies.
Despite allowances for attempts at humor, my crystal ball
proved clouded enough to discourage me from doing any more
crystal ball gazing, for a while at least.

For amusement, let us look at those 1978 predictions
for the coming decade and see how far we have come in these
first five years.

1. "The next decade will be dominated by sequencing
and enzymology." Sequencing, yes; enzymology, no.

2. "We will know as much or more about mechanisms and
control of replication and recombination than about transcription and translation." Possibly true, but that is not saying
much either.

3. "Among the stages in replication, the greatest
clarifications will be about the origin and termination of
chromosome cycles and how they are controlled." We have an
awful lot to do in the remaining five years.

4. "The spatial and temporal orientation of the
replisome and integration of its function in cell growth and
division will be explained." Again it seems we'll need more
than one 5-year plan.

5. "Eukaryotic replication systems will prove to be
interesting variations on the themes of prokaryotic systems."
Right on. No splicing bombshells here. The discoveries of
primases, multisubunit holoenzymes and replication origins
have made this a reasonable prediction.

6. "Studies of phage and plasmid replication will
lead to a cure for cancer." I'll stand by that.

7. "Work with permeable cells, complex lysates, and
folded chromosomes, which may be called '<u>in</u> <u>vivtro</u>,' can
combine the worst or best features of <u>in</u> <u>vivo</u> and <u>in</u> <u>vitro</u>
studies. I predict that '<u>in</u> <u>vivtro</u>' studies will be progressively more informative as they are applied with a better
understanding of cellular behavior and enzyme mechanisms."
This senatorial statement alienates no votes and may even win
a few.

8.   "Finally, if there is a Cold Spring Harbor Symposium on replication in 1988, it will be the last such symposium on this subject." Perhaps there will still be another after all.

In this vein, I will make one more comment about my ability to predict progress in DNA replication. Two years ago, when the 1980 edition of DNA Replication was already about two years old and showing signs of age, I thought I could update it with an appendix of about 30 pages. When the 1982 Supplement appeared last summer it was nearly ten times that size.

In this paper, I will touch on several aspects of duplex DNA replication and then comment on strategies for future work.

1.   Elongation of a DNA chain: E. coli DNA polymerase III holoenzyme.
2.   Initiation of a DNA chain: RNA priming by RNA polymerases, primases and the primosome.
3.   Organization of chain elongation and initiations at the replication fork: the "replisome."
4.   Initiation of a cycle of replication of the E. coli chromosome at its unique origin (oriC).
5.   Termination of replication and segregation of the daughter chromosomes.

I will dwell on predominant patterns, knowing that there are exceptions or multiple variations, even within a single cell. It is disadvantageous or lethal for a cell to lack metabolic alternatives and metabolism of DNA is no exception. As we learn more about DNA metabolism of a single cell, we discover auxiliary polymerases, alternative primases, cryptic origins of replication and many possibilities for suppressing otherwise lethal mutations.

Most of this paper will be about E. coli replication with only parenthetical references to viral and eukaryotic systems. E. coli DNA replication, including that of its viruses and plasmids, and its involvement in many varieties of repair, recombination and transposition, continues to be an experimentally attractive ground for maneuver and to be prototypical for most basic mechanisms.

Elongation of a DNA chain: E. coli DNA polymerase III holoenzyme.

Replication of duplexes is generally semidiscontinuous. One strand (leading) is synthesized continuously; the other strand (lagging) is synthesized in small pieces, that is, discontinuously. The principal synthetic enzyme for both strands in E. coli is the DNA pol III holoenzyme. Some features of the organization and functions of holoenzyme are presented elsewhere in this volume (3). Holoenzyme has a core of 3 subunits with a polypeptide of 140 kd (α subunit) responsible for polymerization and proofreading. Four or more additional subunits endow it with high catalytic efficiency, fidelity and processivity. Animal cell polymerases, as in the case of Drosophila, are also multisubunit holoenzymes with a 180-kd polypeptide doing the polymerization (consult L. Kaguni, et al., elsewhere in this volume). Primase resides in the smaller Drosophila polymerase subunits.

Which of the dozen polypeptide bands commonly seen in highly purified preparations of E. coli DNA polymerase III holoenzyme are genuine and which are adventitious? To make this judgment we rely on several criteria.

First is the persistence of a polypeptide band throughout the purification procedure in an abundance roughly equivalent to one of the bona fide subunits (α, β, ε). Such persistence, in successive procedures approaching a homogeneous preparation, argues against the polypeptide being adventitious.

A second criterion is the functional contribution a polypeptide makes to an in vitro assay. Unfortunately, our assays do not generally impose the stringency of intracellular environments, nor do they elicit the special capacities of an enzyme that expresses its versatility. For example, more than ten years elapsed after the isolation of homogeneous DNA polymerase I before we recognized that a persistent nuclease activity is inherent in the enzyme and is the sum of two discrete exonucleases. One is a 3'→5' exonuclease serving a proofreading function at the polymerizing end of the DNA chain. The other, in a separate domain of the polypeptide, is a 5'→3' exonuclease that excises ultraviolet lesions (e.g., thymine dimer) and performs a vital function in removing RNA that primes DNA synthesis.

Another example is the τ subunit, a product of the dnaZ-X gene locus. McHenry (4) showed that the τ subunit improves the catalytic activity of pol III core on single-stranded templates; it does so by increasing processivity and endowing the core with some holoenzyme features. Yet, separation of the τ subunit from holoenzyme by high pressure liquid chromatography caused no loss in activity in several assay systems that measure holoenzyme function (unpublished

observations by S. Biswas and A. Kornberg). The γ subunit
is very likely processed from the τ subunit (unpublished observations by M. Kodaira, S. Biswas and A. Kornberg) and may
substitute for τ in these particular assays.

A quasi-functional criterion is the ultraviolet-acti-
vated covalent linkage of a nucleoside triphosphate to
holoenzyme subunits (unpublished observations by S. Biswas
and A. Kornberg). ATP (or dATP) is linked to the α, τ, γ
and δ subunits; only dTTP links to the ε subunit. The
selective binding by ε of dTTP, but not of other ribo- or
deoxynucleoside triphosphates, suggests an allosteric function for this subunit. Cox had observed that the mutagenic
activity of mutD (5), the gene now assigned to ε by Echols,
Lu and Burgers (6), requires thymidine in the growth medium.
More recently, thymidine kinase activity was shown to be
required for mutagenesis (7), presumably because a thymidyl-
ate moiety is needed.

Additional and more probing assays of holoenzyme function are always of prime importance. Beyond catalytic
efficiency measured by DNA synthesis, are association and
dissociation rates with various forms of templates and
primers, tests of fidelity, and interactions with other
proteins responsible for template binding, priming, repair
and recombination.

We have mentioned two criteria of subunit legitimacy,
persistence of a polypeptide band during purification and
dependence on a polypeptide for a functional assay in vitro.
A specific inhibitor, such as rifampicin for the β subunit
of RNA polymerase, would be another most important criterion.
Unfortunately we have none, as yet, for pol III holoenzyme.
An antibody specific for a polypeptide would likely be
useful; C. S. McHenry (personal communication) has made a
significant start with antibodies to the α, β and τ subunits.
A powerful alternative is alterations of the polypeptide by
mutations in the gene that encodes it. On this basis, holoenzyme has at least 5 subunits (α, τ, γ, β and ε); it also
seems likely that one of the δ bands is a product of the
dnaX-Z locus. There appear to be several more polypeptides
still lacking a locus.

Initiation of a DNA Chain: RNA Priming by RNA Polymerases,
Primases and the Primosome.

With respect to initiation of chains, no DNA polymerase
has this capacity. With the exceptions of only adenoviruses,

phage φ29 and parvoviruses so far, starts of DNA chains are primed by RNA (consult articles elsewhere in this volume). An oligoribonucleotide is transcribed from the template by a primase or RNA polymerase. Primases differ greatly from each other as well as from RNA polymerases. In some instances the primase expresses a demand or preference for a template sequence; in others the primase counts the number of template residues to be transcribed but is indifferent to their sequence. Some primases also accept a deoxynucleotide in place of a ribonucleotide in extending the initial rNTP residue (generally ATP).

Of considerable mechanistic interest regarding E. coli primase is how the enzyme is organized within a very complex primosome of some 20 polypeptides with a total mass near 10^6 daltons (3). The primosome is translocated on the template in a locomotive-like structure with protein n' as the engine using ATP energy and dnaB protein stabilizing or generating secondary structures suitable for primase transcription. We need to know how these several components of the primosome are related to one another and how the primosome remains processive on the template strand, moving with a polarity that keeps it at the growing fork. With the primosome moving opposite to the direction of elongation, how does the primase within it add residues to elongate the oligonucleotide primer?

Recent studies have examined E. coli primase in its synthesis of a 28-residue primer at the unique origin of phage G4 DNA. These studies show that more than one primase molecule, very likely two, are essential (unpublished observation by M. Stayton and A. Kornberg). Possibly, a primase molecule, unlike polymerases, cannot translocate. Being able to add only one nucleotide within its domain, it may rely on its partner primase molecule to add the next nucleotide, and so on.

We have been concerned also with how a holoenzyme molecule moves from one completed Okazaki fragment to the most nascent primer at the replication fork. Recent results show that transfer of a holoenzyme molecule to another primer on the same template is far more rapid than an intermolecular transfer (8). Should the holoenzyme be capable of rapid diffusion along DNA, as is assumed to be true for RNA polymerase, then one wonders whether there is a polarity to this movement, and whether it is in the anti-elongation direction as observed for the primosome.

Organization of Chain Elongation and Initiations at the
Replication Fork: the "Replisome."

Very likely, the elongation of DNA chains and their
initiations on the lagging strand at the replication fork
are organized in a mechanism more elegantly designed than is
described on paper (9). Although never taken alive, faith
abides in a "replisome", a package that contains the enzymes
responsible for advancing the replication fork (9). This
still hypothetical structure operating at each replicating
fork of an E. coli chromosome might contain a dimeric polymerase, a primosome, and one or more helicases. An even more
complex replisome could be envisioned should both forks of
the bidirectionally replicating chromosome remain contiguous
because of an attachment to the cell membrane by specific
proteins.

This proposal of E. coli DNA polymerase III holoenzyme
functioning as a pair of identical oligomers, each with an
active site, replicating both strands simultaneously, has
several attractive features. Physical and functional linkage of the polymerase to the primosome in a replisome structure would coordinate their actions. Evidence in favor of
a dimeric polymerase is still fragmentary. First is the
stoichiometry of two β subunits, two γ subunits, two primases, and two ATPs per replicating center. Second is the
anomalously large size of pol III' suggestive of a dimeric
pol III core and τ subunits (4). Third are the looped DNA
structures observed in association with the primosome;
looping of DNA is required for concurrent replication and is
topologically feasible.

In the scheme for concurrent replication (9), the
leading strand is always ahead by the length of one nascent
fragment; the regions of the parental template strands
undergoing simultaneous synthesis are therefore not complementary. By looping the lagging strand template 180° (perhaps halfway around the polymerase), the strand has the same
$3' \rightarrow 5'$ orientation at the fork as the leading strand. A
primer generated by the primosome is extended by polymerase
as the lagging strand template is drawn through it. When
synthesis approaches the 5' end of the previous nascent
fragment, the lagging strand template is released and becomes unlooped. Synthesis of the leading strand has, in the
meanwhile, generated a fresh, unmatched length of template
for synthesis of the next nascent fragment for the lagging
strand. By this mechanism, one polymerase molecule not only

copies both templates concurrently but also remains linked
to the primosome whose movement is in the direction opposite
the elongation of the lagging strand.

Initiation of a Cycle of Replication of the E. coli Chromosome at Its Unique Origin (oriC).

Having considered the enzymology responsible for progress of the replication fork, we come now to the basic question of control and regulation of chromosome replication.
In E. coli, the major control is at the commitment to initiate a cycle of replication at the unique chromosomal origin.
What is the molecular anatomy of the switch that determines
whether replication will be active or quiescent?
    Availability of the E. coli chromosomal origin (oriC)
in a functional form as part of a small plasmid, first
prepared by Yasuda and Hirota (10), or as a small phage
chimera, prepared by Kaguni, et al. (11), was crucial to
discovery of the oriC enzyme system (12,13). Our attempts
to resolve the rather complex oriC system and to reconstitute it with pure proteins are described elsewhere in this
volume (14). It took nearly ten man-years of effort before
an enzyme system for oriC replication was achieved. Two
things mattered most. One was proceeding with an inactive
lysate to prepare a narrow and highly concentrated ammonium
sulfate fraction; the second was the serendipitous inclusion
of a high concentration of polyethylene glycol.
    I will digress briefly to tell you about a remarkably
active catenase activity (unpublished observation by R. Low
and A. Kornberg). It depends on 8% polyvinyl alcohol being
present with the supercoiled plasmid DNA in the crude enzyme
fractions. Within four minutes, one nanogram of protein in
the ammonium sulfate fraction converts 2 micrograms of any
supercoiled DNA into a huge catenated mass that cannot enter
an agarose gel and is retained by a Millipore filter. This
is less than 0.1% of the amount of protein needed to obtain
oriC replication by this crude fraction.
    Purification of the catenase disclosed a dependence on:
two distinctive proteins (α and β), Mg^{2+}, hydrophilic polymers, supercoiled DNA and a catalytic amount of RFII (nicked
circular duplex). Protein α proved to be identical with
topoisomerase I and protein β with exonuclease III. The
latter creates gaps in the RFII, thereby providing proper
templates for topoisomerase I action.

To determine whether catenase action is essential for the oriC replication reaction is still technically difficult. Topoisomerase I is essential in the oriC reaction for suppressing nonspecific replication (3); exonuclease III activity contaminates several of the partially purified enzyme fractions.

Termination of Replication and Segregation of the Daughter Chromosomes.

Very little is known about the mechanisms of termination and segregation. Involvement of a topoisomerase activity has been suggested by Varshavsky in studies of SV40 replication (consult articles elsewhere in this volume). The most promising leads are in studies of partition functions of plasmids. For example, partition of F plasmids, which exist as only one or two copies per cell, depends on several factors contributed by both host and plasmid (15).

Strategies and Tactics of Enzymology.

I would like lastly to review the strategies and tactics for advancing our knowledge of the enzymology of DNA operations and consider their adequacy for the future. It is still good advice not to "waste clean thinking on dirty enzymes" and that "time invested in purifying an enzyme is always well spent." To these guidelines, I would add another rule that might seem axiomatic: "Don't waste clean thinking on dirty substrates." Of course, biochemists raised in the traditions of organic chemistry would not make the mistake, often attributed to physical chemists, of making accurate measurements on impure materials.

However, the rule has been relaxed when the substrate has been the huge, vulnerable molecule called DNA. Most studies on DNA replication have, out of apparent necessity, employed fragments of a chromosome, frayed, gapped, and nicked and sometimes battered almost beyond recognition. When offered to crude enzyme fractions, these fragments become the prey of many adventitious reactions. In other studies, repetitive polymers are often used as substitutes for DNA, but they also have many serious limitations. By far the best DNA substrates have been the genomes of small viruses and plasmids. With these DNA molecules we provide substrates clean enough to pass for intact chromosomal DNA and possessing many of the sequence specificities needed to elicit primase recognition of origins and chain starts.

With the recent advances in DNA technology, isolating and engineering suitable DNA substrates has become less formidable. By contrast, assaying and isolating enzymes remains demanding and difficult. The best strategy is to choose an enriched source of the enzyme and if you can identify the gene, to enrich it even further with available recombinant DNA technology. In fact, genetics is commonly substituted entirely for enzymology, often with some arrogance. I recall a departmental research meeting in which a young cloner-sequencer, following a presentation of enzyme purification, presented a slide of his purification table. The slide showed a bench with petri plates and toothpicks. I did not join the laughter that followed because I was more irritated than amused. His recombinant work was carried out with polymerases, ligases, and other enzymes that he purified by a single step -- to my refrigerator.

I would like to mention a few things I have learned about enzyme isolation from my work as well as from that of others.

To begin with, I remain faithful to the conviction that anything a cell can do, a biochemist should be able to do. He should do it even better, being freed from the constraints of substrate and enzyme concentrations, pH, ionic strength, and temperature, and by having the license to introduce novel reagents to drive or restrain a reaction. Put another way, one can be creative more easily with a reconstituted system. One can grapple directly with the molecules instead of trying by remote means to manipulate their structures or levels in the intact cell.

Enzyme purification carries many dangers beyond the well-known exposure of the fragile enzyme to the hostilities of an unfamiliar environment, high dilution, glass containers and a denaturable investigator. I will cite a few examples that trapped even an ardent lover of enzymes.

In the course of purifying the ϕX174 replication enzyme system, we eliminated dUTPase (16) along with a thousand other enzymes. Like the ship that passes in the night, the loss of dUTPase was not detected by our assays of the rate of DNA synthesis or the size and circularity of the product. However, for lack of dUTPase, we were no longer sanitizing our triphosphate mixture of the dUTP invariably produced from the spontaneous deamination of dCTP. The result was that some uracil residues replaced thymine in the DNA product and became the substrate for uracil glycosylase and endonucleases, and the source of considerable confusion for

a long time. I would guess that the unappreciated absence
of dUTPase from purified and reconstituted replication
systems will confuse investigators again and again.

On several occasions when we were pleased with a hand-
some purification of an enzyme by chromatography on some
anionic resin column and reassured by the appearance of a
major band on an SDS gel, we were dismayed to discover later
that the enzyme activity and the prominent polypeptide band
were unrelated. On other occasions, a nuclease was concentrated in the enzyme peak and wiped out our activity.
"Why," I moaned, "should an adventitious intruder, an un-
invited nuclease, pick on me and my enzyme fraction?" The
answer is simple. DNA-binding proteins, be they nucleases,
polymerases or ATPases, are numerous and abundant. They
tend to aggregate and are salted out at relatively low
ammonium sulfate concentrations. They bind to anionic
resins. No wonder we find them in the neighborhood of our
favored replication protein.

Several times after having purified a protein with
considerable effort, we found it completely dispensable when
our assays employed a fresh substrate cocktail or a new
mixture of purified enzymes to replace a previous batch. In
these instances we had purified a protein to neutralize an
inhibitor (e.g., a metal, an oligonucleotide, a nuclease)
that contaminated the earlier reagent or enzyme preparations.

Paradoxically, one of the gravest problems we face with
DNA as our substrate is purifying it too much. In its
intracellular form, DNA is almost invariably a nucleoprotein.
Routinely, we isolate supercoiled viral replicative forms
and plasmids by using detergents, phenol and proteases. As
a result, the DNA is stripped of the conserved primosome,
the eukaryotic histones or the histone-like proteins that
condense the prokaryotic nucleoid. In short, we offer our
enzymes a naked DNA lacking the elements that normally
influence its shape and reactivity, and endow it with personality.

Finally, when we have achieved a soluble, molecularly
dispersed enzyme system, we discover the total absence of
the membranes presumed to be responsible for essential
physiological features of replication. For having done
this, some people accuse us of "membranocide" with a fervor
reserved for supporters of the "right-to-life" movement. In
our defense, I could answer "freedom of choice" but that
would be too glib. A proper answer is that we do what we
can. We can, and we must first isolate the many, key components of the replication system, have them in hand and

know something of their properties and important interactions. Only then, as was true of the fatty acid biosynthetic enzymes, will we be in a good position to do the cytochemistry or discover the subtle assays that measure the influences that membrane orientation very likely has on initiation of replication and segregation of daughter chromosomes.

Sometimes when I feel the weight of the increasing complexity of DNA replication and the vast amount of time it will take to unravel it, I think of the impoverished rabbinical scholar who often appealed to God for help in solving his philosophical and personal problems. On one occasion he said to the Almighty: "I simply cannot fathom a span of a million years." "My child," answered the Lord, "it is just a second." Later the young scholar said, "Lord, I have the same problem thinking about a million dollars." "Very simple," said the Lord, "it is only a penny." At which the scholar brightened and said, "Lord, can I please have a penny?" The Lord replied, "Sure, in just a second."

ACKNOWLEDGMENTS

Understanding of DNA replication has relied on studies of several systems, in particular, phage T7 and T4, the small phages, plasmids, animal viruses and eukaryotic as well as prokaryotic chromosomes. Recent progress with each of these systems will be described in the reports contained in this volume. With regard to contributions from my laboratory, I want to express my indebtedness to the many students and postdoctoral fellows from earlier years who did so much to make the papers in this volume by Dixon, Fuller and their coauthors possible.

REFERENCES

1.   Watson JD, Crick FHC (1953). A structure for deoxyribose nucleic acid. Nature 171:737.
2.   Kornberg A (1978). Aspects of DNA replication. Cold Spring Harbor Symp. Quan. Biol. 43:1.
3.   Dixon NE, Bertsch LL, Biswas SB, Burgers PMJ, Flynn JE, Fuller RS, Kaguni JM, Kodaira M, Stayton MM, Kornberg A (1983). Singlestranded phages as probes of replication mechanisms. This volume.
4.   McHenry CS (1982). Purification and characterization of DNA polymerase III'. Identification of τ as a

subunit of the DNA polymerase III holoenzyme. J Biol Chem 257:2657.

5. Degnen GE, Cox EC (1974). Conditional mutator gene in E. coli: isolation, mapping, and effector studies. J Bact 117:477.

6. Echols H, Lu C, Burgers PMJ (1983). Mutator strains of E. coli, mutD and dnaQ, with defective exonuclease editing by polymerase III holoenzyme. Proc Natl Acad Sci USA, in press.

7. Erlich HA, Cox EC (1980). Interaction of an E. coli mutator gene with a deoxyribonucleotide effector. Mol Gen Genet 178:703.

8. Burgers PMJ, Kornberg A (1983). The cycling of E. coli DNA polymerase III holoenzyme in replication. J Biol Chem, in press.

9. Kornberg A (1982). "1982 Supplement to DNA Replication." San Francisco: W. H. Freeman.

10. Yasuda S, Hirota Y (1977). Cloning and mapping of the replication origin of E. coli. Proc Natl Acad Sci USA 74:5458.

11. Kaguni JM, LaVerne LS, Ray DS (1981). Cloning and expression of the E. coli replication origin in a single-stranded DNA phage. Proc Natl Acad Sci USA 76:6250.

12. Fuller RS, Kaguni JM, Kornberg A (1981). Enzymatic replication of the origin of the Escherichia coli chromosome. Proc Natl Acad Sci USA 78:7370.

13. Kaguni JM, Fuller RS, Kornberg A (1982). Enzymatic replication of the E. coli chromosomal origin is bi-directional. Nature 296:623.

14. Fuller RS, Bertsch LL, Dixon NE, Flynn JE, Kaguni JM, Low RL, Ogawa T, Kornberg A (1983). Enzymes and path-way of initiation of replication at the E. coli chromo-somal origin. This volume.

15. Ogura T, Hiraga S (1983). Partition mechanism of F plasmid: two plasmid gene-encoded products and a cis-cting region are involved in partition. Cell 32:351.

16. Shlomai J, Kornberg A (1978). Deoxyuridine triphos-phatase of E. coli. Purification, properties, and use as a reagent to reduce uracil incorporation into DNA. J Biol Chem 253:3305.

II. DNA STRUCTURE, TOPOISOMERASES, AND HELICASES

Mechanisms of DNA Replication and Recombination, pages 19–28
© 1983 Alan R. Liss, Inc., 150 Fifth Avenue, New York, NY 10011

PROCEDURE USING PSORALEN CROSS-LINKING TO QUANTITATE DNA CRUCIFORM STRUCTURES SHOWS THAT CRUCIFORMS CAN EXIST IN VITRO BUT NOT IN VIVO[1]

Richard R. Sinden, Steven S. Broyles,
and David E. Pettijohn

Department of Biochemistry, Biophysics and Genetics,
University of Colorado Health Sciences Center
Denver, Colorado 80262

A method for detecting cruciform structures in DNA has been developed that is applicable to studies in living cells. A 66-bp perfect palindromic lactose operator DNA sequence cloned into the EcoRI site of pMB9 provides a model system with which we show the palindrome can exist in the linear or cruciform state. In the linear form EcoRI cutting produces a 66-bp linear fragment. When in the cruciform, cutting occurs at EcoRI sites now at the base of the cruciform arms producing 33-bp fragments. Both forms can be stabilized by the introduction of psoralen cross-links into the palindromic DNA. Cross-linking, together with EcoRI cutting, provides an assay to quantitate the fraction of palindromes that exist as cruciforms in vitro or in vivo. In vitro, the fraction in cruciforms was dependent on superhelical density. Relaxed DNA contained no cruciforms. Cruciform structures could not be detected in vivo. Cruciforms were formed during isolation of the plasmid DNA leading to about 90% of palindromes forming cruciforms after isolation. Once formed, cruciforms were quite stable even in DNAs with low superhelical densities. Results suggest that in vivo plasmid DNA either lacks sufficient tension to drive the cruciform transition or that formation is prevented by other molecules interacting with the sequence.

[1]This work was supported by NIH research awards GM 18243-11 to DEP and GM 07703-2 to RRS.

INTRODUCTION

Regions of DNA molecules containing inverted repeated or palindromic sequences can potentially exist in two distinct conformations. The DNA can be in the linear form with base pairing between the complementary strands. Alternatively, palindromes can exist in a cruciform structure where the two strands of the palindromic sequence are involved in intrastrand hydrogen bonding with bases at the center of symmetry at the ends of hairpin loops (see Fig. 1). Since non-hydrogen bonded base pairs should exist at the base and loop of cruciform arms the linear form would be favored thermodynamically (1). However, the free energy of supercoiling in DNA containing unrestrained torsional tension might provide sufficient energy to drive the cruciform transition (2). Cruciforms should be stable in negatively supercoiled DNA since DNA in the cruciform structure is unwound, thus relieving torsional strain (3).

Cruciform structures have been visualized directly by electron microscopy in totally palindromic circular molecules supercoiled by action of DNA gyrase (3). In these molecules the arms of cruciforms were extended until all tension was removed from the core of the circular molecule. Other evidence, suggestive that cruciforms can exist in supercoiled DNA, comes from analysis of double-strand cutting of DNA by single-strand specific nucleases (4-7). These nucleases make cuts in supercoiled DNA specifically at the symmetric center of palindromes, the expected single-strand loops at the ends of cruciform arms.

Palindromic DNA sequences or regions of two-fold rotational symmetry occur at operator and transcription termination regions of genes and at origins of DNA replication. It is clear that this symmetric organization of DNA sequence is important biologically, possibly reflecting binding of multimeric proteins or reflecting RNA secondary structure believed to be important in transcriptional control. Moreover, it has also been suggested that cruciform structures might be important in UV and frameshift mutagenesis (8,9) or in the regulation of gene expression in eukaryotes (10).

Considering the possible significance of cruciform structures in DNA, we have developed a method to detect cruciforms that is applicable to studies in living cells.

RESULTS

The Model System

Betz and Sadler have cloned a perfect 66-bp lac operator DNA sequence into the EcoRI site of pMB9 (11). Including both EcoRI sites and flanking AT:TA sequences, this constitutes a perfect 76-bp palindromic sequence (Fig. 1). This palindromic configuration provides an ideal model system to investigate the existence of cruciform structures. EcoRI digestion of the palindrome in the linear form should produce a 66-bp linear fragment (Fig. 1B). In the cruciform state digestion might release two 33-bp fragments (Fig. 1C). EcoRI digestion of supercoiled pOCE12 plasmid DNA does produce both 33 and 66-bp fragments (Fig. 4A, lane a and see Ref. 12). Tension necessary to form and maintain the cruciform could be relaxed after the first EcoRI cut is made and cruciforms could be lost prior to the second EcoRI cut. Therefore, it was not surprising to find that the relative amount of the palindromic sequence cut out as the 33-bp fragment was dependent on both the DNA and enzyme concentrations (12). When DNA supercoiling was first relaxed by linearization of the plasmid, only the linear 66-bp fragment was produced by EcoRI. Since EcoRI appears to recognize and cut at EcoRI sites at the base of cruciform arms, this suggests that the palindrome sequence exists in the cruciform configuration in supercoiled DNA.

A

B

C

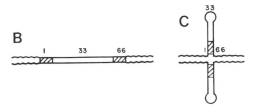

FIGURE 1. (A) Sequence of 66-bp palindrome and flanking regions (11,12). The complete palindromic sequence is 76-bp. Arrow designates center of symmetry. (B) Linear configuration of the palindrome. (C) Cruciform configuration of the palindrome. EcoRI sites (hatched boxes).

Stabilization of Cruciform Structures by DNA Cross-links

If palindromic DNA exists in cruciform structure with hydrogen bonding and base stacking interactions in the cruciform arms, it should be possible to introduce intrastrand cross-links into cruciform arms (12,13). This would fix or stabilize the cruciform configuration and prevent transition to the linear form, even after the loss of DNA supercoiling. Thus, following Me_3psoralen cross-linking of supercoiled DNA and subsequent linearization to relax the DNA, subsequent EcoRI digestion should produce 33-bp molecules only if the cruciform configuration were fixed by a cross-link. As shown in Fig. 2A, lanes a-j, increasing numbers of Me_3psoralen cross-links result in accumulation of 33-bp fragments following linearization and EcoRI digestion. The fraction of the palindromic sequence migrating in electrophoresis as the 33-bp band represents the fraction cross-linked in the cruciform configuration (F_{33}). Following denaturation and rapid renaturation the individual 66-bp strands "snap back" into 33-bp hairpins (11,12). However, an interstrand cross-link will cause "reversible renaturation" of the 66-bp linear fragment. Thus, 66-bp linear fragments remaining after boiling and quick cooling represent the fraction of the palindrome cross-linked in the linear form (F_{66}), Fig. 2A, lanes k-t. F_{66} and F_{33} are shown in Fig. 2B as a function of the average number of cross-links introduced into the palindromic sequence, M_{XL}. The fractions of the palindromic sequence existing in either the cruciform or linear state (P_{33} and P_{66}) were determined from F_{33} and F_{66}. $P_{33} = F_{33}/P_{XL}$ and $P_{66} = F_{66}/P_{XL}$, where $P_{XL} = F_{66} + F_{33}$ (12). These values are also shown in Fig. 2B. From this analysis, 75% of the palindromic sequences were in the cruciform state. This DNA preparation contained about 20% nicked molecules so that the actual P_{33} is greater than 0.75. In addition, it is important to note that P_{33} is constant at all levels of Me_3psoralen photobinding. Thus, unwinding due to Me_3psoralen photobinding does not change the fraction of the palindromic sequence in the cruciform state. Unwinding from Me_3psoralen binding does not affect P_{33} because cruciforms are very stable in DNAs of low superhelical density and because the rate of the cruciform to linear transition is very slow at $0^{\circ}C$ in relaxed DNA (12, RR Sinden & DE Pettijohn, unpublished results).

A

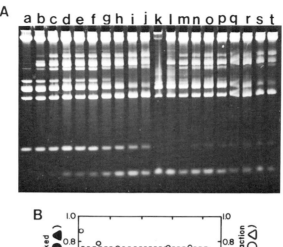

a b c d e f g h i j k l m n o p q r s t

B

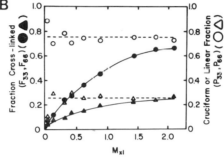

FIGURE 2. Stabilization of palindromic conformations by Me₃psoralen cross-links. (A) Electrophoretic analysis of cross-linked pOCE12 DNA following linearization by treatment with HindIII, BamHI, and SalI and subsequent EcoRI digestion. Direction of electrophoresis is top to bottom. Positions of 33 and 66-bp bands are indicated. Samples were treated with saturating Me₃psoralen and 0, 1.6, 4.0, 8.0, 12, 16, 24, 32, 40, and 48 kJ/m² 360 nm light. a-j, native samples; k-t, denatured samples. Experiment described in detail previously (12). Upper bands, result of HindIII, BamHI, and SalI digestion. (B) Quantitative analysis of gel shown in Fig. 2A. F_{33}, F_{66}, P_{33} and P_{66} are described in text. $M_{XL} = -\ln(1-P_{XL})$.

Effect of Superhelical Density on Cruciform Formation

To examine the effect of superhelical density on cruciform formation, a series of plasmid DNAs of varying superhelical density were made and purified from ethidium bromide–CsCl gradients (12). The high concentration of ethidium bromide by winding in positive superhelical turns drives cruciforms into the linear configuration. After purification, tension is re-established and cruciforms re-equilibrate in the DNA dependent on the level of supercoiling (Fig. 3). Below $\sigma = -0.04$ few cruciforms were detected. Above $\sigma = -0.08$, 98% of the palindrome existed as cruciforms. Additional studies have shown that the midpoint of the transition is temperature dependent (RR Sinden & DE Pettijohn, unpublished results). If plasmid populations were not subjected to high concentrations of ethidium bromide to remove pre-existing cruciforms, cruciform structures were detected in plasmids with σ as low as -0.02, since P_{33} values of 0.25 were obtained. This result suggests that once formed, cruciform structures might be quite stable, even under conditions where little supercoiling exists in the DNA. Results also suggest that cruciforms are not in a rapid equilibrium with the linear state.

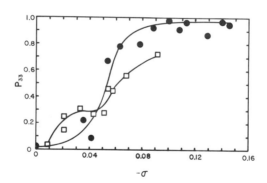

FIGURE 3. Effect of superhelical density on the fraction of palindromes existing in the cruciform configuration. Experiment described in the text and in detail previously (12). (●), cruciforms removed from supercoiled DNAs by incubation in presence of 200 ug/ml ethidium bromide and allowed to reform as tension was re-equilibrated in DNAs upon removal of ethidium bromide. (□), pre-existing cruciforms not removed from DNAs. DNAs relaxed in the presence of topoisomerase I and low concentrations of ethidium bromide to generate DNAs of varying superhelical density and subsequently purified before photobinding.

Conformation of the Palindrome Sequence In Vivo

To investigate the conformation of the palindromic sequence in vivo, we analyzed DNA from cultures of pOCE12 which had been cross-linked in vivo, prior to cell lysis and plasmid isolation, and also in supercoiled plasmid DNA cross-linked after purification from cells (12). The results from a representative experiment are shown in Fig. 4. Photobinding to DNA in vitro, after DNA purification, results in the accumulation of 33-bp fragments diagnostic of the cruciform configuration (Fig. 4A). Analysis of this data showed that at least 90% of the palindromic sequence existed in the cruciform state in purified plasmid DNA. In contrast to this result, no 33-bp band was detectable if the palindrome was cross-linked in vivo, prior to DNA purification (Fig. 4B). Cross-links were formed exclusively in the linear conformation of the palindromic sequence. At maximal cross-linking, greater than 90% of palindromes were

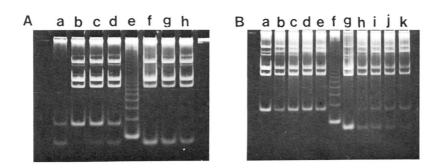

FIGURE 4. Cruciforms exist in vitro but not in vivo. Me3psoralen photobinding to pOCE12 DNA in vitro, following DNA purification (A) or photobinding in vivo, prior to DNA purification (B). Experiment described in text and previously (12). Panel A, lane e and panel B, lane f; 40-bp "ladder", molecular weight marker. Panel A, lane a; supercoiled pOCE12 DNA digested with EcoRI. Lanes b-d; cross-linked in vitro with Me3psoralen and 1.6, 4.0 and 8.0 kJ/m^2 360 nm light. Lanes f-h; denatured and renatured aliquots of samples shown in lanes b-d. Panel B, lanes a-e; samples following permeabilization with EDTA cross-linked in vivo with Me3psoralen and 0, 4.0, 8.0, 12, 16, kJ/m^2 360 nm light. Lanes g-k; denatured and renatured aliquots of DNA samples in a-e.

cross-linked in vivo in the linear form, showing that failure to fix cruciforms was not attributable to inefficient cross-linking. Analysis of this data suggested that less than 2% of the palindromic sequence could exist in the cruciform configuration in vivo. These experiments have been repeated several times with identical results using either chloramphenicol-amplified or non-amplified cells.

DISCUSSION

We have developed a procedure utilizing DNA cross-linking and restriction analysis of DNA to measure quantitatively cruciform structures in DNA in vitro and in vivo. The results show that palidromic lac operator DNA can exist as a stable cruciform structure in purified DNA in vitro but that the cruciform structure does not exist in DNA of living cells. The formation of cruciform structures is dependent on temperature and the level of supercoiling in plasmid DNA. Below $\sigma = -0.04$ few cruciform structures are formed. At high superhelical densities in vitro, nearly all palindromes exist in the cruciform state. Once formed cruciforms are very stable; there may be little transition back into the linear state if minimal tension remains in the DNA. Our results show that the fraction of palindromes in cruciforms does not represent a rapid equilibrium between conformational states.

Results support previous conclusions that cruciforms exist in purified supercoiled DNA (3-7). The cross-linking method seems preferable to other methods for quantitating the fraction of palindromes in cruciforms since at 0°C the photobinding does not appear to alter the equilibrium between states or affect the stability of the cruciform structures. The method also allows quantitation of cruciforms both in vitro and in vivo. This approach should be applicable to any palindromic sequence that harbors a unique restriction fragment consisting exclusively of palindromic DNA. We have also used this approach to identify a cruciform in a palindrome located within a larger restriction fragment. Although quantitation can not be made directly, unique slower migrating bands can be identified diagnostic of cruciform-containing DNA (RR Sinden & PJ Hagerman, unpublished results). The most important advantage of the cross-linking method, however, is that it allows detection of cruciforms in vivo.

Significantly, cruciform structures could not be detected in living cells. One explanation is that the level of torsional tension in plasmid DNA in vivo may be lower than in purified DNA and might remain below a critical transition level. It has been shown that some supercoils can be restrained in prokaryotic cells (14). Alternatively, proteins bound to the DNA might prevent the transition, although there should be enough lac repressor in amplified cells to bind only a small fraction of the palindromic sequence (12). Evidence also suggests that divalent cations and polyamines reduce the rate of cruciform transition in purified DNA (RR Sinden & DE Pettijohn, unpublished results). In addition, the E. coli histone-like HU protein appears to reduce or prevent the linear-to-cruciform transition in supercoiled DNA in vitro (SS Broyles & DE Pettijohn, unpublished results). In vivo there is likely to be a combination of several or all of these possibilities that prevent the cruciform transition. Indeed large cruciform structures may not even be stable in cells as evidenced from the inability to clone long palindromes (3,11,15). Cruciforms may be recognized as Holliday recombinational intermediates and cleaved by recombination nucleases in cells (3,16), or cruciforms may undergo intramolecular recombination. It is clear that there may be subtle differences in the organization and conformation of DNA in vivo compared to in vitro, differences that may be important in the control of reactions involving DNA or in the control of gene expression.

ACKNOWLEDGEMENTS

We thank John R. Sadler and Joan Betz for the gift of strain pOCE12/HB101 and for helpful comments and discussions.

REFERENCES

1. Wang JC (1974). Interactions between twisted DNAs and enzymes: The effect of superhelical turns. J Mol Biol 87:797.
2. Vologodskii AV, Lukashin AU, Anshelevish VV, Frank-Kamenetskii MD (1979). Fluctuations in superhelical DNA. Nuc Acids Res 6:967.

3. Mizuuchi K, Mizuuchi M, Gellert M (1982). Cruciform structures in palindromic DNA are favored by DNA supercoiling. J Mol Biol 156:229.
4. Lilley DMJ (1980). The inverted repeat as a recognizable structural feature in supercoiled DNA molecules. Proc Natl Acad Sci USA 77:6468.
5. Lilley DMJ (1981). Hairpin-loop formation by inverted repeats in supercoiled DNA is a local and transmissible property. Nuc Acids Res 9:1271.
6. Panayotatos N, Wells RD (1981). Cruciform structures in supercoiled DNA. Nature 289:466.
7. Singleton CK, Wells RD (1982). Relationship between superhelical density and cruciform formation in plasmid pVH51. J Biol Chem 257:6292.
8. Todd PA, Glickman BW (1982). Mutational specificity of UV light in Escherichia coli: Indications for a role of DNA secondary structure. Proc Natl Acad Sci USA 79:4123.
9. Ripley LS (1982). Model for the participation of quasi-palindromic DNA sequences in frameshift mutation. Proc Natl Acad Sci USA 79:4128.
10. Groudine M, Weintraub H (1982). Propagation of globin DNAaseI-hypersensitive sites in absence of factors required for induction: A possible mechanism for determination. Cell 30:131.
11. Betz JL, Sadler JR (1981). Variants of a cloned synthetic lactose operator. I. A palindromic dimer lactose operator derived from one strand of the cloned 40-base pair operator. Gene 13:1.
12. Sinden RR, Broyles SS, Pettijohn DE (1983). Perfect palindromic lac operator DNA sequence exists as a stable cruciform structure in supercoiled DNA in vitro but not in vivo. Proc Natl Acad Sci USA 80: in press.
13. Cech TR, Pardue ML (1976). Electron microscopy of DNA crosslinked with trimethylpsoralen: Test of the secondary structure of eukaryotic inverted repeat sequences. Proc Natl Acad Sci USA 73:2644.
14. Pettijohn DE, Pfenninger O (1980). Supercoils in prokaryotic DNA are restrained in vivo. Proc Natl Acad Sci USA 77:1331.
15. Lilley DJM (1981) In vivo consequences of plasmid topology. Nature 292:380.
16. Mizuuchi K, Kemper B, Hays J, Weisberg RA (1982). T4 endonuclease VII cleaves holliday structures. Cell 29:357.

Mechanisms of DNA Replication and Recombination, pages 29–42
© 1983 Alan R. Liss, Inc., 150 Fifth Avenue, New York, NY 10011

IDENTIFICATION OF YEAST DNA TOPOISOMERASE MUTANTS[1]

Stephen DiNardo, Catherine Thrash, Karen A. Voelkel,
and Rolf Sternglanz

Department of Biochemistry, State University of New York
Stony Brook, New York 11794

ABSTRACT Yeast DNA topoisomerase I and II mutants have
been identified. The topoisomerase II mutant is
temperature-sensitive for growth, leads to a
temperature-sensitive enzyme and most probably carries
a mutation in the structural gene for the enzyme.
The mutant is defective in termination of DNA
replication and medial nuclear division. At the
non-permissive temperature 2 micron circle DNA
accumulates as multiply intertwined catenated
dimers. Presumably daughter DNA molecules become
intertwined wherever replication forks approach each
other and DNA topoisomerase II is necessary for their
resolution.
     The topoisomerase I mutant has normal enzymatic
activity at 25° and about 20% of normal activity at
36°. The mutant grows normally at all temperatures.
The mutation leading to the topoisomerase defect has
been mapped; it is tightly linked to the centromere of
chromosome XV at or near the MAK1 gene. Three
previously known mak1 mutants all have less than 1% of
the normal level of topoisomerase I activity,
suggesting that MAK1 is either the structural gene for
DNA topoisomerase I or codes for a protein that
regulates it.

[1]This work was supported by NIH grant GM28220.

INTRODUCTION

DNA topoisomerases are enzymes that catalyze the concerted breakage and rejoining of DNA backbone bonds (1). Topoisomerases can be divided into two classes, type 1 and type 2. Type 1 enzymes make temporary single-strand breaks in DNA and change the linking number (Lk) of closed circular double-stranded DNA in steps of one. Type 2 enzymes make temporary double-strand breaks and change Lk in steps of two. The type 1 enzymes were the first topoisomerases to be isolated, first from E. coli (2), soon thereafter from mouse cells (3), and since then from a great many species (reviewed in ref. 4). The bacterial type 1 topoisomerases differ from their eukaryotic counterparts in two ways: 1) the former can relax only negatively supercoiled DNA while the latter can relax both negatively and positively supercoiled DNA, and 2) the former require Mg^{++} while the latter don't.

The first type 2 topoisomerase to be identified was the bacterial enzyme, DNA gyrase (5). This enzyme requires ATP and can supercoil, catenate and decatenate double-stranded closed circular DNA (6). Eukaryotic type 2 topoisomerases have been discovered only recently. Unlike gyrase, the eukaryotic enzymes cannot supercoil closed circular DNA but can only relax it; they can also catenate, decatenate and unknot double-stranded DNA circles, in all cases in an ATP-requiring reaction (7-9).

Although a great deal is known about the *in vitro* properties of eukaryotic DNA topoisomerases, virtually nothing is known about their *in vivo* roles. It has been thought for a long time that the type 1 enzyme may act as a swivel during DNA replication, serving to relieve positive DNA supercoiling that arises during unwinding of Watson-Crick turns (7). It has been suggested that the type 2 enzyme might be involved in initiation of DNA replication (8), since its enzymatic properties are similar to the bacteriophage T4 topoisomerase (11) and the latter enzyme is thought to be involved in the initiation of T4 DNA replication.

Both DNA topoisomerase I (type 1) and topoisomerase II (type 2) have been partially purified from the yeast, *Saccharomyces cerevisiae* (12-14). The properties of the yeast enzymes are quite similar to their counterparts from mammalian cells. In order to learn more about the role of eukaryotic DNA topoisomerases, we began a search for yeast topoisomerase mutants. We reasoned that a study of the phenotypes of such mutants would lead to a greater understanding of the *in vivo* roles of these enzymes, not only in yeast but also in the DNA metabolism of all eukaryotic cells.

RESULTS

A collection of independently isolated temperature-sensitive mutants of yeast (15), generously provided by C. McLaughlin, was screened for DNA topoisomerase defects. Each mutant was grown at the permissive temperature, shifted to the non-permissive temperature for 20 minutes, harvested, lysed, and assayed at the non-permissive temperature for both DNA topoisomerase I and II. Details of the procedure are in the legend to Fig. 1. The assay for topoisomerase I is the usual one, relaxation of supercoiled DNA. The assay for topoisomerase II is a decatenation assay described previously (10), involving the release of monomer circles from a kinetoplast DNA network. Typical assays for both yeast topoisomerases are shown in Fig. 1.

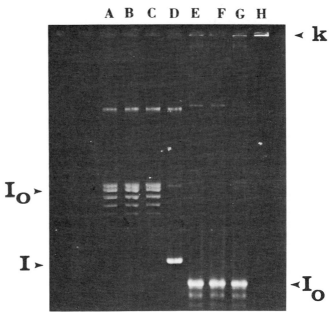

FIGURE 1. DNA topoisomerase I and II assays. Cells were grown in 25ml YPD (15) at 25°. During exponential

growth cultures were shifted to 37° for 20 min,
chilled, centrifuged, washed with cold H_2O and
re-centrifuged. The cell pellet was resuspended in
0.5ml yeast lysis buffer (20mM tris pH=7.0, 1mM
Na_2EDTA, 1mM dithiothreitol, 1mM phenylmethylsulfonyl
fluoride, 500mM KCl, 10% glycerol). One third volume
of glass beads (Sigma, Type IV, 250-300 μm) was added
and the cells lysed by brief sonication. The lysate
was centrifuged for 10 min in a desk top centrifuge.
One μl of the supernatant (undiluted or diluted in
yeast lysis buffer plus 100 μg/ml was used for
topoisomerase assays. DNA topoisomerase I assays were
in 10 μl total volume containing 20 μg/ml supercoiled
pBR322 DNA, 20mM tris HCl pH=7.5, 10mM Na_2EDTA, 1mM
dithiothreitol, 30 μg/ml bovine serum albumin and 150mM
KCl. DNA topoisomerase II assays (10) were in 10 μl
total volume containing 20 μg/ml kinetoplast DNA, 20mM
tris HCl pH=7.5, 10mM $MgCl_2$, 1mM EDTA, 1mM
dithiothreitol, 30 μg/ml bovine serum albumin, 150mM
KCl and 0.4mM ATP. Incubation for both topoisomerase I
and II assays was at 37° for 30 min. Samples were
electrophoresed in 0.7% agarose gels as described (16).
A-D, DNA topoisomerase I assays; E-H, DNA topoisomerase
II assays. A-C and E-G show assays of three different
yeast strains, 1 μl of undiluted supernatant extract
added in each case. D and H, control assays with no
extract added. I indicates the position of supercoiled
pBR322 DNA, I_o the position of relaxed pBR322 DNA (A-C)
or kinetoplast DNA relaxed monomer circles (E-G), and k
the position of the kinetoplast DNA network which does
not enter the gel. In this electrophoresis buffer, the
family of DNA topoisomers centered about fully relaxed
DNA migrates as topoisomers with a few positive
superhelical turns; hence the ladder of bands migrating
faster than position I_o in lanes A-C and E-G.

DNA Topoisomerase II Mutant.

About 200 independent temperature-sensitive mutants
including the well-known cdc mutants (17) were screened for
topoisomerase defects. All the cdc mutants had normal
topoisomerase I and II activity. A topoisomerase II mutant
was identified after screening approximately 150 mutants.
As seen in Fig. 2, strain 14-16 has virtually no
topoisomerase II activity when assayed at 37°. When this
strain is grown at the permissive temperature and assayed at

25° and 37°, it shows almost normal activity at 25° and at
least 10-fold less activity at 37°. This
temperature-sensitive enzymatic activity is also seen when
the mutant enzyme is partially purified about 20-fold on a
hydroxylapatite column (data not shown).

FIGURE 2. DNA topoisomerase II assays of five
independent temperature-sensitive yeast mutants (A-E).
Cells were grown and assayed as in Fig. 1. Strain
14-16 ($top2^{ts}$) is in lane B. Control with no extract
added is in lane F.

The topoisomerase II temperature-sensitive (TS) mutant,
14-16, was mated with a wild-type strain of opposite mating
type and the resulting diploids sporulated and subjected to
standard tetrad analysis. The results showed normal 2:2
segregation of both the TS phenotype and the topoisomerase
II defect, indicating that a single nuclear mutation is
responsible for both phenotypes. This conclusion holds true
for four successive back-crosses of the mutant with
wild-type. Also, temperature-resistant revertants of the
mutant regain enzymatic activity. Taken together, these
data strongly suggest that the mutation originally present
in strain 14-16 is most probably in the structural gene for
topoisomerase II and that the enzyme is essential for
viability.

Phenotypes of the Topoisomerase II Mutant.

Strain SD1-4 from the fourth back-cross described above, and carrying the topoisomerase II mutation was used for phenotype studies. RNA and DNA synthesis were monitored by labeling cells with ^{3}H-uracil. At the non-permissive temperature, RNA synthesis in the mutant strain is normal for about 3 hours and then shuts off completely. DNA synthesis slows down somewhat within 1 hour at 38° but continues at a reduced rate for about 3 hours after which it too shuts off. Only 10% of the mutant cells are viable after 3 hours at 38°. In the microscope two morphological types are seen at this time, approximately equal numbers of single cells with no buds and cells with a large bud. After more than 3 hours at 38°, increasing numbers of aberrantly budded forms are seen. Thus, no single terminal morphology is observed.

The uracil labeling results suggested that the mutant had a possible defect in DNA replication but it was difficult to know if it was a primary phenotype. Experiments with synchronized cells were more informative. The yeast pheremone, α-factor, was used to synchronize both wild-type and mutant a cells in the G1 phase of the cell cycle (18). Removal of the α-factor releases the cells from the G1 block and allows a fairly synchronous cell cycle. Wild-type and mutant cells were released from the block at either 24° or 38° and DNA synthesis was monitored. As seen in Fig. 3, both strains behave identically at 24° with an S phase of DNA synthesis occurring at the expected time. At the non-permissive temperature, 38°, wild-type cells also go through a normal S phase and then continue into a second cell cycle. On the other hand, the mutant at 38° exhibits a normal initial S phase, doubling its DNA, but then it stops DNA synthesis completely.

The data in Fig. 3B show about an 80% increase in DNA and this corresponds well with the fraction of cells observed in the microscope to actually be released from the G1 block. This result is quantitatively reproducible; the topoisomerase II mutant, when synchronously released from a G1 block at 38°, has a normal S phase, doubling its DNA, and then abruptly stops DNA synthesis completely. Examination of the mutant in the microscope at the end of the single S phase shows that it has a striking terminal phenotype; greater than 80% of the cells have a single large bud. Nuclear staining with Hoechst dye indicates that the nuclei are in the neck between the mother cell and its bud. These

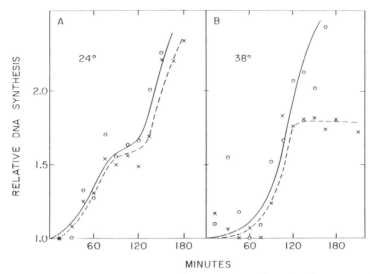

FIGURE 3. DNA synthesis in synchronized cultures of
$TOP2^+$ and $top2^{TS}$ strains of a mating type growing in
YM-5 medium (15) at 25° in the presence of ^{3}H-uracil
(3 μCi/ml) were arrested in G1 by the addition of yeast
α-factor, essentially as described (19). The α-factor
was removed by washing the cells on a filter. Cells
were resuspended in YM-5 medium with 3 μCi/ml ^{3}H-uracil
and incubated at 24° (A) or 38° (B). Samples were
removed at various times and the amount of ^{3}H
incorporated into DNA determined (19). Strain SD2-1
$TOP2^+$ (—o—o—). Strain SD1-4 $top2^{TS}$ (---x---x---).

results suggested that the DNA topoisomerase II mutant is
arrested at medial nuclear division and that it may have a
defect in the termination of DNA replication.

Further evidence for this hypothesis comes from an
examination of the structure of 2 micron circle DNA in the
mutant. The 2 micron circle is a 6kb plasmid present in
most yeast strains at about 50 copies per cell. It is known
that each 2 micron circle is replicated once and only once
per S phase (19). Strain SD1-4 and a wild-type strain were
synchronized with α-factor and released from G1 arrest as in
Fig. 3. After completion of the first S phase, DNA was
extracted, separated by agarose gel electrophoresis, and
subjected to Southern blotting using a 2 micron DNA probe.

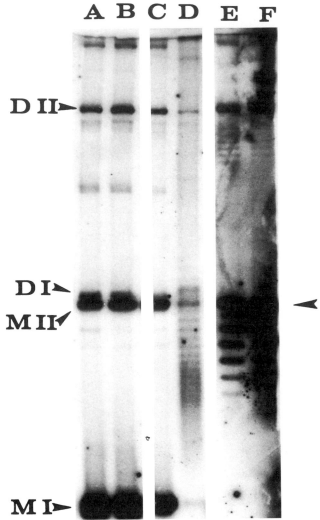

FIGURE 4. Southern blotting of 2 micron DNA from $TOP2^+$ and $top2^{TS}$ strains. Strains were synchronized with α-factor as in Fig. 3 (except no ^{3}H-uracil was added) and total DNA prepared from samples taken at 120 min (24°) or 165 min (38°) after removal of α-factor. The DNA was electrophoresed and analyzed by Southern blot hybridization using ^{32}P-labeled 2 micron DNA as a probe. A: $TOP2^+$, 24°; B: $TOP2^+$, 38°; C: $top2^{TS}$, 24°; D: $top2^{TS}$, 38°; E: As in B but treated with yeast topoisomerase II; F: As in D but treated with yeast topoisomerase II. MI, supercoiled monomer; DI,

supercoiled dimer; MII, nicked monomer; DII, nicked dimer of 2 micron DNA. The arrow to the right of lane F indicates the position cf nicked monomer generated by topoisomerase II treatment. While not easily visible here due to interference from a neighboring lane present in the original gel, the band is clearly visible in shorter exposures of this autoradiogram.

The results in Fig. 4 show that DNA isolated from wild-type cells at 24° and 38° and from the mutant at 24° migrates at the expected positions, mainly monomer and dimer supercoiled 2 micron circles with a small fraction of nicked circles (Fig. 4, A-C). On the other hand, DNA isolated from the mutant at 38° migrates as an unusual series of closely spaced bands, starting around nicked monomer circle and as more rapidly moving species (Fig. 4,D). Very little supercoiled and nicked monomer or dimer is seen. These bands are too closely spaced to be monomer closed circles simply differing in Lk (compare with the bands in Fig. 4E). They also do not have the correct migration for monomer circles containing knots of increasing complexity.

When this DNA (from the mutant at 38°) is treated with partially purified yeast DNA topoisomerase II, the 2 micron DNA is converted mainly into nicked monomer circles and also into closed relaxed monomer circles (Fig. 4, F and data not shown). Treatment of the same DNA with yeast DNA topoisomerase I has no effect. These results, and others to be published elsewhere, lead us to conclude that the 2 micron DNA accumulates as multiply intertwined catenated dimers in the topoisomerase II mutant at the non-permissive temperature.

DNA Topoisomerase I Mutant

A screening of about 200 temperature-sensitive yeast mutants also led to the identification of a DNA topoisomerase I mutant. As can be seen in Fig. 5 the mutant shows normal relaxation activity at 25° but reduced activity at 36°. From this and other experiments we estimate that the mutant has about 20% of the normal level of activity at 36°. Genetic crosses showed that the mutation which causes the topoisomerase I defect is a single nuclear mutation. This mutation can be genetically separated from one or more other mutations present in the original strain that caused that strain to be temperature-sensitive for growth. Strains carrying only the topoisomerase I mutation grow normally at all temperatures; the only phenotype we have detected is the enzyme defect itself.

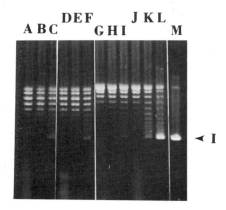

FIGURE 5. DNA topoisomerase I assays of strain A364A (Topo I^+) and 14-1 (Topo I^{TS}). Cells were grown and assayed as in Fig. 1 except that cultures were not shifted to 37°. A–C, A364A assayed at 24°; D–F, 14-1 at 24°; G–I A364A at 36°; J–L, 14-1 at 36°; M, control, no extract added. Each grouping of three represents serial dilutions of a given extract: undiluted, 3.5-fold diluted and 7-fold diluted. The two undiluted extracts were first adjusted to the same protein concentration.

The initial genetic analysis also showed that the topoisomerase I mutation was centromere linked (0/15 second division segregants). In order to map the mutation, further crosses were performed versus a series of strains carrying many centromere linked markers. Fifty-two tetrads from the crosses were analyzed for the topoisomerase I mutation by direct enzymatic assay and for the distribution of the centromere-linked markers. The results confirmed that the topoisomerase I mutation was tightly centromere linked, with only 1 out of 52 second division segregants. Furthermore, direct linkage was seen to a marker on chromosome XV, *PET17*. The ratio of parental, non-parental, and tetratypes for the marker was 7:0:7 indicating that the topoisomerase I mutation is about 25 cM from *PET17*.

Two genes are known near the centromere of chromosome XV, *PHO80* and *MAK1*. Strains carrying mutations in these two genes were obtained from the Yeast Genetic Stock Center and assayed for topoisomerase I. The *pho80* mutant showed normal activity but the *mak1* showed no detectable activity under the assay conditions used. Two other *mak1* mutants were obtained from Reed Wickner, *mak1-2* and *mak1-4*. They too exhibited very low topoisomerase I activity (<1% of normal). Thus, the *MAK1* gene either controls or is the structural gene for DNA topoisomerase I in yeast.

DISCUSSION

The DNA topoisomerase II mutation leads to temperature-sensitive enzymatic activity in crude extracts and after partial purification. Therefore, the mutation is most likely to be in the structural gene for the enzyme. Cloning and mapping of the gene are underway.

The most striking phenotype we have detected in the mutant is a defect in termination of DNA replication. The 2 micron circle DNA accumulates as multiply intertwined catenated dimers, reminiscent of the observations of Sundin and Varshavsky (20) on the terminal stage of SV40 DNA replication. Their thorough studies showed that in hypertonic medium SV40 molecules can replicate but not segregate. They accumulate as multiply intertwined catenated dimers with up to 20-25 intertwinings of one molecular about the other. They speculated that in hypertonic medium DNA replication and DNA segregation are uncoupled with replication permitted but segregation inhibited. They proposed that probably a type 2 DNA topoisomerase is necessary for final separation of daughter molecules. Our findings support their proposal.

We feel that our results are not only applicable to small circular DNA molecules but to the chromosomes themselves. In the mutant at the nonpermisive temperature, cells replicate their DNA only once and then stop. A defect in medial nuclear division is seen. Therefore, it is reasonable to conclude that wherever replication forks approach each other along the chromosomes, the daughter molecules become topologically intertwined and topoisomerase II is necessary for their separation.

The fact that the mutant can go through a complete S phase at 38° (Fig. 3B) strongly suggests that DNA topoisomerase II is not required for initiation of DNA replication or fork movement. In fact, the results of Fig. 3 suggest that topoisomerase II is not needed for any aspect of the cell cycle between the α-factor arrest point in G1 and the end of S phase. If topoisomerase II is required during that part of the cell cycle, more complicated hypotheses must be proposed, such as the mutant enzyme is not inactivated during the first cell cycle at 38° but only in subsequent cycles.

The fact that when an exponentially growing culture of the mutant is raised to the non-permissive temperature half the cells accumulate as single cells with no buds suggests that topoisomerase II is required at another stage of the cell cycle, either late in G2 or early in G1 before the α-arrest point.

The DNA topoisomerase I mutant we identified leads to normal activity in extracts assayed at 25° and reduced activity when assayed at 36°. Strains carrying the topoisomerase mutation, but not the other TS mutations present in the original mutant strain, grow normally at all temperatures. The fact that the topoisomerase I mutation is centromere linked allowed us to map it to the MAK1 gene. Three previously isolated mak1 mutants (21) all have less than 1% of normal topoisomerase I activity and yet they all grow normally. Mak1 mutants were isolated on the basis of their inability to support the maintenance of the double-stranded killer RNA present in many yeast strains. It is unexpected that yeast DNA topoisomerase I is involved in maintenance of a double stranded linear RNA, especially since the RNA is not known to go through a DNA intermediate. It is also surprising that the mak1 mutants grow normally. Perhaps DNA topoisomerase I is not essential for viability in yeast. Or, more likely, our in vitro enzymatic assay is not reflecting the in vivo level or function of topoisomerase I. Further work is needed to clarify the role of the enzyme.

ACKNOWLEDGMENTS

We thank Cal McLaughlin for sending us the collection of temperature-sensitive mutants, Reed Wickner for the mak1 mutants, and Doug Treco and Norman Arnheim for advice.

REFERENCES

1. Wang JC, Liu LF (1979). DNA topoisomerases: Enzymes that catalyze the concerted breakage and rejoining of DNA backbone bonds. In Taylor JH (ed): "Molecular Genetics, Part 3," New York: Academic Press, p. 65.
2. Wang JC (1971). Interaction between DNA and an Escherichia coli protein ω. J Mol Bio 55:523.
3. Champoux JJ, Dulbecco R (1972). An activity from mammalian cells that untwists superhelical DNA - a possible swivel for DNA replication. Proc Natl Acad Sci USA 69:143.
4. Gellert M (1981). DNA topoisomerases. Ann Rev Biochem 50:879.
5. Gellert M, Mizuuchi K, O'Dea MH, Nash HA (1976). DNA gyrase: an enzyme that introduces superhelical turns into DNA. Proc Natl Acad Sci USA 73:3872.
6. Cozzarelli NR (1980). DNA gyrase and the supercoiling of DNA. Science 207:953.
7. Champoux JJ (1978). Proteins that affect DNA conformation. Ann Rev Biochem 47:449.
8. Liu LF, Liu C-C, Alberts BM (1980). Type II DNA topoisomerases: Enzymes that can unknot a topologically knotted DNA molecule via a reversible double-strand break. Cell 19:697.
9. Hsieh T-S, Brutlag D (1980). ATP-dependent DNA topoisomerase from D. melanogaster reversibly catenates duplex DNA rings.
10. Miller, KG, Liu LF, Englund PT (1981). A homogenous type II DNA topoisomerase from HeLa cell nuclei. J Biol Chem 256:9334.
11. Liu LF, Liu C-C, Alberts BM (1979). T4 DNA topoisomerase: a new ATP-dependent enzyme essential for initiation of T4 bacteriophage DNA replication. Nature 281:456.
12. Durnford JM, Champoux JJ (1978). The DNA untwisting enzyme from Saccharomyces cerevisiae. J Biol Chem 253:1086.
13. Badaracco G, Plevani P, Ruyechan WT, Chang LMS (1983). Purification and characterization of yeast topoisomerase I. J Biol Chem 258:2022.

14. Goto T, Wang JC (1982). Yeast DNA topoisomerase II: an ATP-dependent type II topoisomerase that catalyzes the catenation, decatenation, unknotting, and relaxation of double-stranded DNA rings. J Biol Chem 257:5866.
15. Hartwell LH (1967). Macromolecule synthesis in temperature-sensitive mutants of yeast. J Bact 93:1662.
16. Sternglanz R, DiNardo S, Voelkel KA, Nishimura Y, Hirota Y, Becherer K, Zumstein L, Wang JC (1981). Mutations in the gene coding for Escherichia coli DNA topoisomerase I affect transcription and transposition. Proc Natl Acad Sci USA 78:2747.
17. Hartwell LH, Culotti J, Reid B (1970). Genetic control of the cell division cycle in yeast. I. Detection of mutants. Proc Natl Acad Sci USA 66:352.
18. Hartwell LH (1973). Three additional genes required for deoxyribunucleic acid synthesis in Saccharomyces cerevisiae. J Bact 115:966.
19. Zakian VA, Brewer BJ, Fangman WL (1979). Replication of each copy of the yeast 2 micron DNA plasmid occurs during the S phase. Cell 17:923.
20. Sundin O, Varshavsky A (1981). Arrest of segregation leads to accumulation of highly intertwined catenated dimers: Dissection of the final stages of SV40 DNA replication. Cell 25:659.
21. Wickner RB, Leibowitz MJ (1976). Chromosomal genes essential for replication of a double-stranded RNA plasmid of Saccharomyces cerevisiae: the killer character of yeast. J Mol Biol 105:427.

Mechanisms of DNA Replication and Recombination, pages 43–53
© 1983 Alan R. Liss, Inc., 150 Fifth Avenue, New York, NY 10011

BREAKAGE AND REUNION OF DNA HELIX BY MAMMALIAN DNA TOPOISOMERASE II[1]

Leroy F. Liu, Brian D. Halligan, Eric M. Nelson, Thomas C. Rowe, Grace L. Chen and Kathleen M. Tewey

Department of Physiological Chemistry, Johns Hopkins Medical School, Baltimore, Maryland 21205

ABSTRACT Upon addition of protein denaturants to a reaction mixture containing mammalian topoisomerase II and DNA, cleavage of DNA occurs with concomitant attachment of the enzyme to the broken DNA strands. Direct nucleotide sequencing analyses indicate that, similar to bacterial DNA gyrase, mammalian topoisomerase II cuts DNA with a four base stagger and is covalently linked to the protruding 5'-phosphoryl end of each broken DNA strand. The reversibility of this cleavage reaction suggests that it may be a partial reaction of DNA strand passing catalyzed by mammalian topoisomerase II. The interaction between mammalian topoisomerase II and its natural substrate, chromatin, has been investigated using simian virus 40(SV40) minichromosomes as a model system. We have demonstrated that DNA topoisomerase II is associated with the isolated SV40 minichromosomes and the virions both by activity assays and antibody blotting. Furthermore, purified topoisomerase II can promote the catenation and decatenation of purified SV40 chromatin, suggesting a function of topoisomerase II in termination of DNA replication.

[1]This work was supported by an NIH grant GM27731 and the Searle Scholar Program.

INTRODUCTION

Eukaryotic DNA topoisomerase II has been purified from a number of organisms (1-3). All in vitro studies of the enzymatic activities of eukaryotic DNA topoisomerase II indicate that it is similar to the bacteriophage T4 induced DNA topoisomerase (4,5). These similarities include (1) ATP(or dATP)-dependent strand passing reactions, including relaxation, knotting/unknotting and catenation/decatenation. (2) Lack of DNA supercoiling activity. (3) DNA-dependent ATPase activity. The biological function(s) of eukaryotic DNA topoisomerase II is still unknown but the similarities between eukaryotic DNA topoisomerase II and T4 DNA topoisomerase suggest possible similar functions. In order to study the function(s) and mechanism of eukaryotic DNA topoisomerase II, we have purified DNA topoisomerase II from several mammalian organisms to near homogeneity based on the unknotting assay we have developed earlier (6). We report here our studies of the interaction between purified topoisomerase II and DNA. In addition, we have initiated studies on the interaction between topoisomerase II and chromatin using SV40 minichromosomes as a model system.

RESULTS

Cleavage of DNA by mammalian DNA topoisomerase II. At high enzyme to DNA ratio, purified calf thymus DNA topoisomerase II introduces both single and double stranded DNA breaks. Figure I shows such an experiment. With increasing topoisomerase II concentration (lanes B to H), form I plasmid DNA is converted to form II and form III. This cleavage reaction is not ATP-dependent and occurs instantaneously upon incubation at 37°C. Mg (II) stimulates the reaction but is not absolutely required for the cleavage reaction. This cleavage reaction differs from the nuclease reaction in two major aspects: (1) All the cleavage products are covalent protein DNA complexes. We have demonstrated that topoisomerase subunit is linked covalently to the protruding 5'-phosphoryl ends of the broken DNA strands and the other end is a protein free 3'-OH end. Nucleotide sequencing studies of twelve enzyme

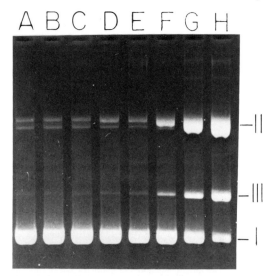

FIGURE 1. Cleavage of DNA by calf thymus DNA topoisomerase II. Plasmid pBR322 DNA was treated with increasing amount of purified calf thymus DNA topoisomerase II. Lane A is the untreated pBR322 DNA. Lanes B to H are pBR322 DNA samples (0.5 μg each) treated with 1.3 ng, 4 ng, 12 ng, 37 ng, 111 ng, 333 ng and 1 μg of calf thymus topoisomerase II respectively. All samples were incubated at 37°C for 3 min in reaction mixtures containing 10 mM Tris, pH 7.2, 1 mM $MgCl_2$, 0.5 mM EDTA and 30 μg/ml BSA. Reactions were stopped by 1% SDS and proteinase K treatment and then analyzed on an agarose gel.

cleavage sites have revealed that the cleavage sites are double strand breaks and the breaks are staggered by four base pairs. No consensus sequence has been deduced from the sequences around the cut sites. (2) The cleavage reaction is reversible. Addition of 0.5 M NaCl to the reaction mixture after the incubation resulted in the instantaneous reversal of the cleavage reaction. Interestingly, all the reversed products resemble the starting materials in that the plasmid DNA is negatively twisted. This experiment demonstrates that the cleavage reaction is dependent upon the protein denaturant treatment. Furthermore, the state of the protein DNA complex is such that the two putative broken ends are tightly held by the topoisomerase subunits and the relative rotation or spontaneous release of the two ends are not allowed. We have also observed that the cleavage reaction can be slowly reversed by lowering the temperature. The reversibility of this cleavage reaction is reminiscent of the cleavage reaction of E. coli topoisomerase I (7). Because of the reversibility of the cleavage reaction, it seems likely that the cleavage reaction may represent a partial reaction of topoisomerase II catalyzed DNA strand passing reaction. It is interesting that the ratio of single-strand and double-strand breaks introduced by calf thymus DNA topoisomerase II varies from preparation to preparation. It is possible that most single-strand breaks are produced by damaged topoisomerase II (e. g. one subunit may be damaged by sulfhydryl oxidation).

Figure 2 shows a schematic drawing of a possible structure of the enzyme DNA complex. Each topoisomerase subunit is covalently linked to the 5'-phosphoryl end of the putative broken DNA strands. The four base stagger may or may not be physically melted for this partial reaction.

We have no information on the detailed structure of this complex. This complex may be at equilibrium with the uncleaved version of the complex. This partial reaction can therefore be considered as a breakage and reunion reaction which does not result in any topological

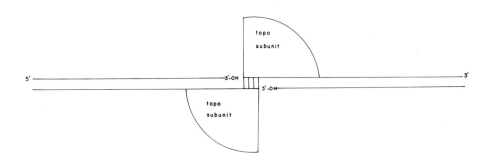

FIGURE 2. A schematic drawing of a possible structure of the topoisomerase II DNA complex revealed by the cleavage reaction..

rearrangement of DNA because of the enzyme bridging. The collision of another DNA segment with this complex may trigger ATP hydrolysis and the active transport of the DNA segment through this putative break by a series of conformational changes of the enzyme-DNA complex. It is interesting to observe that treatment of calf thymus topoisomerase II with 0.2 mM N-ethyl maleimide on ice for 30 min totally abolishes the DNA strand passing activity of topoisomerase II but the cleavage reaction is not significantly affected. It is possible that the ATPase function is more sensitive to inactivation by sulfhydryl reagents.

Association of DNA topoisomerase II with SV40 minichromosomes and isolated virions. We have shown that HeLa topoisomerase II is associated with isolated nucleosomes. Furthermore, we have demonstrated that topoisomerase II is associated with metaphase chromosomes and the protein scaffold derived from metaphase chromosomes (Halligan, R. D., W. Earnshaw and Liu, L. F., unpublished results). To further study the interaction between topoisomerase II and chromatin, we isolated SV40 minichromosomes from SV40 infected BSC cells using sucrose density gradient sedimentation in different salt concentrations. Figure 3 shows the topoisomerase II assays across such a sucrose density gradient.

Topoisomerase II activity is assayed by the P4 unknotting assay (6). The conversion of P4 knotted DNA (the smear in the gel) to the simple P4 circle (the slowest migrating band in gel) parallels the amount of SV40 minichromosomes in the sucrose gradient. Little if any activity is detected near the top of the sucrose gradient. We have also noticed that topoisomerase II completely dissociated from SV40 minichromosomes when the salt in the sucrose gradient was above 0.2 M NaCl. The association of topoisomerase II with SV40 minichromosomes was further demonstrated by immunological staining of the SDS polyacrylamide gel containing samples of SV40 minichromosomes across the sucrose gradient (figure 4).

In this case, it is clear that topoisomerase II is Associated with SV40 minichromosomes. Fractions containing SV40 DNA parallel well with the stained protein band (topoisomerase II) in the SDS polyacrylamide gel. To investigate whether topoisomerase II is also packaged into the virions together with the SV40 chromatin, we have also

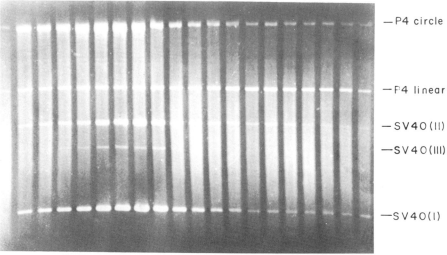

C 5 7 9 11 13 15 17 19 21 23 25 27 29 31 33 35 41 45 47

—P4 circle

—P4 linear

—SV40(II)

—SV40(III)

—SV40(I)

FIGURE 3. Association of topoisomerase II activity with the SV40 minichromosomes. SV40 minichromosomes were isolated following the published procedure (12) with some modifications. SV40 minichromosomes extracted from the infected nuclei at different NaCl concentrations at about 36 hrs post infection were sedimented in a 15–30% sucrose density gradient containing 10 mM Tris, pH 7.2, 1 mM PMSF, 0.5 mM DTT, 0.5 mM EDTA and NaCl (the same concentrations of NaCl were used for nuclei extraction and sucrose gradient sedimentation). Assay of topoisomerase II activity by P4 unknotting reaction was performed as described (6). In this particular experiment, no NaCl was used during nuclei extraction and sucrose sedimentation.

purified SV40 virions from CsCl gradient and blotted the SDS polyacrylamide gel of the virion sample with topoisomerase II antibodies (last lane in gel). Topoisomerase II is packaged into the SV40 virions.

Catenation and Decatenation of SV40 Chromatin by Mammalian DNA Topoisomerase II.

In order to investigate how eukaryotic DNA topoisomerase II manages its natural substrate, chromatin, we have initiated studies using purified SV40 minichromosomes as substrates. As shown in lanes B to E(Fig. 5), SV40 minichromosomes can indeed be

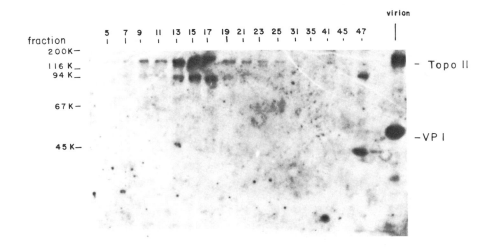

FIGURE 4. Detection of the presence of topoisomerase II in SV40 minichromosomes and SV40 virions by immunological staining. Immunological staining of SDS polyacrylamide gel containing samples of SV40 minichromosomes from the sucrose gradient and purified SV40 virions was done according to the published procedures (14). IgG fraction of the antibodies against calf thymus topoisomerase II and iodinated protein A were used for staining.

catenated by eukaryotic DNA topoisomerase II in vitro. Interlocked dimer, trimer and higher oligomers of SV40 minichromosomes increase with increasing concentrations of topoisomerase II.

In this reaction, no catenation factors were added to the reaction mixtures. The formation of the interlocked minichromosomes must be favored by the aggregation of minichromosomes in the topoisomerase II reaction mixture. At higher enzyme concentrations, cleavage of SV40 DNA into the linear forms (form III) was observed. This cleavage reaction generates linear SV40 DNA with tightly linked proteins. This is seen by comparing lanes E and F. Samples in lanes E and F are identical except that proteinase K treatment was omitted in the sample of lane F. The SV40 linear DNA in lane F (form III') migrated slightly slower than that in Lane E (form III), consistent with

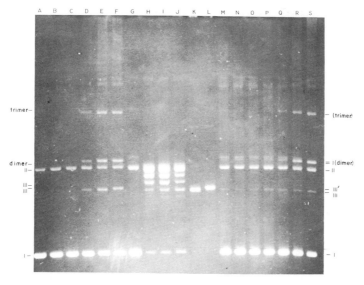

FIGURE 5. Catenation and decatenation of SV40 minichromosomes by calf thymus DNA topoisomerase II. SV40 minichromosomes used in this experiment were purified from salt (0.3 M NaCl) extracted nuclei. Nearly all detectable topoisomerase activities (both type I and type II) were purified away from the minichromosomes. Concentrations of calf thymus topoisomerase II in lanes B to E are 1 ng, 10 ng, 100 ng and 1 µg respectively. All reactions were stopped by 1% SDS. Except samples in lanes F and L, all samples were also treated with proteinase K before loading onto the agarose gel. Electrophoresis was done as usual except that 0.1% SDS was present both in the agarose gel and the electrophoresis buffer.

protein binding to the linearized SV40 DNA. To test whether interlocked SV40 minichromosomes can be decatenated by topoisomerase II treatment, SV40 minichromosomes were pooled near the replicative intermediate region of the sucrose gradient. Interlocked SV40 minichromosome dimers (the doublet in lane M) were enriched in this pool. Treatment of this minichromosome pool with increasing amount of topoisomerase II resulted in the conversion of the doublet dimer band into a singlet dimer band and the formation of trimer minichromosomes (see lanes R and S), suggesting that decatenation and catenation of SV40 minichromosomes were occurring in the reactions. When purified SV40 DNA was used instead of the SV40 chromatin (compare lanes G to K with lanes A to F), relaxation of SV40 DNA occurred rapidly. The cleavage reaction was also more efficient on purified SV40 DNA. The fact that SV40 DNA remained supercoiled after topoisomerase II treatment of SV40 minichromosomes thus indicated that the nucleosomal structure of the minichromosomes was not significantly changed during the incubation.

DISCUSSION

DNA topoisomerases are ubiquitous enzymes (for reviews, see ref. 7-10). Type II DNA topoisomerases from a number of eukaryotic organisms have been purified (1-3). Eukaryotic DNA topoisomerase II is most likely a dimer with a monomer polypeptide molecular weight of about 170,000 daltons. Because of the strand passing reactions catalyzed by type II DNA topoisomerases, it seems most likely that a transient double-stranded DNA break on one of the two crossing DNA double helices may exist as the reaction intermediate. In the case of bacterial DNA gyrase, such a double strand break have been demonstrated and chracterized (9,10). Eukaryotic DNA topoisomerase II has enzymatic properties very similar to baterial DNA gyrase and especially T4 bateriophage encoded DNA topoisomerase (4,5,9,10). Our present studies of the cleavage reaction of calf thymus DNA topoisomerase II further illustrated the similarities among the characterized type II DNA topoisomerases from both prokaryotic and eukaryotic cells. Similar to bacterial DNA gyrase, calf thymus DNA topoisomerase II can be trapped as a covalent protein DNA complex. The structure of the complex is very similiar to that of DNA gyrase. Both enzymes can cut DNA with a four base stagger and are linked covalently to each protruding 5' phosphoryl end of the broken DNA strands. Despite the similarities among all type II DNA topoisomerases, bacterial DNA gyrase appear to be unique in its ability to supertwist DNA. Bacterial DNA gyrase is also unique in its ability to organize a large piece of DNA on its surface in a specific wrapping (11). We have failed to demonstrate that eukaryotic DNA topoisomerase II or T4 DNA topoisomerase can organize such a wrapping using purified plasmid DNAs. Perhaps, DNA wrapping is essential for DNA supercoiling activity by all type II DNA topoisomerases. It is interesting to test whether passive DNA strand passing reactions(e.g. unknotting and decatenation) catalyzed by DNA gyrase result from gyrase DNA interaction which does not involve DNA wrapping. The failure of eukaryotic topoisomerase II and T4 DNA topoisomerase to supertwist DNA may be related to their inability to organize DNA into specific wrap. Whether eukaryotic DNA topoisomerase II has DNA supertwisting activity in vivo is of fundamental importance to our understanding of many genetic functions in eukaryotic cells. Studies of eukaryotic DNA topoisomerase II and T4 DNA topoisomerase in parallel may shed much light on this important problem.

How eukaryotic DNA topoisomerases transect their natural substrate, chromatin, is a totally open area of reserarch. We initiated our studies using the well characterized SV40 minichromosome system. Our results have clearly established that DNA topoisomerase II is physically associated with the SV40 chromatin obtained from either the intracellular SV40 minichromosomes or virions. We have also demonstrated that purified calf thymus DNA topoisomorase II can promote catenation and decatenation of SV40 chromatin. It is not apparent how topoisomerase II may pass chromatin fibers topologically. One possibility for such a reaction is that the isolated SV40 chromatin contains a nucleosome free region which may be the site of topoisomerase II action. In this regard, it is interesting to note that a 400 bp stretch of nucleosome free region has been demonstrated to exist in a significant fraction of isolated SV40 minichomosomes (12). The significance of this in vitro reaction is not clear. It is possible, however, that topoisomerase II may be involved in the segregation of multiply intertwined SV40 minichromosomes during the terminal stage of SV40 DNA replication (13). In view of the multiple functions of E. coli DNA gyrase in various genetic processes, it seems possible that segregation of DNA replication may not be the sole function of eukaryotic DNA topoisomerase II. We are currently mapping the binding site(s) of topoisomerase II on SV40 chromatin using the cleavage reaction we have charaterized in vitro. How multiple eukaryotic DNA topoisomerases regulate the topological states of chromatin and affect various genetic processes in eukaryotic cells remains to be elucidated.

REFERENCES

1. Hsieh TS, Brutlag D (1980) ATP-dependent DNA topoisomerase from Drosophila melanogaster reversibly catenates duplex DNA rings. Cells 21:115.
2. Miller KG, Liu LF, Englund PT (1981) A homogeneous type II DNA topoisomerase from HeLa cell nuclei. J. Biol. Chem. 256:9334.
3. Goto T, Wang JC (1982) Yeast DNA topoisomerase II, an ATP-dependent type II topoisomerase that catalyzes the catenation, decatenation, unknotting, and relaxation of double-stranded DNA rings. J. Biol. Chem. 257:5866.
4. Liu LF, Liu CC, Alberts BM (1979) T4 DNA topoisomerase: a new ATP-dependent enzyme essential for initiation of T4 bacteriophage DNA replication. Nature 281:456.
5. Liu LF, Liu CC, Alberts BM (1980) Type II DNA topoisomerase: enzyme that can unknot a topologically knotted DNA molecule via a reversible double-strand break. Cell 19:697.
6. Liu LF, Davis JL, Calendar R (1981) Novel topologically knotted DNA from bacteriophage P4 capsids: studies with DNA topoisomerases. Nucl. Acids Res. 9:3979.
7. Wang JC, Liu LF (1979) DNA topoisomerases: enzymes that catalyze the breaking and rejoining of DNA backbone bonds, in Molecular genetics, Pt. 3, Taylor JH, ed., academic press, new york, pp 65.
8. Champoux JJ (1978) Proteins that affect DNA conformation. Ann Rev. Biochem. 47:449
9. Gellert M (1981) DNA topoisomerases. Ann. Rev. Biochem. 50:879.
10. Cozzarelli NR (1980) DNA topoisomerases. Cell 22-327.
11. Liu LF, Wang JC (1978) DNA-DNA gyrase complex: the wrapping of the DNA duplex outside the enzyme. Cell 15-979.
12. Varshavsky AJ, Sundin OM, Bohn M (1979) A stretch of 'late' SV40 viral DNA about 400 bp long which includes the origin of replication is specifically exposed in SV40 minichromosomes. Cell 16-453.
13. Sundin O, Varshavsky (1980) Terminal stages of SV40 DNA replication proceed via multiply intertwined catenated dimers. Cell 21:103.
14. Towbin H, Staehelin T, Gordon J (1979) Electrophoretic transfer of proteins from polyacrylamide gels to nitrocellulose sheets: procedure and some applications. Proc. Natl. Acad. Sci. USA. 76:4350.

Mechanisms of DNA Replication and Recombination, pages 55–64
© **1983 Alan R. Liss, Inc., 150 Fifth Avenue, New York, NY 10011**

RECOGNITION OF SUPERCOILED DNA BY
DROSOPHILA TOPOISOMERASE II[1]

Neil Osheroff[2] and Douglas L. Brutlag

Department of Biochemistry, Stanford University
School of Medicine, Stanford, California   94305

The type II topoisomerase from the fly, Drosophila
melanogaster, is able to distinguish the topological
structure of DNA and interacts preferentially with
negatively supercoiled over relaxed molecules.
Furthermore, this study indicates that the enzyme
recognizes points of helix-helix proximity found in
supercoiled DNA and upon binding, condenses its
nucleic acid substrate.

INTRODUCTION

Topoisomerases from prokaryotes are able to distinguish
different topological isomers of DNA and interact preferentially with the form of DNA which serves as the substrate
for their supercoiling-relaxation activities. Thus,
Escherichia coli DNA gyrase, a type II topoisomerase that
induces negative superhelical twists in nucleic acid molecules, displays a 4-fold higher kinetic affinity for substrates which are relaxed or topologically unconstrained
over those which are supercoiled (1). Moreover, preformed
complexes between gyrase and DNA dissociate more rapidly as
superhelical density increases (2). In contrast, E. coli ω
protein, a type I topoisomerase that relaxes nucleic acid
molecules, binds more readily to negatively supercoiled

[1]Supported by Grant GM-28079 from the National
Institutes of Health.
[2]During the course of this study, N.O. was a post-
doctoral fellow of the Helen Hay Whitney Foundation.
Present address: Department of Biochemistry, Vanderbilt
University School of Medicine, Nashville, Tennessee 37232.

DNA (3).

     In order to better elucidate the specificity of
eukaryotic topoisomerases, the interaction between the type
II enzyme from the fly, Drosophila melanogaster, and DNA
was examined. Like the prokaryotic enzymes, the Drosophila
enzyme can discern the topology of DNA. It shows an increased affinity for supercoiled molecules, apparently by
recognizing points of helix-helix proximity which are induced by superhelical twisting. In addition, when bound to
supercoiled molecules, topoisomerase II condenses the DNA
structure. Based on these findings, novel conformational
models for the complexing of DNA by eukaryotic topoisomerase II and implications for the biological function of the
enzyme are presented.

                           METHODS

     All interactions between purified D. melanogaster
topoisomerase II (4) and DNA were at 30°C in 20 µl of 10 mM
Tris (Cl) pH 7.9, 50 mM NaCl, 50 mM KCl, 5 mM MgCl$_2$, 0.1 mM
EDTA, and 15 µg/ml BSA. Detailed procedures for the relaxation and ATPase assays are given in Osheroff et al. (5).

                           RESULTS

Specificity of Topoisomerase II for Supercoiled DNA.

     Three lines of evidence indicate that the type II
topoisomerase from D. melanogaster interacts preferentially
with supercoiled DNA over relaxed molecules.
     Kinetics of relaxation of supercoiled DNA. The time
course for the relaxation of negatively supercoiled closed
circular DNA by the Drosophila enzyme shows an unusually
long linear phase, with as much as 75% of the DNA being relaxed before the velocity of the reaction begins to decrease
(figure 1). This is a strong indication that the kinetic
affinity of the enzyme is several fold higher for supercoiled molecules than it is for relaxed DNA. Interestingly,
the rate of relaxation of positively supercoiled molecules
is comparable to that found for negatively supercoiled DNA
(5). Thus, while the Drosophila protein can discriminate
supercoiled from relaxed molecules, it cannot distinguish
between DNAs with negative or positive superhelical twists.

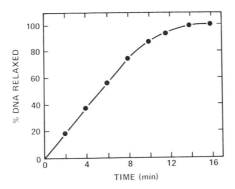

FIGURE 1. Time course for the relaxation of negatively supercoiled closed circular DNA by Drosophila topoisomerase II. Reaction mixtures contained 15 fmol of enzyme and
0.3 μg (0.1 pmol) of pBR322 plasmid DNA. Samples were resolved by electrophoresis on agarose gels in 100 mM Tris-
borate, 2 mM EDTA pH 8.3 and quantitated by scanning densitometry, as previously described (5).

ATPase activity. Although ATP binding to the Drosophila topoisomerase II is sufficient to induce a double-
strand DNA passage event, hydrolysis is required for enzyme
turnover (5,6). When the relaxation of DNA is monitored
under optimal conditions, the reaction is processive and
approximately 4 molecules of ATP are converted to ADP and
Pi for every superhelical twist removed (5). As can be seen
in figure 2, the ATPase activity of topoisomerase II is DNA
dependent, being stimulated 17-fold by the presence of negatively supercoiled DNA. Moreover, the enzyme is sensitive
to the topological state of the nucleic acid substrate, as
supercoiled molecules yield a reaction rate which is 4-fold
higher than generated with relaxed or linear DNA. Despite
the fact that the enzyme is able to accommodate single-
stranded DNA in its active site (5), such structures are
clearly not intermediates in the relaxation process, as
single-stranded DNA is a very poor activator of the protein's
ATPase (figure 2), even at concentrations which are nearly
2500-fold higher than its K_i value ($\sim 10^{-6}$ M) for relaxation.

Competitive binding. In the absence of its ATP cofactor, Drosophila topoisomerase II and DNA form a non-
covalent complex. Although stable to electrophoresis in
agarose gels, this interaction can be dissociated by the
addition of SDS at 0°C. When incubated with an equimolar

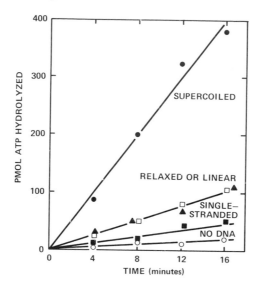

FIGURE 2. ATPase activity of topoisomerase II. Reaction mixtures contained 75 fmol of enzyme, 15 μg (5 pmol) of pBR322 plasmid DNA, and 1 mM ATP. Single-stranded DNA was made by heating linear pBR322 DNA to 70°C for 2 minutes. Activity was followed by monitoring the release of ^{32}Pi from [γ-^{32}P]ATP, as previously described by Osheroff et al. (5).

mixture of negatively supercoiled, relaxed, and nicked circular molecules, under conditions ranging from no addition of detergent to nearly complete disruption with 0.1% SDS, the enzyme always binds 4 to 12-fold more supercoiled than relaxed or nicked DNA (figure 3). Thus, the recognition of topological structures by the enzyme does not require the presence of a nucleotide cofactor.

Catenation of Supercoiled DNA with Stoichiometric Levels of Topoisomerase II.

The catenation of closed circular duplex molecules of DNA by catalytic levels of topoisomerase II requires the presence of a DNA condensing agent such as spermidine (7,8) or histone H1 (6) (figure 4, lane B). However, catenanes can also be produced by incubating the DNA with high levels of Drosophila topoisomerase II in the presence of APP(NH)P (adenyl-5'-yl-imidodiphosphate), an ATP analogue which is

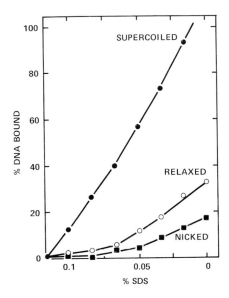

FIGURE 3. Competitive binding of DNA to topoisomerase II. Reaction mixtures contained 2 pmol of enzyme and 0.2 µg (70 fmol) each of supercoiled, relaxed, and nicked pBR322 DNA. Resolution and quantitation of samples are as in figure 1.

sterically similar to ATP (10), but contains no hydrolyzable β-γ phosphonate bond (figure 4, lane D). A comparable re-sult has been found by Hsieh (9), who employed nicked cir-cular molecules, a high ratio of enzyme to DNA, and ATP. These findings strongly suggest that the enzyme itself can condense the DNA. This idea is further supported by the fact that the Drosophila protein, like the type II topo-isomerase from bacteriophage T4 (11), can tie knots in closed circular DNA (unpublished results) (9).

The mechanism of the topoisomerase II induced conden-sation of supercoiled DNA is not known. Although the en-zyme may act via charge neutralization, as is the case for the polycationic agents (6-8), other processes are clearly feasible. As a singular example of another mechanism, poly-vinyl alcohol, which presumably compacts DNA by the process of 'volume exclusion' (12,13) has been successfully employed to promote the enzymic catenation of closed circular DNA (figure 4, lane C).

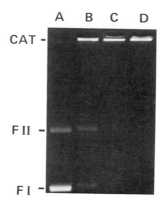

FIGURE 4. Catenation of DNA by topoisomerase II. All reactions contained 0.3 μg (0.1 pmol) of pBR322 closed circular supercoiled DNA. Lanes B and C employed ATP (1 mM) while lane D employed APP(NH)P (1 mM). Lane A, no enzyme; Lane B, enzyme (15 fmol) + histone H1 (0.12 μg); Lane C, enzyme (15 fmol) + polyvinyl alcohol (8%); Lane D, enzyme (2.5 pmol). Prior to electrophoresis on agarose gels (see figure 1), samples were treated with 0.1% SDS - 10 mM EDTA and digested with 1.25 μg of Proteinase K for 20 minutes at 37°C to completely disrupt the binding between protein and DNA. FI, supercoiled DNA; FII, nicked DNA; CAT, catenated networks.

DISCUSSION

The type II topoisomerase from Drosophila melanogaster can discern the topological structure of DNA and interacts preferentially with supercoiled molecules, even in the absence of its ATP cofactor. This specificity for supercoiled DNA may also extend to the type I enzyme from Drosophila, as the addition of relaxed molecules has virtually no effect on the time course for the catalytic removal of negative superhelical twists (unpublished results).

Recognition of Supercoiled DNA by D. melanogaster Topoisomerase II.

As negative superhelical density increases, the structure of closed circular double-stranded DNA changes in many respects: regions with single-stranded character are formed in the underwound double helix (14); the conversion of right-handed B-DNA to left-handed Z-DNA is facilitated (15); cruciform structures are stabilized (16); increased writhing leads to points of helix-helix proximity (17,18). Since the Drosophila topoisomerase II recognizes supercoiled molecules, it is logical that the enzyme can discern one or more of these changes in the helical structure of DNA. Recognition of regions with single-stranded character may be eliminated because 1) single-stranded DNA is a very poor activator of the protein's ATPase (figure 2) and 2) the rate of relaxation of positively supercoiled DNA, which is overwound, is comparable to that found for underwound negatively supercoiled molecules (5). For this second reason, interaction of the enzyme with Z-DNA sequences or cruciform structures may also be excluded, as positive superhelical twisting would destabilize both conformations. However, both negatively and positively supercoiled DNAs contain juxtaposed helices. Therefore, it is likely that topoisomerase II displays a higher affinity for supercoiled over relaxed molecules because it can recognize points of helical proximity in DNA. This conclusion is also consistent with the finding that the Drosophila enzyme can condense DNA (figure 4). As described below, the protein may compact nucleic acids by bridging adjacent helices (see figure 5).

Condensation of Supercoiled DNA by Topoisomerase II.

Negatively supercoiled circular DNA can exist in a number of conformations, ranging from a right-handed interwound structure to a left-handed solenoid (figure 5, A and C). Although these are homeomorphic conformers and have identical linking numbers, the solenoid is clearly the more compact of the two structures. The interwound form, which predominates in nature, is energetically more favorable because it minimizes twist deformation (18). Considering these facts, we suggest that the condensation of supercoiled DNA by stoichiometric levels of the Drosophila topoisomerase II involves the following mechanism. When bound to the interwound form, the enzyme induces a localized conformational transition in the DNA as shown in figure 5B. After several protein molecules are bound, the conformational equilibrium is shifted from the interwound to the solenoid form, thereby condensing the structure of the DNA. This model, which has also been proposed for the interaction

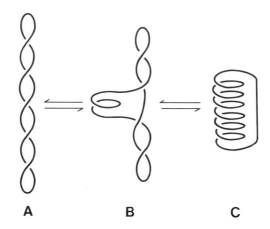

FIGURE 5. Conformation of negatively supercoiled DNA. The right-handed interwound (A) and left-handed solenoid (C) forms are shown. The structure in B, which contains a localized solenoid conformation, depicts a transitional intermediate between these two forms.

of the catabolite activator protein with supercoiled DNA (19), is currently being tested by electron microscopy.

Implications for Biological Activity.

Two major conclusions concerning the in vivo function of eukaryotic type II topoisomerases may be drawn from this study. First, the fact that the Drosophila protein can discern the topological state of DNA is of consequence to its biological activity, because it allows the enzyme to interact specifically with its substrates, yet still turn over rapidly. Second, by recognizing points of helix-helix proximity, topoisomerase II is able to distinguish both supercoiled DNA and interlocked DNA complexes as substrates. Teleologically, this concept is very important. By employing the double-strand DNA passage reaction to resolve interlocked multimolecular structures which are the products of DNA replication, eukaryotic type II topoisomerases may function in such fundamental cellular processes as chromosome segregation (20-22).

ACKNOWLEDGMENTS

We are indebted to Dr. A. Kornberg for his suggestion to employ polyvinyl alcohol as a condensation agent, to Dr. T. Hsieh for sharing his results concerning catenation and knotting prior to publication, to Dr. E.R. Shelton and Dr. C. Glover for many valuable discussions, and to K. Griffin for the preparation of this paper.

REFERENCES

1. Sugino A, Cozzarelli NR (1980). The Intrinsic ATPase of DNA Gyrase. J Biol Chem 255:6299.
2. Morrison A, Higgins NP, Cozzarelli NR (1980). Interaction between DNA Gyrase and Its Cleavage Site on DNA. J Biol Chem 255:2211.
3. Liu LF, Wang JC (1979). Interaction between DNA and Escherichia coli DNA Topoisomerase I. J Biol Chem 254:11082.
4. Shelton ER, Osheroff N, Brutlag DL (1983). DNA Topoisomerase II from Drosophila melanogaster: Purification and Characterization. J Biol Chem, in press.
5. Osheroff N, Shelton ER, Brutlag DL (1983). DNA Topoisomerase II from Drosophila melanogaster: Relaxation of Supercoiled DNA. J Biol Chem, in press.
6. Hsieh T-S, Brutlag DL (1980). ATP-Dependent DNA Topoisomerase from D. melanogaster Reversibly Catenates Duplex DNA Rings. Cell 21:115.
7. Kreuzer KN, Cozzarelli NR (1980). Formation and Resolution of DNA Catenanes by DNA Gyrase. Cell 20:245.
8. Krasnow MA, Cozzarelli NR (1982). Catenation of DNA Rings by Topoisomerases. J Biol Chem 257:2687.
9. Hsieh T-S (1983). Knotting of the Circular Duplex DNA by Type II DNA Topoisomerase from Drosophila melanogaster. J Biol Chem, in press.
10. Yount RG, Babcock D, Ballantyne W, Ojala D (1971). Adenylyl Imidodiphosphate, an Adenosine Triphosphate Analog Containing a P-N-P Linkage. Biochemistry 10:2484.
11. Liu LF, Liu C-C, Alberts BM (1980). Type II DNA Topoisomerases: Enzymes that Can Unknot a Topologically Knotted DNA Molecule via a Reversible Double-Strand Break. Cell 19:697.
12. Tanford C (1961). "Physical Chemistry of Macromolecules." New York: Wiley.

13. Fuller RS, Kaguni JM, Kornberg A (1981). Enzymatic Replication of the Origin of the Escherichia coli Chromosome. Proc Natl Acad Sci USA 78:7370.
14. Wang JC (1974). Interactions between Twisted DNAs and Enzymes: The Effects of Superhelical Turns. J Mol Biol 87:797.
15. Nordheim A, Lafer EM, Peck LJ, Wang JC, Stollar BD, Rich A (1982). Negatively Supercoiled Plasmids Contain Left-handed Z-DNA Segments as Detected by Antibody Binding. Cell 31:309.
16. Mizuuchi K, Mizuuchi M, Gellert M (1982). Cruciform Structures in Palindromic DNA are Favored by DNA Supercoiling. J Mol Biol 156:229.
17. Crick FHC (1976). Linking Numbers and Nucleosomes. Proc Natl Acad Sci USA 73:2639.
18. Bauer WR, Crick FHC, White JH (1980). Supercoiled DNA. Sci Am 243:118.
19. Salemme FR (1982). A Model for Catabolite Activator Protein Binding to Supercoiled DNA. Proc Natl Acad Sci USA 79:5263.
20. Marini JC, Miller K, Englund P (1980). Decatenation of Kinetoplast DNA by Topoisomerases. J Biol Chem 255:4976.
21. Gellert M (1981). DNA Topoisomerases. Ann Rev Biochem 50:879.
22. Sundin O, Varshavsky A (1981). Arrest of Segregation Leads to Accumulation of Highly Intertwined Catenated Dimers: Dissection of the Final Stages of SV40 DNA Replication. Cell 25:659.

Mechanisms of DNA Replication and Recombination, pages 65–76
© **1983 Alan R. Liss, Inc., 150 Fifth Avenue, New York, NY 10011**

FUNCTIONS OF DNA HELICASES IN *Escherichia coli*

G. Taucher-Scholz, M. Abdel-Monem, and H. Hoffmann-Berling

Max-Planck-Institut für medizinische Forschung, Abteilung
Molekulare Biologie, Jahnstr. 29, D-6900 Heidelberg, F.R.G.

ABSTRACT The genes for two DNA helicases of *Escherichia coli* have been identified: Helicase I is specified by the *tra*I gene of the F sex factor while heli-
case II is specified by the bacterial *uvr*D gene. We
further present results of transduction studies which
suggest that *uvr*D mutation is incompatible with *rep*
mutation. We propose that during replication the bacterial chromosome is unwound through the joint action
of helicase II and *rep* helicase. Because of their
different directions of action relative to the chemical polarity of DNA, these enyzmes would act on
opposite arms of the fork. We further propose that one
of the helicases alone is capable of shifting the fork.
Organization of the chromosomal fork with both helicase II and *rep* helicase might be of little advantage
to the cell as long as its DNA is undamaged. Once
damage is inflicted upon the DNA, however, two helicases acting on opposite sides of the fork might gua-
rantee continuation of the replication process beyond
single strand interruptions of the template duplex.

INTRODUCTION

Four enzymes capable of unwinding long stretches of
DNA at the expense of ATP have been identified in *Escherichia coli*, namely the helicases I, II, III, and *rep* protein
(1). Until recently *rep* protein was the only helicase for
which the gene was known. *rep* Mutants are characterized by
inability to replicate the duplex DNAs of some bacteriophages, including ϕX174 and fd (M13), but in other respects
their DNA metabolism seems to be little impaired (2). There
has been a report on a temperature-labile helicase II (3);

however, it was also stated that after the elimination of additional genetic defects the mutant cells were able to grow at elevated temperature. Surprising is also the fact that there is no helicase mutant among the many chromosomal replication mutants which have been characterized. The question thus arises whether the helicases of *E. coli* are capable of substituting each other to overcome mutational defects or whether the bacterial DNA, during replication, is unwound by an unidentified enzyme or DNA binding protein.

IDENTIFICATION OF THE GENE FOR HELICASE I

In studying this problem we have identified the genes for helicase I and helicase II. Helicase I is the product of one of the transfer genes, *tra*I, of the F sex factor (4). The enzyme is thus detectable in F^+ cells and transconjugant cells propagating the F factor, but not F^- cells. The production of helicase I was further induced by a hybrid DNA obtained by cloning *tra*I in a plasmid vector. Using derivatives of this hybrid plasmid prepared in M. Achtman's laboratory (5) we have found that insertion of transposon Tn5 into the cloned *tra*I gene abolishes the production of helicase I and, further, that deletion of part of the cloned gene gives rise to a shortened peptide detectable by antibody. The conclusion that helicase I is encoded by *tra*I implicates a role for this enzyme in bacterial conjugation (6). We assume that helicase I is required to unwind F factor DNA for transferring one of its strands to the recipient cell.

IDENTIFICATION OF THE GENE FOR HELICASE II

The gene for helicase II was found using a modified version (7) of the solid phase radioimmune assay of Broome and Gilbert (8). A representative bank of *E. coli* DNA Sau3A restriction fragments of 10 kb mean length was prepared in the λ L47.1 DNA replacement vector of Loenen and Brammar (9). The packaged DNA was plated on a P2 lysogenic host and secondary plaques of uniform size grown from 4500 of the primary plaques were replicated on a membrane coated with anti-helicase II antibody. Helicase II antigen bound to the membrane was visualized by binding additional radioactive helicase II antibody to the bound antigen followed by autoradiography of the membrane. By limiting the time of contact between plaques and antibody, overproduction of helicase II in a plaque could be distinguished from the background of approximately 5000 helicase II molecules which is liberated per lysed cell.

In this way four phage strains were identified which reproducibly gave an enhanced autoradiographic signal. Two of these phages were found to induce the synthesis of a non-viral protein of 75,000 dalton, the M.W. of helicase II. The DNA insert of one of these phages was shortened with PvuII nuclease, and a 2.9 kb subfragment was cloned into plasmid vector pBR328. The resulting hybrid plasmid was found to contain the full information for the 75,000 dalton protein together with an excess of inserted DNA corresponding to 42% of this information. In cells this plasmid gave rise to a 12 to 15fold overproduction of helicase II (10).

The chromosomal location for the cloned DNA was found using results obtained in three laboratories which have recently cloned the *uvr*D gene of *E. coli* (11-13). All three laboratories agree that the *uvr*D product is a DNA dependent ATPase of M.W. similar to that of helicase II. Exploiting this information we have found that cleavage patterns obtained with various restriction nucleases for helicase II DNA (10) are essentially the same as published for *uvr*D DNA (12). We have also found that extracts prepared from two *uvr*D mutants tested are free of detectable helicase II activity. Finally, helicase II DNA cloned in the high copy pBR328 vector was found to render *uvr*D$^+$ cells sensitive to u.v. irradiation. The same unexplained phenomenon has been observed for *uvr*D$^+$ DNA (12). Since helicase II DNA has not been cloned in a low copy vector we cannot say whether it complements a *uvr*D mutation. With *uvr*D DNA this is the case (12). We have further been informed by S. Kushner (pers. comm.) and H. Arthur and P.T. Emmerson (pers. comm.) that antibody against helicase II prepared in our laboratory is capable of precipitating *uvr*D product and of inactivating its ATPase. Furthermore, recent studies in P.T. Emmerson's laboratory have shown that helicase activity is connected with the *uvr*D product and that this activity, too, is inhibitable by helicase II antibody (14). There is thus little doubt left that helicase II is the *uvr*D product of *E. coli*.

BIOLOGICAL FUNCTIONS OF HELICASE II AND *rep* HELICASE

To discuss these findings we will assume that helicase II is, in fact, encoded by *uvr*D and, accordingly, that *uvr*D mutant cells are such defective for helicase II. In addition we will discuss properties of *rep* mutant cells. As a list of previous observations shows, mutations in *uvr*D and *rep* seem to have consequences both for DNA replication and DNA repair.

1. Mutation in *uvr*D causes a strong increase of the u.v. sensitivity of the cells (11-13). The mutant cells are able to grow at a high rate and to replicate various non-chromosomal DNA molecules which depend on bacterial enzymes, including the DNAs of ColE1-like plasmids and phages λ, φX174 and fd. On the other hand, antibody against helicase II applied to crude subcellular systems interferes with replication of the DNAs of *E. coli*, ColE1 and λ (15). It was also noted in these studies that with the DNAs of *E. coli* and λ this inhibition is only partial. Helicase II antibody does not inhibit replication of the duplex DNA of fd which requires *rep* helicase for separating the template strands (16). In this respect fd DNA behaves like φX174 DNA for which the helicase activity of the *rep* protein was first demonstrated (17).

2. Mutation in *rep* causes a fourfold increase of the u.v. sensitivity of the cells (18) and, in addition, a moderate reduction of the rate of chromosomal growth (19). The mutant cells are able to replicate the DNAs of ColE1 and λ but not those of φX174 and fd, as already mentioned.

It should further be recalled that helicases require a region of single-stranded DNA to initiate unwinding. Helicase II unwinds DNA in the 5'- to 3'-direction of the DNA strand to which it is bound while *rep* helicase unwinds DNA in the 3'- to 5'-direction (1).

To account for these findings we propose that helicase II and *rep* helicase are both involved in shifting the replication fork of the bacterial chromosome. Considering the directions of action of the two helicases relative to the chemical polarity of DNA we expect helicase II to be bound to the parental strand lagging in DNA synthesis, and *rep* helicase to the parental strand leading in DNA synthesis. We further postulate that both helicases are capable of shifting the fork on their own. Helicase II would do so in *rep* cells, and *rep* helicase in *uvr*D cells.

In view of this hypothesis mutation in both *uvr*D and *rep* would be critical to the cell. We have tested this prediction using P1 transduction (R. Moses and H. Hoffmann-Berling, unpublished). *uvr*D and *rep* are located near each other between 84 and 85 min of the chromosomal map, flanked by *met*E and *ilv*A (20,21) which, for transduction, provide selective markers. Using these markers we have attempted to transduce *uvr*D⁻ to a *rep*⁻ recipient and *rep*⁻ to a *uvr*D⁻ recipient. Whereas *uvr*D⁻*rep*⁺ recombinants and *uvr*D⁺*rep*⁻ recombinants were frequent among the 800 transductants which were assayed for non-selective markers, no *uvr*D⁻*rep*⁻ recombinant was found (Fig. 1). There is the remote possibility that the

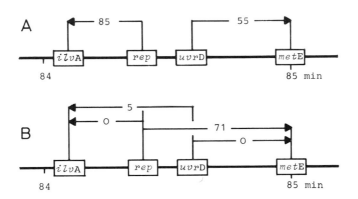

FIGURE 1. P1 transduction (22) using a *rep*⁻ or *uvr*D⁻ donor and a *uvr*D⁻ or *rep*⁻ recipient; absence of *uvr*D⁻*rep*⁻ recombinants among transductants. A (controls). *ilv*A⁺ *rep*⁻*uvr*D⁺ donor, *ilv*A⁻*rep*⁺*uvr*D⁺ recipient or *met*E⁺*uvr*D⁻*rep*⁺ donor, *met*E⁻*uvr*D⁺*rep*⁺ recipient. B. *ilv*A⁺*rep*⁻*uvr*D⁺ donor, *ilv*A⁻*rep*⁺*uvr*D⁻ recipient or *met*E⁺*uvr*D⁻*rep*⁺ donor, *met*E⁻*uvr*D⁺*rep*⁻ recipient. Co-transduction frequencies are given in percent. Arrows point to selective markers. Rep was tested by cross streaking the F⁻ K12/C hybrid bacteria with 10^8 φX174/ml, Uvr by irradiating streaked bacteria at 260 nm with 20-40 J/m².

u.v. sensitivity of a *uvr*D mutant is suppressed by *rep* mutation such that, in the given experiment, *uvr*D⁻*rep*⁻ recombinants were erroneously scored as *uvr*D⁺*rep*⁻. This is unlikely since some 20 of the Uvr⁺*rep*⁻ recombinants obtained retained their Uvr⁺ character after their *rep* defect had been complemented by a *rep*⁺ allele. This was introduced by conjugation using F'uvrD⁻rep^+ donor cells.

We take these results to suggest that uvrD mutation is incompatible with rep mutation. It thus seems justified to associate helicase II and rep helicase with a process of vital importance, i.e. chromosomal replication. Our results also suggest that helicase III does not enable cells to grow when helicase II and rep helicase are both defective. Helicase III will therefore be neglected in the following considerations.

Formation of the fork with both helicase II and rep helicase might be of little advantage to the cell as long as its DNA is undamaged. Once damage is inflicted upon the chromosome, however, two helicases acting on opposite sides of the fork might secure continuation of the replication process beyond single strand interruptions occurring in the template duplex. Attempted replication through a region of discontinuous DNA most likely disrupts the fork (23). A break occurring in the leading strand of the fork where rep helicase acts would cause this arm to detach from the fork together with rep helicase. We expect that in this case helicase II continues to unwind the fork on its own, at the same time creating, beyond the interruption, a single-stranded site for rep helicase to re-adsorb to the DNA and resume its action. Conversely, a break occurring in the lagging strand where helicase II acts would be dealt with by rep helicase continuing its action on the leading strand (Fig. 2). The arm cut off from the fork might subsequently be rescued by recombination of its DNA with that on the intact side of the fork. A replication fork generated at the site of recombination between the two daughter duplexes would enable these duplexes to resume replication.

In a cell defective for helicase II or rep helicase, however, helicase cut off from the fork would be deprived of a site for re-initiating. Accordingly, the replication process would terminate at the site of the interruption. In a uvrD cell the fork would be arrested through a break of the leading strand where rep helicase acts, whereas in a rep cell it would be arrested through a break of the lagging strand where helicase II acts. Replication would continue in a uvrD cell when the break is in the lagging strand, and in a rep cell when it is in the leading strand.

The fate of the chromosome after the arrest of replication through disruption of the fork would depend on whether or not the cell succeeds in restoring the continuity of the double helix such that a subsequent round of replication can pass. Seen under this aspect, the high u.v.

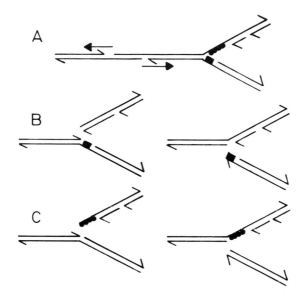

FIGURE 2. Replication of *E. coli* DNA containing single strand interruptions; hypothetical effects of *uvr*D or *rep* mutation. A. Replication fork of a *uvr*D⁺*rep*⁺ cell. Arrows indicate direction of excision re-polymerization of DNA by DNA polymerase I. B. Replication fork of a *uvr*D⁻*rep*⁺ cell following disconnection of the lagging arm (left) or leading arm (right) from the fork. C. Same as B with a *uvr*D⁺*rep*⁻ cell. Hooks indicate 5'-DNA termini. ●, helicase II; ■, *rep* helicase.

sensitivity of a *uvr*D cell as compared to a *rep* cell could mean that, for unknown reasons, it is more difficult to re-pair the leading side of a fork (where in a *uvr*D cell the critical break would occur) than the lagging side. An al-ternative explanation for the high u.v. sensitivity con-nected with *uvr*D mutation would be that *rep* helicase is less effective than helicase II in leading replication through a region of DNA damage. It is therefore of interest to note that, when activated by ATP, *rep* helicase binds tightly to a fork while dissociating from non-branched DNA (24). Moreover, while active helicase II seems to form a stoichiometric complex with its DNA substrate (25), *rep* he-licase acts in catalytic amount (26). For both reasons *rep* helicase might be more frequently lost from a damaged fork than is helicase II. Finally one has to consider the possible occurrence of excision-repair. On the leading side of the replicating duplex depolymerization-repolymerization of DNA would run in the direction opposite to fork movement; on the lagging side it would run in the same direction as fork movement. The fork has thus possibly a greater chance to meet a single strand interruption on its leading side than on its lagging side. Disconnection of the leading arm, however, would be especially critical to a *uvr*D cell. In each of these cases a *uvr*D mutant would be at a disadvantage relative to a *rep* mutant. Furthermore, in each case u.v. sensitivity would be the consequence of impaired replication. The ability of the mutant to deal with the primary u.v. lesion need not be inferior to that of a normal cell.

The behavior of the non-chromosomal DNA molecules which have been mentioned appears to fit into this scheme. The requirement of replicating øX174 DNA or fd DNA specifically for *rep* helicase is, most likely, a consequence of the structure of these DNA molecules at the origin of replication. Following an incision step which activates the DNA for the ensuing rolling circle replication, *rep* helicase is able to bind at the nick (16,25) but probably not helicase II which requires a minimum of approximately 12 unpaired DNA nucleotides to initiate unwinding (26). In the case of øX174 and fd DNA, helicase II can therefore not replace *rep* helicase. On the other hand, the DNAs of ColE1 and λ are known to initiate replication in closed circles (27,28). Following the initiating step which, at the origin of replication, causes the displacement of a loop of single-stranded DNA from the ColE1 or λ DNA duplex, both strands of the DNA are presumably enabled to bind helicase. One can therefore expect that, if either helicase II or *rep* helicase is defective, the other helicase is capable of shifting the replication fork. This would explain why, like the bacterial chromosome, ColE1 and λ DNA can replicate both in *uvr*D cells and *rep* cells.

CONCLUDING REMARKS

As our considerations show, u.v. sensitivity resulting from *uvr*D or *rep* mutation can be interpreted in terms of replication functions of the helicases encoded by these genes. Although we do not associate helicase II and *rep* helicase with the repair of primary u.v. damage, we can ac-

count for the fact that loss of the activity of either helicase results phenotypically in a repair mutant, i.e. not
a chromosomal replication mutant. Reasons can further be
found which explain why the character of a repair-deficient
cell is more strongly expressed with a *uvr*D mutant than a
rep mutant. The fact that these mutants grow almost normally is not inconsistent with the notion of replication defects. In order to rescue replication after the fork has
been damaged, either helicase must be capable of shifting
the fork on its own. This autonomy of function is apparently reflected in the growth properties of a helicase mutant.
The true nature of the defect caused by *uvr*D or *rep* mutation, i.e. its connection with the replication of the chromosome, would only become apparent when the activities of
both helicases are lost. Conditionally defective *uvr*D⁻*rep*⁻
double mutants might be a tool for testing this hypothesis.

Our considerations conform to the view that attempted
replication of a damaged chromosome endangers the cell.
Characteristic of the complications resulting from this
attempt is the phenomenon of liquid holding recovery of u.v.
irradiated bacteria, i.e. the fact that the bacterial titer
increases when, by holding the cells in buffer, time is
allowed for the repair of lesions. While incision mutants
of the *uvr*A,B,C type do not show this phenomenon (29), u.v.
irradiated *uvr*D cells were found to undergo strong recovery
upon liquid holding (G. Taucher-Scholz, unpublished). This
suggests that *uvr*D mutants are capable of repairing primary
u.v. lesions.

Our considerations do not account for phenomena such
as the high rate of spontaneous mutation of a *uvr*D strain
and the fact that *uvr*D mutation is incompatible with *pol*A
polymerase mutation. An additional characteristic of a *uvr*D
mutant, hyperactivity of the cells in general genetic recombination, has recently been shown to be an induced response of the *lex*A *rec*A system of the cells (30). Apparently, *uvr*D mutation can lead to increased expression of *rec*A.
The response of bacteria to *uvr*D mutation is therefore complex.

It should be added that there are other replication
systems which conceivably make use of two DNA unwinding enzymes. Phage T4 specifies so-called DNA polymerase accessory proteins which enable the viral DNA polymerase to perform strand displacement synthesis (31). In addition, by
its gene *dda*, the phage encodes a helicase (1). With phage
T7 the viral gene 4 protein (32) and, in addition, a hypothetical enzyme encoded by the *opt*A gene of the host (33)

have been associated with DNA unwinding (C.C. Richardson, pers. comm.). It would be of interest to know whether in these cases the DNA unwinding enzymes have opposite directions of action on the template DNA, as inferred for helicase II and *rep* helicase, and how far they can substitute for each other. Interestingly enough, loss of the *dda* function is not lethal to T4 (34).

REFERENCES

1. Geider K, Hoffmann-Berling H (1981). Proteins controlling the helical structure of DNA. Ann Rev Biochem 50:233.
2. Denhardt DT (1977). The isometric single-stranded DNA phages. In Fraenkel-Conrat H (ed): "Comprehensive Virology Vol VII," Plenum Press, p 1.
3. Richet E, Kern R, Kohiyama M, Hirota Y (1980). Isolation of DNA-dependent ATPase I mutants of *E. coli*. In Alberts BM, Fox CF (eds): "Mechanistic studies of DNA replication and genetic recombination," ICN-UCLA Symp XIX, New York: Academic Press, p 605.
4. Abdel-Monem M, Taucher-Scholz G, Klinkert M-Q. Identification of *E. coli* DNA helicase I as the *tra*I gene product of the F sex factor. Proc Natl Acad Sci USA, in press.
5. Manning PA, Kuseck B, Morelli G, Fisseau C, Achtman M (1982). Analysis of the promoter-distal region of the F sex factor of *E. coli*. J Bacteriol 150:76.
6. Manning PA, Achtman M (1979). Cell-to-cell interactions in conjugating *E. coli*. In Inouye M (ed): "Bacterial outer membranes," Wiley and Sons, p 409.
7. Taucher-Scholz G (1982). Modifizierter Radioimmuntest zur Identifizierung klonierter Gene: Suche nach dem Gen für Helikase I. PhD thesis, Universität Heidelberg.
8. Broome S, Gilbert W (1978). Immunological screening method to detect specific translation products. Proc Natl Acad Sci USA 75:2746.
9. Loenen WA, Brammar WJ (1980). A bacteriophage lambda vector for cloning large DNA fragments made with several restriction enzymes. Gene 10:249.
10. Taucher-Scholz G, Hoffmann-Berling H. Identification of the gene for DNA helicase II. Manuscript in preparation.
11. Oeda K, Horiuchi T, Sekiguchi M (1982). The *uvr*D gene of *E. coli* encodes a DNA-dependent ATPase. Nature 298:98.

12. Maples VF, Kushner SR (1982). DNA repair in *E. coli*: Identification of the *uvr*D gene product. Proc Natl Acad Sci USA 79:5616.

13. Arthur HM, Bramhill D, Eastlake PB, Emmerson PT (1982). Cloning of the *uvr*D gene of *E. coli* and identification of the product. Gene 19:285.

14. Hickson JD, Arthur HM, Bramhill D, Emmerson PT (1983). The *E. coli uvr*D gene product is DNA helicase II. Mol Gen Genet, in press.

15. Klinkert M-Q, Klein A, Abdel-Monem M (1980). Studies on the functions of DNA helicase I and DNA helicase II of *E. coli*. J Biol Chem 255:9746.

16. Geider K, Bäumel I, Meyer TF (1982). Intermediate stages in enzymatic replication of bacteriophage fd duplex DNA. J Biol Chem 257:6488.

17. Scott JF, Eisenberg S, Bertsch LL, Kornberg A (1977). A mechanism of duplex DNA replication revealed by enzymatic studies of phage ϕX174: Catalytic strand separation in advance of replication. Proc Natl Acad Sci USA 74: 193.

18. Calendar R, Lindqvist G, Sironi G, Clark AJ (1970). Characterization of REP⁻ mutants and their interaction with P2 phage. Virology 40:72.

19. Lane HED, Denhardt DT (1975). The *rep* mutation. IV. Slower movement of replication forks in *E. coli rep* strains. J Mol Biol 97:99.

20. Bachman BJ, Low KB (1980). Linkage map of *E. coli* K-12. Microbiol Rev 44:1.

21. Tessman I, Fassler JS, Benneth DC (1982). Relative map location of the *rep* and *rho* genes of *E. coli*. J Bact 151:1637.

22. Miller JM (1972). Generalized transduction: Use of P1 in strain construction. In Miller JM (ed): "Experiments in molecular genetics," New York: Cold Spring Harbor Lab, p 201.

23. Hanawalt PC (1966). The u.v. sensitivity of bacteria: Its relation to the DNA replication cycle. Photochem Photobiol 5:1.

24. Arai N, Arai K, Kornberg A (1981). Complexes of *rep* protein with ATP and DNA as a basis for helicase action. J Biol Chem 256:5287.

25. Kuhn B, Abdel-Monem M, Hoffmann-Berling H (1979). Evidence for two mechanisms for DNA unwinding catalyzed by DNA helicases. J Biol Chem 254:11343.

26. Eisenberg S, Scott JF, Kornberg A (1978). Enzymatic replication of ϕX174 duplex circles: Continuous synthesis.

Cold Spring Harbor Symp Quant Biol 43:295.

27. Staudenbauer WL (1978). Structure and replication of the Colicin E1 plasmid. Curr Topics Microbiol Immunol 83: 93.

28. Skalka AM (1977). DNA replication - bacteriophage lambda. Curr Topics Microbiol Immunol 78:201.

29. Harm W (1966). The role of host cell repair in liquid-holding recovery of u.v.-irradiated *E. coli*. Photochem Photobiol 5:747.

30. Lloyd RG (1983). *lex*A Dependent recombination in *uvr*D strains of *E. coli*. Mol Gen Genet 189:157.

31. Liu CC, Burke RL, Hibner U, Barry J, Alberts B (1978). Probing DNA replication mechanisms with the T4 bacterio-phage *in vitro* system. Cold Spring Harbor Symp Quant Biol 43:469.

32. Richardson CC, Romano LJ, Kolodner R, LeClerc JE, Tamanoi F, Engler MJ, Dean FB, Richardson DS (1978). Replication of bacteriophage T7 DNA by purified pro-teins. Cold Spring Harbor Symp Quant Biol 43:427.

33. Saito H, Richardson CC (1981). Genetic analysis of gene 1.2 of bacteriophage T7: Isolation of a mutant of *E. coli* unable to support the growth of T7 gene 1.2 mutants. J Virol 37:343.

34. Behme MT, Ebisuzaki K (1975). Characterization of a bac-teriophage T4 mutant lacking DNA-dependent ATPase. J Virol 15:50.

Mechanisms of DNA Replication and Recombination, pages 77–89
© 1983 Alan R. Liss, Inc., 150 Fifth Avenue, New York, NY 10011

STRUCTURE OF CRO REPRESSOR AND ITS IMPLICATIONS
FOR PROTEIN-DNA INTERACTION[1]

D.H. Ohlendorf,[2] W.F. Anderson,[3] Y. Takeda,[4]
and B.W. Matthews[2]

ABSTRACT   The structure of the Cro repressor protein
from bacteriophage lambda has been determined at 2.2Å
resolution. From this structure, model building and
energy refinement techniques have allowed the
development of a detailed model for the presumed
Cro-DNA complex. This model is consistent with the
known affinities of Cro for its six binding sites in
the lambda genome as well as for mutant sites. In
addition, the model suggests a mechanism for sliding
along the surface of DNA duplex prior to the location
of a specific binding site. Co-crystals of Cro with
DNA fragments have been obtained and are being
examined by X-ray techniques.

Comparisons of both the primary and tertiary
structures of Cro with other DNA-binding proteins have
indicated the presence of a homologous region of about
22 amino acids. This region corresponds to the two
consecutive α-helices in Cro, catabolite gene
activator protein and lambda repressor that have been
proposed to interact with DNA. The observed homology
suggests that a bihelical DNA-binding unit is common
to many gene-regulatory proteins.

[1]This work was supported by the NIH (GM20066 to BWM;
GM28138 and GM30894 to YT), the NSF (PCM-8014311 to BWM),
the MRC of Canada through the Group on Protein Structure
and Function (to WFA), the Murdock Trust, and by an NIH
Postdoctoral Fellowship to DHO.
[2]Institute of Molecular Biology and Department of
Physics, University of Oregon, Eugene, OR 97403.
[3]MRC Group on Protein Structure and Function,
University of Alberta, Edmonton, Alberta T6G247, Canada.
[4]Chemistry Department, University of Maryland,
Baltimore County, Catonsville, Maryland 21229.

INTRODUCTION

The genome of phage λ codes for two repressor proteins, "λ repressor" or "cI", and "Cro". The two repressors constitute a molecular switch between the lytic and lysogenic developmental pathways. Both proteins function by recognizing and binding to the same six 17-base-pair sites, each site having approximate two-fold symmetry. The six sites are organized into two operator regions, O_R and O_L, within the λ genome. Both Cro and λ repressor appear to bind to the same side of the DNA and contact many of the same bases. (For review see Refs. 1-3).

RESULTS AND DISCUSSION

Structure of Cro.

The structure of Cro has been determined crystallographically (4) and refined. The current model of the structure includes 2072 protein atoms that comprise the Cro tetramer present in crystal, plus 308 solvent atoms. The crystallographic R-factor for all data to 2.2Å resolution is 0.194 with an rms bond length error of 0.018Å and an rms bond angle error of 2.7°. Except for the five or six C-terminal residues, which are ill-defined in the crystals (thermal factors in excess of 50Å^2), the conformations of the four Cro monomers are very similar, with backbones superimposable within about 0.3Å. This close correspondence, notwithstanding the different environments of the four monomers, suggests that the formation of the crystals does not significantly modify the conformation of the protein (except perhaps in the C-terminal regions).

A monomer of Cro, shown in Figure 1, consists of three α-helices, α_1, α_2 and α_3, packed against a "wall" formed by three strands, β_1, β_2 and β_3, of twisted, antiparallel β-sheet.

Interaction of Cro with DNA.

A striking feature in the structure of a dimer of Cro is the two-fold-related α_3 helices that protrude from the surface of the protein and are 34Å apart. These two helices are oriented such that they can be placed in

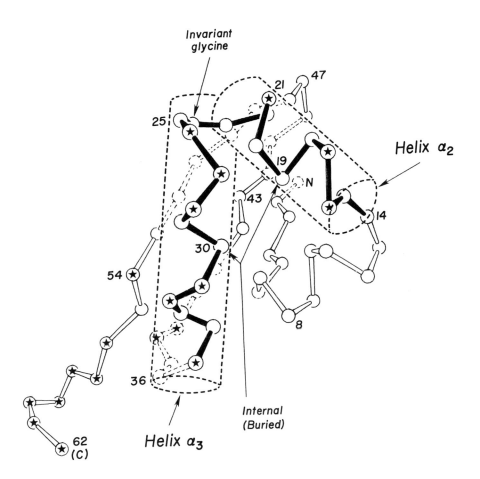

FIGURE 1. Illustration of the backbone conformation of a monomer of Cro. The α_2-α_3 two-helical unit that is common to other DNA-binding proteins is drawn solid, and residues that are presumed to contact the DNA are starred. Residues beyond 62 are ill-defined in the crystals (from Ohlendorf et al. (21)).

successive major grooves of right-handed B-form DNA affording protection to those bases and phosphates involved in the formation of specific Cro-DNA complexes (4).

In order to develop a detailed model for the interaction of Cro with DNA, model building with an MMS-X graphics facility, coupled with energy minimization, was used (5). As a starting point, Cro was aligned with DNA corresponding to O_R3, the site for which Cro has highest affinity. The two-fold axes of Cro and the DNA were required to coincide.

Upon examining the initial complex, three points became clear. First, residues Tyr 26, Gln 27, Ser 28, Asn 31, Lys 32 and His 35 of the α_3 helix were in a position to interact with either the hydrogen bonding groups of the base pairs that were exposed in the major groove, conferring specificity, or with the negatively charged phosphates, promoting general binding. Nearby residues Arg 38 and Lys 39, on strand β_2, were also suitably located to contribute to binding and recognition. Second, there were a number of residues within helix α_2, e.g., Gln 16, Thr 17, and Lys 21, that would contact the phosphate backbone of the DNA if either the DNA was bent or if the two monomers of Cro underwent a "hinge-bending" motion relative to one another. Third, the β_3 strand was positioned such that the subsequent flexible carboxyl terminal residues could be placed along the surface of the minor groove of the DNA. In this position Ser 60, Asn 61, Lys 62, Lys 63, Thr 64 and Thr 65 could make sequence-independent interactions with the phosphate backbone.

Utilizing these ideas it was possible to obtain a detailed atomic model for the Cro-DNA complex (5). The presumed sequence-specific hydrogen-bonding between Cro and O_R3 is shown in Figure 2 using a schematic representation based on that proposed by Woodbury et al. (6). In Figure 3 the bend built into the DNA is apparent.

We chose to bend the DNA, for which there is some precedence (7-9), rather than assume an arbitrary conformational change in the Cro dimer. However, it is to be noted that a "hinge bending" motion of Cro seems feasible, and might occur in other DNA-binding proteins as well (5,10). Such flexibility may be important in initial, sequence independent, binding and in "sliding" along the DNA to desired target sites.

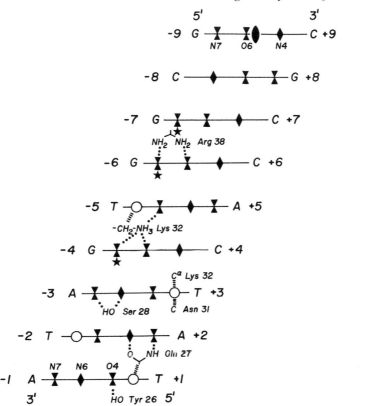

FIGURE 2. Schematic representation, following Woodbury et al. (14), of the presumed sequence-specific interactions between Cro and the parts of the base pairs exposed within the major groove of the DNA. The direction of view is imagined to be directly into the major groove of the DNA with the base pairs seen edge-on. The dyad symbol within the topmost base pair indicates the center of the overall 17-base pair binding region. The symbols are as follows: ⊥ , hydrogen bond acceptor; ✦ , hydrogen bond donor; ⊖ , methyl group of thymine; ✶ , guanine N7 which is protected from methylation when Cro is bound (5,16). Presumed hydrogen bonds between Cro side chains and the bases are indicated (....). Apparent van der Waals contacts between Cro and the thymine methyl groups are shown (ⅢⅢ). Other van der Waals contacts are not shown (from Ohlendorf et al. (4).

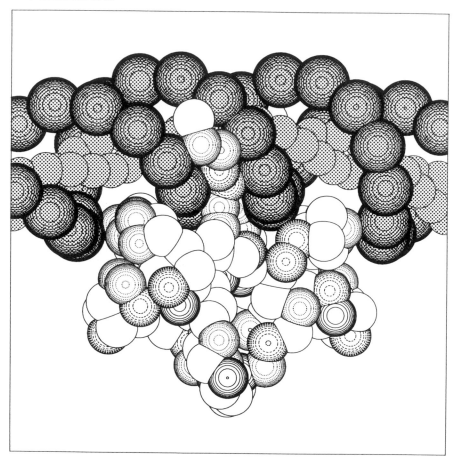

FIGURE 3. Stylized drawing showing the
complementarity between the structure of Cro repressor and
DNA. In the presumed sequence-specific complex, the
protein is assumed to move about 4Å closer to the DNA with
the α₃ helices penetrating further into the major grooves
of the DNA than is shown in the figure. The DNA is
represented stylistically by large dotted spheres centered
at the phosphate positions and small dotted spheres that
follow the bottom of the major groove. In the protein,
each residue is represented by a single sphere. Acidic
residues have solid concentric circle shading, basic
residues have broken circle shading, uncharged hydrophilic
residues have dotted circle shading and hydrophobic
residues have no shading (after (4,5)).

Supporting Evidence for the Model.

Johnson and coworkers (11,12) investigated the
relative affinities of Cro for its six naturally occurring
binding sites plus a number of mutant binding sites (see
also Ref. 13). The proposed model for Cro bound to DNA is
consistent with the postulate that the reduction in
binding affinity of the different sites, relative to O_R3,
is due to the loss of one or more sequence specific
hydrogen bonds (5). O_R3 itself is thought to make about
20 sequence-specific hydrogen bonds with a dimer of Cro,
as well as many other sequence-independent interactions
(5).

The model is supported by the observation that removal
of four or five residues from the carboxyl terminus of Cro
with carboxypeptidase A reduces the affinity of Cro for
λ DNA (14) (see Fig. 3). Also, a frame shift mutation
that changes the carboxyl-terminal sequence from
Thr-Thr-Ala to Glu-Glu-His-Lys results in poor growth,
consistent with the participation of this region of Cro
with DNA binding (M. Daniels, D. Schultz and G. Smith,
personal communication). The presumed involvement in DNA
binding of six lysine residues (Fig. 2; also see Fig. 4 of
Ref. 5) has been confirmed by chemical protection
experiments (14). Also, crystallographic studies show
that there are a number of anion binding sites on the
surface of the Cro molecule in the vicinity of the
presumed DNA phosphate binding sites (15). Lu and
coworkers (16,17) have demonstrated a pronounced quenching
of tyrosine fluorescence during formation of the specific
Cro-DNA complex. Finally, the striking structural and
sequence similarities between Cro and other DNA binding
proteins also support the proposed interaction of Cro with
DNA (see below).

Crystals of Cro complexed with both six- and
nine-base-pair segments of O_R3 have been obtained (18),
and are currently being analyzed both by isomorphous
replacement methods and by "molecular replacement" with
the native Cro structure. It is hoped that these crystals
will eventually allow direct visualization of Cro-DNA
complexes.

Comparison with Other DNA-binding Proteins.

In order to ascertain whether the structures of other
DNA-binding proteins might contain features in common with

Cro, comparisons of primary sequences (19-21) and, where available, tertiary structures (22,23) have been made. The sequence comparisons suggested that a number of proteins were homologous with Cro and with each other (Fig. 4) (see also Refs. 24,25).

The region of homology includes residues 14-35 of Cro. This region corresponds to the bihelical α_2-α_3 unit that contains many of the residues that are presumed to interact with DNA (Fig. 1). Examination of the Cro structure (Fig. 1) indicates that the residues that are most strongly conserved are important in the establishment of the bihelical geometry (21). First, the "invariant" glycine at position 24 of Cro, in the elbow bend between the α_2 and α_3 helices, has the conformation characteristic of a residue in a left-handed α-helix. This conformation is very uncommon for any residue other than glycine. Second, residues at Cro positions 19, 23, 25, and 30 are generally hydrophobic and always uncharged. This is also consistent with a bihelical geometry because in Cro these residues are almost completely buried. Third, there are no prolines at positions that would disrupt the hydrogen bonding within the two helices. Fourth, residue 20 is nearly always a glycine or an alanine. In Cro this residue is wedged in a surface crevice between the α_2 and α_3 helices. Because of the close contacts around the beta carbon, only residues with non-beta-branched side chains seem possible at this location.

Two of the proteins included in Figure 4 are the λ repressor and the catabolite gene activator protein ('CAP'), for which the tertiary structures have been determined (26,27). In both cases the region of these proteins presented as homologous in Figure 4 fold into two consecutive alpha helices (α_2-α_3 for λ repressor, α_E-α_F for CAP). When the alpha carbon coordinates of the α_2-α_3 helices of Cro and the α_E-α_F helices of CAP were compared, a high degree of structural homology was seen. For the 24 homologous residues, the rms error between Cro and CAP was 1.1Å (22). A similar comparison for Cro and λ repressor found an rms error of 0.7Å for 23 homologous residues (23).

Although there is, therefore, a remarkable conservation of the two-helical unit in the three DNA-binding proteins, this unit does not seem to interact with the DNA in exactly the same way in all three cases. For Cro and λ repressor, the proposed model of alignment of their two-helical units on DNA are similar but not

|← —— α₂ —— →| |← —— α₃ —— →|

⬒○★★○○⬒⬒★○○⬒○○●★★★★○⬒⬒●★★●⬒○★

λ Cro 14 Phe-Gly-Gln-Thr-Lys-Thr-Ala-Lys-Asp-Leu-Gly-Val-Tyr-Gln-Ser-Ala-Ile-Asn-Lys-Ala-Ile-His 35

434 Cro 16 Met-Thr-Gln-Thr-Glu-Leu-Ala-Thr-Lys-Ala-Gly-Val-Lys-Gln-Gln-Ser-Ile-Gln-Leu-Ile-Glu-Ala 37

P22 Cro 11 Gly-Thr-Gln-Arg-Ala-Val-Ala-Lys-Ala-Leu-Gly-Ile-Ser-Asp-Ala-Ala-Val-Ser-Gln-Trp-Lys-Glu 32

λ Rep 31 Leu-Ser-Gln-Glu-Ser-Val-Ala-Asp-Lys-Met-Gly-Met-Gly-Gln-Ser-Gly-Val-Gly-Ala-Leu-Phe-Asn 52

434 Rep 16 Leu-Asn-Gln-Ala-Glu-Leu-Ala-Gln-Lys-Val-Gly-Thr-Thr-Gln-Gln-Ser-Ile-Glu-Gln-Leu-Glu-Asn 37

P22 Rep 19 Ile-Arg-Gln-Ala-Ala-Leu-Gly-Lys-Met-Val-Gly-Val-Ser-Asn-Val-Ala-Ile-Ser-Gln-Trp-Glu-Arg 40

λ cII 24 Leu-Gly-Thr-Glu-Lys-Thr-Ala-Glu-Ala-Val-Gly-Val-Asp-Lys-Ser-Gln-Ile-Ser-Arg-Trp-Lys-Arg 45

434 cII 24 Leu-Gly-Thr-Glu-Lys-Thr-Ala-Glu-Ala-Val-Gly-Val-Asp-Lys-Ser-Gln-Ile-Ser-Arg-Trp-Lys-Arg 45

P22 cI 23 Arg-Gly-Gln-Arg-Lys-Val-Ala-Asp-Ala-Leu-Gly-Ile-Asn-Glu-Ser-Gln-Ile-Ser-Arg-Trp-Lys-Gly 44

LacR 4 Val-Thr-Leu-Tyr-Asp-Val-Ala-Glu-Tyr-Ala-Gly-Val-Ser-Tyr-Gln-Thr-Val-Ser-Arg-Val-Val-Asn 25

CAP 167 Ile-Thr-Arg-Gln-Glu-Ile-Gly-Gln-Ile-Val-Gly-Cys-Ser-Arg-Glu-Thr-Val-Gly-Arg-Ile-Leu-Lys 188

GalR 2 Ala-Thr-Ile-Lys-Asp-Val-Ala-Arg-Leu-Ala-Gly-Val-Ser-Val-Ala-Thr-Val-Ser-Arg-Val-Ile-Asn 23

TrpR 66 Met-Ser-Gln-Arg-Glu-Leu-Lys-Asn-Glu-Leu-Gly-Ala-Gly-Ile-Ala-Thr-Ile-Thr-Arg-Gly-Ser-Asn 87

Matα1 115 Lys-Glu-Lys-Glu-Glu-Val-Ala-Lys-Lys-Cys-Gly-Ile-Thr-Pro-Leu-Gln-Val-Arg-Val-Trp-Val-Cys 136

FIGURE 4. Homologous amino acid segments from 14 gene regulatory proteins. Solid, broken or dotted underlines are used to distinguish different groups of identical residues. The locations of the α_2-α_3 helices in Cro protein are indicated; this also corresponds to the α_2-α_3 helices of λ repressor and the α_E-α_F helices of CAP. The symbols above the first row indicate the location of that residue in the Cro protein structure; ○ : external residue with side chain fully exposed to solvent; ⬒ : surface residue which is only partly exposed; ● : internal buried residue; ★ : residue presumed to interact with the DNA when the protein is bound. Proteins included in this list are taken from Refs (21) and (24).

identical (4,26). A detailed comparison of the structures of Cro and λ repressor indicates that Cro cannot be readily adjusted to fit on to the DNA in exactly the same way as is proposed for λ repressor; nor can λ repressor be aligned with DNA exactly as is proposed for Cro (23). Therefore, even though these two structures are, in one respect, very similar, and recognize the same sites on the DNA, their detailed interactions with the DNA must be somewhat different in the two cases.

REFERENCES

1. Ptashne M, Jeffrey A, Johnson AD, Maurer R, Meyer BJ, Pabo CO, Roberts TM, Sauer RT (1980). How the λ repressor and cro work. Cell 19:1.

2. Johnson AD, Poteete AR, Lauer G, Sauer RT, Ackers GK, Ptashne M (1981). λ repressor and cro-components of an efficient molecular switch. Nature 294:217.

3. Ohlendorf DH, Matthews BW (1983). Structural studies of protein-nucleic acid interactions. Ann Rev Biophys Bioeng 12:259.

4. Anderson WF, Ohlendorf DH, Takeda Y, Matthews BW (1981). Structure of the cro repressor and its interaction with DNA. Nature 290:754.

5. Ohlendorf DH, Anderson WF, Fisher RG, Takeda Y, Matthews BW (1982). The molecular basis of DNA-protein recognition inferred from the structure of cro repressor. Nature 298:718.

6. Woodbury CP Jr., Hagenbuchle O, von Hippel PH, (1980). DNA site recognition and reduced specificity of the Eco RI endonuclease. J Biol Chem 255:11534.

7. Dickerson RE, Drew HR, Conner BN, Wing RM, Fratini AY, Kopke ML (1982). The anatomy of A-, B-, and Z-DNA. Science 216:475.

8. Klug A, Rhodes D, Smith J, Finch JT, Thomas JO (1980). A low resolution structure for the histone core of the nucleosome. Nature 287:509.

9. Schellman JA (1974). Flexibility of DNA. Biopolymers 13:217.

10. Matthews BW, Ohlendorf DH, Anderson WF, Fisher RG, Takeda Y (1983). Cro repressor and its interaction with DNA. 47th Cold Spring Harbor Symposium on Quantitative Biology. Structures of DNA. In press.

11. Johnson AD, Meyer B, Ptashne M (1979). Interactions between DNA-bound repressors govern regulation by the λ phage repressor. Proc Natl Acad Sci USA 76:5061.

12. Johnson AD (1980). Mechanism of action of the lambda cro protein. Ph.D. thesis. Harvard University.

13. Takeda Y (1979). Specific repression of in vitro transcription by the cro repressor of bacteriophage lambda. J Mol Biol 127:177.

14. Takeda Y, Cadag CG, Davis D, Stier E. In preparation.

15. Anderson WF, Ohlendorf DH, Cygler M, Takeda Y, Matthews BW (1983). Cro repressor and its interaction with DNA. UCLA Symposium on Molecular and Cellular Biology, New Series, Vol. 8, "Gene expression". In press.

16. Boschelli F, Arndt K, Nick H, Zhang Q, Lu P (1982). Lambda phage cro repressor: DNA sequence-dependent interactions seen by tyrosine fluorescence. J Mol Biol 162:251.

17. Boschelli F (1982). Lambda phage cro repressor: non-specific DNA binding. J Mol Biol 162:267.

18. Anderson WF, Cygler M, Vandonselaar M, Ohlendorf DH, Matthews BW, Kim J, Takeda Y (1983). Crystallographic data for complexes of the cro repressor with DNA. J Mol Biol (submitted).

19. Anderson WF, Takeda Y, Ohlendorf DH, Matthews BW (1982). Proposed α-helical super-secondary structure associated with protein-DNA recognition. J Mol Biol 159:745.

20. Matthews BW, Ohlendorf DH, Anderson WF, Takeda Y (1982). Structure of the DNA-binding region of lac repressor inferred from its homology with cro repressor. Proc Natl Acad Sci USA 79:1428.

21. Ohlendorf DH, Anderson WF, Matthews BW (1983). Many gene-regulatory proteins appear to have a similar α-helical fold which binds DNA. J Molec Evol. In press.

22. Steitz TA, Ohlendorf DH, McKay DB, Anderson WF, Matthews BW (1982). Structural similarity in the DNA-binding domains of catabolite gene activator and cro repressor proteins. Proc Natl Acad Sci USA 79:3097.

23. Ohlendorf DH, Anderson WF, Lewis M, Pabo CO, Matthews BW (1983). Comparison of the structures of cro and repressor proteins from bacteriophage λ. J Mol Biol. In press.

24. Sauer RT, Yocum RR, Doolittle RF, Lewis M, Pabo CO (1982). Homology among DNA-binding proteins suggests use of a conserved super-secondary structure. Nature 298:447.

25. Weber IT, McKay DB, Steitz TA (1982). Two helix binding motif of CAP found in lac repressor and gal repressor. Nucleic Acids Res 10:5085.

26. Pabo CO, Lewis M (1982). The operator-binding domain of λ repressor: structure and DNA recognition. Nature 298:443.

27. McKay DB, Steitz TA (1981). Structure of catabolite gene activator protein at 2.9Å resolution suggests binding to left-handed B-DNA. Nature 290-744.

III. BACTERIOPHAGE DNA REPLICATION

Mechanisms of DNA Replication and Recombination, pages 93–113
© 1983 Alan R. Liss, Inc., 150 Fifth Avenue, New York, NY 10011

SINGLE-STRANDED PHAGES AS PROBES OF REPLICATION MECHANISMS[1]

N. E. Dixon, L. L. Bertsch, S. B. Biswas, P. M. J. Burgers,[2]
J. E. Flynn, R. S. Fuller, J. M. Kaguni, M. Kodaira,[3] M. M.
Stayton[4] and A. Kornberg

Department of Biochemistry, Stanford University School of
Medicine, Stanford, California 94305

ABSTRACT   Conversion of single-stranded circular DNA
of M13, G4 and φX174 to their duplex replicative forms
relies exclusively on host enzymes. These reactions
are attractive probes for events in progress of the
replication fork of duplex DNA [e.g., E. coli chromosome and plasmids that bear the chromosomal origin
(oriC)]. Two aspects are considered: (i) The subunits and functions of DNA polymerase III holoenzyme
responsible for continuous synthesis of the leading
strand and extension of RNA transcripts that initiate
nascent (Okazaki) fragments in discontinuous synthesis
of the lagging strand. (ii) Mechanisms of RNA priming
of Okazaki fragments with particular attention to the
action of primase on G4 DNA. With regard to holoenzyme, the ε subunit has been identified as the
product of the dnaQ (mutD), γ as the product of the
dnaZ, and τ as the product of the dnaX+Z genes.   ATP
forms a binary complex with holoenzyme, activating it
for formation of an initiation complex at a primer
terminus. The bound ATP is hydrolyzed on initiation
complex formation. ATP is not required for subsequent
elongation. Although dissociation of holoenzyme from
the 3' end of a nascent strand requires ATP and is

[1]This work was supported by grants from the National
Institutes of Health and the National Science Foundation.
      [2]Present address: Department of Biological Chemistry,
Washington University School of Medicine, St. Louis, MO.
      [3]Present address: Department of Genetics, Osaka Uni-
versity Medical School, Osaka, Japan.
      [4]Present address: Advanced Genetic Sciences, Inc.,
Manhattan, KS 66502.

slow, a mechanism does exist for rapid intramolecular transfer to a new primer terminus. With regard to RNA priming, evidence is presented for the need for a dimeric primase possibly closely associated with holoenzyme in its action.

INTRODUCTION

Studies of in vitro DNA replication of the single-stranded (SS) phages M13, G4 and φX174 have been used to probe events that occur during chromosomal replication in E. coli. Attention has been focused on these particular phage DNAs because their replication relies almost completely on the replication machinery of the host bacterium. Insofar as the various host-derived proteins serve a function in replication of the phage DNA analogous to their roles in chromosomal replication, these simpler reactions model the more complex events occurring at a replication fork (1,2).

Replication at the fork occurs semidiscontinuously. DNA polymerase III holoenzyme (holoenzyme) is responsible both for uninterrupted DNA synthesis on the leading strand, and also for synthesis of Okazaki fragments on the lagging strand. During discontinuous replication, the small number of holoenzyme molecules present in the cell must repeatedly and rapidly recycle from the ends of newly synthesized Okazaki fragments to new primer termini at the replication fork (1,2).

DNA POLYMERASE III HOLOENZYME: SUBUNITS AND FUNCTIONS

DNA polymerase III holoenzyme, as isolated, is a complex of as many as 13 separate polypeptides. Three of these (α, ε and θ) make up the core. The core itself will function as a DNA polymerase on DNase I-activated (gapped) DNA, but its action does not show the rapidity, fidelity or processivity characteristic of holoenzyme. At least four other polypeptides (β, γ, δ and τ) are required in association with the core for true holoenzyme action (Table 1).

Subunits of Pol III Core.

Since the core itself has polymerase activity, it must bear the active site responsible for DNA synthesis; this

TABLE 1
SUBUNITS OF POL III HOLOENZYME

Subunit	Mol.wt. (kdal)	Gene	Function	Ref.
Core subunits				
α	140	dnaE	Polymerization, proofreading	3,4
ε	25	dnaQ, mutD	Control of proof-reading	3,5
θ	10			3
Accessory subunits				
β	37	dnaN	Holoenzyme properties	6
γ	52	dnaZ	Holoenzyme properties	7,8
δ	32	"dnaX"	Holoenzyme properties	8,9
τ	78	dnaX + Z	Holoenzyme properties	8

site and the 3'→5' exonuclease activity required for proof-reading are on the large α subunit (10). Its structural gene has been cloned and is now securely identified as dnaE (4).

The ε subunit of pol III core is encoded by the dnaQ (mutD) gene (5). MutD is one of the strongest mutator loci in E. coli. Mutants in mutD have a 100,000-fold higher spontaneous mutation rate than wild-type cells. It is now known that dnaQ mutants, with a quick-stop phenotype in DNA synthesis, also show high mutator activity; both mutD and dnaQ mutations map in the same gene. Two-dimensional gel electrophoresis demonstrated that the protein overproduced from the dnaQ gene on a thermoinducible plasmid has the same charge and size as the ε subunit of holoenzyme. The mutator activity of the dnaQ-mutant holoenzyme is due to a deficiency in its 3'→5' exonuclease activity. Since ε itself does not contain the exonuclease active site, it must exert some controlling influence on proofreading by holoenzyme (5).

Other Holoenzyme Subunits.

The action of holoenzyme is distinguished from that of core and other subassemblies by its rapidity, processivity, fidelity, and insensitivity to inhibition by salt. Elongation of a primer by intact holoenzyme proceeds at about 400 nucleotides/second at 30°C, about as fast as the movement of a replication fork in vivo. A single holoenzyme molecule will synthesize a segment of DNA at least 5000 nucleotides long without dissociating. Its action is essentially error-free, and it works best at physiological salt concentrations. In order to achieve these characteristics, at least four subunits in addition to core are needed (Table 1). That these subunits are necessary for holoenzyme action has been established by functional assays and genetic studies.

The dnaX and dnaZ mutations map together at 10.4 min on the E. coli chromosome, and both show a quick stop in DNA replication after a temperature shift. The dnaX gene product was believed to be the δ subunit of holoenzyme (9) and the dnaZ protein was shown to be γ (7). Both products have since been reexamined (8). Both genes have been cloned and the inserts minimized by Bal31 nuclease digestion. Deletion mapping showed that the two genes are located together on a 2.2-kb piece of DNA. Further deletions from one end of this fragment complemented the dnaZ but not the dnaX mutation, while small deletions from the other produced clones that complemented neither mutation.

The products of the cloned genes have been examined using the maxicell technique. Those clones that complemented both dnaX and dnaZ produced two distinctive proteins (78 and 52 kdal). Deletions that complemented dnaZ but not dnaX produced the 52-kdal dnaZ protein, but only truncated versions of the 78-kdal protein (8). Similar results have been obtained independently by another group using other dnaZ and dnaX plasmids (11).

When maxicell extracts were run in parallel with holoenzyme on an SDS gel, the 52-kdal protein comigrated with the γ subunit, confirming the earlier assignment of dnaZ to γ. The 78-kdal species comigrated with the τ subunit of holoenzyme, suggesting that the dnaX mutation lies within the region coding for τ. This was confirmed by two-dimensional gel electrophoresis of the maxicell extracts and holoenzyme, in which the 78-kdal species was identical to τ both in charge and size (8).

Since the two genes are encoded by a 2.2-kb fragment, whose total coding capacity is little more than 78 kdal, the coding region for γ must lie within that for τ. This is consistent with the deletion mapping since the small deletions that complement neither dnaX nor dnaZ produce neither τ nor γ in maxicells. This possibility was examined by peptide mapping of the τ and γ subunits cut from SDS gels of maxicell extracts. The proteins were partially digested in situ with V8 protease at three different levels and the resulting fragments were run on an SDS gel. The dnaB gene product from maxicells transformed with a plasmid carrying that gene was digested in parallel as a control. Digestion products of the τ and γ bands showed extraordinary homology, to be contrasted with the dnaB control. These data indicate that τ and γ must at least in part be transcribed from the same gene in the same reading frame (8). It is not yet certain whether γ is a product of post-translational processing of τ, or if it is produced from a shorter transcript. Genetic, functional, and structural studies of holoenzyme and its subassemblies indicate that both τ and γ are required for holoenzyme action. Thus the presence of γ in preparations of holoenzyme is not simply an artifact of proteolysis.

In order for holoenzyme to act in replication, it must be activated by ATP or dATP. This is difficult to demonstrate with natural templates which require dATP as a substrate for elongation. The requirement for ATP is apparent, however, when poly(dA)·oligo(dT) is used as template-primer. In the absence of ATP on poly(dA)·oligo(dT), holoenzyme will synthesize poly(dT), but its activity is completely inhibited by 100 mM salt and its processivity is no greater than 40 nucleotides. In the presence of ATP, and especially ATP and single-strand binding protein (SSB), its action is less sensitive to salt and its processivity is at least 20-fold greater (13).

The function of the β subunit, product of the dnaN gene, in holoenzyme action has also been probed. Chromatography of holoenzyme on phosphocellulose dissociates β and produces a subassembly termed pol III*, which contains all the other subunits. Mixing of pol III* with purified β results in reconstitution of both the structure and activities of native holoenzyme (6,12,13). In the presence or absence of ATP, pol III* behaves like holoenzyme in the absence of ATP, displaying marked salt sensitivity and low processivity. Reconstitution of holoenzyme by addition of β to pol III* restores both salt resistance and high processivity, suggesting the respective roles of ATP and the β subunit in holoenzyme action are intimately related.

MECHANICS OF HOLOENZYME ACTION

The interaction of ATP with holoenzyme has been used to probe the mechanism of its action in the conversion of SS G4 DNA to the duplex replicative form (RF). While the work with poly(dA)·oligo(dT) shows that ATP is required in replication, it does not show where it acts. ATP could be involved only in initiation complex formation by holoenzyme at a primer terminus, only during elongation, or in both processes. RNA-primed SSB-coated G4 DNA has been used to address this problem. G4 DNA was specifically primed by the action of primase and isolated by gel filtration after treatment with SDS.

Holoenzyme associates with the primed DNA in the presence of ATP to form a stable initiation complex which can be isolated by gel filtration. Addition of dNTP's to the initiation complex results in elongation to full-length RFII without dissociation of holoenzyme. Elongation is very rapid, requiring 15 to 20 seconds (12). The synthetic RFII is left with a gap no longer than one or a few nucleotides at the 5' terminus of the primer (14).

ATP is Required for Formation of an Initiation Complex.

A key feature of these experiments is the use of the nonhydrolyzable β-γ imido derivative of dATP (dAMP-PNP) in the elongation assays. dAMP-PNP is incorporated efficiently into DNA in place of dATP, but will not activate holoenzyme as does ATP or dATP. Its use has allowed the separation of the requirement for ATP in activation of holoenzyme from the requirement for dATP in elongation (12). With dATP replaced by dAMP-PNP, the RNA-primed DNA is extended efficiently by holoenzyme, but only if ATP is also present. If, on the other hand, the isolated initiation complex is used in the same experiment, elongation proceeds in the absence of additional ATP. The requirement for ATP in holoenzyme action is therefore only at the level of initiation complex formation. Elongation from the initiation complex does not require ATP (12).

This was confirmed by an examination of lag times at very early stages in replication of the primed template. As the ATP concentration was increased, the lag time decreased without affecting the rate of elongation. Elongation from the isolated initiation complex occurred at exactly the same rate, though without a lag. The lag, therefore, reflects the time required for events that precede elongation, including initiation complex formation. Even at saturating levels of ATP, these events take about 7 seconds. In the presence of 4 µM ATP, ATPγS competes with ATP for binding to holoenzyme, and inhibits formation of the initiation complex (12).

When holoenzyme was preincubated with ATP in the absence of DNA before its addition to an assay containing 4 µM ATP, no ATP, or 10 µM ATPγS, replication of the primed SS G4 DNA occurred at the same rate after a lag of 3 sec, a rate identical to that with the preformed initiation complex. These data suggest that during preincubation without DNA, holoenzyme had already formed a complex with ATP. The preformed holoenzyme·ATP complex then produced an initiation complex with primed DNA in a reaction that consumed about 3 sec. For its action, the holoenzyme·ATP complex no longer required further addition of ATP, and it was no longer able to bind ATPγS (12).

DNA synthesis from an initiation complex is a rapid and efficient process, the product being full-length RFII. The extent of synthesis can therefore be used to quantify initiation complexes. The stability of the ATP·holoenzyme complex could thus be measured by its competence to form an initiation complex in a coupled replication assay. ATP and holoenzyme were preincubated; excess ATP was then rapidly removed with hexokinase. The mixture was sampled at time intervals and added to an assay containing no ATP. The dissociation of the binary complex followed first-order kinetics with a half-life of 38 seconds at pH 7. At pH 8 the complex dissociated about 10 times more rapidly (12).

The ATP·holoenzyme complex was thus sufficiently stable at pH 7 to be studied using a filter binding assay. The ATP binding curve showed saturation effects. A Scatchard plot indicated that holoenzyme bound 2-3 molecules of ATP at equilibrium with a dissociation constant of 0.8 µM. Under these conditions neither pol III* nor the β subunit bound ATP appreciably, but holoenzyme reconstituted from pol III* and β behaved similarly to native holoenzyme (12). The complex bound on the nitrocellulose filter was allowed to dissociate and the released nucleotides were chromatographed on PEI plates. Of the recovered nucleotide, 80% co-chromatographed with ATP, indicating that ATP was not hydrolyzed during activation of holoenzyme (12).

Hydrolysis of ATP during Formation of an Initiation Complex.

Holoenzyme binds 2 to 3 molecules of ATP in the absence of DNA. The complex is competent in producing an initiation complex on RNA-primed DNA. The question arises: What is the fate of the bound ATP during formation of the initiation complex?

When preformed holoenzyme·ATP complex was incubated with unprimed SSB-coated φX DNA, a low level of DNA-dependent ATPase activity was observed. If instead of unprimed DNA, multiply-primed φX DNA was used, a burst of ATP hydrolysis occurred within 4 seconds, superimposed on the background rate of ATP hydrolysis. Addition of an excess of unlabeled ATP competed out the background reaction without affecting the burst. These data suggest that holoenzyme-bound ATP is hydrolyzed during the formation of the initiation complex (13).

Similar burst-hydrolysis experiments were carried out using singly-primed G4 DNA. The number of ATP molecules hydrolyzed in the burst was determined and compared with the number of initiation complexes formed, as assessed by a coupled replication assay. The ratio of the two values obtained over a range of holoenzyme and ATP concentrations indicated that 2 molecules of ATP were hydrolyzed to ADP for each initiation complex formed (13).

The Initiation Complex May Contain Two Holoenzyme Molecules.

Since the number of initiation complexes in solution can be determined, it should be possible to determine the stoichiometry of holoenzyme subunits in the isolated initiation complex. So far this has been done with holoenzyme reconstituted with labeled β and γ subunits and isolated by gel filtration. The initiation complex eluted in the void volume; free holoenzyme was included. The stoichiometry of β in the void fractions in relation to initiation complexes as assessed by the coupled replication assay was close to 2:1 (13). The same stoichiometry has been obtained with γ. Since scanning of SDS gels makes it appear that the ratio of β and γ to core in holoenzyme is near 1:1, these results suggest that holoenzyme functions in an initiation complex as a dimer.

Dissociation of Holoenzyme from Replicated DNA.

The mechanism of dissociation of holoenzyme from the end of a nascent fragment and its reassociation with a new primer terminus has been probed. It appears that dissociation also requires the binding of ATP. The isolated initiation complex of holoenzyme on RNA-primed G4 DNA was extended to full length after addition of dNTP's. In the presence of an excess of primed DNA, holoenzyme dissociates from the completed RFII, forms a new initiation complex with the excess primed DNA and proceeds to replicate it.

The recycling of holoenzyme to the new primer terminus requires ATP. In the absence of ATP, the first round of replication is completed within 2 minutes and no further synthesis ensues (Figure 1). If ATP is present from the outset, recycling to the new primer terminus occurs and the excess primed DNA is replicated. The second round (proceeding from 45 pmol of nucleotide incorporated to 130 pmol), takes about 4 minutes. The difference between the time required for the first and second rounds (about 2 minutes) is the time required for recycling. Since reassociation at the new primer terminus takes only a few seconds, most of this transfer time must be taken up by the slow dissociation of holoenzyme from RF II. If ATP is added after 10 min, subsequent rounds occur at the same rate as when ATP is present throughout. These data show that dissociation is slow and that it requires binding of ATP (13).

During recycling, one role of ATP is therefore to promote the dissociation of holoenzyme from the 3' terminus of the completed nascent strand. Although an isolated initiation complex is indefinitely stable to spontaneous dissociation, nonhydrolyzable analogs of ATP (ATPγS, AMP-PNP) will still bind holoenzyme in place of ATP to promote rapid dissociation. Because ATP facilitates reassociation, its effect in promoting dissociation is masked since the equilibrium lies heavily in favor of the associated complex. This is demonstrated when both ATP and ATPγS are present. A new equilibrium with 30% of holoenzyme in solution as the ATPγS complex, and 70% bound in the initiation complex is rapidly established. This situation would predict that the initiation complex would catalyze a steady-state turnover of ATP to ADP, and this is in fact observed. At 4 μM ATP, the turnover number of the initiation complex acting as an ATPase is 25 min^{-1} (15).

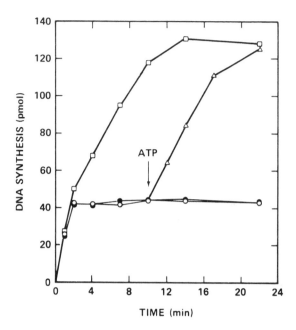

FIGURE 1.   Cycling of holoenzyme after reactivation by
ATP (13).   Reactions contained 45 pmol (as nucleotide) of
initiation complex per time point aliquot and dAMP-PNP
instead of dATP.   Additions were: (o) none; (●) 85 pmol
(as nucleotide) of RNA-primed G4 DNA per time point aliquot;
(□) 85 pmol (as nucleotide) of RNA-primed G4 DNA per time
point aliquot and 0.5 mM ATP; (Δ) 85 pmol (as nucleotide)
of RNA-primed G4 DNA per time point aliquot and, after 10
min of incubation (indicated by the arrow), 0.5 mM ATP.

Dissociation of holoenzyme labeled in the β subunit
from the initiation complex was measured more directly by
gel filtration (15);   labeled initiation complexes eluted
in the void volume while free holoenzyme was included.
Although incubation of the initiation complex alone in
buffer for 90 sec led to very little dissociation, a 90-sec
incubation with ATPγS released 69% of the bound holoenzyme.
Similarly, incubation with excess unlabeled holoenzyme did
not exchange the labeled holoenzyme out of the complex
unless ATP was also present.   ATP therefore does promote
dissociation.   Under these conditions, it is a slow process

with a half-life near 90 sec. The holoenzyme is released
as an ATP complex already activated for immediate binding
to a new primer terminus.

The rate of transfer of holoenzyme from one primed DNA
to another was measured directly using an initiation complex
prepared with an RNA-primed M13 chimera containing the G4
origin and excess RNA-primed G4 DNA (15). In the presence
of ATP, the transfer from the larger to the smaller DNA
occurred with a half-life again of 90 sec. Thus dissociation of holoenzyme from the 3'-end of a nascent strand,
even in the presence of ATP, takes about 2 minutes. In
sharp contrast with this slow intermolecular transfer of
holoenzyme, intramolecular transfer is facile and rapid
(see below).

MECHANISMS OF RNA PRIMING

Holoenzyme by itself cannot initiate replication on
single-stranded DNA. It requires an RNA or DNA primer with
a free 3' hydroxyl group which it can extend. In vitro
studies with the single-stranded phages have revealed 4
different priming mechanisms (1).

One mechanism operates in the absence of SSB. All
three phage DNAs (M13, G4, φX174) can be primed by primase,
the dnaG protein, assisted only by the dnaB protein (16,17).
Under these circumstances, priming occurs multiply and
relatively randomly.

The other three mechanisms depend on the action of SSB
as a specificity protein, allowing the secondary structures
of each of the different complementary strand origins to be
specifically recognized.

In the presence of SSB, priming of M13 occurs at the
unique origin. It requires only RNA polymerase holoenzyme
(18).

SSB-coated φX DNA is primed by a complex of 7 proteins
including primase and six pre-priming proteins. This
complex, termed the primosome, is assembled at a unique
site, but subsequently moves along the template in the
antielongation direction laying down primers at intervals
(19-25).

G4 has evolved a simpler method for specific priming.
It occurs in vitro at the G4 complementary strand origin
and requires only primase. In the absence of coupled DNA
synthesis, primase recognizes secondary structure at the G4
origin and traverses the first of three predicted hairpin

loops in synthesizing a unique 28-nucleotide primer (26).
Priming, a rather slow process in the uncoupled reaction,
is much more efficient when coupled to elongation by holoenzyme, which may reflect some interaction between the two
proteins. Full-length primers are never made in coupled
synthesis since holoenzyme supervenes over primase at an
early stage to extend primers as short as 2 nucleotides
(27).

Primase Acts as a "Dimer" on G4 SS DNA.

        The dependence on primase concentration of the steady-
state rates of primer synthesis in an uncoupled reaction,
and of DNA synthesis in the normal coupled G4 SS→RF assay
was examined. In the coupled system, the reaction is more
efficient in the sense that it requires less primase. The
DNA synthesis rate is saturated near 5 primase molecules
per circle of DNA while the rate of primer synthesis is
still in the linear range with an input of 15. Both rates
show a biphasic dependence on the concentration of primase,
suggesting the cooperative involvement of two or more
enzyme molecules in the priming reactions (28).
        In the presence of SSB and Mg^{2+} ions, primase binds
specifically to DNA containing the G4 origin. The complex
formed is stable and isolable by gel filtration. At high
levels of input primase, the isolated complex contained up
to two molecules of primase per DNA molecule and was competent for priming in a subsequent coupled SS→RF assay. The
isolated complexes were active without addition of more
primase, and maximum activity was seen with the 2:1 complex.
At lower levels of bound primase, the replication activity
paralleled closely the primase content of the complexes.
In the normal coupled G4 SS→RF assay, a rather high input
of primase is required for maximum efficiency. Two molecules per circle give a very feeble reaction. The reaction
is saturated with primase at greater than 10 molecules per
circle. The rate of synthesis by the isolated complex
containing 2 molecules of primase is comparable with that
at saturation in the normal assay (28).
        Studies of a 1:1 complex of primase bound to G4 DNA
suggest that primase acts as a dimer. The stability of the
isolated 1:1 complex and its activity in primer synthesis
were examined by gel filtration after preincubation under
various conditions. As with holoenzyme complexes, the
primase-DNA complex eluted in the void volume while free
primase was included. The 1:1 complex is itself stable to

refiltration, but pretreatment with an excess of unlabeled primase led to exchange of labeled primase out of the complex. With α-[^{32}P]rNTP's, active primase would synthesize a labeled primer on the DNA and [^{32}P] would elute in the void volume with the primase·DNA complex; not so with the 1:1 complex, which was clearly inactive in primer synthesis even though labeled primase was retained in the complex. When excess unlabeled primase was added to the 1:1 complex so that a 2:1 complex could form, again in the presence of rNTP's to allow primer synthesis, the results were strikingly different. Not only was the complex thus formed now active in primer synthesis, but the labeled primase bound to the DNA could no longer be exchanged with excess unlabeled enzyme (28).

Many mechanistic consequences of these data remain to be probed, but they are most consistent with the involvement of more than one molecule of primase in primer synthesis.

The Primosome in Replication of ϕX DNA.

The action of primase in priming G4 SS DNA, even in the coupled reaction, is somewhat distant from its role in priming lagging strand synthesis in chromosomal replication.

Those events are better modeled by priming of SSB-coated ϕX DNA, which requires not only primase, but also the dnaB and dnaC proteins, proteins i, n, n' and n". A unique site on the ϕX viral strand is recognized by n', a site-specific DNA-dependent ATPase. It directs the assembly at this site of a complex of all seven proteins, termed the primosome. The primosome moves along the DNA in the anti-elongation direction, using the energy of ATP hydrolysis as fuel. At intervals, primase lays down primers for extension by holoenzyme (2,19-25).

Originally, a complex of ϕX DNA with the primosome could be isolated which was competent for further reaction. It was retained on the synthetic ϕX RFII (21). Since that time, further progress in studying primosome action has been limited by some inconsistency in obtaining a primase-independent complex. A complex of proteins bound to DNA can still be isolated, and it still contains one molecule of labeled primase. However, it is inactive in priming unless more primase is added. Perhaps we are now lacking a previously unidentified factor or condition essential to maintenance of the integrity of the primosome. It is

expected that replication of oriC templates (29,30) will
show greater stringency than does φX in its requirement for
the primosomal proteins, and will eventually prove to be a
better system for probing the structure and function of the
primosome in replication of duplex DNA.

EVENTS AT A REPLICATION FORK

The probing of chromosomal replication with phage DNA
has allowed the development of a consistent model of events
during elongation at a replication fork (Fig. 2).
Holoenzyme is involved in continuous replication on
the leading strand, progressively extending a primer laid
down by RNA polymerase at the origin of replication. On
the lagging strand, the primosome moves in the anti-elongation direction and at intervals synthesizes RNA primers
which are extended to full-length Okazaki fragments by
holoenzyme. The primers are excised and the gaps filled by

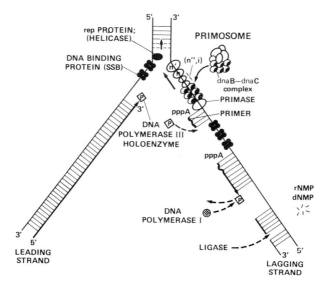

FIGURE 2. Scheme for enzymes operating at one of the
forks in the bidirectional replication of an E. coli chromosome or of a plasmid (oriC) employing the unique origin of
the E. coli chromosome (2).

DNA polymerase I and the remaining nicks are sealed by ligase.

The intermolecular transfer of holoenzyme from the 3' end of a nascent fragment to a new primer terminus in the model systems described is a very slow process taking about 2 minutes. Since an E. coli cell contains less than 10 molecules of holoenzyme per replication fork and Okazaki fragments are produced at a rate of one per second, how can lagging strand synthesis be efficient enough to keep pace with synthesis on the leading strand? A single holoenzyme molecule would need, effectively, to recycle from the end of an Okazaki fragment to a new primer terminus every second or two.

Intramolecular Recycling of Holoenzyme.

The attractive possibility of a facile mechanism for intramolecular transfer of holoenzyme has been probed by comparing the kinetics of DNA synthesis from initiation complexes on singly- and multiply-primed DNAs (15).

The multiply- and randomly-primed templates contained on the average up to 10 primers, each about 30 nucleotides long; replication was carried out with primed DNA in 4-fold molar excess over the preformed initiation complexes. On addition of dNTP's to the mixture of initiation complex and excess multiply-primed DNA, the initiation complex would be expected rapidly to synthesize DNA to the 5' end of the next primer. If it were necessary for holoenzyme to dissociate into solution and form a new initiation complex at any other primer terminus either on this molecule or another, further replication would be very slow, and all of the multiply-primed DNA present would be replicated at the same slow rate.

This was not the case. Even though nine reassociations at primer termini occurred in replication of the molecule of DNA with 10 primers, it was replicated almost as fast as singly-primed DNA (Fig. 3). For both the singly- and multiply-primed DNA circles the reactions were biphasic. The input initiation complexes (80 pmol of nucleotide), were all extended to full length within 30 seconds. Only then did holoenzyme dissociate to form new initiation complexes on the excess primed DNA and begin to replicate those molecules. During the first round, the time taken to replicate a circle with 10 primers was little more than that required for replication of singly-primed DNA (15).

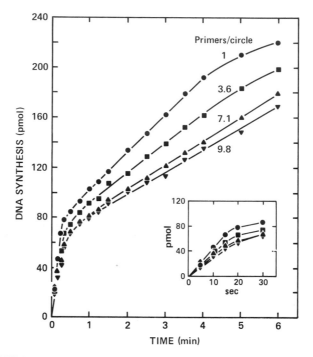

FIGURE 3. Kinetics of replication of singly- and
multiply-primed DNAs (15). The 700-µl incubations contained SSB-covered singly- or multiply-primed G4 DNA circles (5300 pmol of nucleotide), 0.5 mM ATP and 0.45 µg of
holoenzyme (0.24 pmol). After a preincubation at 30°C for
1 min to allow complete formation of initiation complexes,
the reaction was started by addition of [3H]dNTPs; 42-µl
aliquots were processed.

Agarose gel electrophoresis of the DNA products at
various early times in these reactions confirmed that the
major product produced in the first round on multiply-
primed DNA is full-length RFII (15).
The data show that a facile mechanism for intramolecular transfer of holoenzyme from the end of a nascent
fragment to a new primer does exist, and appropriate calcu-
lations show that the intramolecular transfer takes only 2
to 5 seconds. The mechanism and polarity of this intramolecular transfer deserve further study.

SUMMARY

Our current understanding of DNA synthesis at a replication fork has evolved from studies of the replication of single-stranded phage DNA.

The subunit structure of pol III holoenzyme and roles of subunits in its action are still being investigated. The ε subunit of the core appears to have some controlling role in proofreading since it is the mutD(dnaQ) gene product. Both the γ and τ subunits are encoded in part by the same segment of DNA. Their separate existence and roles in holoenzyme action imply a novel post-translational processing mechanism in E. coli, or a mechanism involving occasional premature termination of transcription.

Contrary to earlier reports, primase appears to act as a dimer in priming SS G4 DNA and perhaps also in the primosome on φX DNA. Perhaps it does not act as does a normal transcriptase. The two active sites on the two monomeric units could, for example, act alternately in tandem, each extending a primer by one nucleotide and then dissociating in favor of the other. Both a dimeric primase and a dimeric holoenzyme could associate to form the core of the elusive replisome.

Work on the cycling of holoenzyme is summarized in Figure 4. Holoenzyme in solution forms a binary complex with ATP. It is thereby activated for formation of a stable initiation complex at a primer terminus. This reaction takes 3 seconds and results in hydrolysis of the bound ATP. Dissociation of holoenzyme from the initiation complex or from the 3' end of a nascent fragment is a slow process and requires binding of more ATP. The dissociated holoenzyme is already activated for formation of a new initiation complex. Elongation from the initiation complex proceeds at about 400 nucleotides/second, which is similar to the rate of movement of a replication fork in vivo.

Finally, transfer of holoenzyme from the end of a nascent strand to a new primer terminus on the same molecule does not involve dissociation of holoenzyme into solution and occurs at a rate almost sufficient to account for the behavior of holoenzyme in the frequent strand initiations in lagging strand synthesis at a replication fork.

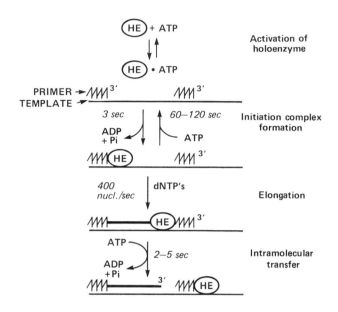

FIGURE 4. Scheme for the actions of holoenzyme dependent on ATP (15).

ACKNOWLEDGMENTS

We are indebted to Drs. H. Echols and J. R. Walker for the communication of results prior to publication. N.E.D. is supported by a C. J. Martin Fellowship of the NH and MRC (Australia) and a Fulbright Award.

REFERENCES

1. Kornberg A (1980). "DNA Replication." San Francisco: W. H. Freeman.
2. Kornberg A (1982). "1982 Supplement to DNA Replication." San Francisco: W. H. Freeman.

3. McHenry C, Crow W (1979). DNA polymerase III of
    Escherichia coli. Purification and identification of
    subunits. J Biol Chem 254:1748.
4. Welch MM, McHenry CS (1982). Cloning and identifica-
    tion of the product of the dnaE gene of Escherichia
    coli. J Bact 152:351.
5. Echols H, Lu C, Burgers PMJ (1983). Mutator strains
    of E. coli, mutD and dnaQ, with defective exonuclease
    editing by polymerase III holoenzyme. Proc Natl Acad
    Sci USA, in press.
6. Burgers PMJ, Kornberg A, Sakakibara Y (1981). The
    dnaN gene codes for the β subunit of DNA polymerase
    III holoenzyme of Escherichia coli. Proc Natl Acad
    Sci USA 78:5391.
7. Hübscher U, Kornberg A (1980). The dnaZ protein, the
    γ subunit of DNA polymerase III holoenzyme of Escher-
    ichia coli. J Biol Chem 255:11698.
8. Kodaira M, Biswas SB, Kornberg A (1983). The dnaX
    gene encodes the DNA polymerase III holoenzyme τ
    subunit, precursor of the γ subunit, the dnaZ gene
    product. Submitted for publication.
9. Hübscher U, Kornberg A (1979). The δ subunit of
    Escherichia coli DNA polymerase III holoenzyme is the
    dnaX gene product. Proc Natl Acad Sci USA 76:6284.
10. Spanos A, Sedgwick SG, Yarranton GT, Hübscher U, Banks
    GR (1981). Detection of the catalytic activities of
    DNA polymerases and their associated exonucleases
    following SDS-polyacrylamide gel electrophoresis.
    Nucleic Acids Res 9:1825.
11. Mullin DA, Woldringh CL, Henson JM, Walker JR (1983).
    Cloning of the Escherichia coli dnaZX region and
    identification of its products. Submitted for publica-
    tion.
12. Burgers PMJ, Kornberg A (1982). ATP activation of DNA
    polymerase III holoenzyme of Escherichia coli. I.
    ATP-dependent formation of an initiation complex with
    a primed template. J Biol Chem 257:11468.
13. Burgers PMJ, Kornberg A (1982). ATP activation of DNA
    polymerase III holoenzyme from Escherichia coli. II.
    Initiation complex: stoichiometry and reactivity. J
    Biol Chem 257:11474.
14. Stayton MM, Kornberg A. Unpublished results.
15. Burgers PMJ, Kornberg A (1983). The cycling of E.
    coli DNA polymerase III holoenzyme in replication. J
    Biol Chem, in press.

16. Arai K, Kornberg A (1979). A general priming system employing only dnaB protein and primase for DNA replication. Proc Natl Acad Sci USA 76:4308.

17. Arai K, Kornberg A (1981). Mechanism of dnaB protein action. IV. General priming of DNA replication by dnaB protein and primase compared with RNA polymerase. J Biol Chem 256:5267.

18. Kaguni JM, Kornberg A (1982). The σ subunit of RNA polymerase holoenzyme confers specificity in priming M13 viral DNA replication. J Biol Chem 257:5437.

19. Arai K, Kornberg A (1981). Unique primed start of phage φX174 DNA replication and mobility of the primosome in a direction opposite chain synthesis. Proc Natl Acad Sci USA 78:69.

20. Arai K, Low RL, Kornberg A (1981). Movement and site selection for priming by the primosome in phage φX174 DNA replication. Proc Natl Acad Sci USA 78:707.

21. Low RL, Arai K, Kornberg A (1981). Conservation of the primosome in successive stages of φX174 DNA replication. Proc Natl Acad Sci USA 78:1436.

22. Arai K, Low R, Kobori J, Shlomai J, Kornberg A (1981). Mechanism of dnaB protein action. V. Association of dnaB protein, protein n', and other prepriming proteins in the primosome of DNA replication. J Biol Chem 256:5273.

23. Arai K, McMacken R, Yasuda S, Kornberg A (1981). Purification and properties of Escherichia coli protein i, a prepriming protein in φX174 DNA replication. J Biol Chem 256:5281.

24. Low RL, Shlomai J, Kornberg A (1982). Protein n, a primosomal DNA replication protein of Escherichia coli. Purification and characterization. J Biol Chem 257:6242.

25. Kobori JA, Kornberg A (1982). The Escherichia coli dnaC gene product. II. Purification, physical properties, and role in replication. J Biol Chem 257:13763.

26. Bouché J-P, Rowen L, Kornberg A (1978). The RNA primer synthesized by primase to initiate phage G4 DNA replication. J Biol Chem 253:765.

27. Rowen L, Kornberg A (1978). A Ribo-deoxyribonucleotide primer synthesized by primase. J Biol Chem 253:770.

28. Stayton MM, Kornberg A (1983). Complexes of E. coli primase with the replication origin of G4 phage DNA. Submitted for publication.

29. Fuller RS, Kaguni JM, Kornberg A (1981). Enzymatic
     replication of the origin of the Escherichia coli
     chromosome. Proc Natl Acad Sci USA 78:7370.
30. Kaguni JM, Fuller RS, Kornberg A (1982). Enzymatic
     replication of the E. coli chromosomal origin is
     bidirectional. Nature 296:623.

Mechanisms of DNA Replication and Recombination, pages 115–124
© 1983 Alan R. Liss, Inc., 150 Fifth Avenue, New York, NY 10011

RECONSTITUTION OF φX174–SYNTHESIZING SYSTEM FROM PURIFIED COMPONENTS[1]

Akira Aoyama[*], Robert K. Hamatake[+] and Masaki Hayashi[*]

Departments of Biology[*] and Chemistry[+]
University of California at San Diego
La Jolla, California 92093

ABSTRACT. An in vitro system capable of synthesizing infectious φX174 phage was reconstituted from purified components. Synthesis required φX174 supercoiled replicative form DNA, φX174 gene A, C, J proteins, prohead, Escherichia coli DNA polymerase III holoenzyme, rep protein and dUTPase. Phage production was coupled to the synthesis of viral single-stranded DNA. More than 70% of the synthesized particles sedimented at the position of mature phage in a sucrose gradient and associated with the infectivity. The simple requirement of the host proteins suggests that the mechanism of viral strand synthesis in the phage synthesizing reaction resembles that of viral strand synthesis during the replication of replicative form DNA.

INTRODUCTION

During infection of Escherichia coli by the single-stranded (ss) DNA bacteriophage φX174, the circular ssDNA is replicated via three successive stages of DNA synthesis (for review, see ref. 1). Stage I is the conversion of ss DNA to double-stranded (ds) replicative form (RF) DNA and requires at least 13 host proteins but no phage-encoded

[1]This work was supported by U.S. Public Health Service grant GM12934 from National Institute of Health and grant PCM8011741 from the National Science Foundation to M.H.

proteins (2). Stage II is the semiconservative replication of RF DNA and is carried out by the proteins involved in Stage I and two additional proteins: E. coli rep protein and φX174 gene A protein (3). Stage III is the synthesis of viral circular ssDNA using RF DNA as template. This process is tightly coupled to the formation of mature phage and requires the functions of the φX174-encoded proteins A, B, C, D, F, G, H, and J, as well as host-originated proteins (for review, see ref. 4). Previous studies have indicated that φX174 gene A protein initiates stage III DNA synthesis by nicking supercoiled RF (RFI) DNA at a specific site [ori(+)] on the viral strand of RF DNA to form open circular RF (RFII) DNA whose 5'-terminus is covalently attached to gene A protein (5-10). The RFII-gene A protein complex then associates with prohead (phage head precursor composed of phage-structural proteins F, G, and H and nonstructural proteins B and D) (11,12) to form the replication assembly sedimenting at 50S when centrifuged in sucrose gradients (50S complex) (11). Gene C protein is thought to be required for the association of the prohead and the template (13). Stage III DNA synthesis proceeds in the 50S complex by a rolling circle mechanism (14,15) with the displaced viral strand packaged into the prohead. Gene J protein, a capsid protein, is incorporated into phage particles during these processes. After one round of replication is completed, gene A protein cleaves the viral strand at ori(+) and joins the two ends to form a circular genome packaged in the phage particle (16).

To study this whole process in detail, we have been developing in vitro systems capable of stage III DNA synthesis leading to the formation of mature phage (17,18). Previously, we described an in vitro system composed of purified phage-encoded protein components and an unfractionated E. coli protein fraction (18). In this system, we found that phage synthesis requires φX174 RFI DNA, φX174 gene A, C, J proteins and prohead. Recently, we also showed that a fragment of RF DNA containing ori(+) carries the necessary information for initiating the stage III reaction in vitro by cloning this DNA fragment into plasmid DNA molecules (19). The requirement of phage-encoded proteins and the necessary DNA sequence for stage III reaction was thus established in vitro. The host protein requirement, however, remained to be elucidated. The identities and roles of host proteins in stages I and II have been determined by reconstituting these reactions _in vitro_ with purified proteins (2, 3). Such an approach was not previously possible for the stage III reaction because of the requirement for many purified viral components. With the viral components now purified, we have initiated the analysis of the host proteins involved in the stage III reaction. We have reconstituted an _in vitro_ system that is capable of stage III reaction by substituting the host protein fraction of the previous _in vitro_ system (18) with purified proteins (20).

RESULTS AND DISCUSSION

In Vitro Stage III System Reconstituted from Purified Components.

TABLE 1

Omission	DNA synthesis pmole	Infectivity PFU/ml
None	200	2.0×10^{10}
Template	1	$< 10^{3}$
A protein	6	4.0×10^{8}
C protein	1	4.0×10^{7}
J protein	124	$< 10^{3}$
Prohead	1	$< 10^{3}$
PolIII holo	1	8.5×10^{6}
rep protein	4	6.0×10^{8}
dUTPase	160	1.5×10^{9}

Table 1. Requirements for Stage III reaction _in vitro_. The complete reaction mixture (25 µl) was 50 mM Tris–HCl pH 7.3, 10 mM 2-mercaptoethanol, 20 mM MgCl$_2$, 0.1 mM dATP, 0.1 mM dGTP, 0.1 mM dCTP, 0.1 mM ^{3}H–TTP, 0.8 mM ATP, 0.1 mg/ml BSA, 0.1 pmole RFI DNA, 280 ng A protein, 75 ng C protein, 48 ng J protein, 20 µg prohead, 7 ng rep protein, 120 ng PolIII holo, and 3.2 units of dDTPase. Incubation was at 30°C for 30 min. DNA synthesis was

measured by incorporation into TCA insoluble cpm. Phage synthesis was determined by assaying for infectivity on E. coli HF4704.

The complete system was composed of ϕX174 RFI DNA, ϕX174-encoded proteins A (10), C (21), and J (22), prohead (18) E. coli DNA polymerase III holoenzyme (23), rep protein (24), dUTPase (25), four deoxyribonucleoside triphosphates, ATP and $MgCl_2$. The electrophoretic patterns of the purified host and viral protein components are shown in (20). The complete system was capable of synthesizing DNA and infectious phage using ϕX174 RFI DNA as template (Table 1). DNA sythesis continued for 60 min and produced circular ssDNA (more than 70% of the total synthesized DNA) packaged into proheads and sedimenting at 114S (infectious phage particles) and 70S (fractions 16 to 24 in Fig. 1; defective particles with no infectivity). The 70S materials, which are also observed in vivo (26), may occur by inactivation of infectious particles during incubation, or by packaging ssDNA into defective proheads. The remaining DNA synthesized in this system sedimented at 50S and 20S in a sucrose gradient. The 50S material contained dsDNA with a single-stranded tail as observed in vivo. This indicates that ssDNA synthesis occurs in the 50S complex by a rolling circle mechanism and is coupled to packaging of the displaced viral DNA into prohead. The 20S material contained RFII DNA which represents the end product of one or several rounds of replication.

Requirements of Host Components.

The system required DNA polymerase III holoenzyme, rep protein, and dUTPase as host components for stage III reaction (Table 1). No other host factors stimulating the stage III reaction could be detected during the purifications of these host proteins under our experimental conditions (unpublished results). This simple requirement of host proteins implies that the mechanism of viral DNA synthesis during stage III reaction resembles the looped rolling circle mechanism occurring in the stage II(+) system of Eisenberg et al. (27,28). The synthesis of viral strand during stage II(+) requires ϕX174 RFI DNA, gene A protein, E. coli DNA polymerase III holoenzyme, rep protein, single-stranded DNA binding (ssb) protein, $MgCl_2$,

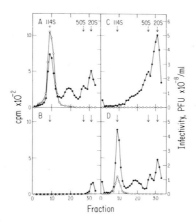

FIGURE 1. Sucrose gradient centrifugation of Stage III reaction products. Reaction conditions are described in the legend to Table 1. Reaction mixtures were centrifuged in sucrose gradients and acid-insoluble radioactivity (o) and infectivity (o) determined (18). A. Complete, B. Minus C protein, C. Minus J Protein, D. Minus dUTPase.

dNTPs and ATP and produces circular ssDNA covered with ssb protein (3,28). In this system, gene A protein nicks the RFI DNA at ori(+) on the viral strand to generate a 3'-hydroxyl group. DNA chain growth from the 3'-hydroxyl end at ori(+) is carried out by DNA polymerase III holoenzyme, while the parental viral strand of RF DNA is displaced at the replication fork by rep protein. The ssb protein is required for this strand separation by covering the displaced viral strand DNA. On the other hand, strand separation during stage III DNA synthesis occurs by the packaging of the displaced viral strand into the prohead. This reaction requires two more viral components; gene C and J proteins (see below).

E. coli dUTPase was required for maximum infectivity of the product (Table 1). There were no differences between the kinetics of DNA synthesis (data not shown) or sedimentation patterns of the products in sucrose gradients (Fig. 1A, D) made in the presence or the absence of dUTPase. The 114S material made in the absence of dUTPase

contained circular ssDNA (data not shown) as in the case
of the complete system. These results suggest that dUT-
Pase is not necessary for stage III reaction directly but
is required to sanitize the system by preventing the
incorporation of uracil instead of thymine into DNA.
Although the source of dUTP is not clear, a similar
requirement of dUTPase was observed in the stage II reaction described by Shlomai and Kornberg (25). In their
system, dUTP is not discriminated against by DNA polymerase III holoenzyme and is incorporated into DNA. The
uracil-containing DNA is attacked and degraded by uracil-
excising enzymes present in the partially purified proteins necessary for stage I DNA synthesis. Such degradation of DNA was not observed in our stage III system. The
uracil-containing ssDNA might be inactivated in vivo during infection.

Requirements of Viral Components – Gene C Protein, Prohead, and Gene J Protein.

In E. coli cells infected with gene C-deficient
ΦX174, proheads (108S particles) accumulate and newly synthesized FF molecules are rapidly converted to RFI
molecules associated with RNA (RFI*) (13). Based on these
observations, Fujisawa and Hayashi proposed that gene C
protein is involved in the maintenance of template DNA in
a form required for viral DNA replication in stage III.
Gene C protein was clearly required for DNA synthesis in
the stage III system (Table 1). The correlation between
the absence of gene C protein and the non-utilization of
prohead, i.e. the non-formation of the 50S complex, is
also observed in vitro (Fig. 1). This correlation may be
related to the DNA binding properties of gene C protein.
Gene C protein is able to bind ssDNA but has little affinity for supercoiled dsDNA (21). These observations suggest that the form of DNA required for viral DNA replication is one that has bound gene C protein. Gene C protein
may mediate the association of RFII-gene A protein complex
and prohead by binding to single-stranded regions around
ori(+).

A 108S protein complex observed in E. coli cells
infected with gene C-deficient ΦX174 as well as in cells
infected with gene A-deficient ΦX174 or cells infected
with a lysis-deficient mutant and starved for thymidine
during the period of stage III DNA synthesis contained

gene F, G, and H protein in the same proportions as those found in mature phage but had no DNA (11). After release from thymidine starvation, the 108S particle is converted to infectious phage _in vivo_, indicating that the 108S particle is a prohead structure of φX174. This was first shown by using an _in vitro_ complementation reaction. When added to the extract prepared from gene B-mutant-infected cells, which are deficient in the 108S particles and incapable of stage III reaction, the 108S particles caused stage III DNA synthesis (12). In the purified stage III system, the prohead was clearly required for not only the production of phage but also the synthesis of DNA, indicating that DNA synthesis and the packaging of ssDNA into prohead are coupled (Table 1).

Gene J codes for a phage structural protein. Amino acid sequence analysis of gene J protein places its map position between gene D and gene F of the nucleotide sequence of the φX174 genome (22). However, no mutants in gene J have been found. Consequently, it is not known whether or not gene J protein is involved in stage III DNA synthesis _in vivo_. Gene J protein, in the purified system as well as the previous system (18), was apparently not needed for initiating DNA synthesis (Table 1). The products made in the absence of gene J protein sedimented between 20S and 70S, which contained rolling-circle type DNA (data not shown), but no mature phage was found in the 114S position (Fig. 1C). These results suggest that gene J protein is required for the proper packaging of ssDNA into proheads to produce infectious phage particles.

Requirement of Template DNA and φX174 Gene A Protein.

During semiconservative replication of RF DNA (stage II), φX174 gene A protein initiates DNA synthesis by introducing a nick at ori(+) on the viral strand of RFI DNA to produce a 3'-hydroxyl group (nucleotide residue 4305). The 30 nucleotide sequence around ori(+) (nucleotide residue 4299 to 4328), which is highly conserved among the φX174-related phages, is crucial for the nicking at ori(+) (29). The sequence may also act as a signal for terminating viral strand DNA synthesis after one round of replication. In our purified stage III system, φX174 RFI DNA acts as template and gene A protein is required for DNA synthesis (Table 1). The RFII-gene A protein complex isolated by sucrose sedimentation after incubating RFI DNA

and gene A protein acts as template in the stage III system (data not shown). These results indicate that the
ori(+) sequence of the stage II reaction is also used for
stage III DNA synthesis in vitro. In fact, a DNA fragment
of φX174 RF DNA containing ori(+) region has been shown to
carry the necessary information for the entire stage III
reaction (19). Plasmid DNAs carrying the third largest
HindII-fragment of φX174 RF DNA in either direction were
as active as φX174 RFI DNA in supporting stage III DNA
synthesis and produced infectious phage particles containing plasmid ssDNA of either strand depending on the orientation of the insert in the template DNA. The synthesized
and packaged ssDNA is converted to dsDNA after infection,
and the dsDNA is maintained in the cell as evidenced by
its ability to confer drug resistancy to the host cells.

Size of Template DNA

    The plasmid-packaging system also provides a way to
determine the minimal and maximal length of DNA able to be
packaged into viable phage particles. Derivatives of
plasmid pBR322 or pBR325 of various sizes which carry a
fragment containing the ori(+) sequence of φX174 RF DNA at
EcoR1 sites were used as templates in the in vitro stage
III system. All chimeric DNA tested were as active as
φX174 RFI DNA in supporting stage III DNA synthesis (data
not shown). However, only template DNA whose sizes were
from 4.2 to 5.4 kb (80 to 100% of φX174 RF DNA) produced
infectious phage particles as effectively as φX174 RFI DNA
(unpublished data).

CONCLUSION

    The system described here provides a sensitive means
for investigating the complex interactions between viral
and host components during the development of φX174. In
addition to analyzing the Stage III reaction, this system
is also useful for studying the mechanism by which Stage
II DNA synthesis is converted to Stage III DNA synthesis.
Some proteins not necessary for our Stage III system could
be required for this conversion. For example, Eisenberg
and Ascarelli proposed that φX174 A* protein has some role
in this conversion (30). Hamatake et al. observed the
requirement for DNA gyrase subunit A in Stage III DNA

synthesis occurring in vivo as well as in a crude in vitro system (31), although the purified DNA gyrase subunit A had no effect on our purified Stage III system (data not shown). Recently, Wolfson and Eisenberg (32) reported that a partially purified host fraction was required for stage III DNA synthesis in their system composed of a crude ϕX174-infected cell extract. The requirements for DNA gyrase subunit A and the Wolfson and Eisenberg factor were both observed in crude systems which may have contained the enzymes necessary for stages I and II DNA synthesis. DNA gyrase subunit A and the Wolfson and Eisenberg factor may have a role in the conversion to or maintenance of stage III DNA synthesis in the presence of the other replicative enzymes. Alternatively, the crude systems may contain inhibitors that when removed circumvents the need for these proteins. Once the mechanism of the stage II to stage III conversion is determined, it might be possible to couple the stage I, II and III reactions in vitro with purified components in order to examine the complete cycle of ϕX174 development occurring in the cell.

REFERENCES

1. Dressler D, Hourcade D, Koths K, Smith J (1978). In Denhardt DT, Dressler D, Ray DS (eds): "The Single-Stranded DNA Phages," Cold Spring Harbor, New York, Cold Spring Harbor Laboratory p 187.

2. Shlomai J, Polder L, Arai K, Kornberg A (1981). J Biol Chem 256:5233.

3. Arai N, Polder L, Arai K, Kornberg A (1981). J Biol Chem 256:5239.

4. Hayashi M (1978). In Denhardt DT, Dressler D, Ray DS (eds): "The Single-Stranded DNA Phages," Cold Spring Harbor, New York: Cold Spring Harbor Laboratory, p 531.

5. Franke B, Ray DS (1972). Proc Natl Acad Sci USA 69:475.

6. Baas PD, Jansz HS, Sinsheimer RL (1976). J Mol Biol 102:633.

7. van Mansfeld ADM, Lengeveld SA, Weisbeek PJ, Baas PD, van Arkel GA, Jansz HS (1978). Cold Spring Harbor Symp Quant Biol 43:331.

8. Ikeda J-E, Yudelevich A, Hurwitz J (1976). Proc

Natl Acad Sci USA 73:2669.
9. Ikeda J-E, Yudelevich A, Shimamoto N, Hurwitz J (1979). J Biol Chem 254:9416.
10. Eisenberg S, Kornberg A (1979). J Biol Chem 254:5328.
11. Fujisawa H, Hayashi M (1977). J Virol 24:303.
12. Mukai R, Hamatake RK, Hayashi M (1979). Proc Natl Acad Sci USA 76:4877.
13. Fujisawa H, Hayashi M (1977). J Virol 21:506.
14. Fujisawa H, Hayashi M (1976). J Virol 19:409.
15. Koths K, Dressler D (1980). J Biol Chem 255:4328.
16. Fujisawa H, Hayashi M (1976). J Virol 19:416.
17. Mukai R, Hayashi M (1977). Nature (London) 270:364.
18. Aoyama A, Hamatake RK, Hayashi M (1981). Proc Natl Acad Sci USA 78:7285.
19. Aoyama, A, Hayashi M (1982). Nature (London) 297:704.
20. Aoyama, A, Hamatake RK, Hayashi M (1983). Proc Natl Acad Sci, in press.
21. Aoyama A, Hamatake RK, Mukai R, Hayashi M (1983). J Biol Chem, in press.
22. Freymeyer D, Shank PR, Vanaman T, Hutchison CA III, Edgal MH (1977). Biochemistry 16:4550.
23. McHenry C, Kornberg A (1977). J Biol Chem 252:6478.
24. Scott JF, Kornberg A (1978). J Biol Chem 253:3292.
25. Shlomai J, Kornberg A (1978). J Biol Chem 253:3305.
26. Spindler KR, Hayashi M (1979). J Virol 29:973.
27. Eisenberg S, Scott JF, Kornberg A (1976). Proc Natl Acad Sci USA 73:3151.
28. Eisenberg S, Griffith J, Kornberg A (1977). Proc Natl Acad Sci USA 74:3198.
29. Godson GN, Fiddes JC, Barrell BG, Sanger F (1978). In Denhardt DT, Dressler D, Ray DS (eds): "The Single-Stranded DNA Phages," Cold Spring Harbor, New York: Cold Spring Harbor Laboratory, p 51.
30. Eisenberg S, Ascarelli R (1981). Nucleic Acids Res 9:1991.
31. Hamatake RK, Mukai R, Hayashi M (1981). Proc Natl Acad Sci USA 78:1532.
32. Wolfson R, Eisenberg S (1982). Proc Natl Acad Sci USA 79:5768.

Mechanisms of DNA Replication and Recombination, pages 125–134
© 1983 Alan R. Liss, Inc., 150 Fifth Avenue, New York, NY 10011

THE INTERACTION OF E. COLI REPLICATION FACTOR Y WITH ORIGINS OF DNA REPLICATION

W. Soeller, J. Greenbaum, P. Abarzua and K.J. Marians

Department of Developmental Biology and Cancer
Albert Einstein College of Medicine, Bronx, N.Y. 10461

ABSTRACT   The interaction of E. coli replication factor Y with its DNA effector site from the L-strand of pBR322 has been studied. DNA footprinting indicates that factor Y can protect 50-70 nucleotides of DNA from nuclease attack. The phenotypes of point mutated Y sites suggests that under some conditions factor Y's ATPase activity may not be directly connected to its role in the initiation of DNA replication.

INTRODUCTION

Replication factor Y from E. coli has been shown to interact with discrete segments of DNA from the viral (+) strand of ØX174 and from both the H and L strands of pBR322 and ColEl (1-3). The DNA segments from these extrachromosomal elements serve as effectors for the ATPase activity of factor Y (1,2) and also function as origins of complementary strand synthesis in an SS→ RF pathway characteristic of ØX174 DNA replication (3).

Our interest is to define the manner with which factor Y interacts with its effector region in what is one of the initial events in the formation of the multienzyme priming complex (the primosome) (4). Utilizing the L strand Y site from pBR322, which has been cloned into an fl-phage vector, we have performed a series of experiments designed to identify areas of this effector sequence which are necessary for factor Y recognition and function.   Single base changes have been introduced into this region via controlled mutagenesis with sodium bisulfite, and the effects of the mutations on factor Y ATPase activity and SS→RF synthesis have been evaluated.   In addition, footprinting studies with pancreatic DNase and methylation-protection experiments have revealed an array of points where factor Y makes contact with this site.

RESULTS

Isolation of Base-Substitution Mutant Factor Y Effector Sites

Construction of the Heteroduplexes and Mutagenesis. In order to target the mutagenic agent (sodium bisulfite) to the L strand Y site, a heteroduplex was formed beteween DNA of the wild-type parent (μYEl) and the respective deletion mutant (YEdl) (see Fig. 1). This approach is similar to the deletion loop mutagenesis procedure described by Peden and Nathans (5). μYEl is a recombinant fl phage whose inserted Eco RI DNA fragment (nuc. 2383-2520 of pBR322) contains the pBR322 L strand Y site. The vector into which the DNA was inserted is a microplaque derivative of flR229 (see below). YEdl is a recombinant fl phage whose inserted Eco RI fragment contains a deletion within the pBR322 L strand Y site, from nucleotide position 2450-2484 (6). Its vector DNA is flR229 (7). These heteroduplexes were constructed in two different orientations (Fig. 1). In orientation A, μYEl ss(c) DNA was annealed to YEdl RF DNA which had been digested with Asu I and then heat denatured. In orientation B, YEdl ss(c) DNA was annealed to Bam HI-linearized, denatured, μYEl RF DNA. In this fashion, approximately 35 bases of the 3' half of the L strand Y site sequence (orientation A) or its complementary sequence (orientation B) were exclusively single-stranded and could be exposed to directed sodium bisulfite mutagenesis (8). Mutagenized heteroduplexes were then used for transformation into the E. coli K746 (ung) (P. Model, Rockefeller Univ.).

Selection for Y Site Mutants. The screen for mutations in the factor Y effector site is based on plaque morphology. M.H. Kim et al (9) isolated Ml3 phage containing deletions within the Ml3 complementary strand origin of replication which were viable but formed morphologically faint, turbid plaques and gave reduced phage yields. Ray et al (10) used these mutants as vectors for isolating complementary strand origin sequences, since any DNA sequence which acted as an origin of DNA replication restored the Ml3 mutant vector to a normal plaque morphology. We used this approach in screening for Y site mutants. Deletions were constructed in the fl complementary strand origin as described for Ml3 (9). These fl deletion mutants, similar to the case of Ml3 described above, give tiny, faint plaques which we call "microplaques". Insertion of an Eco RI fragment containing the pBR322 L strand Y site from flYE3 (3) into the fl microplaque vector (flR229-5B) restored plaque morphology to normal (unpublished results). Conversely it should follow that any mutations which inactivate this Y site's origin function will revert the phenotype to one of a microplaque. Thus, after transformation of K746(ung) cells with mutagenized

MUTAGENESIS & SELECTION

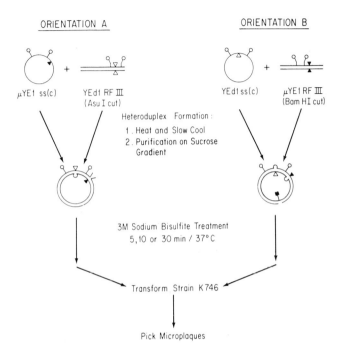

FIGURE 1. Construction of heteroduplex DNA, mutagenesis and selection. Heteroduplexes were formed and treated with sodium bisulfite as described by Shortle and Nathans (8). ——O, EcoRI and ——■ , Bam HI, restriction sites. The deletion within YEdl is indicated by △ ; the deletion in the fl origin is indicated by ▲ .

heteroduplexes, microplaques were picked and plaque-purified. Preparations of ss(c) DNA stocks from these phage were sequenced by the dideoxy sequencing procedure (11). Figure 2 shows the sequences of the strand (H) complementary to the strand (L) containing the Y site of two isolates and the parental wild-type DNA. Isolate B5.27, a derivative of an orientation B heteroduplex, had a single mutation (G → A transition in the Y site sequence) at position 2358. Isolate A30.l, a derivative of an orientation A heteroduplex, had a single mutation (C → T in the Y site sequence) at position 2374. Other single hit mutants have been identified in this manner; they are in the process of being phenotypically characterized.

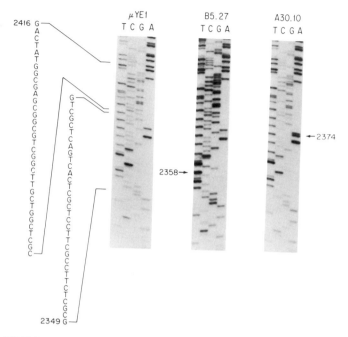

FIGURE 2. DNA sequence of wild-type and mutant L strand Y sites. The sequence shown is that of the strand complementary to the L strand Y site. Nucleotide positions are numbered according to Sutcliffe (14).

Characterization of Mutant Phenotypes

Factor Y ATPase Effector Activity of the Mutant ss(c) DNAs The ss(c) DNAs of the parent (μYEl), the vector (flR2295B), and the mutants B5.27, and A30.l0 were tested for their ability to induce factor Y's ssDNA-dependent ATPase activity (Fig. 3). μYEl exhibited a level of P_i release close to that previously reported for flYE3 (3,6) as does one of the mutants, A30.l0. Mutant B5.27 was completely inactive under the conditions of the assay, as an effector for the factor Y ATPase. The microplaque vector (flR2295B) is also inactive. Thus a single nucleotide change, depending on its location in the Y site sequence, is sufficient to reduce factor Y ATPase effector activity whereas other single nucleotide changes (i.e., A30.l0) do not affect this activity at all.

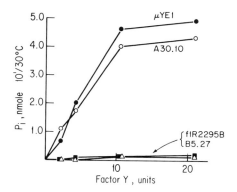

FIGURE 3. ATPase effector activity of recombinant phage ss(c) DNAs bearing wild-type and mutant L strand Y sites. Factor Y ATPase assay has been previously described (2).

Reconstitution of DNA Synthesis in vitro   The template activity of the mutant ss(c) DNAs was determined using purified E. coli DNA replication proteins in a reconstitution assay (Table 1). Each of the templates supported ss(c) → RF DNA synthesis in a ϕX-like manner, even though one of them was inactive as an effector for the factor Y ATPase activity. Our studies with these DNAs has indicated the presence of a weak, secondary, complementary strand origin on fl DNA which operates with a ϕX-like mechanism. We assume that this origin is responsible for propogation of the microplaque phage DNA. When our values were corrected for the level of synthesis from this origin we found that B5.27 DNA consistently exhibited 50% of the activity seen for μYEl DNA. Increasing the level of factor Y concentrations in the assay to 10-fold higher than that needed to saturate μYEl-directed DNA synthesis did not alter the ratio of B5.27 to μYEl-directed DNA synthesis. These results suggest that, under certain conditions, factor Y's ss DNA dependent ATPase activity may not directly bear on its role in initiation of DNA replication.

Determination of Contact Points Between Factor Y and its Effector Site

In order to identify the areas of the L strand DNA sequence which interact with factor Y, pancreatic DNase footprinting and methylation-protection experiments were performed.   The substrates employed in these studies were the single strands of a

TABLE 1

DNA SYNTHESIS[a]

Protein omitted	pmol dCMP incorp. 15'/30°C				
	ØX174	F1R2295B	μYE1	A30.10	B5.27
dnaB	< 0.01	< 0.01	0.12	< 0.01	< 0.01
dnaC	0.17	0.20	0.20	0.02	0.08
dnaG	< 0.01	< 0.01	< 0.01	0.02	< 0.01
i	0.17	0.63	1.67	0.68	1.25
n + n''	0.05	< 0.01	0.09	0.07	0.05
Y(n')	1.18	1.18	5.24	2.97	4.37
Factor I + pol III*	< 0.01	< 0.01	< 0.01	< 0.01	< 0.01
Complete	19.1	5.60	26.9	24.8	17.1
% of μYE1 template activity[b]	89.7	-	100	90.1	54.0

[a]Conditions for DNA synthesis and the purification of the replication proteins will be published (Soeller and Marians, in preparation).
[b]The activity due to the microplaque vector background was subtracted.

restriction fragment bearing a Y site. Factor Y was allowed to bind to the recognition site under the standard conditions employed in assaying for its ATPase activity. Pancreatic DNase, at a concentration and incubation time which gave the best distribution of cleavage sites within the template DNA, was then added to the reaction mix. The digested DNA was then electrophoresed on a polyacrylamide gel in parallel with L strand Y site DNA which had been subjected to the Maxam-Gilbert sequencing reactions (Fig. 4). Within its effector site (12), factor Y protected virtually every region which was susceptible to the action of pancreatic DNase. When the complementary (non-effector) region on the H strand of pBR322 was similarly treated, protection by factor Y was not observed (Fig. 4).

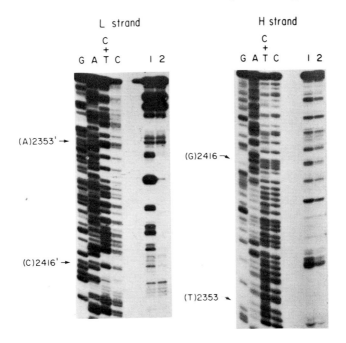

FIGURE 4. Footprinting of the L strand factor Y site. G,A,C + T, and C are Maxam-Gilbert sequencing reactions. L strand DNA or its H strand complement was digested with pancreatic DNase in the absence (1) or presence (2) of factor Y. Arrows denote the boundaries of the effector sequence. Complete conditions will be published (Greenbaum and Marians, in preparation).

Along similar lines, the methods of Maxam and Gilbert were used to determine which G residues of the L strand site were methylated, by dimethylsulfate, in the presence and absence of factor Y (Fig. 5).

DISCUSSION

The isolation of point mutations in the pBR322 L strand factor Y effector site and analysis of their phenotype with respect to factor Y ATPase activity and origin function are allowing us to determine the active configuration of this DNA replication origin. The wild-type level of activity reported here

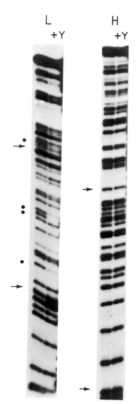

FIGURE 5. Methylation-Protection Experiments. L = L strand; H = H strand. Guanine sequencing ladders were produced by incubation of the DNA with dimethylsulfate and subsequent cleavage at methylated sites. (·) Denotes protected base.

for the A30.10 mutant, both in the ATPase and replication assays, suggests the presence of nonessential bases within the 72 base Y site sequence (12). Perhaps these serve as spacer sequences which separate flanking recognition sites by a critical distance. Such spacer sequences have already been observed in the E. coli ori C region by Oka et al (13).

The B5.27 mutation falls within the hexanucleotide sequence 5'-A-A-G-C-G-G-3' previously shown to be common to the ØX174 Y site and the two pBR322 Y sites. This mutant's inability to function as an effector for factor Y's ATPase activity suggests an association between factor Y recognition of this

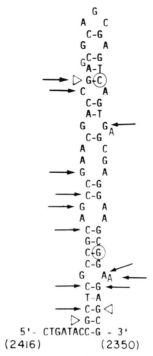

FIGURE 6. Possible secondary structure of the L-strand effector site. → Pancreatic DNase cleavage sites protected by factor Y. ▷ Methylation sites protected by factor Y.○ mutated bases.

hexanucleotide sequence and factor Y ATPase function. Characterization, now in progress, of other mutations will answer this question.

The finding that B5.27 DNA supports 50% of the wild-type level of DNA synthesis upon addition of the appropriate replication proteins suggests that factor Y's ATPase function may be separable from at least one of its roles in replication. Alternatively, conditions extant during DNA replication may serve to activate the site. This is currently under investigation.

All of the known factor Y effector sites can be drawn as well paired helical hairpins. The L strand Y site from pBR322 could form a similar structure (Fig. 6) which is folded upon itself. Compared to the effector regions on ØX174, Col El, and the H strand of pBR322, the structure of the pBR322 L strand site is less stable; it contains considerably fewer hydrogen bonded bases. In this light the inactivating effect of the mutation at position 2358 may be traced to the loss of a critical base pairing. The mutation, a G to A transition, would eliminate one of the four consecutive base pairs which start at position 2357. Conceivably, in a poorly stabilized hairpin, this alteration could lead to a significant loss of secondary structure.

That the factor Y recognition site is folded upon itself is a view that arises from the interpretation of the footprinting and methylation-protection data. The regions of the structure which are attacked by pancreatic DNase (and which are protected by bound factor Y) can be grouped into three areas. The guanine residues which are protected from methylation by dimethylsulfate lie within these same areas: bases 2351-2356; 2360-2364; 2370-2374. It is possible that these three domains are folded so that they 1) face exposure to the action of pancreatic DNase, and 2) lie in juxtaposition and hence are bound to and protected by factor Y. Model building and studies with the other known effector sites of factor Y are in progress to further evaluate the validity of this view.

ACKNOWLEDGEMENTS

These studies were supported by grants GM26410 and GM31624 from the NIH and a Sinsheimer Scholar Award to K.J.M.. W.S. and J.G. were supported by NIH Training Grants 5T32CA09060 and 2T32GM07491

REFERENCES

1.   Shlomai, J. and Kornberg, A. PNAS 77, 799-803 (1980)
2.   Zipursky, S.L. and Marians, K.J. PNAS 77, 6521-6525 (1980).
3.   Zipursky, S.L. and Marians, K.J. PNAS 78, 6111-6115 (1981).
4.   Arai, K.I. and Kornberg, A. PNAS 78, 69-73 (1981).
5.   Peden, K.W.C., and Nathans, D. PNAS 79, 7214-7217 (1982).
6.   Soeller, W.C. and Marians, K.J. PNAS 79, 7253-7257 (1982).
7.   Zinder, N.D. and Boeke, J.E. Gene 19, 1-10 (1982).
8.   Shortle, D. and Nathans, D. PNAS 75, 2170-2174 (1978).
9.   Kim, M.H., Hines, J.C., and Ray, D.S. PNAS 79, 6784-6788 (1981).
10. Ray, D.S. et al Gene 18, 231-238 (1982).
11. Smith, A.J.H. Methods in Enzymology 65, 560-580 (1980).
12. Marians, K.J., Soeller, W. and Zipursky, S.L., J. Biol. Chem., 257, 5656-5662 (1982).
13. Oka, A. et al Gene 19, 59-69 (1982).
14. Sutcliffe, J.G. CSHSQB 43, 77-90 (1978).

Mechanisms of DNA Replication and Recombination, pages 135–151
© 1983 Alan R. Liss, Inc., 150 Fifth Avenue, New York, NY 10011

ENZYMATIC MECHANISMS OF T7 DNA REPLICATION[1]

Steven W. Matson,[2] Benjamin B. Beauchamp,
Michael J. Engler,[3] Carl W. Fuller, Robert L. Lechner,
Stanley Tabor, John H. White and Charles C. Richardson

Department of Biological Chemistry
Harvard Medical School
Boston, MA 02115

ABSTRACT   Studies with bacteriophage T7 have
revealed the minimal requirements for the
initiation and accurate replication of a duplex DNA
molecule.   The primary origin of T7 DNA
replication, determined by deletion mapping, is
located at position 15 on the physical map.
Sequence analysis of the primary origin reveals two
T7 RNA polymerase promoters and a 61 base-pair
AT-rich region. Inactivation of the RNA polymerase
promoters by in vitro mutagenesis forces DNA
replication to initiate at secondary origins in
vivo. Using purified proteins we have obtained
initiation of DNA replication at the primary origin
of the T7 DNA molecule. In addition to T7 DNA
polymerase, initiation requires T7 RNA polymerase
and the 4 ribonucleoside 5'-triphosphates. Using
plasmids into which the primary origin has been
cloned we have shown that DNA synthesis initiates
at the primary origin but proceeds unidirectionally
to the right in the direction of transcription from
the ∅1.1A and ∅1.1B promoters. Analysis of the

[1]This work was supported by United States Public
Health Service Grant AI-06045, American Cancer Society,
Inc. Grant NP-1L and Jane Coffin Childs Fund for Medical
Research Fellowship 61-578 to S.W.M.
[2]Present address: Department of Biology, University
of North Carolina at Chapel Hill.
[3]Present address: Department of Biochem. and
Molecular Biol., Univ. of Texas Medical School at Houston.

newly synthesized DNA molecules reveals that they are covalently linked to RNA at their 5'-termini. These primer RNAs consist of two species, each having a unique 5'-terminus as a result of transcription from the Ø1.1A and Ø1.1B promoters. They are heterogeneous in length, however, their 3'-terminus being determined by the transition to DNA synthesis.

Two proteins, T7 DNA polymerase and T7 gene 4 protein account for most of the reactions occuring at the replication fork. The gene 4 protein has multiple activities: it is a single-stranded DNA dependent nucleoside 5'-triphosphatase, a helicase and a primase. The NTPase activity is coupled to unidirectional 5' to 3' translocation of the gene 4 protein along a single-strand of DNA, a reaction required for both helicase and primase activities. The helicase activity of the gene 4 protein can be demonstrated directly in that it will unwind a polynucleotide annealed to a single-stranded DNA molecule, provided the fragment has a single-stranded 3'-tail. Using a preformed replication fork we have shown that the gene 4 protein specifically stimulates T7 DNA polymerase on duplex templates. The rate of fork movement catalyzed by the two proteins is 300 bases/sec. at 30°C. As a primase the gene 4 protein catalyzes the synthesis of tetraribonucleotide primers on single-stranded DNA. The predominant recognition sites for primer synthesis by the gene 4 protein are 3'-CTGGG/T-5'; synthesis at these sites gives rise to primers having the sequences pppACCC/A.

INTRODUCTION

Bacteriophage T7 codes for most of its replication proteins, the consequence of which has been the development of an extremely efficient mechanism for the replication of its chromosome. In vivo, DNA replication proceeds through three well defined stages: (i) initiation of DNA replication at a specific site on the T7 genome, (ii) bidirectional DNA synthesis from this site and (iii) the formation and processing of DNA concatemers several times the unit length of T7 DNA. The enzymes that catalyze the first two stages have been purified and used to reconstitute the essential biochemical reactions responsible for initiation at the chromosomal origin and for the elongation reactions that occur at the replication fork.

DNA replication is initiated, in vivo, at a specific site on the T7 genome giving rise to replication bubbles located approximately 17% of the distance (position 17) from the genetic left end of the chromosome (1,2). As discussed below, nucleotide sequence analysis, genetic analysis and in vitro studies have revealed a direct role of T7 RNA polymerase in the initiation of T7 DNA replication. T7 RNA polymerase, the product of gene 1, is a 98,000 dalton protein that accounts for intermediate and late gene expression by initiating transcription at 17 promoters, each consisting of a conserved 23 base-pair sequence (3). Earlier studies had, in fact, shown that T7 RNA polymerase is essential for T7 DNA replication in vivo (4).

Once DNA synthesis has initiated, the two replication forks proceed bidirectionally to the ends of the linear DNA molecule. Genetic analysis originally indicated a role for the products of genes 4 and 5 in the elongation reaction (5,6). The product of gene 5 is an 84,000 dalton polypeptide that combines with the host thioredoxin, a 12,000 dalton polypeptide, to form the T7-induced DNA polymerase (7,8). The product of gene 4, first purified by complementation assay (9,10), is both a primase (11-13) and a helicase (14-16). Both activities are coupled to the single-stranded DNA dependent NTPase activity of the gene 4 protein, an activity that provides the energy for unidirectional 5' to 3' translocation of the protein along single-stranded DNA (17).

As a primase, the gene 4 protein catalyzes the synthesis of tetraribonucleotides at specific recognition sites in single-stranded DNA (17). These tetraribonucleotides can then be extended by T7 DNA polymerase and, as such, serve as primers for lagging strand DNA synthesis (11-13). As a helicase, the gene 4 protein unwinds duplex DNA ahead of the advancing replication fork using the chemical energy derived from the hydrolysis of nucleoside triphosphates (14). Recently, we have demonstrated the helicase activity of the gene 4 protein in the absence of other proteins using a DNA substrate that consists of a short DNA fragment bearing a single-stranded 3'-tail annealed to circular, single-stranded DNA (18).

Using a novel DNA substrate that resembles a replication fork we have reconstituted leading strand DNA synthesis using the T7 DNA polymerase and the gene 4 protein. DNA synthesis on this template is highly processive and proceeds at in vivo rates. In these reactions the gene 4 protein functions as a helicase to unwind the duplex DNA allowing polymerization of nucleotides by the T7 DNA polymerase. It is remarkable that a single enzyme, the gene 4 protein, serves as both primase and helicase in DNA synthesis. This dual role is representative of the economy with which phage T7 accomplishes DNA replication.

RESULTS

Initiation of DNA Replication at the Primary Origin

Mapping and Genetic Organization of the Primary Origin of Replication. As described above, during the first round of T7 DNA replication in vivo, DNA synthesis initiates at a specific site giving rise to replication bubbles located at approximately position 17, and synthesis then proceeds bidirectionally (1,2). Viable deletion mutants lacking the entire region between positions 14.5 and 21.8 on the physical map are known (19,20); these mutants initiate DNA replication at one of several secondary replication origins (21). Using a series of deletion mutants extending into this region, an essential sequence for initiation from the primary origin has been located within a 120 base-pair segment of DNA (21). Sequence analysis (22) has shown that this region of the T7 chromosome contains several genetic elements that may be involved in the initiation of DNA replication (Fig. 1). These elements include two tandem T7 RNA polymerase promoters and an AT-rich region within which is a single gene 4 protein primase recognition site. A single RNase III cutting site lies between the two promoters.

We are currently analyzing the mechanism of initiation at the primary origin by two approaches. First, we are mutagenizing the various genetic elements within the primary origin using in vitro mutagenesis in order to

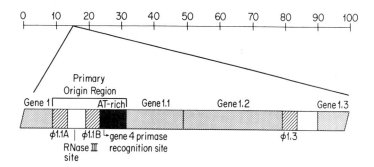

Figure 1. Genetic organization of the primary origin region of the T7 chromosome. The primary origin is located approximately 15% of the distance from the genetic left end of the chromosome (22). The insert shows a segment of the T7 genetic map containing the primary origin of DNA replication and identifies several genetic elements located within this sequence.

determine the role each of these elements has in initiation _in_ _vivo_. Second, we are reconstituting the initiation reaction _in_ _vitro_ using purified proteins in order to investigate the manner in which the genetic elements interact with the replication enzymes to achieve proper initiation.

_In_ _Vitro_ Mutagenesis of the _Primary_ Origin. Using _in_ _vitro_ mutagenesis techniques we have constructed T7 mutants defective in each of the two T7 RNA polymerase promoters in the primary origin, and in the single gene 4 protein primase recognition site within the AT-rich region. We have altered each of the two T7 RNA polymerase promoters by inserting a synthetic octomer within the sequence of each of the promoters. Analysis of the transcripts generated _in_ _vitro_ using T7 RNA polymerase shows that these insertions inactivate the promoters. We have determined, using electron microscopy, the ability of these T7 mutants to initiate DNA replication _in_ _vivo_ at the primary origin. Mutants in which both promoters are rendered inactive initiate DNA replication exclusively at secondary origins (18, S. Tabor and C.C. Richardson, unpublished results). Thus the two T7 RNA polymerase promoters within the primary origin region are essential for initiation of DNA replication. This result supports earlier studies which showed that T7 RNA polymerase is essential for T7 DNA replication (4). When either of the two T7 RNA polymerase promoters are inactivated individually, initiation of DNA replication occurs with approximately equal frequencies from the primary and secondary origins. Thus inactivation of each promoter individually reduces but does not eliminate initiation at the primary origin.

Using a synthetic oligonucleotide, a single base change has been introduced into the gene 4 protein primase recognition site within the AT-rich region (Figure 1). The recognition sequence, 3'-CTGGG-5', has been changed to 3'-CCGGG-5', a sequence not recognized for priming by the gene 4 protein (17). We are currently carrying out studies with this mutant to determine if the absence of the gene 4 primase recognition site affects initiation in vivo. Specifically, does initiation still occur exclusively at the primary origin?

Reconstitution of the Initiation Reaction In Vitro. Site specific initiation of DNA synthesis has been achieved in vitro using wild type T7 DNA, T7 DNA polymerase, T7 RNA polymerase and the four dNTPs and four rNTPs (23). There is no absolute requirement for gene 4 protein for in vitro initiation at the primary origin. Electron microscopy shows that DNA synthesis is initiated within the origin region as evidenced by the appearance of replication bubbles. Restriction enzyme analysis reveals that DNA synthesis initially proceeds unidirectionally in the same direction as transcription from the Ø1.1A and Ø1.1B promoters (see Fig. 2). The site of initiation in vitro has been mapped precisely by initiating DNA replication, using purified proteins, on a plasmid containing the primary origin (18). The 5'-termini of the newly synthesized DNA molecules are covalently linked to RNA. These primer RNAs result from transcription from both the Ø1.1A and Ø1.1B promoters; the transition to DNA synthesis occurs at numerous sites, many located within the AT-rich region. No initiation from the gene 4 protein primase recognition site located within the AT-rich segment of the primary origin region has been observed (18). We conclude that, in vitro, RNA transcripts originating at either of the T7 RNA polymerase promoters serve as primers for DNA

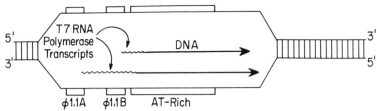

Figure 2. Model for the initiation of DNA replication in vitro. Transcription catalyzed by the T7 RNA polymerase from either promoter Ø1.1A or Ø1.1B provides the primer required by the T7 DNA polymerase for the initiation of DNA synthesis (18). The length of the primer is found to be heterogeneous in vitro.

synthesis by T7 DNA polymerase. The site at which the transition from RNA to DNA takes place is heterogeneous in these reactions. However, the possibility exists that the transition from RNA to DNA in vivo may be more specific, depending on additional proteins. For example, the role of neither the RNase III recognition site nor the single gene 4 protein primase recognition site located within the primary origin region has yet been determined.

Movement of the Replication Fork

Leading Strand DNA Synthesis. Reconstitution of the essential reactions of leading strand DNA synthesis has been accomplished using a novel DNA substrate (Fig. 3) and two highly purified proteins, T7 DNA polymerase and gene 4 protein. This DNA molecule resembles a replication fork, in that it contains a single-stranded 5'-tail, a structure analagous to the lagging strand at a replication fork. Branch migration, a reaction that would destroy the topological characteristics of the preformed fork cannot occur since the 5'-tail is not complementary to the DNA in the circular DNA template.

In an effort to determine the role of the gene 4 protein during elongation, we have compared the rate of DNA synthesis catalyzed by the combined action of the T7 DNA polymerase and the gene 4 protein using both this template and a nicked, circular duplex DNA template (Fig. 4). T7 DNA polymerase has been purified in two forms (24,25). Form II appears to be the predominant form of the T7

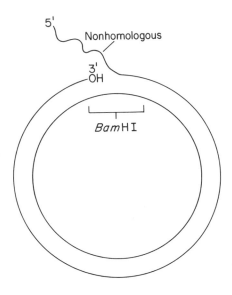

Figure 3.   A preformed and topologically stable
replication fork (26).   This DNA substrate contains a 3'–OH
and a single-stranded, nonhomologous 5'–tail.   The circular
DNA molecule is 7200 bp in length.

DNA   polymerase   in   the   phage   infected-cell   and   has
characteristics that would be expected of a true elongation
polymerase;   namely   high   levels   of a 3'   to 5'   exonuclease
activity   and the inability to catalyze strand displacement
synthesis in the absence of other proteins.
     As   shown   in   Figure   4,   T7   DNA   polymerase,   in the
absence   of   gene   4   protein,   catalyzes   little or no DNA
synthesis on either a nicked, circular DNA substrate or on
the   preformed replication fork.   In fact, with the nicked,
circular primer–template there is little DNA synthesis even
in   the   presence   of   gene 4 protein. If,   however,   the
preformed   replication   fork is used as the substrate there
is   rapid and extensive synthesis in the presence of gene 4
protein.   After 10 minutes at 30°C there have been greater
than   6   rounds   of   "rolling   circle"   DNA   synthesis   on
molecules bearing a preformed replication fork (Fig. 4).

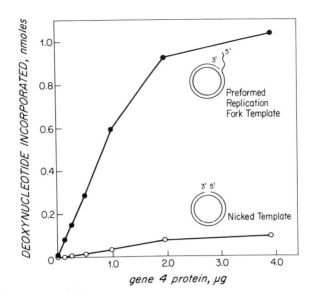

Figure 4. Measurement of the rate of DNA synthesis catalyzed by the T7 DNA polymerase on duplex DNA substrates as a function of increasing gene 4 protein concentration (26). DNA synthesis was measured by the incorporation of [^{3}H]dTMP into DNA using either nicked, circular duplex DNA or the preformed replication fork DNA as substrate. Reactions contained 0.15 nmoles of DNA substrate, incubation was at 30° for 10 minutes.

DNA synthesis in this reaction is accompanied by and dependent on the hydrolysis of dNTPs, a reaction catalyzed by the gene 4 protein (14). While the gene 4 protein will hydrolyze most NTPs, the K_m for dTTP (0.4mM) is the lowest; the K_m value for rATP is ten times greater (S.W. Matson and C.C. Richardson, unpublished results).

Gene 4 protein is unable to stimulate DNA synthesis catalyzed by other DNA polymerases on the preformed replication fork (26), suggesting that the gene 4 protein interacts specifically with the T7 DNA polymerase. The single-stranded 5'-tail is presumably required to provide a site for the gene 4 protein to bind and then to translocate 5' to 3' (17) until it reaches the replication fork at

which time its helicase activity is manifest (15, 18).
    Recently, we have been able to demonstrate the
helicase activity of the gene 4 protein directly using the
DNA substrate depicted in Figure 5. The substrate is made
by annealing a labelled, single-stranded polynucleotide to
the circular DNA of phage M13mp7. The fragment is 115
nucleotides long; the 20 nucleotides at the 3'-end are not
homologous to the M13 DNA and will not base-pair with the
single-stranded circle. Displacement of the labelled
fragment from the single-stranded circle is easily and
quantitatively detected using polyacrylamide gel
electrophoresis (Table 1).

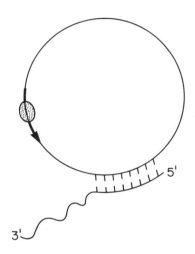

    Figure 5.   Substrate used for demonstrating and
characterizing the helicase activity of the gene 4 protein.

    When this DNA substrate is incubated with the gene 4
protein and an NTP up to 80% of the fragment is displaced
within 10 minutes at 37°C. If the NTP is omitted or an
inhibitor of NTP hydrolysis such as β,γ-methylene dTTP is
added the fragment is not displaced. Fully base-paired
polynucleotides are not displaced by the gene 4 protein,
but fragments having 7 or more non-complementary, and
therefore single-stranded nucleotides at the 3'-end will be
displaced.
    It is interesting to compare the structure of this
substrate (Fig. 5) with the structure of a replication

TABLE 1
Helicase Reaction Requirements

Reaction components	Fragment displaced
	(%)
Complete	81
$-MgCl_2$	<3
+25mM EDTA	<3
$-dTTP$	<3
$+\beta,\gamma-$methylene dTTP	<3

Helicase reaction mixtures contained 120ng of gene 4 protein and 1.0mM dTTP. Incubation was at 37°C. for 10 minutes.

fork. The non-complementary region at the 3'-end of the fragment is oriented in such a way that the circular strand is analagous to the displaced (lagging) strand at a replication fork and the fragment is analagous to the leading strand template. The gene 4 protein binds to the single-stranded circle and translocates unidirectionally 5' to 3' (17) to the annealed fragment. Here it unwinds the duplex DNA just as duplex DNA at a replication fork must be unwound.

Using the preformed replication fork and purified proteins, it is possible to obtain a measurement of the rate of fork movement. Our ability to accurately determine fork speed is aided by the unique characteristics of the preformed replication fork. As shown in Figure 6, DNA synthesis initiates on all DNA molecules at the unique 3'-OH terminus located within the single BamHI restriction endonuclease recognition sequence. Thus, the ^{32}P-labelled newly synthesized DNA has a known starting point with synthesis occuring via a "rolling circle" mechanism. Analysis of the length distribution of newly synthesized DNA is facilitated by the BamHI site that is generated after extension of the 3'-OH terminus. Cleavage by BamHI after short incubations with T7 DNA polymerase generates labelled DNA fragments of two types. Some fragments will be of an indeterminate length less than unit length while other fragments will have discrete lengths equal to two times unit length or higher integral multiples of unit length depending on the number of times the fork has

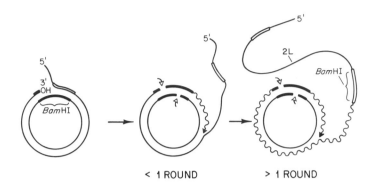

Figure 6. Scheme for determining the rate of movement of a replication fork. DNA synthesis on the replication fork DNA substrate is stopped after brief incubations by adding EDTA and heating to 70°C. for 10 minutes. The DNA is then cleaved using BamHI restriction endonuclease, denatured and the products analyzed by electrophoresis through an alkaline agarose gel. This provides a determination of the size of DNA product made during the incubation.

circumnavigated the circular template. These fragments are easily identified and quantitated using electrophoresis through an alkaline agarose gel. Such an experiment is presented in Figure 7A. After just 40 seconds of incubation at 30° fragments two times unit length are observed; in 60 seconds fragments three times unit length appear. After longer times of incubation fragments that are eight times unit length and longer are seen. The radioactive content of each band (nucleotide incorporated) is plotted versus the time of incubation in Figure 7B. DNA synthesis, as measured by the incorporation of labelled precursor into discrete DNA fragments, is linear with time. Extrapolation of the curves for each fragment size to the time axis provides a measure of the minimum time required to synthesize a molecule of a specific length, and therefore provides a determination of the maximum rate of fork movement. In these experiments, such an analysis for the 2L band (containing 7200 nucleotides of newly synthesized DNA) and the 3L band (containing 14,400 nucleotides of newly synthesized DNA) indicates that the

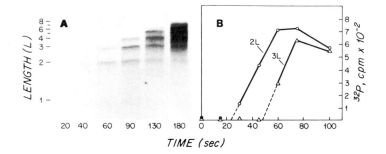

Figure 7. (A) Analysis of the products of DNA synthesis on the preformed replication fork DNA substrate on an alkaline agarose gel after cleavage with BamHI restriction endonuclease and denaturation (26). Unit length (L) is 7200 nucleotides. (B) A graph of radioactivity incorporated into newly synthesized DNA of a specific length versus time of incubation at 30°C. The bands from a gel similar to that shown in (A) were cut out and their radioactivity determined. The 2L line extrapolates to the time axis at 23 seconds and the 3L line at 45 seconds.

replication fork is moving at 300 nucleotides per second at 30°C. This impressive rate of DNA synthesis, which is similar to the rate calculated for in vivo DNA synthesis, is achieved using only two enzymes, the gene 4 protein and the T7 DNA polymerase.

Since cleavage by BamHI provides a sensitive assay for even short extensions by DNA polymerase, it is possible to estimate the percentage of available template molecules extended by the DNA polymerase. These measurements suggest that less than half of the available DNA molecules have been used as primers after 120 seconds of incubation. Since polymerization of nucleotides in this time period results in several revolutions of the fork around the circle, we conclude that DNA synthesis is extremely processive. When the DNA polymerase concentration is decreased 5-fold and similar incubations are performed, total DNA synthesis is reduced and the percentage of available primer-template utilized is decreased. However, the product lengths are not appreciably altered (R.L. Lechner and C.C. Richardson, unpublished results). At this time we have not determined an upper limit for the processivity of DNA synthesis catalyzed by T7 DNA polymerase and gene 4 protein. A lower limit based on these results suggests the processivity of polymerization is greater than 3600 nucleotides, a value in the same range as the processivities measured for both E. coli DNA polymerase III holoenzyme (27) and the bacteriophage T4 replication complex (28).

Lagging Strand DNA Synthesis. As the gene 4 protein moves along the displaced strand at a replication fork it also catalyzes the synthesis of oligoribonucleotides at specific primase recognition sites (17). These sites consist of the pentanucleotide sequences, 3'-CTGGG/T-5'. The gene 4 protein will catalyze the synthesis of tetraribonucleotides that are complementary to the four nucleotides at the 5' end of the primase recognition site giving rise to the sequences 5'-pppACCC/A-3'. These primers serve to initiate DNA synthesis on the lagging strand (11-13,17). Although the gene 4 protein binds randomly to single-stranded DNA, it translocates unidirectionally in the 5' to 3' direction (17). The translocation of the protein along single-stranded DNA requires concomitant hydrolysis of NTPs and ensures that the gene 4 protein will encounter primase recognition sites directly. Thus, the two diverse activities of the gene 4 protein, helicase and primase, are both coupled to the DNA dependent NTPase activity of the enzyme. This activity provides the energy necessary for unidirectional translocation along the lagging strand, a process which simultaneously locates primase recognition sites and opens the DNA helix ahead of the advancing T7 DNA polymerase.

DISCUSSION

The initiation and elongation stages of DNA replication have been reconstituted in vitro using enzymes purified from T7-infected cells. T7 RNA polymerase and T7 DNA polymerase, together, are able to initiate DNA synthesis in vitro at the primary origin on either wild type T7 DNA or on plasmids containing the T7 origin region. RNA transcripts originating from either of the two T7 RNA polymerase promoters in the primary origin provide the primer necessary for initiation of DNA synthesis. We have presented a model for the in vitro initiation reaction that reflects these observations (see Fig. 2). Data obtained from experiments carried out in vivo substantiate this mechanism. When the tandem RNA polymerase promoters located within the primary origin are inactivated using site directed in vitro mutagenesis techniques, DNA replication no longer initiates at the primary origin. We conclude that transcription catalyzed by the T7 RNA polymerase from the RNA polymerase promoters in the primary origin is essential in initiating DNA replication in vivo and in vitro and that, at least in vitro, the RNA transcripts themselves are the initiating primers.

Using a novel DNA substrate we have shown that purified T7 DNA polymerase and gene 4 protein together catalyze the elongation phase of DNA synthesis in vitro at a high microscopic rate. Our studies suggest that the gene 4 protein acts as a helicase using the chemical energy supplied by NTP hydrolysis to promote unwinding of the duplex DNA template for the advancing DNA polymerase. The helicase activity of the gene 4 protein requires single-stranded DNA on both sides of the duplex to be unwound (18), a condition that may exist at a replication fork. As the protein translocates unidirectionally 5' to 3' along the displaced strand it will encounter primase recognition sites at which a tetraribonucleotide primer is synthesized by the gene 4 protein (17). These primers provide the 3'-OH termini necessary to initiate discontinuous lagging strand DNA synthesis. Together the T7 DNA polymerase and gene 4 protein are capable of synthesizing DNA on both the leading and lagging strand sides of the replication fork. That these two proteins, in the absence of other proteins, are capable of carrying out this complicated process reflects the economy of the mechanism that T7 has developed for the replication of its DNA.

REFERENCES

1. Wolfson, J., Dressler, D. and Magazin, M. (1972) Proc. Natl. Acad. Sci. U.S.A. 69, 499–504.
2. Dressler, D., Wolfson, J. and Magazin, M. (1972) Proc. Natl. Acad. Sci. U.S.A. 69, 998–1002.
3. Dunn, J.J. and Studier, F.W. (1983) J. Mol. Biol., in press.
4. Hinkle, D.C. (1980) J. Virol. 34, 136–141.
5. Grippo, P. and Richardson, C.C. (1971) J. Biol. Chem. 246, 6867–6873.
6. Studier, F.W. (1972) Science 176, 367–376.
7. Modrich, P. and Richardson, C.C. (1975) J. Biol. Chem. 250, 5515–5522.
8. Mark, D.F. and Richardson, C.C. (1976) Proc. Natl. Acad. Sci. U.S.A. 73, 780–784.
9. Stratling, W. and Knippers, R. (1973) Nature 245, 195–197.
10. Hinkle, D.C. and Richardson, C.C. (1975) J. Biol. Chem. 250, 5523–5529.
11. Romano, L.J. and Richardson, C.C. (1979) J. Biol. Chem. 254, 10476–10482.
12. Romano, L.J. and Richardson, C.C. (1979) J. Biol. Chem. 254, 10483–10489.
13. Hillenbrand, G., Morelli, G., Lanka, E. and Scherzinger, E. (1978) Cold Spring Harbor Symp. Quant. Biol. 43, 449–460.
14. Kolodner, R. and Richardson, C.C. (1977) Proc. Natl. Acad. Sci. U.S.A. 74, 1525–1529.
15. Kolodner, R. and Richardson, C.C. (1978) J. Biol. Chem. 253, 574–584.
16. Kolodner, R., Masamune, Y., LeClerc, J.E. and Richardson, C.C. (1978) J. Biol. Chem. 253, 566–573.
17. Tabor, S. and Richardson, C.C. (1981) Proc. Natl. Acad. Sci. U.S.A. 78, 205–209.
18. Fuller, C.W., Beauchamp, B.B., Engler, M.J., Lechner, R.L., Matson, S.W., Tabor, S., White, J.H. and Richardson, C.C. (1983) Cold Spring Harbor Symp. Quant. Biol. 48, 669–679.
19. Simon, M.N. and Studier, F.W. (1973) J. Mol. Biol. 79, 249–265.
20. Studier, F.W., Rosenberg, A.H., Simon, M.N. and Dunn, J.J. (1979) J. Mol. Biol. 135, 917–937.
21. Tamanoi, F., Saito, H. and Richardson, C.C. (1980) Proc. Natl. Acad. Sci. U.S.A. 77, 2656–2660.

22. Saito, H., Tabor, S., Tamanoi, F. and Richardson, C.C. (1980) Proc. Natl. Acad. Sci. U.S.A. 77, 3917–3921.

23. Romano, L.J., Tamanoi, F. and Richardson, C.C. (1981) Proc. Natl. Acad. Sci. U.S.A. 78, 4107–4111.

24. Fischer, H. and Hinkle, D.C. (1980) J. Biol. Chem. 255, 7956–7964.

25. Engler, M.J., Lechner, R.L. and Richardson, C.C. (1983) J. Biol. Chem. 258, in press.

26. Lechner, R.L. and Richardson, C.C. (1983) J. Biol. Chem. 258, in press.

27. Johanson, K.O. and McHenry, C.S. (1981) in ICN–UCLA Symposia on Molecular and Cellular Biology, Vol XXII, 425–436 ed. by D.S. Ray (Academic Press).

28. Sinha, N.K., Morris, C.F. and Alberts, B.M. (1980) J. Biol. Chem. 255, 4290–4303.

Mechanisms of DNA Replication and Recombination, pages 153–171
© 1983 Alan R. Liss, Inc., 150 Fifth Avenue, New York, NY 10011

DYNAMIC AND STRUCTURAL INTERACTIONS
IN THE T4 DNA REPLICATION COMPLEX[1]

Peter H. von Hippel, John W. Newport[2], Leland S. Paul,
Frederic R. Fairfield, Joel W. Hockensmith,
and Mary K. Dolejsi

Institute of Molecular Biology
and Department of Chemistry
University of Oregon
Eugene, OR 97403, U.S.A.

ABSTRACT. In this paper we examine molecular aspects
of the protein-protein and protein-DNA interactions
involved in processive DNA synthesis by the five
protein T4-coded system, carrying out template-
directed DNA replication at model (oligo dT-poly dA)
or "real" (natural DNA) primer-template junctions.
Binding of the T4 DNA polymerase and of the polymerase
accessory proteins at the primer-template junction is
studied, and preliminary results of laser-induced
protein-DNA crosslinking experiments on the replication complex are considered. Based on these results,
a tentative molecular model of the processive
synthesis "cycle" is proposed.

[1] This work was supported in part by USPHS Research
Grants GM-15792 and GM-29158, by USPHS Training Grants
GM-00715 and GM-07759, and by USPHS Postdoctoral
Fellowship GM-08252 (to FRF).

[2] Present Address: Department of Biochemistry and
Biophysics, University of California Medical Center,
San Francisco, California 94143

INTRODUCTION

Considerable progress has been made in isolating and purifying the protein components of various DNA replication systems. Several prokaryotic systems, notably (in order of increasing complexity) those coded by bacteriophages T7 and T4 and by E. coli, have been successfully reconstituted in vitro, and have been shown to carry out the elongation phase of DNA replication at approximately in vivo rates and fidelities (for recent overviews see ref. 1 and various articles in this volume).

The bacteriophage T4-coded DNA replication system, which has been developed primarily by Alberts, Nossal and their coworkers (2,3), is well suited for mechanistic studies. Effective leading and lagging strand synthesis is carried out in vitro by a seven-protein system. The proteins required are: (i) gene 43 protein (DNA-dependent DNA polymerase), which catalyzes both template-directed DNA synthesis and "editing" -- the latter function via an intrinsic 3' → 5' exonuclease; (ii) gene 32 protein (single-stranded DNA binding protein), which functions specifically with the T4 polymerase by stabilizing the transient single-stranded DNA sequences formed at the replication fork; (iii) genes 44/62 and 45 proteins (polymerase accessory proteins), which feature a DNA-stimulated ATPase activity that is required for strand displacement and to maintain the processivity of the system under physiological conditions (see below); and (iv) genes 41 and 61 protein, which carry out RNA priming of lagging strand synthesis, perform helicase functions, and couple leading and lagging strand synthesis (for an up-to-date summary, see Jongeneel et al., this volume).

In the work to be reported here we will focus on current studies in our laboratory on molecular aspects of the central process of DNA replication; that is, the processive template-directed elongation (and/or editing) of the primer strand at the primer-template junction (Figure 1).

To this end we will discuss first the binding and function of the polymerase at the primer-template locus, and then consider the roles of genes 32, 44/62 and 45 proteins ("the five-protein system") in achieving fully processive synthesis under physiological conditions.

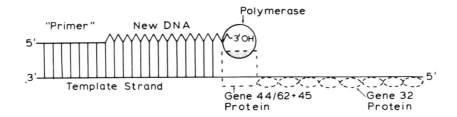

FIGURE 1. The primer-template DNA replication model system. The required polymerase is shown as a solid figure; the other proteins of the five protein complex are only required under some conditions (see text) and are thus shown as dotted outlines.

PROCESSIVITY

The elementary process carried out by the T4 DNA polymerase is the (template-directed) addition of a deoxyribose nucleoside triphosphate (dNTP) to the 3'-OH of a pre-existing primer, with the concomitant release of a pyrophosphate molecule:

$$\text{Primer} + \text{dNTP} \underset{\leftarrow}{\overset{Mg^{2+}}{\rightarrow}} \text{Primer-dNMP} + \text{PP}_i \quad (1)$$

In the editing mode, the equivalent elementary process is:

$$\text{Primer-dNMP} \underset{\leftarrow}{\overset{MG^{2+}}{\rightarrow}} \text{Primer} + \text{dNMP} \quad (2)$$

As written above, these one-step equations describe the basic chemical reactions catalyzed by the polymerase. However these equations do not reflect the fact that a large number of these steps (especially in synthesis, as described by eq. 1) must occur during each polymerase binding event in order to achieve the observed in vivo rates of synthesis of several hundred to a thousand nucleotide residues incorporated per second per functioning polymerase (4). (We estimate that the maximum rate of primer extension achievable at physiological polymerase concentrations, if the enzyme were to dissociate from the

primer-template junction after each nucleotide incorporation step -- i.e., if synthesis were fully dispersive --
would be about one nucleotide residue per second, assuming
polymerase rebinding proceeds at diffusion-controlled
rates.)   Thus the establishment and maintenance of
<u>processive</u> synthesis (the template-directed incorporation
of more than one nucleotide residue per polymerase binding
event) represents a central functional requirement for the
incorporation of polymerase into a working DNA replication
system.   It is the nature of such processive synthesis,
and the role of the various components of the five-protein
system in establishing it, that we examine in this paper.

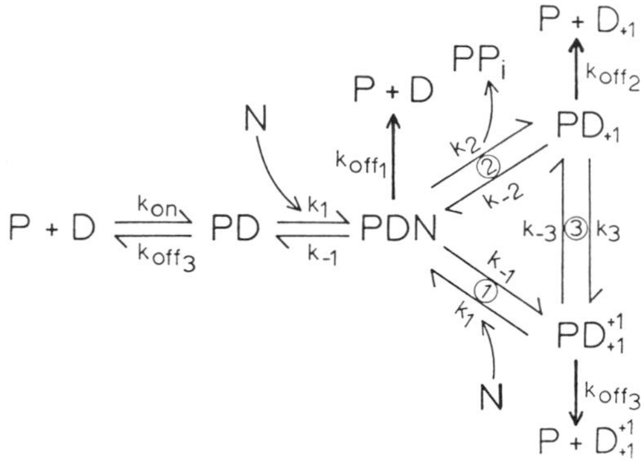

FIGURE 2.   Model for processive synthesis of DNA.   P
here represents the polymerase, D is the DNA primer-
template complex which has <u>not</u> been elongated, N a
nucleoside triphosphate and PP_i is pyrophosphate.   The
subscripts on D represent the number of nucleotides that
have been added to the primer, and the superscripts the
number of translocation events that have occurred; for
example D_{+5}^{+4} would represent a primer-template complex that
has been elongated by 5 nucleotide residues, and has
undergone 4 polymerase translocations.   The "processivity
cycle" (steps 1,2 and 3) continues until the polymerase
dissociates from the primer-template complex at any one of
the three cyclic steps (Figure from ref. 5.).

Definitions and Model.

We define the microscopic processivity parameter, P, as the probability that a polymerase positioned at the primer-template junction at template position n will <u>not</u> dissociate from the DNA in translocating to position n + 1 in the polymerizing mode (or to position n - 1 in the editing mode). Thus for fully processive synthesis P = 1, for fully dispersive synthesis P = 0, and for partially processive synthesis 0 < P < 1.

In these terms, the elementary elongation process can be represented by a three step mechanism (5,6), as shown in Figure 2. These steps are: (i) binding of a template complementary nucleoside triphosphate to the polymerase-primer-template complex; (ii) covalent linking of this unit (as the nucleoside monophosphate) to the 3'-primer terminus; and (iii) translocation of the polymerase forward along the template by one nucleotide residue to achieve correct alignment at the new primer terminus. The overall (mean) processivity of this synthesis depends on the average number of such three-step cycles the polymerase completes before dissociating from the primer-template complex. Figure 2 shows that, in principle, dissociation can occur at each of the steps of the cycle.

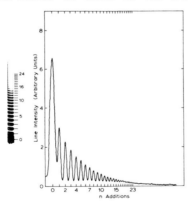

FIGURE 3. Processivity assay for the oligo dT-poly dA system containing reaction buffer (10 mM Tris, 0.1 mM EDTA, 2.5 mM $MgCl_2$, 0.5 mM DTT, 200 μg/ml BSA, pH 7.8), 95 mM NaCl, 300 μM dTTP, 3 x 10^{-7} M $(dT)_{16}$ (5'-labelled with ^{32}P to 10^5 cpm), 4 x 10^{-5} M poly dA (as M nucleotide phosphate) and 5 x 10^{-9} M T4 polymerase. The reaction was run for 30 sec at 37°C. (Figure from ref. 5)

Assay.

The processivity of T4 polymerase under various
conditions was investigated by examining the distribution
of extended primers after reaction with polymerase
(5,7,8). Either a $5'-{}^{32}$P-labelled oligo dT primer
annealed to a poly dA template, or a 5'-labelled M13
primer annealed to a variety of single-stranded M13
templates, was used. After reaction, the extended primers
were separated by polyacrylamide gel electrophoresis and
assayed by quantitative autoradiography. In general,
experiments were run under conditions of considerable
primer excess, so that each extended primer represents the
result of a single polymerase binding event.

The results of a typical assay, carried out under
conditions of moderate processivity with the oligo dT-poly
dA model system, are shown in Figure 3. The autoradiogram
of the separated primers is shown on the left, and the
corresponding densitometer tracing is displayed on the
right. The peak areas under the densitometer trace follow
accurately a single exponential decay; the value of P (the
microscopic processivity parameter) for this system can be
obtained by plotting log (n_x/n_t) versus x, where n_x is the
number of primers extended by exactly x residues, and n_t
is the total number of extended primers (5,7). For the
experiment presented in Figure 3, P = 0.83, meaning that
under these conditions the mean primer extension per
polymerase binding event is about 6 residues calculated as
$1/(1-P)$.

PROCESSIVITY STUDIES WITH THE OLIGO dT-POLY dA SYSTEM

In this section we briefly summarize some processivity
studies carried out with the oligo dT-poly dA primer-
template system by Newport (5), in order to provide a
background for the more recent and continuing studies on
which we wish to present a progress report in this paper.
Some results of these model studies have been described
previously (7); a complete report will be presented
elsewhere (Newport and von Hippel, manuscript in
preparation).

Conclusions.

P is Independent of Primer Length. The fact that
n_x/n_t decreases exponentially with increasing length of
extended primer under a variety of experimental conditions
(5,7; for example, see Figure 3) shows that P (the microscopic processivity parameter) is a constant for the oligo
dT-poly dA system. That is, the decay of this primer-
template-polymerase complex follows first order kinetics.
P is Dependent on Salt Concentration. Experiments
like the one presented in Figure 3 were carried out at a
variety of concentrations of added salt. The measured
value of P falls markedly with increasing salt
concentration; from $P \simeq 0.97$ at 35 mM NaCl to $P \simeq 0$ at
salt concentrations greater than about 180 mM (5,7). This
result shows that while DNA synthesis in the presence of
polymerase alone can be very processive (at low salt), it
becomes fully dispersive at salt concentrations equivalent
to those present in vivo (see ref. 9 for discussion of
physiological salt concentration in vivo).
Processive Synthesis Can be Restored at High Salt
Concentrations by Adding the Other Components of the
Five-Protein DNA Replication System. It was shown that
fully processive synthesis with the oligo dT-poly dA model
system at 170 mM NaCl could be restored by adding gene 32
protein, gene 44/62 and 45 proteins, and ATP (> 300 mM) to
the primer-template-polymerase system. All components
were required to make an appreciable change in either the
rate or the processivity of polymerization (5,7). Thus,
to a first approximation, the other proteins of the five-
protein system (and ATP) serve to restore the processivity
of the system to values comparable to those achieved by
the polymerase alone at low salt.
ATP is Required for the Assembly of the Five-Protein
Replication Complex, Not for its Translocation. Piperno
and Alberts (10) have demonstrated that ATP must be
hydrolyzed in order for the gene 44/62 and 45 proteins to
stimulate DNA synthesis. These workers also showed that
less than one ATP is hydrolyzed per ten deoxyribonucleotides incorporated into DNA. We (5,7) have shown that the
number of primers elongated per unit time by the five-
protein system is dependent on ATP concentration, but the
length to which these primers are elongated is not. This
suggests that ATP hydrolysis is required to form
(assemble) the five protein replication complex, but not

for its translocation along the template. Experiments in which the rate of DNA synthesis was reduced by limiting the concentration of dNTP substrates suggested that the lifetime of the assembled five protein replication complexes is about 45 seconds.

PHYSICAL AND ENZYMATIC STUDIES ON INTERACTIONS IN THE FIVE-PROTEIN REPLICATION COMPLEX

Binding of T4 Polymerase to the Primer-Template Junction.

A variety of direct and binding competition approaches have been used to define the binding interactions of T4 polymerase with the primer-template junction. Some results have been presented (5,7); full details will be presented elsewhere (Newport, Paul and von Hippel, manuscript in preparation). The processivity (synthesis) experiments outlined above suggest that polymerase binding is very salt dependent; this conclusion is confirmed by direct binding measurements of polymerase to single-stranded (ϕX-174) DNA and by examining the rate of the $3' \rightarrow 5'$ exonuclease action of the polymerase on single-stranded DNA. All three approaches suggest that free energy of polymerase binding to either single-stranded DNA or to the oligo dT-poly dA primer-template junction is fully electrostatic in nature, and (based on the treatment of Record et al., 1976) involves the formation of $\sim$ 8 polymerase-DNA charge-charge interactions between basic residues in the polymerase binding site and DNA phosphates. Furthermore, in this electrostatic binding mode polymerase appears to be binding with comparable affinities to single-stranded DNA and to primer-template junctions. A model has been presented suggesting that the DNA phosphates of the primer-template involved in charge-charge interactions with the polymerase are located in two clusters over a distance of $\sim$ 11 base pairs along the primer side of the primer-template junction (5,7). The recent determination that the site size for T4 DNA polymerase binding to polyethenoadenylic acid (poly rϵA) is $\sim$ 12 nucleotide residues (11) is consistent with such binding models.

Assembly of the Polymerase Accessory and Gene 32 Proteins at the Primer-Template Junction.

Processivity studies with the five-protein system have suggested that a functional polymerase accessory protein complex can assemble, independently of polymerase, at the primer-template junction (5,7). This has been further investigated (11, and Paul and von Hippel, manuscript in preparation) by examining the behavior of the DNA-stimulated ATPase activity of the gene 44/62 and 45 protein system when complexed at this locus.

The protein products of genes 44 and 62 are isolated as a tight multi-subunit complex (12). These proteins carry an ATPase activity that is stimulated by polynucleotide binding; the presence of gene 45 protein greatly enhances the ATPase activity of the gene 44/62 protein complex. Previous studies have shown that single-stranded DNA is much more effective than double-stranded DNA at stimulating this ATPase activity (12). In order to better define the DNA binding specificity of the polymerase accessory proteins, we have investigated further the nucleic acid cofactor requirement of this ATPase system.

Most Single-Stranded Deoxyribopolynucleotides Stimulate Accessory Protein ATPase Activity at Low Salt Concentration. By monitoring ATPase activity of the polymerase accessory proteins as a function of polynucleotide concentration under low salt concentration conditions (6 mM $MgCl_2$, 20 mM NaCl), it has been shown that most single-stranded deoxyribopolynucleotides (poly dT, poly dA, poly dI) are effective ATPase cofactors. In contrast, most single-stranded ribopolynucleotides do not support substantial ATPase activity. Since T4 DNA polymerase can bind effectively to single-stranded RNA (5), the accessory proteins may provide some of the binding specificity (for DNA over RNA) of the five-protein replication complex that seems to be lacking in the polymerase alone.

In addition, we have examined the ATPase rates supported by the binding of single-stranded deoxyribo-oligonucleotides of defined length. These studies have shown that oligonucleotides at least 12 residues in length are required to stimulate the ATPase of the accessory proteins, and full (polynucleotide level) stimulation requires oligonucleotides that are ~ 20 residues in length.

<u>The Polymerase Accessory Proteins Bind Specifically to</u>
<u>the Primer-Template Junction</u>. We tested the effectiveness
of oligo dT-poly dA (or oligo dA-poly dT) primer-template
models as cofactors for the ATPase activity of the accessory protein complex. It was found that these primer-
template models are better cofactors than either of the
single-stranded DNA cofactors working separately, and
furthermore that at moderate salt concentrations (95 mM
NaCl) the primer-template complexes continue to stimulate
with appreciable effectiveness, while stimulation by
single-stranded polynucleotides falls virtually to zero.
Thus it appears likely that the primer-template junction
represents the target site for the binding of the accessory protein complex under physiological salt
concentrations. Preliminary experiments have also shown
that RNA primers can serve as well as DNA primers in
activating the ATPase activity of the accessory protein
complex. This is consistent with the presumed requirement
for activity of this complex at RNA primers in lagging
strand DNA synthesis.

<u>Interactions with Gene 32 Protein</u>. The effect of gene
32 protein on the utility of the above nucleic acid
systems as cofactors of the ATPase activity of the polymerase accessory proteins has also been examined (11).
With single-stranded DNA (poly dT or poly dA) titration by
gene 32 protein progressively lowers the observed ATPase
activity, until the DNA-dependent ATPase activity of the
accessory protein complex is abolished at DNA-saturating
levels of gene 32 protein. This suggests that gene 32
protein displaces the accessory protein complex from these
single-stranded DNA molecules.

In contrast, gene 32 protein has little effect on
ATPase activity stimulated by primer-template binding,
suggesting that this single-stranded DNA binding protein
is unable to displace the polymerase accessory protein
complex from the primer-template junction. In addition,
at saturating levels of gene 32 protein, the ATPase activity of the accessory proteins that is stimulated by
primer-template junction binding is totally salt insensitive (at least up to 110 mM NaCl). Thus it appears that
gene 32 protein stabilizes the accessory proteins-primer-
template complex, as well as preventing nonspecific
binding of the accessory proteins (and of polymerase as
well) to single-stranded regions of the replication
complex.

Model for Binding of the Five-Protein Replication
Complex at the Primer-Template Junction. This and related
evidence is summarized in Figure 4 to provide a model of
the five-protein complex assembled at the primer-template
junction. The model suggests that the accessory proteins
must carry binding subsites that can recognize single- and
double-stranded DNA sequences. Gene 32 protein binds to
single-stranded regions of the primer-template, and probably interacts directly with the polymerase as well (13).
The binding interactions of the polymerase have been
described above (see also ref. 7). These results also
show that the accessory proteins can assemble properly at
the primer-template junction in the absence of polymerase,
as inferred previously from more indirect experiments
(5,7).

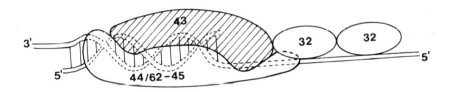

FIGURE 4. Model of the five protein complex assembled
at the primer-template junction.

Crosslinking of the Constituents of the T4 DNA Replication
Complex.

In order to function in processive DNA synthesis, the
T4 DNA replication proteins must interact with the primer-
template, and probably with one another. In an attempt to
establish the relative positions and contacts of these
components in the functioning five-protein replication
complex, we have examined these interactions using
protein-protein and protein-DNA crosslinking techniques.
Protein-Protein Crosslinking. Efforts at protein-
protein crosslinking were pursued using chemical crosslinking reagents and SDS-polyacrylamide gel electrophoresis techniques (14-16). This approach is applicable

to the establishment of chemical proximity and relative
distances between polypeptide chains. However, this
approach proved unusable here. Both intermolecular and
intramolecular crosslinking occur, with the intramolecular
type resulting in altered mobilities of the individual
polypeptides on SDS-polyacrylamide gels. However, under
typical conditions used for DNA synthesis reactions, substantial intermolecular crosslinking was observed only for
the protein components known from other work to have
strong interactions: the 44/62 complex crosslinks well
and gene 32 protein crosslinks to form oligomers when it
is bound to DNA. The difficulty in detecting other intermolecular crosslinking events in the replication complex
may result from the dynamic state of the polypeptide
interactions on the time scale of the chemical crosslinking reaction (minutes).

<u>Protein-DNA Crosslinking</u>. Protein-DNA interactions
were probed using photochemical crosslinking induced by
broad-band ultraviolet irradiation (17), and also by laser
irradiation (18). These crosslinks are of "zero-length"
and thus should reveal points of contact between DNA and
proteins. Samples contained a radiolabelled nucleic acid
oligomer, and one or more purified gene products under
conditions typically used for DNA synthesis <u>in</u> <u>vitro</u>.
After crosslinking, the samples were denatured and run on
a SDS-polyacrylamide gel, followed by autoradiography of
the gel. The oligonucleotide migrates more rapidly than
any of the gene products unless crosslinking has occurred.
If the oligonucleotide is crosslinked to a gene product,
the resulting complex will be essentially the same size as
the gene product itself. Consequently, the crosslinked
complex will migrate more slowly than the oligomer; i.e.,
at the position of the gene product alone.

The broad-band UV irradiation studies were performed
with a 30 watt germicidal lamp with a primary output at
254 nm, and demonstrated that the products of genes 43,
32, 44/62, and 45 can all be crosslinked to a single-
stranded DNA oligomer after five minutes of irradiation.
This crosslinking is sensitive to the presence of added
salt, suggesting a relationship between DNA binding and
crosslinking. While broad-band UV crosslinking can be
used to crosslink proteins to DNA, it has a number of
disadvantages. Large numbers of UV damaged molecules
(photoproducts) are generated and cannot be entirely
eliminated, even at low temperatures. Again, the time

scale of crosslinking is long relative to the lifetime of
the protein-DNA interactions that are being trapped. This
long time of irradiation permits conformational changes in
the proteins and protein movement to new sites to occur,
and also allows long-lived photoproducts and UV-damaged
molecules to form spurious crosslinks.

Such spurious crosslinking complicates interpretations,
but multiple protein systems are difficult to interpret
for other reasons as well. This method of irradiation
introduces too few photons per second into the sample and
too large a dispersion of light occurs to permit the
crosslinking of all of the proteins in a complex before
any particular protein can dissociate. Also the multiple
species of proteins cause the generation of photoproducts
to become very complex. The difficulties with the time
frame of conventional UV irradiation have led us to the
use of laser pulses to induce protein-DNA crosslinking.

Laser irradiation was carried out with an KrF excimer
laser with a wavelength of 249 nm, a pulse length of 9
nanoseconds, and a power delivery of 115 mJ ± 10% per
pulse to the sample. The laser crosslinking technique
offers a number of striking advantages over traditional
broad-band UV crosslinking. The time scale of irradiation
for a single pulse is very short, and consequently, few
protein or nucleic acid conformational changes can take
place. This decreases the probability of crosslinking of
photoproducts that have bound to the DNA in an abnormal
way. The nine nanosecond laser pulse effectively
"freezes" intermediates, since dissociation and reassociation cannot occur within this time frame.

For RNA polymerase, there appear to be two classes of
reactive species, one is short-lived and one is
long-lived. The long-lived species can be quenched with
2-mercaptoethanol, thereby avoiding artifactual crosslinking (18). Linear dependence of crosslinking with
laser intensity suggests that a single photon process is
occurring. And lastly, simultaneous crosslinking of
several proteins in a complex is possible.

Laser crosslinking at 249 nm leads to a specific
crosslinking of protein to DNA. We have not observed any
protein-protein or primer-template crosslinking using the
T4 DNA replication system. Additionally, no protein
degradation has been detected at the 115 mJ input level.
Photodegradation can be induced using multiple (ten) laser
pulses, and leads to the generation of discrete bands in a

silver-stained polyacrylamide gel.

The protein products of T4 genes 32, 43, 44, and 45 all crosslink to a single-stranded DNA oligomer, while the 62 gene product does not. This correlates well with the broad-band UV studies, with the exception of the 62 gene product result. The crosslinking can be altered by the addition of certain chemicals typically found in DNA synthesis systems in vitro. The addition of Mg^{2+} to a sample containing gene 43 protein greatly increases the crosslinking of that protein to the DNA, suggesting an increase in DNA binding by the gene 43 product. Alternatively, Mg^{2+} decreases the crosslinking of gene 45 protein. In a similar manner, added ATP decreases the crosslinking of the gene 44 protein to the DNA oligomer. Using multiple protein systems, we found that the crosslinking of gene 32 protein to a single-stranded oligomer is decreased by the addition of gene 43 protein or gene 45 protein. Crosslinking of the 34,000 molecular weight subunit of the gene 44/62 proteins is enhanced by the presence of the gene 32 protein, but is diminished by the addition of gene 43 protein or gene 45 protein. Gene 32 protein or gene 44/62 proteins diminish crosslinking of gene 45 protein, while gene 43 protein totally eliminates gene 45 protein crosslinking. Further studies will be needed in order to integrate the crosslinking studies with the other physical-biochemical studies employed in this paper. However, it appears that laser irradiation represents an excellent technique for crosslinking proteins to DNA, and we are currently investigating the crosslinking of the individual protein and mixed protein systems to a DNA primer-template complex.

PROCESSIVITY AT NATURAL DNA PRIMER-TEMPLATE JUNCTIONS

In conclusion, we report briefly on some approaches we are currently taking to define the processivity of T4 DNA polymerase at natural DNA primer-template junctions, in order to further investigate the nature of processive synthesis itself, and to compare the effects of DNA sequence on processivity (if any) with the picture of processivity developed with model (homopolymer) primer-template systems as summarized above. A detailed discussion of the work will be presented elsewhere (Fairfield, Dolejsi and von Hippel, manuscript in preparation).

The M13 Primer-Template System.

These studies have been carried out using various circular M13 single-stranded viral DNA molecules as template, and forming a primer-template locus at which to initiate polymerization by annealing specific natural oligomeric primer sequences to this template. Since the M13 sequence is totally known (19,20), and in addition all sorts of variants of the template sequence are available (e.g., M13 mp7, M13 mp8, M13 mp9, etc.), this system provides an ideal substrate to test the effects of (known) DNA sequence and sequence alterations on processivity. Quantitative gel electrophoresis and densitometric analyses were carried out much as described for the oligo dT-poly dA system (above).

Initial results immediately showed that the situation is complex, and that processive synthesis with T4 DNA polymerase is far more likely to "terminate" at some DNA sequences than at others. Other groups have obtained similar results with other polymerase systems; e.g., (21,22,23), and De Pamphilis et al. (this volume). In terms of our model system for processive DNA synthesis (Figure 2), this means that P (the microscopic processivity parameter) is <u>not</u> a constant at every template position; rather P is very dependent on sequence at or near the primer-template junction.

Initial analyses of our results using the M13 primer-template system suggest the following: (i) P (the probability of <u>not</u> dissociating at any particular synthesis step) is very dependent on sequence; (ii) the sequence dependence of P is fairly extensive (e.g., the value of P a particular locus does not simply depend on the nature or sequence of one or two residues right at the polymerization site); (iii) stable stem-loop structures in the single-stranded DNA template do represent strong "termination" sites (as previously shown with ϕX174 templates by Huang et al., 24); (iv) many other specific strong (and weak) "termination" sites are not located in any obviously relevant proximity to stem-loop structures, suggesting that other elements of sequence, either on the single- or the double-stranded side of the primer-template junction, may also be relevant.

In a preliminary effort to establish the positions at the primer-template locus that effect P, we have used the Perceptron computer algorithm procedure employed by Stormo et al. (25) to find common sequence elements in mRNA at ribosome binding sites. For our purposes we wished to determine whether preferred sequence elements exist on either the double-stranded or single-stranded sides of the primer-template junction at loci of preferred termination (i.e., at sites characterized by low P values). Analysis of ~ 40 preferred termination sites over ~ 200 nucleotides of the M13 primer showed that positions located within about 8 base pairs of the primer-template junction on the double-stranded (primer) side, and within ~ 3 residues of the junction on the single-stranded side, appear to exert specific effects on local termination (polymerase dissociation) probabilities. We note this "site size" for the primer-template bound polymerase is very consistent with that estimated by other techniques described above. Sequences further from the primer-template locus had no effect above the random "noise" in the analysis. More experiments carried out under a variety of conditions will be analyzed to assign specific preferred "low P" sequences at the primer-template junction with confidence (Fairfield, Dolejsi and von Hippel, in preparation).

Thoughts on the Nature of Processivity.

Present results of these studies, and those of others, provide certain "facts" on the nature of the processive DNA replication process at a single-stranded template-primer locus.

(i) The potential to carry out processive synthesis is inherent in the polymerase itself, since substantial processivity is observed (at least at low salt) with both the oligo dT-poly dA and natural DNA primer-template junctions.

(ii) The polymerase accessory proteins (with gene 32 protein) can assemble specifically at the primer-template junction to form a "platform" for polymerase that permits processive synthesis at physiological salt concentrations (at least at oligo dT-poly dA primer-template junctions).

(iii) ATP binding and hydrolysis is required to stabilize this complex, but not as a source of energy to translocate it along the template.

(iv) Polymerase can bind to DNA and at the primer-template junction in at least two modes (conformations?). In one mode binding is totally electrostatic and sequence non-specific. In the other mode, reflected in sequence specific preferential termination (or dissociation) sites, binding affinities at the primer-template junction must be at least partially sequence dependent.

Tentatively, we propose the following model for the movement of T4 polymerase (and the five-protein replication complex?) around the three-step "processivity cycle" (Figure 2). A specific binding event occurs to position the polymerase at the primer-template junction in a proper arrangement to bind the incoming (template-complementary) nucleoside triphosphate and to catalyze its covalent linkage to the 3'-OH of the primer, with release of pyrophosphate. We speculate that the formation of this bond is coupled with a conformational change in the polymerase, which "snaps" it back into the nonspecific electrostatic binding mode. In this mode the polymerase can "slide" along the DNA, in a fashion similar to that recently proposed for lac repressor and other genome regulatory proteins (26,27). This sliding is a one-dimensional diffusion process, and is driven only by thermal fluctuations (kT). Such sliding carries the polymerase (and the accessory protein platform), on a random walk basis, to the new (displaced by one residue) primer-template junction, at which the polymerase conformation reverts to the specific primer-template binding mode, thus positioning itself to catalyze the next template-complementary primer elongation event. Additional studies are obviously needed to test and modify this proposal further.

ACKNOWLEDGEMENTS

We are very grateful to our laboratory colleagues for discussions as this work developed, and to Bruce Alberts and his associates for much advice and help in learning how to purify and handle the components of the T4 DNA replication system. We also acknowledge the interest and help of Ed Herbert, George Sprague and their colleagues in setting-up the M13 primer-template system for our purposes.

REFERENCES

1. Kornberg A (1980). "DNA Replication" and "1982 Supplement to DNA Replication." San Francisco: WH Freeman and Co.

2. Alberts BM, Barry J, Bedinger P, Burke RL, Hibner U, Liu C-C, Sheridan R (1980). Alberts, B (ed) "Mechanistic Studies of DNA Replication and Genetic Recombination" (ICN-UCLA Symposia on Molecular and Cellular Biology, Vol 19) New York: Academic Press, p. 449.

3. Silver LL, Venkatesan M, Nossal NG (1980). Alberts, B (ed) "Mechanistic Studies of DNA Replication and Genetic Recombination" (ICN-UCLA Symposia on Molecular and Cellular Biology, Vol 19) New York: Academic Press, p. 475.

4. McCarthy D, Minner C, Bernstein H, Bernstein C (1976). J Mol Biol 106:963.

5. Newport JW (1980). PhD Thesis, University of Oregon.

6. Detera SD, Wilson SH (1982). J Biol Chem 257:9770.

7. Newport JW, Kowalczykowski SC, Lonberg N, Paul LS, von Hippel PH (1980). Alberts, B (ed) "Mechanistic Studies of DNA Replication and Genetic Recombination" (ICN-UCLA Symposia on Molecular and Cellular Biology, Vol 19) New York: Academic Press, p. 485.

8. McClure WR, Chow Y (1980). Meth Enzymol 64:277.

9. Kao-Huang Y, Revzin A, Butler AP, O'Connor P, Noble D, von Hippel PH (1977). Proc Natl Acad Sci USA 74:4228.

10. Piperno JR, Alberts BM (1978) J Biol Chem 253:5174.

11. Paul L (1983). PhD Thesis, University of Oregon.

12. Piperno JR, Kallen RG, Alberts BA (1978) J Biol Chem 253:5180.

13. Huberman JA, Kornberg A, Alberts BM (1971). J Mol Biol 62:39.

14. Smith RJ, Capaldi RA (1977). Biochemistry 16:2629.

15. Smith RJ, Capaldi RA, Muchmore D, Dahlquist F (1978). Biochemistry 17:3719.

16. Wang K, Richards FM (1974). J Biol Chem 249:8005.

17. Park CS, Hillel Z, Wu CW (1980). Nuc Acids Res 8:5895.

18. Harrison CA, Turner DH, Hinkle DC (1982). Nuc Acids Res 10:2399.

19. Messing J, Crea R, Seeburg P (1981). Nuc Acids Res 9:309.

20. van Wezenbeek PMGF, Hulsebos TJM, Schoenmakers JGG (1980). Gene 11:129.

21. LaDuca R, Crute J, Fay P, Chuang C, Johanson K, McHenry C, Bambara R (1983). J Cellular Biochemistry Supplement 7B:93.

22. Nossal NG (1983). J Cellular Biochemistry Supplement 7B:103.

23. Weaver DT, DePamphilis ML (1982). J Biol Chem 257:2075.

24. Huang C-C, Hearst JE, Alberts BM (1981) J Biol Chem 256:4087.

25. Stormo GD, Schneider TD, Gold L, Ehrenfeucht A (1982). Nuc Acids Res 10:2997.

26. Winter RB, Berg CG, von Hippel PH (1981). Biochemistry 20:6961.

27. Berg O, Winter RB, von Hippel PH (1982). Trends in Biochemical Sciences 7:52.

Mechanisms of DNA Replication and Recombination, pages 173–186
© 1983 Alan R. Liss, Inc., 150 Fifth Avenue, New York, NY 10011

GENE EXPRESSION AND INITIATION OF DNA REPLICATION OF
BACTERIOPHAGE T4 IN PHAGE AND HOST TOPOISOMERASE MUTANTS

Gisela Mosig, Paul Macdonald, Gene Lin,
Margaret Levin and Ruth Seaby,

Vanderbilt University, Department of Molecular Biology,
Nashville, Tennessee 37235

ABSTRACT We show that altered torsional stress due
to phage and/or host topoisomerase mutations has
major effects on T4 gene expression and consequently
on packaging of the DNA. The well-known DNA-delay
phenotype of T4 topoisomerase mutants is largely
due to a delay in expression of certain recombination
genes. Additional inhibition of host gyrase had
little effect on overall DNA synthesis in the T4
mutants. It did, however, influence the selection
of T4 origins from which primary replication started.

INTRODUCTION

There is now much evidence that torsional stress in
DNA affects transcription and, in a less well defined way,
DNA replication (1). Since initiation at many replication
origins requires transcription, for "transcriptional
activation" i.e. to open the origin sequences, and/or for
priming of leading strand DNA synthesis, it is an attractive
hypothesis to consider effects of altered torsional stress
on both transcription and replication as consequences of
altered promoter recognition and utilization (2).
DNA molecules must be covalently closed in order
to measure topological stress by the degree of supercoiling
or of solenoid condensation in vitro (3). It has been
demonstrated, however, that linear T4 DNA molecules can

This work was supported by NIH grants GM 13221,
RR07201 and T32 GM07319.

be supercoiled in vivo, presumably because they are topologically constrained by other means (4). We have therefore
investigated effects on phage T4 gene expression and DNA
replication when torsional tension is altered by changes
in unwinding and relaxing activities of two topoisomerases,
E. coli gyrase and T4 topoisomerase. E. coli gyrase, a
type II topoisomerase, can negatively supercoil or relax
DNA; a different form of this enzyme that contains a
truncated peptide of the gyrB subunit lacks the supercoiling but retains the relaxing activity (1,5). Phage T4
type II topoisomerase whose subunits are encoded in genes
39, 52, and 60 can relax supercoiled DNA but cannot supercoil
DNA in vitro (6,7,8). Both host and phage enzymes can knot
and unknot DNA rings (9, 10).

When one or all of the T4 topoisomerase subunit genes
are mutated, overall DNA replication and progeny particle
production are delayed and reduced, but not abolished
(11,12,13). These defects are less severe at higher than
at lower temperatures.

Bernstein's group has shown that the T4 mutants are
slow in initiating replication forks, but not in DNA
elongation (14). Since all initiation was thought to
occur at origins, it appeared paradoxical at first, that
the T4 mutants showed little, if any delay in their first
round of replication (11,12), which is presumably initiated
at origins, but that they were largely delayed in subsequent
^{3}H-thymidine incorporation.

This paradox could be resolved, since it has become
evident that phage T4 uses at least two modes to initiate
replication forks: (i) origin initiation which requires
functional RNA polymerase for priming leading strand
synthesis (15) and (ii) subsequent initiation from recombinational intermediates (15). The latter mode is necessary
for T4 growth, because the RNA polymerase dependent priming
from origin promoters is turned off as a consequence of
the T4 developmental program (15,16).

Accordingly, the T4 topoisomerase mutants are capable
of origin initiation, but subsequent initiation from
recombinational intermediates and consequently major
^{3}H-thymidine incorporation is delayed (13).

On the other hand, it has been suggested that the T4
topoisomerase is an origin specific gyrase which has to
unwind T4 replication origins prior to initiation (7).
In this model, the host gyrase can partially substitute
for origin unwinding, when the T4 topoisomerase is inactive.
This model is based on the report that inhibitors of host

gyrase abolish the residual replication of T4 topoisomerase
mutants mentioned above (17).

Several replication origins have been found in T4
(13,18,19,20,21,22, see Fig. 1). Most of these origins

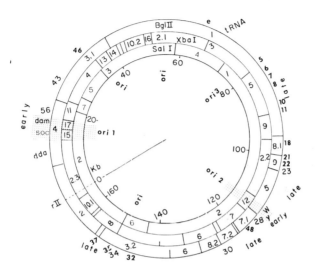

Figure 1: T4 map showing relative positions of
DNA fragments cut by restriction enzymes SalI, XbaI,
and BglII, and the positions of the origins and
genes (23) discussed here.

are located in early regions, as predicted (15) since
origin initiation requires early transcription. One,
however, near gene 5 appears to be in a late region. It
remains to be seen whether this origin region really contains only late promoters and, in fact, whether all described
origins initiate replication de novo and use RNA polymerase
for priming. It appears that "firing" of different origins
depends on different growth and labeling conditions, i.e.,
ultimately on the activity of certain genes (20,22).

To clarify these issues we have investigated the
effects of phage and host topoisomerase mutations on
T4 gene expression, origin selection and overall DNA
synthesis.

RESULTS

Inhibiting the host gyrase with novobiocin completely inhibited progeny production of the T4 topoisomerase mutants, both at 25° and at 42°. In contrast to McCarthy (17), we found, however, less than 50% inhibition of DNA replication by novobiocin under these conditions. (We infected exponentially growing cells, whereas McCarthy (17) had starved the infected cells for an extensive period.) Acrylamide gels of T4 proteins labeled under these conditions, showed drastic changes in the pattern of late T4 proteins in the mutants (data not shown), suggesting that inactivating the host gyrase largely altered late gene expression and consequently prevented DNA packaging in the T4 topoisomerase mutants. Since inhibitors might also act on other proteins, we investigated growth of the T4 mutants in the ts gyrB host LE 316 (24), kindly supplied by E. Orr. In this host, overall DNA synthesis of the T4 mutants was not significantly reduced as compared with the gyr+ parent host (Fig. 2). However, the pattern of late T4 proteins

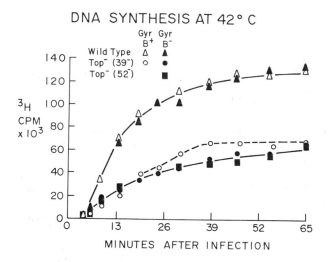

Figure 2: DNA synthesis (continuous labeling) of wild type T4 and gene 39 or gene 52 mutants in E. coli LE 234 (gyr+) and LE 316 (gyr⁻) hosts.

was grossly distorted (Fig. 3). Certain important late T4 gene products were definitely underproduced (as compared with wild type T4), e.g., gp 23 (major head protein),

gp 18 (tail protein), and gp 37 (tail fiber protein). Therefore, the DNA could not be packaged, a sufficient reason for

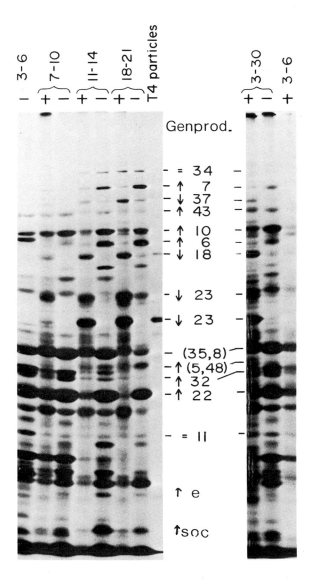

Figure 3:
Autoradio-
gram of
^{14}C-pro-
teins pulse
labeled at
the indi-
cated times
after
infection of
the ts gyrB
host LE 316
at 42°C with
+ = wild
type T4,
- = T4 gene
39 mutant
N116.

the failure to produce progeny particles. Other gene
products were definitely overexpressed, e.g., gp 6, gp 7,
gp 22, gp soc (small outer capsid protein), and probably

gp e (lysozyme). Expression of still other late genes
(e.g., 34, tail fiber protein) was essentially unaltered.
(For map positions of these genes see Fig. 1).

By comparison, effects on early T4 gene expression
were smaller and less dependent on the host gyrase activity,
but nevertheless sufficient to explain the DNA delay
phenotype. Expression of at least two early recombination
genes, gp 30 (ligase) and gp 46 (recombination nuclease)
was delayed in the T4 mutants both in gyr^+ and gyr^- hosts
(data not shown). Since these proteins are required for
recombinational initiation of replication, their delayed
appearance must contribute to the DNA-delay phenotype
of the T4 topoisomerase mutants.

Although overall DNA synthesis of the T4 toposomerase
mutants was indistinguishable in gyr^+ and gyr^- hosts
(Fig 2), we observed a remarkable difference in particular
replication origins used. To map origin regions we have
isolated early replicated T4 DNA labeled in vivo and have
used it as probes for Southern hybridization (Southern
1975) to T4 restriction fragments cut from cytosine containing T4 DNA (26,27). Hybridization of replicated ^{32}P
Hydroxymethylcytosine containing T4 DNA (isolated early
after infection) to these immobilized restriction fragments
revealed a strong origin region within a 680 bp EcoRI
fragment spanning the junction of XbaI restriction fragments
15 and 17 between the nonessential T4 genes dam and dda
(see Fig. 1). At 25°C, replication of wild type T4 or
of recombination deficient gene-46 mutants began at first
from this region unidirectionally in the counterclockwise
direction (the major direction of early T4 transcription).
Soon thereafter, replication became bidirectional. In a
T4 gene-39 (topoisomerase⁻) mutant, replication was
initiated in the same region but, in contrast to wild
type T4, replication remained unidirectional until late
after infection. This showed that unidirectional initiation in the major direction of early T4 transcription
did not depend on T4 topoisomerase but that bidirectional
replication in the opposite direction does require this
function. Gratuitously, this observation allowed more
precise mapping of the origin regions at the transition
between the most and the least frequently labeled restriction
fragments (22).

The strong origin between dda and dam was used predominantly, when ^{3}H thymidine was used for labeling in vivo,
and the DNA initiated de novo was isolated in a density
shift experiment and then labeled in vitro by nick translation

(Fig. 4). In addition to this strong origin, we found
another origin in wild type T4 and in T4 mutants (shaded
area near <u>uvsY</u> in Fig. 1) when a high level of ^{32}P was
present before and during infection. This is probably
the same as the minor origin near gene 25 (19, 20).

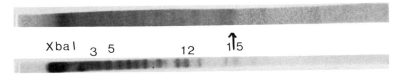

Figure 4: Autoradiogram of Southern blot of early wild
type T4 DNA replicated at 25°C, in <u>E</u>. <u>coli</u> B isolated
from a Cs_2SO_4 density gradient (upper panel) and
T4 DNA (lower panel) to an XbaI digest of total T4 DNA.
XbaI 15 maps at ori 1 in Fig. 1.

At 25°, in wild type hosts, we have not detected the
strong origin near gene 5 (19). In contrast, at 42°, this
origin was preferentially used, specifically by wild type
T4 in a <u>gyr</u>⁻ host or by the T4 39⁻ mutant in a <u>gyr</u>⁺ host.
Surprisingly, however, when both gyrase and T4 topoisomerase
were defective, the other two origins were again labeled (Fig. 5).

DISCUSSION

We have shown that changing the relative activities of
two topoisomerases, <u>E</u>. <u>coli</u> gyrase and T4 topoisomerase,
dramatically altered the pattern of T4 late gene expression.
Likewise, different origins were used under these conditions,
although the effects on the overall DNA synthesis were far
less severe. The DNA delay phenotype of the T4 topoisomerase mutants which is largely caused by delayed recombinational initiation (13) must be due, in part, to the
delayed expression of certain recombination genes. Most
likely, T4 topoisomerase has additional direct effects
on stimulating DNA replication (28), and perhaps by facilitating recombinational initiation (29).

Most of these effects are readily explained by differential sensitives of certain promoters, including origin
promoters to torsional stress. What factors might contribute
to such differential sensitivities?

We consider three plausible causes which are not
mutually exclusive: (i) GC content, (ii) secondary structure,
(iii) distance from topoisomerase binding sites

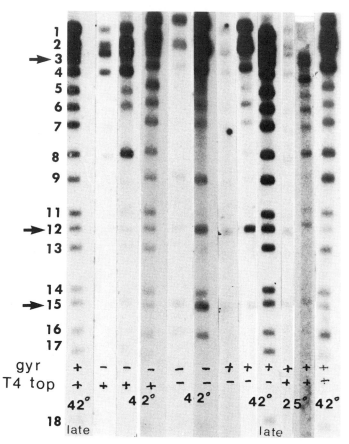

Figure 5: Autoradiograms of Southern blots of early wild type T4 DNA and gene 39 mutant (T4 topoisomerase⁻) DNA to XbaI digests of total T4 DNA. The XbaI fragments 15, 12 and 3 are diagnostic for the origins labeled 1, 2, and 3 respectively in Figure 1.

(i) The promoter for gene 23 which is grossly under-expressed in a T4 topoisomerase mutant in the gyr⁻ host has the most GC pairs of any known late T4 promoters. On the other hand, the promoter for gene 22, which is over-expressed, under these conditions contains only ATs (30).

It is tempting to speculate that "tight" GC rich promoters have to be activated by unwinding and that they are therefore less active, when both phage and host topoisomerases are defective. On the other hand, "open" AT rich promoters may become too open and thus be less well recognized by RNA polymerase when the two topoisomerases are active, therefore such promoters appear stimulated when these functions are defective.

(ii) As extensively discussed in this volume, palindromes or sequences that tend to form left handed helices may go into alternative configurations dependent on torsional stress. Thus, promoters in or near such sequences should be particularly sensitive to supercoiling. The putative late promoters for lysozyme (31) and for gp soc (Macdonald, Kutter and Mosig manuscript in preparation, Fig. 6) are

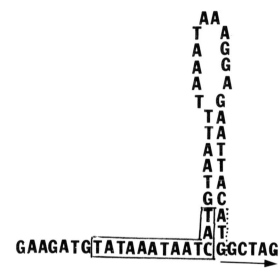

Figure 6: Sequence and secondary structure of the late promoter (boxed) for soc.

located in palindromes. Both of these gene products are overproduced when topoisomerases are inactive. Perhaps, RNA polymerase can only recognize these sequences when they are not involved in hairpins.

(iii) It is conceivable that the proximity of preferential binding sites of one or the other of the two topoisomerases may have additional effects on promoter utilization in general and on origin recognition specifically. We note, however, that the two promoters for genes 22 and 23 (30) are in close proximity on the chromosome but they respond in opposite ways to torsional stress. Perhaps promoters have to be located between topoisomerase binding sites, to be affected.

Whereas expression of certain late genes appeared most sensitive to alterations of the two topoisomerase activities, there was little effect on most early genes. This may be so because the modifications of RNA polymerase by accessory proteins during T4 development (16) renders promoter recognition by the "late" RNA polymerase more sensitive to torsional stress. This hypothesis could explain certain effects of additional mutations on late gene expression in T4 topoisomerase mutants. The topoisomerase mutants appeared completely defective in all late gene expression, when they contained additional mutations in alc and when the host gyrase was inactivated by coumermycin (33). We suspect that a functional T4 alc gene, which prevents late transcription on cytosine containing T4 DNA (26) is essential for the topoisomerase effects that we observed.

In light of these findings on gene expression, we can rationalize the major effects of the topoisomerase mutations on DNA synthesis in terms of promoter recognition as follows: (i) the delayed appearance of certain T4 recombination enzymes, such as DNA ligase and recombination nuclease must delay recombinational initiation, and, therefore, overall thymidine incorporation. (ii) Altered recognition of origin promoters due to torsional tension would alter the utilization of specific origins. (iii) Uni- vs. bidirectional initiation from certain origins might depend on the recognition of primer promoters facing in different directions. For example, at 25°, the T4 topoisomerase mutants initiated from the origin between dam and dda in the counterclockwise direction only, whereas wild type T4 initiated in both directions (though initiation

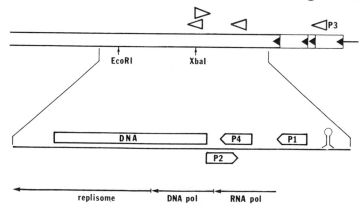

Figure 7: The arrangement of early promoters in the origin region between dda and dam (Macdonald and Mosig, manuscript in preparation).

in the clockwise direction occurs somewhat later). Interestingly, this origin region contains at least four closely spaced promoters (Fig. 7, Macdonald and Mosig, manuscript in preparation). Three of them face in the counterclockwise direction, which is the major direction of early transcription. The other one points in the clockwise direction. Possibly, the latter promoter, which could be a primer promoter, cannot be used in the absence of T4 topoisomerase and therefore no primer RNA for leading strand synthesis in that direction is made.

All of these results, taken together, support the idea that the T4 topoisomerase has indeed gyrase-like activity in vivo (7) and suggest in addition that such activity is not restricted to origin sequences. Obviously, if this is the case, at least one component must be missing in the current T4 topoisomerase preparations. One intriguing speculation is that the product of the alc gene mentioned above might be such a missing component.

REFERENCES

1. Gellert M (1981). DNA topoisomerases. Ann Rev Biochem 50:879.
2. Filutowicz M, Jonczyk P (1981). Essential role of the gyrB gene product in the transcriptional event coupled to dnaA-dependent initiation of Escherichia coli chromosome replication. Mol Gen Genet 183:134.
3. Bauer M, Crick FHC, White JH (1980). Supercoiled DNA. Sci Am 243:118.
4. Sinden RR, Pettijohn DE (1982). Torsional tension in intracellular bacteriophage T4 DNA: evidence that a linear DNA duplex can be supercoiled in vivo. J Mol Biol 162:654.
5. Cozzarelli NR (1980). Topoisomerases. Cell 22:327.
6. Stetler GL, King GJ, and Huang WM (1979). T4 DNA-delay proteins, required for specific DNA replication, form a complex that has ATP-dependent DNA topoisomerase activity. Proc Natl Acad Sci USA 76:3737.
7. Liu LF, Liu C-C, Alberts BM (1979). T4 DNA topoisomerase: a new ATP dependent enzyme essential for initiation of T4 bacteriophage DNA replication. Nature 281:456.
8. Seasholtz AF, Greenberg GR (1983). Identification of bacteriophage T4 gene 60 product and a role for

this protein in DNA topoisomerase. J Biol Chem
258:1221.

9.   Liu LF, Liu C-C, Alberts BM (1980). Type II DNA
     topoisomerases: Enzymes that can unknot a topologically
     knotted DNA molecule via a reversible double-stranded
     break. Cell 19:697.

10. Forterre P, Assairi L, Duguet M (1983). Topology,
     type II DNA topoisomerases and DNA replication in
     prokaryotes and eukaryotes. In de Recondo AM (ed.):
     "New Approaches in Eukaryotic DNA Replication,"
     New York and London: Plenum Press, p 123.

11. Yegian CD, Mueller M, Selzer G, Russo V, Stahl FW
     (1971). Properties of the DNA-delay mutants of
     bacteriophage T4. Virol 46:900.

12. Mosig G, Luder A, Garcia G, Dannenberg R, Bock S
     (1979). In vivo interactions of genes and proteins
     in DNA replication and recombination of phage T4.
     Cold Spring Harbor Symp Quant Biol 43:501.

13. Mosig G, Luder A, Rowen L, Macdonald P, Bock S (1981).
     On the role of recombination and topoisomerase in
     primary and secondary initiation of T4 DNA replication.
     In Ray D (ed.): "The Initiation of DNA Replication,"
     New York: Academic Press, Inc., p 277.

14. McCarthy D, Minner C, Bernstein H, Bernstein C (1976).
     DNA elongation rates and growing point distributions
     of wild-type phage T4 and a DNA-delay amber mutant.
     J Mol Biol 106:963.

15. Luder A, Mosig G (1982). Two alternative mechanisms
     of initiation of DNA replication forks in bacteriophage
     T4: priming by RNA polymerase and by recombination.
     Proc Natl Acad Sci USA 79:1101.

16. Rabussay D (1982). Changes in Escherichia Coli
     RNA Polymerase After Bacteriophage T4 Infection.
     ASM News 48: 398-403.

17. McCarthy D (1979). Gyrase-dependent initiation
     of bacteriophage T4 DNA replication: interactions
     of Escherichia coli gyrase with novobiocin, coumermycin
     and phage DNA-delay gene products. J Mol Biol 127:265.

18. Marsh RC, Breschkin AM, Mosig G (1971). Origin
     and direction of bacteriophage T4 DNA replication.
     II. A gradient of marker frequencies in partially
     replicated T4 DNA as assayed by transformation.
     J Mol Biol 60:213.

19. Halpern ME, Mattson T, Kozinski AW (1979). Origins
     of phage T4 DNA replication as revealed by hybridiza-
     tion to cloned genes. Proc Natl Acad Sci USA 76:6137.

20. Morris CF, Bittner (1981). Effect of temperature and specific labelling conditions on the identification of T4 bacteriophage replication origins. J. Supramolecular Structure and Cellular Biochemistry, Supplement 5, 334.

21. King GJ, Huang WM (1982). Identification of the origins of T4 DNA replication. Proc Natl Acad Sci USA 79:7248.

22. Macdonald PM, Seaby RM, Brown W, Mosig G (1983). Initiator DNA from a primary origin and induction of a secondary origin of bacteriophage T4 DNA replication. In Schlessinger D (ed.): "Microbiology," Washington, D.C., ASM, p. 111.

23. Mosig G (1982). Genetic Map of Bacteriophage T4. In O'Brien S (ed.): "Genetic Maps 2," Frederick, Maryland, National Cancer Institute, p. 15.

24. Orr E, Fairweather NF, Holland IB, Pritchard AH (1979). Isolation and characterization of a strain carrying a conditional lethal mutation in the cou gene of Escherichia coli K12. Molec Gen Genet 177:103-112.

25. Southern EM (1975). Detection of specific sequences among DNA fragments separated by gel electrophoresis. J Mol Biol 98:503.

26. Snyder L, Gold L, Kutter E (1976). A gene of bacteriophage T4 whose product prevents true late transcription on cytosine-containing T4 DNA. Proc Natl Acad Sci USA 73:3098.

27. Wilson GG, Tanyashin VI, Murray NE (1977). Molecular cloning of fragments of bacteriophage T4 DNA. Molec Gen Genet 156:203.

28. Huang WM (1978). Positive regulation of T-even-phage DNA replication by the DNA-delay protein of gene 39. In "Cold Spring Harbor Symposia on Quantitative Biology, Volume XLIII". Cold Spring Harbor Laboratory, Cold Spring Harbor, p. 495.

29. Dannenberg R, Mosig, G (1983). Early intermediates in bacteriophage T4 DNA replication and recombination. J Virol 45:813.

30. Christensen AC, Young ET (1982). T4 late transcripts are initiated near a conserved DNA sequence. Nature 299:369.

31. Owen JE, Schultz DW, Taylor A, Smith GR (1983). Nucleotide sequence of the lysozyme gene of bacteriophage T4. J Mol Biol 165:229.

32. Young ET, Menard RC, Harada J (1981). Monocistronic and polycistronic bacteriophage T4 gene 23 messages. J Virol 40:790.

33. Jacobs K, Geiduschek EP (1981). Regulation of expression of cloned bacteriophage T4 late gene 23. J Virol 32:46.

Mechanisms of DNA Replication and Recombination, pages 187–201
© 1983 Alan R. Liss, Inc., 150 Fifth Avenue, New York, NY 10011

REPLICATIVE TRANSPOSITION OF BACTERIOPHAGE MU IN VITRO[1]

N. Patrick Higgins[t][2], Dino Moncecchi[t], Martha M. Howe[§]
Purita Manlapaz-Ramos[*] and Baldomero M. Olivera[*]

[*]Department of Biology, University of Utah,
Salt Lake City, UT 84112;

[§]Department of Bacteriology, University of
Wisconsin, Madison, Wis. 53706;

[t]Department of Biochemistry, University of
Wyoming, Laramie, WY 82070

ABSTRACT   Replicative transposition of Mu was studied
using the cellophane film technique. In vitro, Mu
replication is initiated at the very left (or c) end,
and is restricted to Mu (or mini-Mu) sequences. Mu
transposition functions control replication initiation
and termination. A biochemical pathway for Mu
transposition is suggested leading either to
unreplicated insertions or replicated cointegrates.

INTRODUCTION

Genetic transposition is the movement of discrete DNA
sequences to new loci (see ref. 1 for a recent review).
Such DNA sequences, generally called transposons or
transposable elements, have a number of common features.
All autonomously transposing elements have two ends which
define the DNA which can hop; all DNA found between the two
ends jumps as a single unit. There is at least one open
reading frame between the two ends. This open reading frame

[1]This work was supported by Grants GM26107 (NIH) and
PCM 8004689 (NSF).
[2]Present address: Department of Biology, University of
Alabama, Birmingham, Alabama

codes for a protein that must be expressed before the sequence can transpose. It has become custom to call the gene product necessary for hopping the "transposase".

Despite the intensive scrutiny that transposons have received in recent years, the biochemical events that take place during the DNA transposition process have not yet been characterized. One problem is that most transposable elements present in nature transpose relatively infrequently ($< 10^{-4}$ transposition events per cell per generation). Such a low transposition frequency would be expected to be the rule in natural transposable elements. As the frequency of transposition increases in a cell, the probability that the transposon will land in an essential gene and impair host cell growth and division becomes high. Therefore, high frequency transposition would be selected against.

High frequency transposons. There are elements that have evolved a mechanism to rapidly amplify their number through transposition. Such elements also can become autonomous from the host cell, i.e., they are viruses. One well characterized example of this type of transposable element is the bacteriophage Mu (2). Upon Mu infection or induction of a Mu prophage, the virus undergoes a lytic cycle in which essentially lethal levels of transposition take place (3,4). Over 100 transposition events occur per cell during the lytic cycle of Mu; the high frequency of Mu transposition makes it a favorable system for the biochemical study of transposition.

The in vitro system for Mu replicative transposition. We recently established an in vitro system for transposition (5-7). Mu replicative transposition can be efficiently carried out using film lysates on cellophane discs (5, 8-9). One advantage of this system is that the macromolecular components necessary for DNA synthesis are present at close to intracellular concentrations. Furthermore, the presence of the intact folded host chromosome in the cellophane disc system may be essential for a high frequency of replicative transposition in vitro.

Using the film lysate system, we can differentiate between replication of Mu sequences by a replication fork under host control, and replication of Mu sequences by a replication fork under Mu replicon control (see Figure 1); the latter events result in transposition. The salient difference between these two types of replication forks is diagrammatically shown in the figure. The essential point is that while a host replication fork replicates all the bordering DNA sequences semi-conservatively, Mu controlled

replication forks only replicate the Mu sequences. Such a
replication pattern is what would be expected of replicative
transposition. Newly transposed Mu sequences appear in Mu-
sized "replication patches" as shown in Figure 1b. This was
observed by density shift experiments in the film lysate
system (7).

FIGURE 1. Normal replication versus transposition
replication of Mu sequences. The thick lines represent
newly replicated DNA. The Mu sequences (stippled) can be
replicated by a normal host fork (A); in this case the host
regions around it (solid) are similarly replicated. Thus,
in a density transfer experiment, this entire piece of
duplex DNA would band at a hybrid density. However, if
transposition replication took place, only Mu sequences are
replicated (B). Thus, a DNA molecule of this type would
band at a density intermediate between light and hybrid.

The second criterion for Mu replicative transposition
events is that newly replicated Mu sequences should only
appear if Mu "transposase" is continuously synthesized.
Pato and Reich (10) showed that a brief period of protein
synthesis inhibition prevents further Mu transposition in
vivo. We have shown that protein synthesis inhibition
causes a similar selective disappearance of Mu controlled
replicative transposition in vitro (7). Replication events
of the type shown in Figure 1a continue after a brief
protein synthesis inhibition period, but events of the type
shown in Figure 1b are no longer seen. Concomitant with the
disappearance of replication events in "Mu sized patches" is
a drastic decrease in the total level of Mu sequences being
replicated in vitro. This sensitivity to brief periods of
protein synthesis inhibition is consistent with a continued

requirement for new Mu proteins, most likely the MuA gene. There is increasing evidence that most transposases are used stoichiometrically in the transposition process, hence the need for continued protein synthesis.

The establishment of the film lysate system for Mu replicative transposition has also revealed that the Mu controlled replication forks have the basic characteristics of host replication forks, except that Mu has overlaid a special initiation and termination event at Mu sequence boundaries. The in vitro studies show that Mu replication is semi-conservative, and semi-discontinuous. Furthermore, the semi-discontinuous replication fork traverses Mu DNA sequences predominantly from the left end of Mu (also called immunity or c end) to the right end of Mu (also called the S end). This was shown by the asymmetry of semi-discontinuous DNA replication in the in vitro system (5,6). When DNA sealing was completely inhibited in vitro, half of the DNA was synthesized as Okazaki pieces and half as continuously synthesized DNA. Mu Okazaki pieces hybridized predominantly to the Mu light strand, and the continuously synthesized Mu DNA hybridized predominantly to the Mu heavy strand. These results are consistent with Mu proteins capturing a host replication fork, which then moves through Mu sequences from left to right. Termination occurs at the right end of Mu. This special initiation and termination process connects a new locus in the host chromosome to the newly replicated Mu sequences.

RESULTS

The replicative transposition event in mini-Mus. As discussed above, replicative transposition events result in replicated Mu DNA sequences surrounded by unreplicated host DNA. Therefore, the replication of "mini-Mus" (11,12), defective Mu genomes which are much smaller than intact Mu prophage, should form even smaller "replication patches" than intact Mu. The experimental results comparing a density shift analysis of both intact Mu and mini-Mu is shown in Figure 2. Replication in vitro was carried out in the presence of dBUTP and $[\alpha^{32}P]dGTP$. With full length Mu, a broad band between hybrid and light DNA is seen in addition to the replicated host sequences. We previously showed (7) that this anomalously banding material is enriched in Mu, and has the structure shown in Figure 1b. In this experiment, the DNA was sheared to a molecular weight of approximately 80 kb. If the DNA is sheared to a much lower molecular weight (ca. 1 kb), then all the Mu sequences banded at the hybrid position.

Figure 2 (lower panel) shows a density shift experiment for λP mini-Mu (12). Since the mini-Mus are smaller than Mu (16 kb vs. 35 kb), when the DNA is sheared to an 80 kb size, semi-conservatively replicated mini-Mus would shift the density much less than an intact transposing Mu sequence. Newly replicated mini-Mu sequences are most enriched (at least five-fold) in a peak banding much closer to the light density in a CsCl gradient. This is consistent with the diagram shown at the bottom of Figure 2.

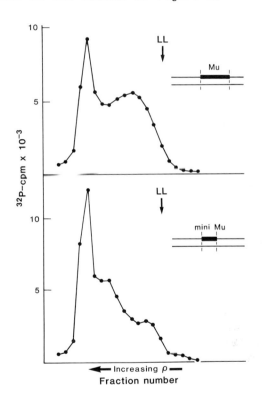

FIGURE 2. Density transfer experiments with intact Mu and mini-Mu. The top panel shows a density gradient analysis of newly replicated DNA from an induction of intact, full length Mu, while the bottom panel shows an infection using a λp mini-Mu (N33). In both cases, the lytic cycle was allowed to proceed for half an hour at 42°C,

cells were harvested by centrifugation and lysed on cellophane discs. DNA replication was carried out in vitro for 45 min using a reaction mixture containing 20 µM dBUTP (replacing dTTP) and ^{32}P-dGTP and the DNA analyzed by isopycnic centrifugation using a CsCl gradient as previously described (7). The DNA was sheared to an average length of 80 kb. The Mu lysogen used was E. coli strain W3110::MuCts. The infection with λp mini-Mu was carried out on E. coli strain HB101. These strains were grown to an A_{650} of 0.2 at the time of induction or infection. The position of light DNA (LL) was determined by a pulse of ^{3}H-thymidine just before harvesting the cells.

In the induced (intact) Mu lysogen, the intermediate density region is highly enriched for Mu sequences. In the infection with the mini-Mu, the peak closest to light (LL) is enriched five fold in Mu sequences. The identity of the shoulder to the hybrid peak, also somewhat enriched in Mu sequences is currently under study. Each graph includes the proposed structure of DNA pieces containing fully replicated Mu or mini-Mu sequences. Since the mini-Mu is smaller, a piece of DNA encompassing hybrid mini-Mu sequences would band at a lighter density.

Initiation of Mu replicative transposition.   The asymmetry of Okazaki piece hybridization previously described (5,13) showed that the predominant direction of replication through Mu sequences was from left to right. However, since whole Mu separated strands were analyzed in those experiments, the origin of DNA replication at the left end of Mu was not precisely localized. The basic strategy for a more precise analysis to pinpoint the origin of replication on the left end is shown in Figure 3. Instead of analyzing separated strands of whole Mu (35 kb), we have analyzed the separated strands of the left-most 1.5 kb segment of Mu. If the origin of replication were to the right of the segment, then the hybridization of Okazaki pieces would be reversed compared to whole Mu. If the origin of replication were in the middle of the segment, then equal hybridization of Okazaki pieces to the two separated strands should be observed. If the origin of replication were at the very left end of Mu, then the same strand asymmetry observed for whole Mu strands should be observed with the left-most 1.5 kb separated strands.

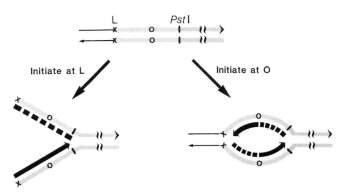

FIGURE 3. Two alternatives for initiation of replication at the left end of Mu. This diagram shows semi-discontinuous replication if initiation had occurred in the middle of the left end fragment of Mu (upper panel) or at the very left end of Mu. Given the first alternative, Okazaki pieces would be found hybridizing evenly to both strands of Mu. The second alternative predicts that even for the left end fragments, Okazaki pieces would only hybridize to the light strand.

As shown in Figure 4, the strand asymmetry for the left-most 1.5 kb segment is the same as for separated strands of whole Mu; the Okazaki pieces hybridize predominantly to the light strand. This experiment, although not accurate to the exact nucleotide, nevertheless indicates that the origin of replication is very near the left genetic end of Mu. Since 1.5 kb is about the size of an Okazaki piece in the in vitro system, the most straightforward interpretation is that the left genetic end of Mu is the origin of replication.

How is replication initiated at the very left end of Mu? Preliminary experimental evidence suggests that initiation of DNA replication at the left end involves priming of the continuous Mu strand with bacterial DNA. An induced Mu Ats lysogen shifted to the nonpermissive temperature in vivo was harvested, and incubated in vitro at the permissive temperature. We have previously shown that initiation-like events occur on the film lysate system under these conditions (6). Such an in vitro initiation was carried out in the presence of NMN to completely inhibit DNA sealing. Even with a brief in vitro incubation, (5 min), the continuously synthesized Mu strand is very large (it

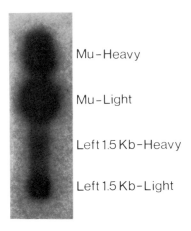

Mu-Heavy

Mu-Light

Left 1.5 Kb-Heavy

Left 1.5 Kb-Light

FIGURE 4. Hybridization of Okazaki pieces to separated strands of the 1.5 kb left end of Mu and intact Mu. A Mu cts Ats infected HB101 culture was shifted to 42° to stop transposition in vivo. Cellophane discs were prepared and DNA synthesis was carried out at 30° in vitro with [α^{32}P]-dGTP, in the presence of NMN. Under these conditions new rounds of replicative transposition are initiated in vitro (16). Okazaki pieces were isolated from an alkaline sucrose gradient and then hybridized to Mu strands. The heavy and light strands of full length Mu were separated by centrifugation on poly UG containing CsCl gradients by the method of Szybalski (14). The left 1.5 kb Mu fragments were separated into H and L strands after they were cleaved from plasmid pPH03 (6) and then separated in a strand separating agarose gel (15). 0.2 µg dots were prepared and hybridization was carried out as described (6). The autoradiograph was developed after 1 month.

sediments with the bulk of the prelabeled chromosomal DNA). This indicates that during the initiation event, the continuously synthesized strand is primed by chromosomal sized host DNA.

A simple model consistent with these results is shown in Figure 5. In this model initiation of DNA replication takes place at the very left end of Mu. Replicative transposition is initiated by breaking the phosphodiester linkage between the left end of the Mu light strand and host DNA and joining the 5' end of Mu to a new chromosome location. The break at the genetic left end leaves a 3' hydroxyl terminal to prime the continuously synthesized strand. It should be noted that the data do not eliminate the possibility of heavy Mu strand transfer with priming at the acceptor site, but this seems a priori less likely.

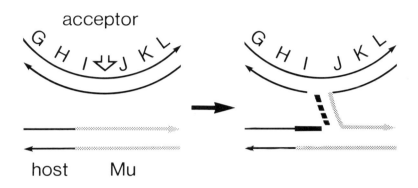

FIGURE 5. Initiation model. The initiation of replicative transposition is postulated to occur by a break at the genetic left end of the light strand of Mu, and transfer of the 5' end of the Mu light strand to an acceptor site; this generates a replication fork-like structure where the 3' terminus of the host DNA can serve as primer for the continuous strand. The data for the origin being at the very left end, as well as continuous strand priming is discussed in the text.

DISCUSSION

The role of the Mu transposase complex in replicative transposition. The establishment of an in vitro system for Mu replicative transposition has allowed us to define a set of rules which govern the process. It is now clear from the in vitro work that during the lytic cycle, Mu sequences are replicated by Mu controlled replication forks which very likely contain the components normally found in a host replication fork (DNA polymerase III holoenzyme, the primisome complex, etc.). The major difference between a Mu controlled replication fork and a host replication fork is that a special initiation event takes place at the left end of the Mu sequences, the semi-conservative, semi-discontinuous (basically host) replication fork then traverses the Mu sequences from left to right, and a special termination event occurs such that a new locus in the host DNA is now connected to Mu sequences. Depending on how such connections are made, the results will either be a simple insertion or a cointegration event (17). Thus, in replicative transposition it is the initiation and termination events that the transposase complex defines.

It should be noted that not all Mu transposition events may involve replicative transposition. There is increasing evidence that upon infection, the first jump into the chromosome may involve a slightly different mechanism. In contrast to most Mu transposition events that have been observed in the lytic cycle, the first hop is a simple insertion (18,19). Furthermore, there is evidence that such hops are conservative, and not a product of replication (18). However, we are confident that by elucidating the role of the transposase complex in replicative transposition, considerable insight into the mechanism of simple insertions will be obtained. Since the factors necessary to achieve simple insertions are a subset of the Mu factors necessary to achieve replicative transposition, the mechanism of these two processes are likely to involve common intermediates.

The biochemical events in transposition: speculation and future directions. Although the work in vitro has defined the role of the transposase complex in general terms, the biochemical details need to be defined. There are, however, a number of experimental results in the literature that make it possible to delineate a probable biochemical pathway for transposition.

Our experimental results in vitro, as well as the in vivo results of others are condensed into the working model shown in Figure 6. We propose that attachment of both the left and right 5' ends of Mu to the target site occurs prior to Mu DNA replication. If these attachment sites on the target are 5 nucleotides apart, when transposition is complete the host DNA will contain a 5 base duplication. In the specific model shown in Figure 6, after replication of Mu sequences, a cointegration event will have occurred. This is consistent with the data in the literature regarding replicative transposition during the lytic cycle.

Although conclusive evidence for when attachment occurs to the target site is not available, in vivo evidence suggests immediate attachment to the chromosomal nucleoid well before DNA synthesis is completed (20,21). The finding of "key structures" in several studies is consistent with

attachment to the target sites before replication of Mu
sequences is complete (22). In addition, <u>in vivo</u> labeling
studies suggest that both ends of Mu are <u>required</u> for any
DNA replication to occur at all (M. Pato, personal
communication). These experimental results make it seem

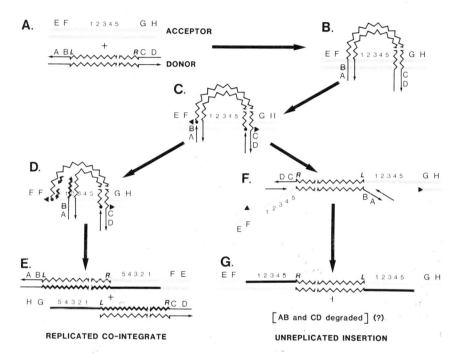

FIGURE 6. Replicative and non-replicative
transposition from a common synaptic intermediate. Panel A
shows the donor site with the parental Mu insert, and the
target acceptor site, including the five base duplication
which is indicated by the numbers. Panel B shows a
configuration just before transposition synapsis has
occurred. Panel C shows the postulated structure at
transposition synapsis; in this version, we postulate that
covalent connections to both strands are made simultaneously
across the five base duplication. As discussed in the text,
a topoisomerase mechanism seems most attractive; a covalent
protein DNA complex would therefore be expected at the loci
represented by triangles. Figures D and E show the
replication pathway. In Figure D, a semi-discontinuous
replication fork is shown beginning at the left end. When

replication is completed, and all discontinuous pieces have been sealed, a structure as shown in Figure E is generated, in this case a "co-integrate structure". This result is analogous to a model proposed by Shapiro (17). Figures F and G show a postulated mechanism for simple insertion. In the absence of replication, unfolding of the five base duplication takes place. If the regions AB and CD are linear, these are susceptible to attack by a variety of nucleases. If regions EF and GH are part of the circular chromosome, these would be resistant. Trimming of the linear regions would generate Mu sequences which are able to prime repair synthesis by DNA polymerase I, generating the five base duplication as shown in Panel G.

likely that attachment of both ends of Mu to the target site occurs as DNA replication is being inititiated on Mu sequences. Such a prediction can be tested using the in vitro system.

How does DNA attachment occur? It could occur by a coupled nuclease-ligase mechanism, although preliminary in vitro experiments where ligase is completely inhibited suggest that ligase may not be involved in parental strand attachment (our unpublished results). There are numerous precedents for a mechanism involving a covalent protein DNA complex, i.e., a DNA topoisomerase-type mechanism. Initiation of ϕX Rf $\rightarrow$ Rf replication, which is formally analogous, proceeds by the covalent attachment of the cisA protein to the origin of replication; the resulting strand interruption is then used to initiate continuous strand synthesis at the 3' hydroxyl end, with discontinuous synthesis proceeding on the opposite strand (23). This is analogous to the mechanism in Figure 6. In addition, the specific insertion of λ DNA into its attachment site through the activity of the int protein (24) and the resolution reaction of Tn3 and γ-σ make a site-specific DNA topoisomerase mechanism attractive (25,26).

The diagrams in Figures 5 and 6 incorporate the features of a number of previously proposed models of transposition (17,27-29; for a review of these and earlier models, see 1,30). The biochemical events proposed in Figure 6 are consistent with all of our in vitro data so far. This scheme shows a common synaptic intermediate (6C) which is converted either to replicated cointegrates or simple insertions in the absence of replication. Such a model is useful because it makes specific predictions regarding strand connections, sites of covalent protein DNA complexes, and when simple insertions might occur (i.e., during transposition from linear DNA). Clearly, the ability to carry out replicative transposition in vitro provides an experimental system for investigating the biochemical details that are predicted by mechanistic transposition schemes of this type.

REFERENCES

1. Kleckner N (1981). Transposable elements in prokaryotes. Ann Rev Genet 15:341.
2. Taylor AL (1963). Bacteriophage-induced mutation in Escherichia coli. Proc Natl Acad Sci USA 50:1043.
3. Toussaint A, Resibois A (1983). Phage Mu: transposition as a lifestyle. In Shapiro J (ed): "Mobile Genetic Elements," New York: Academic Press, p 105.
4. Bukhari A (1976). Bacteriophage Mu as a transposition element. Ann Rev Genet 10:389.
5. Higgins NP, Moncecchi D, Manlapaz-Ramos P, Olivera BM (1983). Mu DNA replication in vitro. J Biol Chem 258:4293.
6. Higgins NP, Olivera BM (1983). Mu DNA replication in vitro: criteria for initiation. Mol Gen Genet, in press.
7. Higgins NP, Manlapaz-Ramos P, Gandhi T, Olivera BM (1983). Bacteriophage Mu: A transposing replicon. Cell, in press.
8. Schaller H, Otto B, Nusslein U, Huf J, Herrmann R, Bonhoeffer F (1972). Deoxyribonucleic acid replication in vitro. J Mol Biol 63:183.
9. Olivera BM, Bonhoeffer F (1972). Discontinuous DNA synthesis in vitro. Nature New Biol 240:233.
10. Pato ML, Reich C (1982). Instability of transposase activity--evidence from bacteriophage Mu DNA replication. Cell 29:219.
11. Resibois A, Toussaint A, Gijsegem F, Faelen M (1981). Physical characterization of mini-Mu and mini-D108. Gene 14:103.
12. Schumm JW, Howe, MM (1981). Mu-specific properties of λ phages containing both ends of Mu depend on the relative orientation of Mu end DNA fragments. Virology 114:429.
13. Goosen T (1977). Replication of bacteriophage Mu:

Direction and possible location of the origin. In Molineaux I, Kohiyama M (eds.): "DNA Synthesis," New York, London: Plenum Press, p 121.

14. Szybalski W, Kubinsky H, Hradeena A, Summers WC (1971). Analytical and preparative separation of the complementary DNA strands. Methods Enzymol 21:383.

15. Maxam AM, Gilbert W (1980). Sequencing end-labeled DNA with base specific chemical cleavages. Methods Enzymol 65:499.

16. Kafatos FC, Jones CW, Efstradiatis A (1979). Determination of nucleic acid sequence homologies and relative concentrations by a dot hybridization procedure. Nucleic Acids Res 7:1541.

17. Shapiro JA (1979) Molecular model for the transposition and replication of bacteriophage Mu and other transposable elements. Proc Natl Acad Sci USA 76:1933.

18. Liebart JC, Ghelardini P, Paolozzi L (1982). Conservative integration of bacteriophage Mu DNA into pBR322 plasmid. Proc Natl Acad Sci USA 79:4362.

19. Chaconas G, Kennedy DL, Evans D, Bukhari AI (1983). Predominant integration end-products of infecting bacteriophage Mu DNA are simple insertions with no preference for integration of either Mu strand. Virology in press.

20. Bialy H, Waggoner BT, Pato ML (1980). Fate of plasmids containing Mu DNA: chromosome association and mobilization. Molec Gen Genet 180:377.

21. Chaconas G, Harshey R, Bukhari A (1980). Association of Mu-containing plasmids with the Escherichia coli chromosome upon prophage induction. Proc Natl Acad Sci USA 77:1778.

22. Harshey RM, Bukhari AI (1981). A mechanism of DNA transposition. Proc Natl Acad Sci USA 78:1090.

23. Kornberg A (1980). " DNA Replication." San Francisco: W. H. Freeman Co, p 508

24. Nash, HA (1981). Integration and excision of bacteriophage λ: the mechanism of conservative site-specific recombination. Ann Rev Genet 15:143.

25. Krasnow MA, Cozzarelli NR (1983). Site specific relaxation and recombination by the Tn3 resolvase: recognition of the DNA path between oriented res sites. Cell 32:1313.

26. Reed RR (1981). Transposon-mediated site-specific recombination: a defined in vitro system. Cell 25:713.

27. Hershey RM, Bukhari AI (1981). A mechanism of DNA transposition. Proc Natl Acad Sci USA 78:1090.
28. Grindley ND, Sherratt DJ (1978). Sequence analysis at IS1 insertion sites: Models for transposition. Cold Spring Harbor Symp. Quant. Biol. 43:1257.
29. Galas DJ, Chandler M (1981). On the molecular mechanisms of transposition. Proc Natl Acad Sci USA 78:4858.
30. Bukhari AI (1981). Models of DNA transposition. Trends Biol. Sci. 6:56.

Mechanisms of DNA Replication and Recombination, pages 203–223

REQUIREMENTS FOR THE INITIATION OF PHAGE Ø29 DNA
REPLICATION IN VITRO PRIMED BY THE TERMINAL PROTEIN

Margarita Salas, Juan A. García, Miguel A. Peñalva,
Luis Blanco, Ignacio Prieto, Rafael P. Mellado
José M. Lázaro, Ricardo Pastrana,
Cristina Escarmís and José M. Hermoso

Centro de Biología Molecular (CSIC-UAM)
Universidad Autónoma, Canto Blanco,
Madrid 34, Spain.

ABSTRACT   Incubation of extracts from Ø29-infected
B. subtilis with $\{\alpha\text{-}^{32}P\}$dATP and ATP in the presence
of Ø29 DNA-protein p3 as template gave rise to the
formation of a covalent complex between protein p3 and
5'dAMP. Proteinase K-treated Ø29 DNA was not active as
template for the formation of the p3-dAMP complex suggesting that the parental protein in the DNA is an essential requirement for the initiation reaction. Protein-containing DNA fragments from the left or right
ends were active in the formation of the p3-dAMP complex, although there was a size effect. In addition to
the template, protein p3, dATP and ATP, the product of
gene 2 is also required for the initiation reaction.
Gene 3 has been sequenced and cloned in a pBR322 derivative plasmid under the control of the P_L promoter of
phage lambda. Protein p3, which is overproduced in E.
coli, has been highly purified and is active in the
formation of the p3-dAMP initiation complex. The effect of two mutations at the carboxyl end of protein
p3 on the initiation reaction has been studied.

[1] This work was supported by research grant 1 R01 GM-
27242-03 from the National Institutes of Health and by
grants from the Comisión Asesora para el Desarrollo de la
Investigación Científica y Técnica and Fondo de Investiga-
ciones Sanitarias.

INTRODUCTION

The B. subtilis phage Ø29 contains a linear, double-stranded DNA of about 18,000 base pairs (1) with a viral protein product of gene 3, p3, covalently linked to the two 5' ends (2-5) by a phosphoester bond between the OH group of a serine residue in the protein and 5'dAMP (6). An inverted terminal repetition six nucleotides long is present at the ends of Ø29 DNA (7,8). Genes 2, 3, 5, 6 and 17 are involved in Ø29 DNA replication (9-11). Shift-up experiments using ts mutants showed that genes 2 and 3 are involved in an initiation step, and genes 5 and 6 in elongation or maturation steps (12).

Bacteriophage Ø29 replication starts at either DNA end and proceeds by strand displacement (13-15). To explain the initiation at the 5' ends of the linear Ø29 DNA, a model, similar to that proposed for the replication of adenovirus DNA which also has a protein linked at the 5' ends (16), has been postulated by which a molecule of free protein p3 interacts with the parental protein p3, with the inverted terminal repetition or with both and primes replication by reaction with dATP and formation of a protein p3-dAMP covalent complex that provides the free 3'OH group needed for elongation (13,14). In agreement with this model, protein p3 is present at the ends of the parental and daughter DNA strands in Ø29 replicative intermediates (15). Moreover, a covalent complex between protein p3 and dAMP has been obtained by incubation of extracts from Ø29-infected B. subtilis with dATP, in the presence of ATP and Ø29 DNA-protein p3 as template (17-19). Similarly, a covalent complex between the 80K precursor of the adenovirus terminal protein and 5'dCMP, the nucleotide at both 5' ends of the DNA, has been found (20-24).

The template requirements for the formation of the p3-dAMP initiation complex in vitro are presented here. Isolated protein-containing left or right DNA fragments are active as template for the initiation reaction although there is a size effect. A terminal fragment 26 base pairs long is still active as template while a 10 base pairs long terminal fragment is very little active (25). In agreement with in vivo results (2) the parental protein p3 in the DNA is an essential requirement for the initiation reaction.

In addition to protein p3, dATP, ATP and the Ø29 DNA-protein p3 template, the product of gene 2 is required for the formation of the p3-dAMP initiation complex (26).

Gene 3 has been sequenced (27,28) and cloned (29) in the pBR322 derivative plasmid pKC30 (30) under the control of the P_L promoter of phage λ. After heat induction, protein p3 is overproduced in E. coli (29). Protein p3 has been obtained in a highly purified form from E. coli transformed with the recombinant plasmid containing gene 3 (31). The purified protein is active in the formation of the initiation complex when supplemented with extracts from sus3-infected B. subtilis. The effect of two mutations at the carboxyl end of protein p3 on the initiation of φ29 DNA replication in vitro has been studied.

RESULTS

a) Formation of a Covalent Complex between the Terminal Protein, p3, and 5'dAMP.

When extracts from φ29-infected B. subtilis containing endogenous φ29 DNA-protein p3 complex were incubated with {α-^{32}P}dATP in the presence of ATP, a ^{32}P-labeled protein with the electrophoretic mobility of p3 was found (17). The reaction with {α-^{32}P}dATP was strongly inhibited by anti-p3 serum. Incubation of the ^{32}P-labeled protein with piperidine under conditions in which the φ29 DNA-protein p3 linkage is hydrolyzed released 5'dAMP indicating that the reaction product is a protein p3-dAMP covalent complex. When the endogenous φ29 DNA-protein p3 complex was removed from the extracts by DEAE-cellulose chromatography, the formation of the protein p3-dAMP complex was dependent on the presence of φ29 DNA-protein p3 as template; no reaction took place with proteinase K-treated φ29 DNA or in the absence of DNA (Fig. 1). These results are in agreement with transfection experiments carried out in parallel with φ29 DNA-protein p3 complex or proteinase K-treated φ29 DNA. As shown in Table 1 the DNA-protein p3 complex was able to transfect competent B. subtilis while the protease-treated DNA was inactive. Since protease-treated DNA is able to enter into the cell, as shown by marker rescue (32), the inability of this DNA to transfect is probably because it cannot initiate replication.

The protein p3-dAMP initiation complex formed in the presence of {α-^{32}P}dATP could be elongated by addition of dNTPs. Treatment with piperidine of the product elongated in the presence of ddCTP released the expected oligonucleotides, 9 and 12 bases long (17). Figure 2 shows the effect of the concentration of dATP in the elongation

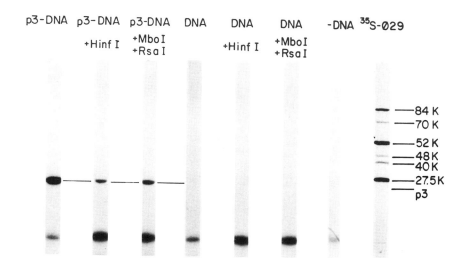

FIGURE 1. Formation of the protein p3-dAMP initiation complex with Ø29 DNA-protein p3 as template before or after treatment with restriction endonucleases or proteinase K. The standard incubation mixture for the initiation reaction contained, in a final volume of 0.05 ml, 50 mM Tris-HCl, pH 7.5, 10 mM MgCl$_2$, 1 mM ATP and 0.25 µM {α-^{32}P}dATP (5 µCi) (26). 75 µg of extracts from B. subtilis infected with mutant sus3(91) (33), passed through DEAE-cellulose to remove endogenous DNA (17) and 25 µg of extracts from E. coli transformed with the gene 3-containing recombinant plasmid pKC30 A1 (29), were added. When indicated, 0.5 µg of Ø29 DNA-protein p3 (p3-DNA) (17) or of proteinase K-treated Ø29 DNA (DNA) (32), either untreated or treated with the restriction endonucleases Hinf I or Mbo I + Rsa I, were added. After treatment with micrococcal nuclease the samples were subjected to 10% acrylamide gel electrophoresis in the presence of SDS (34). ^{35}S-labelled phage Ø29 proteins were used as molecular weight marker in this and the remaining electrophoresis, except when indicated otherwise.

reaction. In the presence of 0.5 µM {α-^{32}P}dATP there was very small amount of elongation, which was greatly increased when 40 µM {α-^{32}P}dATP was used, indicating that a high concentration of dATP is needed for elongation.

TABLE 1

TRANSFECTION OF COMPETENT B. SUBTILIS WITH Ø29 DNA-
PROTEIN p3 COMPLEX OR PROTEINASE K-TREATED Ø29 DNA

---

	pfu/μg DNA
a. Ø29 DNA-protein p3 complex	1.7×10^5
b. Proteinase K-treated Ø29 DNA	<10
c. <u>sus</u>7(614)	8.4×10^3
d. <u>sus</u>7(614)+proteinase K-Ø29 DNA	1.0×10^6

Competent B. subtilis Mu8u5u1 (0.5 ml), prepared as described (35),were incubated with 1 μg of
either Ø29 DNA-protein p3 complex (a) or proteinase
K-treated Ø29 DNA (b). After incubation for 30 min
at 37ºC, 5 μg of DNase I were added and, after 10
min at 37ºC, the cells were plated on B. subtilis
110NA try‾spoA‾ (33). Mutant sus7(614) (33) was
added to the competent B. subtilis at a multiplici-
ty of infection of 20 in the absence (c) or presence of proteinase K-treated Ø29 DNA (d). After 30
min at 37ºC, anti-Ø29 serum was added together with
the DNase I and, after 10 min at 37ºC, the cells
were plated as indicated above.

b) Template Requirement for the Formation of the Initiation
   Complex.

To study whether or not an intact Ø29 DNA molecule is
needed as template for the initiation reaction, the Ø29 DNA-
protein p3 complex was treated with several restriction endonucleases. Treatment with Bst EII, Cla I, Bcl I, Eco RI
or Hind III (see physical map in Fig. 3) essentially did not
affect the amount of protein p3-dAMP complex formed (25).
Treatment with Hinf I, which produces protein-containing
left and right terminal fragments 158 and 59 base pairs
long, respectively, or with a mixture of Mbo I and Rsa I
producing left and right terminal fragments 73 and 10 base
pairs long, respectively, gave rise to the formation of
high amounts of initiation complex (Fig. 1). As a control,
proteinase K-treated Ø29 DNA digested with the same enzymes
was completely inactive.

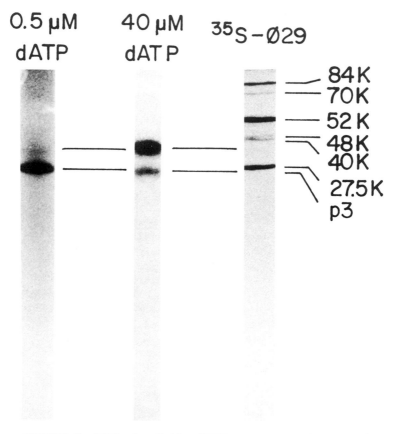

FIGURE 2. Effect of the dATP concentration on the elongation of the protein p3-dAMP initiation complex. About 35 µg of PEG-fraction prepared from Ø29-infected B. subtilis (17) were incubated in the presence of 40 µM dGTP, 40 µM dTTP and 100 µM ddCTP and either 0.5 µM or 40 µM {α-^{32}P} dATP (10 µCi) as described (17). After 20 min at 30°C the samples were subjected to 20% acrylamide gel electrophoresis in the presence of SDS.

To test whether a single terminal fragment from the left or right end of Ø29 DNA could act as template for the formation of the initiation complex, protein-containing fragments Eco RI A (9000 bp) and Eco RI C (1800 bp) as well as the internal fragment Eco RI B (5490 bp) were isolated by sucrose gradient centrifugation. As shown in Fig. 4

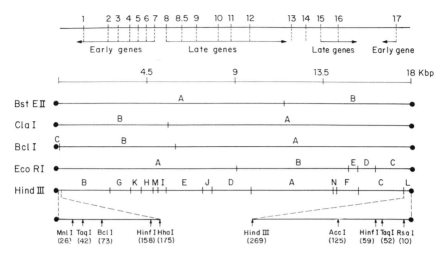

FIGURE 3. Genetic and physical map of Ø29 DNA. The genetic map and Eco RI restriction map were taken from Sogo et al. (1), the Bst EII, Cla I, Bcl I and Hind III restriction maps from Yoshikawa and Ito (28) and the restriction maps at the left and right ends of Ø29 DNA from Escarmís and Salas (7) and Yoshikawa et al. (8).

fragments Eco RI A and C were quite active in the formation of the initiation complex, the values of protein p3-dAMP complex formed being 98% and 54%, respectively, relative to the control with intact Ø29 DNA-protein p3. As expected, the internal fragment Eco RI B was completely inactive. To study further the size effect, the left fragment Eco RI A was cut with the restriction nucleases Hha I, Hinf I, Bcl I, Taq I or Mnl I which produce terminal fragments 175, 158, 73, 42 and 26 base pairs long, respectively (see Fig. 3). As shown in Table 2 the amount of protein p3-dAMP formed directed by those templates was 74%, 47%, 51%, 32% and 27%, respectively, of that obtained with fragment Eco RI A. The right fragment Eco RI C was cut with the restriction nucleases Hind III, Acc I, Hinf I, Taq I or Rsa I which produce terminal fragments 269, 125, 59, 52 and 10 base pairs long, respectively (see Fig. 3). The values obtained for the formation of the initiation complex were 181%, 157%, 36%, 29% and 14% of that obtained with fragment Eco RI C, the last value being only slightly higher than the background obtained in the absence of DNA (Table 2) (25).

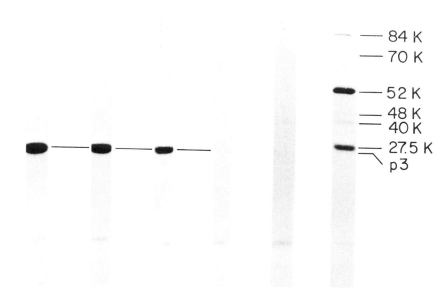

FIGURE 4. Formation of the protein p3-dAMP initiation complex with isolated terminal fragments Eco RI A and C. Ø29 DNA-protein p3 was digested with the restriction endonuclease Eco RI and the fragments were separated by sucrose gradient centrifugation (2). The left and right terminal fragments Eco RI A and C, respectively, as well as the internal fragment Eco RI B were used as template for the formation of the initiation complex as described in Fig. 1 except that the DNA-free extracts from sus3-infected B. subtilis were prepared as described (18). After treatment with micrococcal nuclease, the samples were subjected to 10% acrylamide gel electrophoresis, in the presence of SDS.

To study whether the low level of initiation obtained when the 10 base pairs long right terminal Rsa I fragment was used as template was due to the lack of specific DNA sequences or to the small size of the fragment, an excess of lambda DNA digested with Rsa I was ligated to the mixture of five fragments produced after digestion of fragment

TABLE 2

EFFECT OF THE SIZE OF THE TERMINAL FRAGMENTS ON THE FORMATION OF THE PROTEIN p3-dAMP INITIATION COMPLEX

Terminal left fragment	Base pairs	Amount of p3-dAMP, %	Terminal right fragment	Base pairs	Amount of p3-dAMP, %
Eco RI A	9000	100	Eco RI C	1800	100
+ Hha I	175	74	+ Hind III	269	181
+ Hinf I	158	47	+ Acc I	125	157
+ Bcl I	73	51	+ Hinf I	59	36
+ Taq I	42	32	+ Taq I	52	29
+ Mnl I	26	27	+ Rsa I	10	14
- DNA	-	4	- DNA	-	9

The terminal fragments Eco RI A and C were isolated from φ29 DNA-protein p3 complex by sucrose gradient centrifugation (2), treated with each of the restriction nucleases indicated above and used as template for the formation of the initiation complex as described in Fig. 4. After treatment with micrococcal nuclease the samples were subjected to 10% acrylamide gel electrophoresis in the presence of SDS. The amount of the protein p3-dAMP band was determined by densitometry of the autoradiographs.

Eco RI C with Rsa I. A marked stimulation in the initiation reaction as compared to the unligated control mixture was obtained (25).

c) Requirement of Protein p2 for the Formation of the Protein p3-dAMP Covalent Complex.

To study the protein requirements for the in vitro initiation reaction, extracts from B. subtilis infected with Ø29 mutants in genes 2, 3, 5, 6 and 17, involved in DNA synthesis, were tested in the formation of the protein p3-dAMP complex. As expected, extracts from B. subtilis infected with a sus3-mutant were completely inactive in the formation of the initiation complex (Fig. 5). Extracts from B. subtilis infected with mutant sus2(513) were also inactive in the formation of the p3-dAMP complex but they could be complemented by addition of the sus3 extracts (Fig. 5). The same results were obtained with mutant sus2(515) (26) showing that, not only protein p3, but also protein p2 is essential for the initiation reaction. Similar experiments using mutants in genes 5, 6 and 17 suggested that the products of these genes are not required for the initiation reaction (26).

d) Cloning and Expression of Gene 3 in E. coli. Purification of Protein p3.

A Ø29 DNA fragment containing genes 3, 4, 5 and most of 6 was cloned in the pBR322 derivative plasmid pKC30 (30) under the control of the P_L promoter of bacteriophage λ, resulting in the recombinant plasmid pKC30 A1 (29). Four specific polypeptides of M_r 27,000, 18,500, 17,500 and 12,500 were labeled with ^{35}S-methionine after heat inactivation of the λ repressor. The protein with a M_r 27,000, accounting for about 3% of the de novo protein synthesis, was characterized as protein p3 by radioimmunoassay (29). The protein p3 synthesized in E. coli was active in the in vitro formation of the initiation complex when supplemented with extracts from B. subtilis infected with a sus3 mutant (29). The protein of M_r 12,500 was characterized as p4, involved in the control of late transcription (1),and those of M_r 17,500 and 18,500 are probably p5 and the fusion product of p6 and the amino-terminus of the N protein of phage λ. Protein p3 has been purified from E. coli harbouring the gene 3-containing recombinant plasmid pKC30 A1 (29).

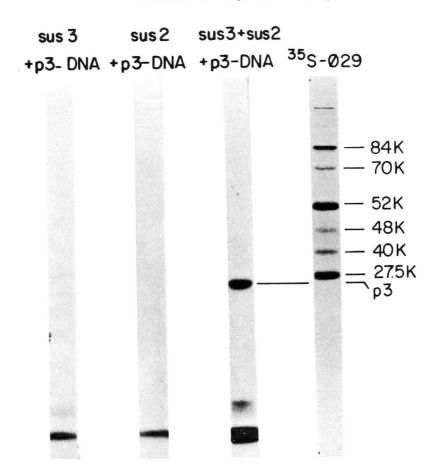

FIGURE 5. Effect of sus mutations in genes 2 or 3 on the formation of the protein p3-dAMP complex. Extracts were prepared from B. subtilis infected with mutants sus3(91) or sus2(513) (33). The sus3-extracts were subjected to DEAE-cellulose chromatography and the sus2-extracts were extracted with PEG-dextrane as described (17). Incubation was carried out as described in Fig. 1 in the presence of Ø29 DNA-protein p3 as template, except that the E.coli extracts containing protein p3 were not added. After treatment with micrococcal nuclease the samples were subjected to 10% acrylamide electrophoresis in the presence of SDS.

Extracts were passed through a DEAE-cellulose column at a
salt concentration of 0.3 M. Protein p3, detected by radioimmunoassay, eluted at 0.7 M NaCl, together with nucleic
acid and free of most of the protein. By precipitation with
0.2% polyethyleneimine in the presence of 1 M NaCl, the
bulk of the nucleic acid was removed. Protein p3, remaining
in the supernatant, was further purified by elution from a
phosphocellulose column at 0.6 M NaCl (Fig. 6). The protein
was essentially homogeneous and it was active in the formation of the initiation complex in vitro when supplemented
with extracts from sus3-infected B. subtilis. It did not
contain DNA polymerase or ATPase activities (31).

e) In vitro Mutagenesis of Gene 3. Effect of Mutations at
   the Carboxyl End on the in vitro Formation of the Initiation Complex.

   The Ø29 DNA Hind III G fragment contains gene 4 and
most of gene 3, except the last 15 nucleotides coding for
the five carboxy-terminal amino acids (27). This fragment
was cloned in the Hind III site of pBR322 (37) giving rise
to a recombinant in which the information for the five carboxy-terminal amino acids of protein p3 (ser-leu-lys-gly-
phe) has been replaced by another one which would produce a
protein with the sequence ser-phe-asn-ala-val-val-tyr-his-
ser at the carboxyl end. To obtain a high level of expression of the mutated protein p3 (p3'), the Eco RI-Bam HI
fragment from the above recombinant was cloned in the Eco
RI-Bam HI sites of plasmid pPLc28 (38), under the control
of the P_L promoter of phage λ,resulting in the recombinant
plasmid pRMw51 (37). Three specific polypeptides were labeled with ^{35}S-methionine after heat induction. One of them,
which accounted for about 6% of the de novo protein synthesis was characterized as p3' by its electrophoretic mobility in SDS-polyacrylamide gels, by radioimmunoassay and by
the fact that it dissappeared when the Hind III G fragment
cloned was from a sus3 mutant (37). Protein p3' was active
in the in vitro formation of the initiation complex, although
the specific activity was about 10% of that obtained with
intact protein p3 (Fig. 7), when corrected for the relative
production of proteins p3 and p3'.
   The Ø29 DNA Hind III G fragment was also cloned in
plasmid pKTH 601 which has the Hind III-Bam HI fragment of
pBR322 replaced by a synthetic stop oligonucleotide with
Hind III-Bam HI linkers (kindly donated by Dr. R.F.Pettersson).

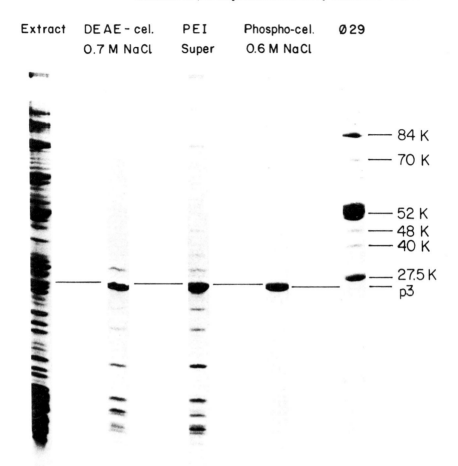

Extract   DEAE-cel.   PEI   Phospho-cel.   Ø29
          0.7 M NaCl Super 0.6 M NaCl

— 84 K
— 70 K
— 52 K
— 48 K
— 40 K
— 27.5 K
— p3

FIGURE 6. SDS-polyacrylamide gel electrophoresis of the protein present in the different purification steps of protein p3. Between 5 and 100 μg of protein were precipitated with 10% trichloroacetic acid in the presence of tRNA as carrier (200 μg/ml) and subjected to slab gel electrophoresis in a 10-20% acrylamide gradient in the presence of SDS as described (10). Phage Ø29 proteins were used as molecular weight marker. After electrophoresis the gel was stained as described (36).

In the recombinant obtained, the information for the last three carboxy-terminal amino acids of protein p3, lys-gly-phe,

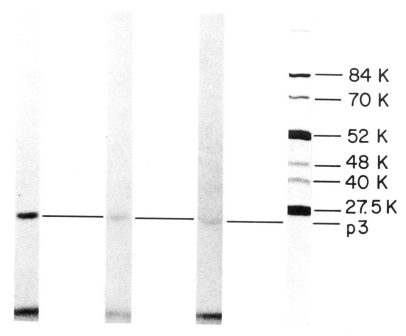

pKC 30 A1 : .. ser-leu-lys-gly-phe

pRM w 51 : .. ser-phe-asn-ala-val-val-tyr-his-ser

pRM t 121 : .. ser-leu-leu-ile-asp

FIGURE 7. Effect of specific mutations at the carboxyl end of protein p3 on the formation of the initiation complex. About 12 μg of protein from extracts from E. coli transformed with the recombinant plasmids pKC30 A1 (29), pRMw51 (37) or pRMt121 (39) were prepared as described (29), and they were incubated with extracts from sus3-infected B. subtilis prepared as described in Fig. 4, in the presence of Ø29 DNA-protein p3. After treatment with micrococcal nuclease the samples were subjected to 10% acrylamide gel electrophoresis in the presence of SDS. The protein p3-dAMP band was quantitated by densitometry of the autoradiographs.

should have been replaced by leu-ile-asp. The Eco RI-Bam HI fragment from the above recombinant was cloned in plasmid pPLc28 under the control of the P_L promoter of phage λ. When E. coli transformed with the recombinant plasmid, pRMt121, was labeled with ^{35}S-methionine after heat induction, a protein with the electrophoretic mobility of p3 was found, accounting for about 9% of the total de novo protein synthesis. This protein was active in the in vitro formation of the initiation complex, although the specific activity was about 7% of that obtained with intact protein p3 (Fig. 7) when corrected for the relative production of both proteins (39).

f) Cloning of Genes 2 and 6.

The Hind III B fragment of Ø29 DNA, containing gene 2, was cut with Bcl I (see Fig. 3) and the resulting fragment was cloned in the Hind III-Bam HI sites of plasmid pBR322. To obtain a high expression of protein p2, the Eco RI-Pst I fragment of the recombinant plasmid with the Ø29 DNA insert was cloned into the Eco RI-Pst I sites of pPLc28, under the control of the P_L promoter of phage λ. A protein of M_r 68,000, the size expected for protein p2 (28), was synthesized after heat induction, accounting for about 2% of the de novo protein synthesis.

Gene 6 protein is probably involved in an elongation (or maturation) step in Ø29 replication (12). We undertook the cloning of gene 6 to overproduce the protein and study its role in elongation. Fragment Hind III H, containing gene 6 as well as the strong promoters A2 and A3 (1,40), could not be cloned in the Hind III site of plasmid pPLc28. However, after digestion with Hinf I which removes the Ø29 promoters (40), the resulting fragment could be cloned in the Hind III site of pPLc28, after filling the single-stranded tails. A protein of M_r 13,500, the size expected for protein p6 (40) was synthesized after heat induction, accounting for about 12% of the de novo protein synthesis.

Purification of proteins p2 and p6 from E. coli harbouring the recombinant plasmids is being carried out.

DISCUSSION

A covalent complex between protein p3 and 5'dAMP is formed when extracts from Ø29-infected B. subtilis are incubated with dATP and ATP (17). This complex can be elonga-

ted in vitro, supporting the proposed model in which the
terminal protein acts as a primer in the initiation of Ø29
DNA replication. The formation of the protein p3-dAMP covalent complex requires the presence of Ø29 DNA-protein p3 as
template; no reaction occurs with proteinase K-digested Ø29
DNA. These results indicate that the parental 5' linked
protein p3 is an essential requirement for the initiation
of Ø29 DNA replication. This conclusion is in agreement
with the fact that competent B. subtilis is transfected
with Ø29 DNA-protein p3 and not with protease-treated DNA,
as first shown by Hirokawa (41) and also with the results
of mixed infection of B. subtilis at 42°C with mutants ts2
(98) and ts3(132) where most of the progeny had the ts2 genotype (2).

An intact Ø29 DNA-protein p3 template is not a requirement for the initiation reaction since treatment of the
latter with different restriction endonucleases essentially
did not affect the formation of the initiation complex.
Even the presence of the left and right terminal fragments
together is not necessary since isolated protein-containing
fragments Eco RI A (left) and C (right) can function as
template for the formation of the initiation complex. As
expected, an internal fragment (Eco RI B) was completely
inactive. To further determine the minimal size of the terminal fragments required, isolated fragments Eco RI A and C
were cut with different restriction endonucleases. A 26 base
pairs long left fragment, produced by treatment with Mnl I,
was still active in the initiation reaction. At the right
side, a Taq I fragment, 52 base pairs long, was active but
a Rsa I fragment, 10 base pairs long, was essentially inactive. However, the activity was restored when the latter
fragment was ligated to an excess of lambda DNA digested
with Rsa I, suggesting that a specific DNA sequence after
the 10 terminal base pairs at the right end is not required
for the initiation reaction.

In addition to the protein-containing template as described above and free protein p3, the formation of the initiation complex requires the presence of the gene 2 product.
The product of genes 5, 6 and 17 do not seem to be needed
for the initiation reaction. These results are in agreement
with in vivo shift-up experiments in which it was shown that
the products of genes 2 and 3 are involved in the initiation
of replication and that of genes 5 and 6 act in elongation
(or maturation) steps (12). Fig. 8 shows a summary of our
knowledge of the process of initiation of Ø29 replication.
A free molecule of protein p3 interacts with the parental

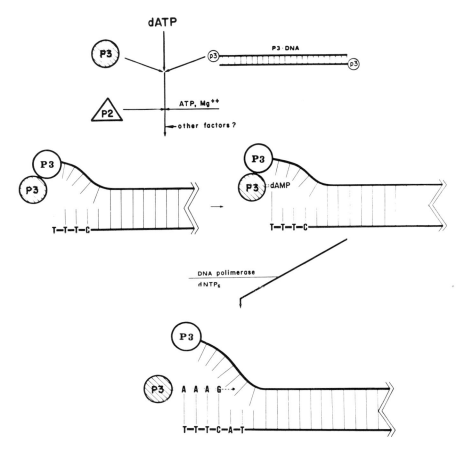

FIGURE 8. Model for the initiation of phage Ø29 DNA replication.

protein at the 5' ends of the DNA and may be also with the inverted terminal repetition and starts replication by reaction with dATP in the presence of ATP and protein p2. A protein p3-dAMP covalent complex is formed, and it provides the 3'OH group needed for elongation. Whether or not other protein factors are required in the initiation reaction remains to be determined.

To further study the initiation and elongation processes of Ø29 replication we have cloned genes 2, 3, 5 and 6 to overproduce the proteins for its purification. Protein p3 has been obtained in a highly purified form, which is active in the formation of the initiation complex. The purified protein does not contain DNA polymerase or ATPase activities. It has been reported that the initiation of adenovirus replication requires a viral-coded DNA polymerase activity, in addition to the precursor of the terminal protein (42,43). Whether the gene 2 product of phage Ø29 is a DNA polymerase activity is being studied. Since proteins p5 and p6 are also overproduced in E. coli transformed with the corresponding recombinant plasmids we are purifying the proteins to study their role in replication. Cloning of gene 17, the other known cistron involved in DNA replication, and purification of the protein will allow a study of its function.

We have started the in vitro mutagenesis of gene 3 to study the effect of specific mutations on the protein function. We have obtained two mutated proteins with changes at the carboxy-terminal end. The two mutants are still active in the formation of the initiation complex although their specific activity is about 10% of that obtained with wild-type protein p3. We intend to produce deletions at the carboxyl end to determine how many amino acids can be eliminated without loosing completely the protein activity in the initiation reaction. The effect of mutations at other sites of protein p3 will be also studied. We are presently determining the serine residue through which the linkage to 5'-dAMP occurs to introduce mutations in the neighbouring region by oligonucleotide-directed mutagenesis and study their effect on the protein function.

REFERENCES

1. Sogo JM, Inciarte MR, Corral J, Viñuela E, Salas M (1979). RNA polymerase binding sites and transcription map of the DNA of Bacillus subtilis phage Ø29. J Mol Biol 127:411.

2. Salas M, Mellãdo RP, Viñuela E, Sogo JM (1978). Characterization of a protein covalently linked to the 5' termini of the DNA of Bacillus subtilis phage Ø29. J Mol Biol 119:269.

3. Harding NE, Ito J, David GS (1978). Identification of the protein firmly bound to the ends of bacteriophage Ø29 DNA. Virology 84:279.

4. Yehle CO (1978). Genome-linked protein associated with the 5' termini of bacteriophage Ø29 DNA. J Virol 27:776.

5. Ito J (1978). Bacteriophage Ø29 terminal protein : Its association with the 5' termini of the Ø29 genome. J Virol 28:895.
6. Hermoso JM, Salas M (1980). Protein p3 is linked to the DNA of phage Ø29 through a phosphoester bond between serine and 5'-dAMP. Proc Natl Acad Sci USA 77: 6425.
7. Escarmís C, Salas M (1981). Nucleotide sequence at the ends of the DNA of Bacillus subtilis phage Ø29. Proc Natl Acad Sci USA 78: 1446.
8. Yoshikawa H, Friedmann T, Ito J (1981). Nucleotide sequences at the termini of Ø29 DNA. Proc Natl Acad Sci USA 78:1336.
9. Talavera A, Jiménez F, Salas M, Viñuela E (1972). Temperature-sensitive mutants of bacteriophage Ø29. Virology 46:586.
10. Carrascosa JL, Camacho A, Moreno F, Jiménez F, Mellado RP, Viñuela E, Salas M (1976). Bacillus subtilis phage Ø29 : Characterization of gene products and functions. Eur J Biochem 66:229.
11. Hagen EW, Reilly BE, Tosi ME, Anderson DL (1976) Analysis of gene function of bacteriophage Ø29 of Bacillus subtilis : Identification of cistrons essential for viral assembly. J Virol 19:501.
12. Mellado RP, Peñalva MA, Inciarte MR, Salas M (1980). The protein covalently linked to the 5' termini of the DNA of Bacillus subtilis phage Ø29 is involved in the initiation of DNA replication. Virology 104:84.
13. Inciarte MR, Salas M, Sogo JM (1980). Structure of replicating DNA molecules of Bacillus subtilis bacteriophage Ø29. J Virol 34:187.
14. Harding NE, Ito J (1980). DNA replication of bacteriophage Ø29 : Characterization of the intermediates and location of the termini of replication. Virology 104: 323.
15. Sogo JM, García JA, Peñalva MA, Salas M (1982). Structure of protein-containing replicative intermediates of Bacillus subtilis phage Ø29 DNA. Virology 116:1.
16. Rekosh DMK, Rusell WC, Bellett AJD (1977). Identification of a protein linked to the ends of adenovirus DNA. Cell 11:283.
17. Peñalva MA, Salas M (1982). Initiation of phage Ø29 DNA replication in vitro : Formation of a covalent complex between the terminal protein, p3, and 5'-dAMP. Proc Natl Acad Sci USA 79:5522.

18. Shih M, Watabe K, Ito J (1982). In vitro complex formation between bacteriophage Ø29 terminal protein and deoxynucleotide. Biochem Biophys Res Commun 105:1031.

19. Watabe K, Shih MF, Sugino A, Ito J (1982) In vitro replication of bacteriophage Ø29 DNA. Proc Natl Acad Sci USA 79:5245.

20. Lichy JH, Horwitz MS, Hurwitz J (1981). Formation of a covalent complex between the 80,000 dalton adenovirus terminal protein and 5'-dCMP in vitro. Proc Natl Acad Sci USA 78:2678.

21. Pincus S, Robertson W, Rekosh D (1981). Characterization of the effect of aphidicolin on adenovirus DNA replication : Evidence in support of a protein primer model of initiation. Nucl Acids Res 9:4919.

22. Challberg MD, Ostrove JM, Kelly TJ Jr (1982). Initiation of adenovirus DNA replication : Detection of covalent complexes between nucleotide and the 80 kilodalton terminal protein. J Virol 41:265.

23. Tamanoi F, Stillman BW (1982). Function of adenovirus terminal protein in the initiation of DNA replication. Proc Natl Acad Sci USA 79:2221.

24. De Jong PJ, Kwant MM, Van Driel W, Jansz HS, Van der Vliet PC (1982). The ATP requirements of adenovirus type 5 DNA replication and cellular DNA replication. Virology 124:45.

25. Garcia JA, Peñalva MA, Blanco L, Salas M. Template requirements for the initiation of phage Ø29 DNA replication in vitro. In preparation.

26. Blanco L, García JA, Peñalva MA, Salas M (1983). Factors involved in the initiation of phage Ø29 DNA replication in vitro : requirement of the gene 2 product for the formation of the protein p3-dAMP complex. Nucl Acids Res 11:1309.

27. Escarmís C, Salas M (1982). Nucleotide sequence of the early genes 3 and 4 of bacteriophage Ø29. Nucl Acids Res 10:5785.

28. Yoshikawa H, Ito J (1982). Nucleotide sequence of the major early region of bacteriophage Ø29. Gene 17:323.

29. García JA, Pastrana R, Prieto I, Salas M (1983). Cloning and expression in Escherichia coli of the gene coding for the protein linked to the ends of Bacillus subtilis phage Ø29 DNA. Gene 21:65.

30. Shimatake H, Rosenberg M (1981). Purified λ regulatory protein CII positively activates promoters for lysogenic development. Nature 292:128.

31. Prieto I, Lázaro JM, García JA, Hermoso JM, Salas, M. Purification of the terminal protein of bacteriophage Ø29 DNA. In preparation.
32. Inciarte MR, Lázaro JM, Salas M, Viñuela E (1976). Physical map of bacteriophage Ø29 DNA. Virology 74: 314.
33. Moreno F, Camacho A, Viñuela E, Salas M (1974). Suppressor-sensitive mutants and genetic map of Bacillus subtilis bacteriophage Ø29. Virology 62:1.
34. Laemmli UK (1970). Cleavage of structural proteins during the assembly of the head of bacteriophage T4. Nature 227:680.
35. Bott KF, Wilson GA (1968). Metabolic and nutritional factors influencing the development of competence for transfection of Bacillus subtilis. Bacteriol Rev 32: 370.
36. Steven AC, Couture E, Aebi U, Showe MK (1976). Structure of T4 polyheads. II. A pathway of polyhead transformations as a model for T4 capsid maturation. J Mol Biol 106:187.
37. Mellado RP, Salas M (1982). High-level synthesis in Escherichia coli of the Bacillus subtilis phage Ø29 proteins p3 and p4 under the control of phage lambda P_L promoter. Nucl Acids Res 10:5773.
38. Remaut E, Stanssens P, Fiers W (1981). Plasmid vectors for high-efficiency expression controlled by the P_L promoter of coliphage lambda. Gene 15:81.
39. Mellado RP, Salas, M. Effect of mutations at the carboxyl end on the activity of the terminal protein of Ø29 DNA. In preparation.
40. Murray CL, Rabinowitz JC (1982). Nucleotide sequences of transcription and translation initiation regions in Bacillus phage Ø29 early genes. J Biol Chem 257:1053.
41. Hirokawa H (1972). Transfecting deoxyribonucleic acid of Bacillus bacteriophage Ø29 that is protease sensitive. Proc Natl Acad Sci USA 69:1555.
42. Lichy JH, Field J, Horwitz MS, Hurwitz J (1982). Separation of the adenovirus terminal protein precursor from its associated DNA polymerase : Role of both proteins in the initiation of adenovirus DNA replication. Proc Natl Acad Sci USA 79:5225.
43. Stillman BW, Tamanoi F, Mathews MB (1982). Purification of an adenovirus-coded DNA polymerase that is required for initiation of DNA replication. Cell 31:613.

Mechanisms of DNA Replication and Recombination, pages 225–243
© 1983 Alan R. Liss, Inc., 150 Fifth Avenue, New York, NY 10011

TERMINAL PROTEIN-PRIMED INITIATION OF φ29 DNA REPLICATION
IN VITRO

Kounosuke Watabe, Meng-Fu Shih and Junetsu Ito

Department of Molecular and Medical Microbiology

University of Arizona, College of Medicine
Tucson, Arizona 85724

ABSTRACT

We have been studying the mechanism of linear DNA
replication by using Bacillus bacteriophage φ29 as a model
system. We have developed a cell-free replication system.
A cell-free extract prepared from φ29-infected Bacillus
subtilis catalyzed the semiconservative replication of φ29
DNA, but only if exogenous φ29 DNA-protein complex is used
as the template. This template consists of linear duplex
DNA with a 30,000-dalton terminal protein attached covalently to both 5' ends. The extract also catalyzes the specific
binding between dATP and the φ29 terminal protein. We have
used this system for the isolation of replication activity
associated with gene 3 protein (terminal protein) from
φ29-infected cells. We utilized two assay systems:
(i) DNA replication dependent on φ29 DNA of which 5' end
covalently link to terminal protein (DNA-protein), and
(ii) the complex formation between the terminal protein and
dAMP. Both the DNA replication and the complex forming
activities were copurified through all steps of purification. The complex of terminal protein and dAMP formed in
the purified fraction was shown to serve as an effective
primer for successive chain elongation in the presence of
dNTP by a pulse-chase experiment. The analysis of DNA
synthesized in vitro revealed that the replication starts
at both termini of the φ29 genome. The protein fraction
purified from cells infected with a temperature-sensitive
φ29 mutant in gene 3 was thermolabile compared to the wild-
type activity in the assay system for complex formation.
This shows that the purified fraction having replication

activity includes the gene 3 product of $\phi29$. Both the DNA replication and complex formation activities are highly specific for $\phi29$ DNA protein as template. We have prepared two distinct antibodies against $\phi29$ terminal protein (anti-TP) and synthetic peptide corresponding to the carboxy-terminus of the terminal protein (anti-gp3c). Our immunochemical studies indicated that the anti-gp3c react with native $\phi29$ terminal protein. Furthermore, it was found that both antibodies inhibit specifically complex forming reaction between the terminal protein and dAMP. These results are consistent with the basic elements of the protein-priming model for the initiation of linear DNA synthesis.

INTRODUCTION

Bacteriophage $\phi29$ provides a useful model system for the study of the initiation of linear DNA synthesis. The genome of $\phi29$ is a relatively small linear nonpermuted duplex DNA of 18 kilobases long (about half the size of the DNA of adenovirus of phage T7)(1,2). A good deal is known about genetics of the $\phi29$ (3). The novel features of $\phi29$ DNA are that it contains short inverted terminal repeats 6 base pairs long (4,5) and that the 5'-ends of both DNA strands are covalently linked to a protein of molecular weight 30,000 (6-8). Genetic studies indicated that this terminal protein is required for $\phi29$ DNA synthesis (9,10). Studies on in vivo $\phi29$ DNA synthesis have shown that replication is started at either end of the DNA and proceeds in the 5' to 3' direction by an asymmetric strand-displacement mechanism (11,12).

Ever since it was discovered that DNA polymerases are incapable of initiating the synthesis of new chains, replication of linear DNA has been an intriguing problem (13,14). Although various models have been proposed for replication of linear DNA molecules, such as adenovirus DNA (15) and $\phi29$ DNA (11,12), the protein-priming model proposed originally by Rekosh et al. (16) is most consistent with the experimental results described to date (11,12). In this model, the terminal protein binds covalently with the first 5'-nucleotide and thus provides the 3'-OH end that can be used as primer for subsequent chain elongation by DNA polymerase (16).

The development of an in vitro $\phi29$ DNA replication system would facilitate the functional analysis of the terminal protein and will enable the factors involved in the φ29 DNA replication to be isolated and their function characterized. To this end, we have developed an in vitro DNA replication system for phage φ29 (17). Cell-free extracts prepared from φ29-infected Bacillus subtilis catalyze the initiation and the semiconservative replication of the φ29 genome. This cell-free system also catalyzed the formation of a covalent complex between φ29 terminal protein and dAMP, the 5'-terminal nucleotide of φ29 genome (18). Our in vitro φ29 DNA replication system requires specifically φ29 DNA covalently linked to a 30,000-dalton protein at each 5'-terminus (φ29 DNA-protein) as template, the four dNTPs and Mg^{++} (17,18). A similar complex formation between φ29 terminal protein and dAMP has recently been described independently by others (19).

In this paper, we describe the isolation and characterization of a protein fraction that catalyzes both terminal protein-dAMP complex formation and subsequent chain elongation reaction. We also describe the preparation of two distinct antibodies against the φ29 terminal protein and the chemically synthesized peptide corresponding to the C-terminal portion of the φ29 terminal protein. All our results are consistent and verify the protein-priming mechanism for the initiation of the φ29 DNA replication.

RESULTS

Isolation of the replication activity associated with the terminal protein.

The enzymatic activity of the terminal protein can be assayed by using a system to detect the complex formation between the terminal protein and dAMP, the first 5'-terminal nucleotide of the φ29 genome. During the course of the purification of the φ29 terminal protein, we found that the fraction active in the complex formation between the terminal protein and dAMP is also active in DNA replication. Thus, we have been able to isolate this DNA replication activity by using two assay systems: φ29 DNA synthesizing activity and complex forming activity between the terminal protein and dAMP. Elution profiles for both the φ29 DNA synthesizing activity and terminal protein-dAMP complex forming activity as well as total protein from a phosphocellulose column chromatography are shown in Fig. 1. As indicated, both activities have rather high affinity for phosphocellulose and are eluted with 0.5M NaCl

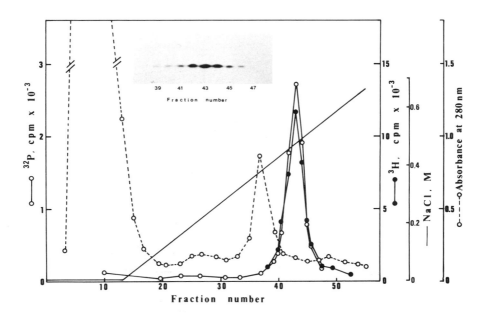

<u>Figure 1</u>. Phospho-cellulose chromatography. The fraction
II was chromatographed on a phospho-cellulose column.
Inserted figure shows the autoradiogram of product of
complex formation between terminal protein and ^{32}P-labeled
dAMP. The band was cut and counted for radioactivity
(o———o). ●———●, replication activity for φ29 depend
on DNA-protein as template. o–––––o, absorbance of A_{280}.

coincidentally in the same fraction. On the other hand, host DNA polymerase III was eluted near the void volume and host DNA polymerase II was eluted around 0.2M NaCl (data not shown). Some residual amount of DNA synthesizing activity using calf thymus DNA as template was also coincidental with the ϕ29 DNA replication activity. Generally, the ϕ29 DNA replication activity increased significantly after phosphocellulose column chromatography, suggesting the removal of some inhibitors or proteolytic enzyme activities. The results of a typical purification are summarized in Table I. Throughout these purifications, both ϕ29 DNA replication activity and complex forming activity were copurified. The purified fraction was stable at 0°C for at least 2 weeks, although dilution of the fraction resulted in the rapid loss of activity.

Thermosensitivity of the complex forming activity isolated from B. subtilis infected with a gene 3 ts mutant.

Gene 3 of ϕ29 encodes for the terminal protein (2). To confirm that gene 3 protein is indeed responsible for the complex forming activity in the purified fraction, we prepared the active fraction through phospho- and DNA-cellulose column chromatography from cells infected with ϕ29 temperature-sensitive mutant ts3. As a control, the active fraction was also prepared from cells infected with wild type. Fractions were preincubated at various temperatures and were then examined for their ability to support the complex formation between the terminal protein and dAMP at 25°C. Figure 2 shows temperature inactivation of complex forming activity in the purified fraction prepared from cells infected with ts3 and wild-type ϕ29, respectively. The fraction purified from the ts3-infected cells was far less stable at 35°C than those from wild-type infected cells. These results indicate that protein labeled with dAMP is indeed the gene 3 protein, the same gene product previously shown to be the terminal protein of the ϕ29 DNA-protein (20).

Properties of the purified fraction

The properties of the purified fraction V are shown in Table II, both DNA synthesizing activity and complex forming activity depended on the addition of Mg^{++} and ϕ29 DNA-protein. The activities in both reactions are specific for template. The ϕ29 DNA-protein cannot be

Table I Purification of the activity of DNA replication and complex formation

Fraction	Total protein (mg)	Complex forming activity		Replication activity		The ratio of (A) to (B)
		Total units[1]	Specific activity(A) (units/mg)	Total units[2]	Specific activity(B) (units/mg)	
I. Crude extracts	2365	–	–	–	–	–
II. $(NH_4)_2SO_4$	1340	1136	0.8	63.0	0.05	–
III. Phospho-cellulose	171	1732	10.1	36.6	0.21	48
IV. DEAE-cellulose	2.0	1234	617	26.0	13.0	47
V. Denatured DNA-cellulose	0.9	472	514	7.8	8.6	59

[1] One unit incorporates the 1 pmole of ^{32}P-dAMP into terminal protein–dAMP complex.

[2] One unit catalyzes the incorporation of 1 nmole of ^3H–dTMP.

Table 1.   Purification of the activity of replication and
complex formation.   A crude extract (Fraction I) was
prepared from a 12 l culture of B. subtilis cells infected
with φ29 as described previously (17).   After passing
through the DEAE cellulose column, ammonium sulfate was
added to 50% saturation and the resultant precipitate was
suspended with 20 ml of buffer I (20 mM Tris-HCl, pH 7.0,
20% glycerol, 5 mM $MgCl_2$, 60 mM $(NH_4)_2SO_4$, and 1 mM mer-
captoethanol).   The suspension was dialyzed against the
same buffer for 5 hours (Fraction II 23.5 ml).   Fraction II
was diluted by the same volume of buffer I and applied to
a phosphocellulose column (100 ml) activated as described
by (26).   The column was washed and eluted with 600 ml
linear gradient of NaCl (0-0.7 M) in buffer I.   Fractions
containing the activity (0.5 M NaCl) were pooled, adjusted
at pH 8.0 with NaOH and dialyzed against the buffer II
(20 mM Tris-HCl, pH 8.0, 20% glycerol, 5 mM $MgCl_2$, 60 mM
$(NH_4)_2SO_4$ and 1 mM DTT) for 3 hours (Fraction III 5 ml).
Fraction III was applied to a DEAE-cellulose column (6 ml)
equilibrated with buffer II.   The column was washed and
then eluted with 30 ml linear gradient of NaCl (0-0.6M) in
buffer II.   Active fractions (0.4-0.45 M NaCl) were pooled
and dialyzed against the buffer I for 3 hours (Fraction IV
4.4 ml).   Fraction IV was applied to denatured DNA-cellulose
column (1.2 ml) equilibrated with buffer I.   The column
was washed with buffer I containing 0.2 M NaCl extensively.
The activity was eluted in a stepwise fashion with buffer I
containing 0.7 M NaCl (Fraction V 1.4 ml).

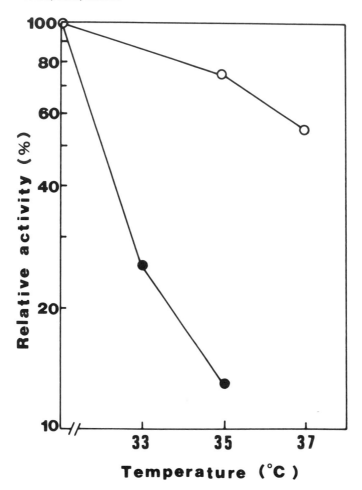

Figure 2. Thermolability of gene 3 product purified from cells infected with temperature-sensitive mutant. The activity for complex formation was purified through phospho- and DNA-cellulose chromatography as described in Table I from cells infected with φ29 wild type, or a temperature-sensitive mutant ts3. Each fraction was pre-incubated at various temperatures for 10 min and tested for their ability to support the complex formation of terminal protein dAMP at 25°C. Actual count for radio-activity at 100% is 6202 cpm (wild type o———o), 2497 cpm (ts3 •———•).

TABLE 2   Properties of the purified fraction.

Assay condition	Complex formation (f moles)	(%)	DNA replication (p moles)	(%)
Complete	2.0	100	11.2	100
-MgCl$_2$	0.08	4	0.7	6
-ATP	0.88	44	12.1	107
-(NH$_4$)$_2$SO$_4$	0.86	43	-	-
-DNA·protein	0.02	1	1.0	8
-DNA·protein +DNA *	0.1	5	2.0	17
-DNA·protein +heat denatured DNA *	0.1	5	-	-
-DNA·protein +M13 RF DNA	0.04	2	0.8	7
-DNA·protein +M13 SS DNA	0.02	1	0.8	7
-DNA·protein +BstNI A	0.22	11	6.7	60
-DNA·protein +BstNI B	0.28	14	3.9	35
+N.E.M.(10 mM)	1.2	60	5.4	48
+BuPG(0.2 mM)	2.6	130	10.7	95
+Pol III immune serum	1.9	95	9.7	80

* Deproteinized ϕ29 DNA was used as DNA template.

substituted with de-proteinized DNA even though the DNA was piperidine treated or heat denatured. Single- and double-stranded DNA of the E. coli phage M13 had no activity as template. Likewise, φX174 DNA did not work as template (data not shown).

To examine if the whole genome structure of φ29 DNA-protein is necessary, the DNA-protein was cleaved with BstN-1 which cuts φ29 DNA only once and generates two fragments, a 15.3 Kb fragment (right end) and a 2.7 Kb fragment (left end) (8). These fragments containing the terminal protein at one end were purified through sucrose gradient centrifugation and then tested for their template activity. Both fragments showed a significant amount of activity, though the efficiency of complex formation appeared to be lower than that of replication. These results suggest that the intact structure of the φ29 DNA-protein may not be necessary for the template activity.

It was of considerable interest to determine whether DNA polymerase III of the host cell is involved in φ29 DNA synthesis. Therefore, we have tested anti-DNA polymerase III serum and the BuPG, a specific inhibitor of B. subtilis DNA polymerase III (21) using the purified fraction. Neither inhibited the activity, suggesting the DNA polymerase III of the host cell is not involved in φ29 DNA replication in vitro.

Antibodies, anti-TP and anti-gp3c precipitate the TP-dAMP complex

To facilitate further functional analysis of the terminal protein, we have prepared two distinct antibodies against the φ29 terminal protein. One antibody was raised against sonicated φ29 DNA-protein complex isolated from phage virions (anti-TP). The other antibody was prepared by immunization of rabbits with a conjugate of bovine serum albumine and the synthetic peptide corresponding to the carboxy-terminus of the φ29 terminal protein, pre-dicted from the nucleotide sequence of φ29 DNA (anti-gp3c). Our results indicated that both antibodies inhibited specifically the complex forming reaction between the terminal protein protein and dAMP (Shih, et al. in prepa-ration).

We also examined if both antibodies were capable of precipitating with the TP-dAMP complex. The φ29 terminal protein was labeled with α-^{32}P-dAMP in vitro complex forming reaction. The reaction products were mixed with anti-gp3c, anti-TP or preimmune sera, respectively, and then precipitated with 10% suspension of Staphylococcus aureus cells. The bound labeled protein were released by heating to 100°C for 3 min in sample buffer and analyzed by 10% polyacrylamide gel electrophoresis. The anti-gp3c serum (Fig. 3, Lane 2) and anti-TP serum (Fig. 3, Lane 4) were found to precipitate a 30 Kd TP-dAMP complex, this complex was not present when preimmune serum was used (Fig. 3, Lanes 1 and 3).

Anti-TP inhibits in vitro ϕ29 DNA replication

To confirm that antibody raised against the terminal protein inhibits ϕ29 DNA replication, the following experiments were performed. First, extracts prepared from ϕ29 infected cells were mixed with anti-TP or preimmune sera as the control and incubated for 12 hrs at 0°C. The preincubated extracts were then assayed for ϕ29 DNA synthesizing activity. As shown in Figure 4, anti-TP completely inhibits in vitro DNA synthesis, whereas no inhibition was observed with preimmune serum. These findings strongly support the notion that in vitro ϕ29 DNA replication is indeed de novo synthesis depending on the exogenously added ϕ29 DNA-protein template, but is not a repair-type of DNA synthesis.

Can the ϕ29 terminal protein-dAMP complex serve as a primer for replication?

The following experiments were designed to test the protein-primer model for the initiation of DNA replication directly. The ϕ29 DNA sequences are 5'-AAAGTAAGC (left end), and 5'-AAAGTAGGGTAC (right end)(4,5). Accordingly, if the terminal protein-dAMP complex serves as primer, the addition of chain-termination analogue ddGTP instead of dGTP in the reaction mixture for replication should result in the termination of chain elongation at the 4th oligonucleotides, and ddCTP should yield two species of complexes; terminal protein-9 oligonucleotides (left end) and terminal protein-12 oligonucleotides (right end).

In the first stage, the terminal protein-dAMP complex was formed in the standard assay mixture for DNA replication in the absence of dNTP with the purified protein fraction. Then, in the second stage, one ddNTP together with the remaining three dNTP was added and the reaction mixture was further incubated. The formation of these

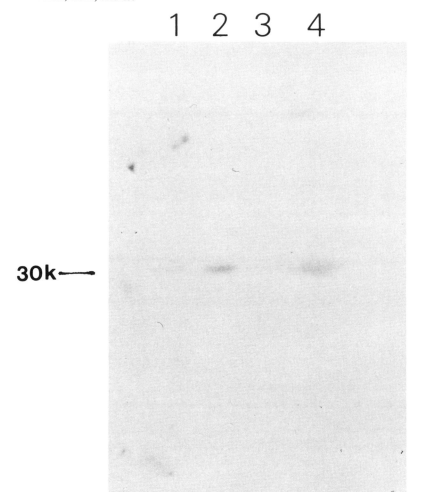

Figure 3. Immunoprecipitation of protein-dAMP complex with antibodies. The terminal protein was labeled specifically with α-^{32}P-dATP as in vitro complex formation condition. The reaction products were suspended in 100 μl extraction buffer and incubated with 20 μl of anti-gp3C (Lane 2), anti-TP serum (Lane 4) and preimmune serum (Lane 1 and 3) for 18 hours at 0°C. After a further 1 hr of incubation (0°C) with 40 μl of a 10% suspension of S. aureus, the precipitates were processed, subjected to electrophoresis through a 10% SDS-polyacrylamide gel and autoradiographed.

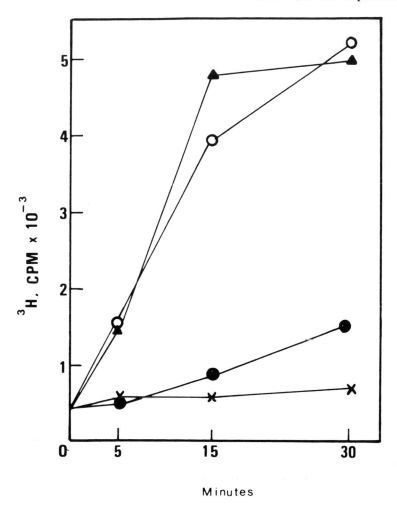

Figure 4. Effects of antisera on in vitro DNA replication.
Cell extract (6 μl) was incubated with 6 μl of anti-TP or
preimmune serum at 0°C for 12 hrs. The preincubated
extracts were then assayed in a condition for DNA replication in vitro. The reaction was carried out at 30°C for
the time as indicated, and the acid-insoluble radioactivity
was measured. (○——○, control reaction: DNA-protein as
template. ●——●, control reaction: Deproteinized DNA
as template. ▲——▲, preimmune serum. X—X, anti-TP).

products was analyzed by acrylamide gel electrophoresis as shown in Figure 5. When ddCTP was added in the second stage (Lane 2), more than 90% of the terminal protein-dAMP complex was elongated into product of molecular weight of 35K as expected. The product of molecular weights of 31.5K was obtained when ddGTP or ddTTP was used in the second stage (Lane 3,4). Moreover, we chased the elongation product for a short time (10-80 seconds) after addition of dNTP (without ddNTP) in the second stage (Lane 5-8). With increasing the time of second incubation, the 30K dalton band was decreased and shifted to relatively high molecular weight band (35K-40K dalton) increased with time. These results indicate that the complexes of terminal protein-dAMP formed in the purified protein fraction serve as primer for successive chain elongation of φ29 DNA in vitro.

DNA product synthesized in the purified fraction

In vivo studies have shown that φ29 replication starts from either end of the DNA and proceeded by a strand displacement mechanism (11,12). The same results were obtained with in vitro system using the purified fraction. The mode of replication in the purified fraction was analyzed by restriction enzyme digestion followed by agarose gel electrophoresis. After a 20 or 40 min incubation of the reaction mixture, the DNA product was treated with Proteinase K, extracted and digested with MspI. Figure 6 shows the distribution of the radioactivity incorporated predominantly into both the terminal ends. The mode of distribution of radioactivity was basically the same after 5 min incubation (data not shown). The rate of the chain elongation was quite low after the initial 15-20%, although the reaction was linear even after 40 min. This is probably due to the fact that some factors required for elongation in the purified fraction are missing.

DISCUSSION

We have described the isolation of replication activity for φ29 DNA from phage infected cells. In all the steps of purification, the activity of complex formation between terminal protein and dAMP is associated with the activity of DNA replication. The purified protein fraction prepared from cells infected with a temperature-

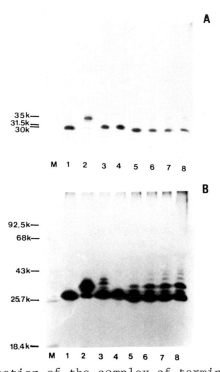

Figure 5. Function of the complex of terminal protein-dAMP as a primer for chain elongation. Purified fraction (3 µl) was incubated in the reaction mixture for DNA replication without dNTP in the presence of 0.5 µM of ^{32}P-dATP for 20 min at 30°C (stage I). Then 40 µM of dCTP, dTTP, ddCTP and 0.8 nM of dATP (2), 40 µM of dCTP, dGTP and ddTTP and 0.8 nM of dATP (3), or 40 µM of dCTP, dTTP and ddGTP and 0.8 nM of dATP (4) was added in the reaction mixture. After the addition of 3 µl of fresh protein fraction, each reaction mixture was further incubated at 30°C for 30 min (stage II). (1) is the control only for stage I. For the pulse-chase experiment without using a ddNTP, 40 µM of dGTP, dCTP and dTTP, 0.8 nM of dATP and 3 µl of protein fraction were added in the reaction mixture after the reaction of stage I (5-8). Then reaction mixtures were incubated at 25°C for 10 sec (5), 20 sec (6), 40 sec (7), or 80 sec (8), respectively. After the reaction of stage II, all the reaction mixtures were processed and run through 10% SDS-polyacrylamide gel electrophoresis and autoradiogramed for 2 hr (A) or 6 hr (B) with intensifying screen. Lane (M) shows the protein markers; phosphorylase B (92.5 k dalton), BSA (68 k), ovalbumine (43 k), α-chymotrypsinogen (25.7 k) and β-lactoglobulin (18.4 k).

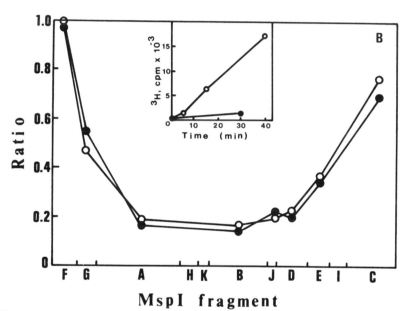

MspI fragment

<u>Figure 6</u>. Product analysis of DNA replication. After
20 min (upper lane) or 40 min (lower lane) of incubation
of reaction mixture, DNA was treated with Proteinase K and
extracted with phenol. DNA was digested by <u>Msp</u> I and run
through a 1.4% agarose gel in TBE buffer for 5 hr with
5 V/cm. Gels were dried and autoradiogramed (A). Each
band was cut and counted for radioactivity by liquid
scintillation counter. The ratio of radioactivity to the
DNA length of each fragment were plotted (B). The ratio
of the fragment F was arbitrarily set as 1.0 and the other
values were normalized to it. The axis of abscissa represents the <u>Msp</u>I restriction map of ϕ29 DNA. The actual
radioactive count for fragment F was 4933 cpm (20 min,
●————●) and 73282 cpm (40 min, o————o). The inserted
figure shows the kinetics of ^{3}H-dTTP incorporation with
DNA-protein (o————o) or with deproteinized DNA (●————●)
as template.

sensitive mutant of gene 3 was heat labile in contrast to a protein fraction from wild-type infected cells. Moveover, immune serum against the terminal protein which was prepared from φ29 DNA-protein inhibited the activity of purified fraction both in complex formation and in DNA replication. These results indicate that the gene 3 product is included in the purified fraction and that the protein labeled with dAMP in the assay system for complex formation is indeed the terminal protein which is also necessary for DNA replication in vitro.

According to the protein-primer model, first step of the replication is to make a covalent complex of terminal protein-dAMP (stage I). This complex should provide a 3'-OH for the subsequent chain elongation (stage II). Thus, one of the most direct evidence to prove the model is to show the coupling of both reactions. Purified protein fraction catalyzed the reaction of both stages separately in each assay system. Moreover, pulse-chase experiments showed that most of the product of stage I (90%) was able to serve as a substrate for the stage II reaction. These results indicate that progeny terminal protein of φ29 indeed functions as a primer for DNA replication.

Recently, Ikeda et al. (23) reported that when adenovirus terminal protein (ad-pTP), prepared from adenovirus-infected cells, was incubated with ΦX174 single-stranded DNA, complex formation can take place between the ad-pTP and dCMP, the terminal nucleotide of the adenovirus genome. More recently, Tamanoi et al. (24) reported that adenovirus DNA deproteinized by piperidine treatment and heat denatured DNA treated with protease supported the covalent linking between the ad-pTP and dCMP when partially purified enzyme fraction was used. Thus, they suggested the possibility that the newly synthesize terminal protein recognized a specific sequence but not parental protein which linked to the DNA. In contrast, neither φX174 or M13 DNA nor heat denatured or piperidine -treated Φ29 DNA treated with Proteinase K was able to support the covalent linking between the φ29 terminal protein and dAMP. Only the φ29 DNA-protein complex isolated from the phage particle has been found to be active. Moreover, in vivo studies showed that mixed infection at a nonpermissive temperature with ts2 (gene 2) and ts3 (gene 3) mutants of φ29 produce only ts2 progeny (25). However, mixed infection with nonsense mutants of sus2 and sus3 in a nonpermissive host yielded both types of progeny. Thus, it is likely that the parental terminal protein bound to the template DNA is necessary for recognition by the newly synthesized terminal protein via protein-protein interaction.

Previous in vivo studies suggested that replication of the φ29 DNA starts at both ends, not simultaneously, and occurs by a mechanism of strand displacement (11,12). The data presented here obtained by analysis of the DNA product synthesized in the purified fraction are in good agreement with in vivo data. However, the rate of chain elongation is quite low. The distribution of radionucleotide incorporated into DNA is predominantly at both ends even after 40 min incubation of the reaction mixture (15-20% of the length at both ends). Possibly, one or more factors required for further elongation are missing in the purified fraction. The factor(s) appear to be missing already in the early stage of purification (Fraction II), since the mode of replication was basically the same when fraction II was used for analysis of the DNA product (data not shown). A complementation assay to stimulate DNA replication should enable us to purify this unknown factor(s).

In summary, we have shown that φ29 DNA replicating activity can be copurified with the terminal protein (gp3 protein). This DNA polymerase activity is distinct from all known B. subtilis DNA polymerases and exhibited a remarkable template specificity. It requires φ29 DNA-protein as template. The origin and the nature of this unique DNA polymerase is under investigation.

ACKNOWLEDGMENTS

We are grateful to May-Fen Shih and Laura Weber for their invaluable assistance. This research was supported by National Instute of Health Grant GM28013.

REFERENCES

1. Anderson, D.L., Hickman, D.D., and Reilly, B.E. (1966) J. Bacteriol. 91, 2081-2089.
2. Geiduschek, E.P., and Ito, J. (1982) in Molecular Biology of Bacillus (Dubnau, D., Ed.) Academic Press, New York, pp. 203-245.
3. Mellado, R.P., Moreno, F., Vinuela, E., Salas, M., Reilly, B.E., and Anderson, D.L. (1976) J. Virol. 19, 495-500.
4. Yoshikawa, H., Friedmann, T., and Ito, J. (1981) Proc.

Natl. Acad. Sci. U.S.A. 78, 1336–1340.

5. Escarmis, C., and Salas, M. (1981) Proc. Natl. Acad. Sci. U.S.A. 78, 1446–1450.

6. Salas, M., Mellado, R.P., Vinuela, E., and Sogo, J.M. (1978) J. Mol. Biol. 119, 269–291.

7. Yehle, C.D. (1978) J. Virol. 27, 776–783.

8. Ito, J. (1978) J. Virol. 28, 895–904.

9. Yonofsky, S., Kawamura, F., and Ito, J. (1976) Nature (London) 259, 60–63.

10. Mellado, R.P., Penalva, M.A., Inciarte, M.R., and Salas, M. (1980) Virology 104, 84–96.

11. Harding, N.E., and Ito, J. (1980) Virology 104, 323–338.

12. Inciarte, M.R., Salas, M., and Sogo, J.M. (1980) J. Virol. 34, 187–199.

13. Kornberg, A. (1980) In DNA Replication, W.H. Freeman and Co., San Francisco.

14. Watson, J.D. (1972) Nature New Biol. 239, 197–201.

15. Winnacker, E.L. (1978) Cell 14, 761–773.

16. Rekosh, D.M.K., Russel, W.C., Bellett, A.J.D., and Robinson, A.J. (1977) Cell 11, 283–295.

17. Watabe, K., Shih, M., Sugino, A., and Ito, J. (1982) Proc. Natl. Acad. Sci. U.S.A. 79, 5245–5248.

18. Shih, M., Watabe, K., and Ito, J. (1982) Biochem. Biophys. Res. Comm. 105, 1031–1036.

19. Penalva, M.A., and Salas, M. (1982) Proc. Natl. Acad. Sci. U.S.A. 79, 5522–5526.

20. Hermoso, J.M., and Salas, M. (1980) Proc. Natl. Acad. Sci. U.S.A. 77, 6425–6428.

21. Wright, G.E., Baril, E.F., Brown, V.M., and Brown, N.C. (1982) Nucleic Acid Res. 10, 4431–4440.

22. Shih, M.F., Watabe, K., and Ito, J. (Virology, submitted).

23. Ikeda, J.E., Enomoto, T., and Hurwitz, J. (1982) Proc. Natl. Acad. Sci. U.S.A. 79, 2442–2446.

24. Tamanoi, F., and Stillman, B.W. (1982) Proc. Natl. Acad. Sci. U.S.A. 79, 2221–2225.

25. Salas, M., Mellado, R.P., Vinuela, E., and Sogo, J.M. (1978) J. Mol. Biol. 119, 269–291.

26. Green, P.J., Betlach, M.C., Boyer, H.W., and Goodman, H.M. (1974). In Methods in Molecular Biology (R. Wickner, Ed.), Vol. 9, pp. 87–111.

Mechanisms of DNA Replication and Recombination, pages 245–254
© 1983 Alan R. Liss, Inc., 150 Fifth Avenue, New York, NY 10011

BACTERIOPHAGE N4 DNA REPLICATION[1]

J.K. Rist, D.R. Guinta, A. Sugino[2,3]
J. Stambouly, S.C. Falco[4] and L.B. Rothman-Denes

Departments of Biophysics and Theoretical Biology
and of Biochemistry, The University of Chicago,
Chicago, IL 60637 and
Laboratory of Genetics, National Institute of
Environmental Health Sciences, Research Triangle Park,
NC 27709[2]

ABSTRACT E. coli and phage genes required for bacteriophage N4 DNA replication have been identified.
N4 DNA replication is independent of the host replication genes dna A,B,C,E and G and requires the dna F,
gyr A and B, lig and pol A exo activities. N4 DNA
replication requires the expression of N4 early and
middle genes. At least five N4 gene products, dbp,
dnp, dns, exo and vrp are directly required for N4
DNA synthesis. We have developed an in vitro system
to study N4 DNA replication. DNA synthesis catalysed
by extracts of N4 infected cells is dependent upon
exogenous N4 DNA. Extracts of cells infected with N4
mutants in the dnp, dbp and exo genes do not synthesize DNA unless they are complemented with the missing
wild type protein. This system has been used to purify the N4 coded DNA polymerase (dnp), DNA binding
protein (dbp) and exonuclease (exo). The in vitro
replication system is specific for N4 DNA and synthesis begins at or near the ends of the genome proceeding toward the middle. Its relationship to in vivo
N4 replication is discussed.

[1]This work was supported by grants AI 12575 and CA
19265 from the National Institutes of Health to L.B. R-D.
[2]Present address: Department of Molecular Genetics,
University of Georgia, Athens, Georgia
[3]Present address: E. I. DuPont de Nemours Central
Research, Wilmington, Delaware.

INTRODUCTION

The genome of coliphage N4 is a double-stranded, linear DNA molecule, 71 Kb in length containing 400 base pair direct terminal repeats (1). The left end of the molecule has a seven base 3' overhang (CCATAAA, L. Haynes, unpublished) while the right end is not unique. Five major ends, differing from each other by approximately 10 bases in length have been detected (L. Haynes, unpublished).

N4 is unique among double-stranded DNA containing bacteriophages in that transcription of its genome requires the activity of three distinct DNA-dependent RNA polymerases (2). Early N4 transcription is carried out by a rifampicin-resistant DNA-dependent RNA polymerase which is present in virions and injected into the host cell with the DNA (3,4). This enzyme, composed of one polypeptide 320000 MW, is responsible for transcription of the leftmost 10% of the genome (2,5). In vitro, the N4 virion RNA polymerase transcribes only denatured N4 DNA (6) and recognizes its promoters on this template with in vivo specificity (L. Haynes, unpublished). N4 middle transcription requires the synthesis and activity of three N4 early proteins of molecular weights 40,000, 30,000 and 15,000 (2,7). The 40,000 and 30,000 molecular weight polypeptides constitute a second rifampicin-resistant RNA polymerase (8). Finally, transcription of the late region of the N4 genome requires the activity of the E. coli RNA polymerase in an, as yet, unelucidated form (2).

Soon after infection of E. coli K12 strains, N4 shuts off host DNA replication through a mechanism that requires phage protein synthesis. N4 DNA replication starts 7 to 10 min post infection at 37° (9).

Towards the aim of elucidating the mode of replication of the N4 genome, we have determined the in vivo host and phage requirements, developed an in vitro system and have made use of this system to purify phage coded proteins required for replication of N4 DNA.

RESULTS

Host and Phage Gene Products Required for N4 DNA Replication.

In order to determine the host requirements for N4 DNA replication, various E. coli dna ts mutants were infected at both the permissive and restrictive temperatures and N4

DNA synthesis was measured by pulse labeling the cultures with [^{3}H] thymidine. Table 1 summarizes the results obtained.

TABLE 1
HOST AND PHAGE REQUIREMENTS FOR N4 DNA SYNTHESIS[a]

Host	Phage	^{3}Hthymidine incorporation into N4 DNA	N4 progeny production
wild type	N4$^+$	+	+
dnaA	N4$^+$	+	+
dnaB	N4$^+$	+	+
dnaC	N4$^+$	+	+
dnaE	N4$^+$	+	+
dnaF	N4$^+$	–	–
dnaG	N4$^+$	+	+
polA1	N4$^+$	+	+
polAex1	N4$^+$	+	–
lig	N4$^+$	–	–
gyrB	N4$^+$	–[b]	nd
wild type	N4dnsam12	–	–
wild type	N4dnpam25	–	–
wild type	N4dbp33am7	–	–
wild type	N4exoD11	+[c]	reduced[d]

[a][^{3}H]thymidine incorporation into N4 DNA was determined as described in ref. 9. [b]Cells were shifted to non permissive temperature 15 min after replication had started. [c][^{3}H]thymidine incorporation is normal up to 60 min after infection at 37°C, then DNA synthesis stops. [d]Burst is 30 while wild type is 1000.

N4 DNA replication does not require the activity of the E. coli dnaA, dnaB, dnaC, dnaE and dnaG gene products. In contrast, no N4 specific thymidine incorporation was observed after infection of E. coli strains carrying ts alleles of dnaF, lig and gyrB at the restrictive temperature, indicating that these functions are required for N4 DNA replication. The requirement for E. coli DNA gyrase activity was corroborated by the use of coumermycin and oxolinic acid in a wild type strain. Both drugs inhibited thymidine incorporation when added after the onset of N4 DNA replication (not shown).

The involvement of polI was studied by measuring thymidine incorporation, burst size and by alkaline-sucrose gradient analysis of the newly synthesized DNA. While the pattern of thymidine incorporation was normal after infection of polAl, polAexl and polAl2 strains, progeny phage were produced only in polAl infected cells. Further analysis of the fate of the newly synthesized DNA in polAexl and polAl2 infected cells showed that small DNA fragments accumulated in cultures shifted to the non-permissive temperature after DNA synthesis had begun (not shown).

We have isolated N4 mutants defining several genes required for replication. No DNA synthesis was detected in cells infected with mutants in three of these genes, dns, dbp and dnp. The function of dns is unknown while, as will be shown below, dnp codes for the N4 DNA polymerase and dbp for a DNA binding protein (Table 1).

The involvement of two other N4 genes in N4 DNA replication is more complex. When cells are infected at the permissive temperature with a conditional-lethal, temperature-sensitive mutant in the N4 virion RNA polymerase, expression of N4 early genes and N4 DNA synthesis occurs normally. However, if cells are shifted to the restrictive temperature after the onset of N4 DNA replication, N4 DNA synthesis shuts off indicating that the N4 virion RNA polymerase is required for N4 DNA replication. The activity of the N4 virion RNA polymerase is not simply required for the continuous synthesis of an early gene product since addition of chloramphenicol does not affect replication. Finally, N4 mutants in a $5' \rightarrow 3'$ exonuclease show a temperature-dependent DNA arrest phenotype (10).

Development of an in vitro N4 DNA replication system.

To determine and characterize the components required and to study the mechanism of N4 DNA replication, we have developed a cell-free system using N4 infected polAl E. coli. As seen in Table 2, the system is dependent on the addition of exogenous DNA and is highly specific for native N4 DNA.

TABLE 2
REQUIREMENTS FOR IN VITRO N4 DNA REPLICATION[a]

Conditions	%
Complete	100
- N4 DNA	3
- Mg Cl$_2$	0.6
- dNTPs	3.4
- N4 DNA + λ DNA	14
- N4 DNA + φX174RF DNA	3

[a]E. coli D110 (polA1) is infected with N4, and cells are collected 20 min later by centrifugation. Cells are lysed gently with lysozyme, 0.5% Brij 58, and 1 M NaCl. The lysate is centrifuged 100,000 x g 45 min, and $(NH_4)_2SO_4$ is added to 50% saturation. The precipitate is pelleted, resuspended in 1/500 original volume Buffer A (50 mM Tris pH 8, 10 mM βME, 1 mM EDTA, 10% glycerol), dialysed against Buffer A for 2 hr, and stored at -70°C. The reaction mixture (0.1 ml) contained: 66 mM Tris-HCl pH8, 25 mM KCl, 10 mM MgCl$_2$, 2 mM DTT, 1 mM spermidine-HCl 0.1 mg/ml BSA, 5 mM ATP, 0.01 mM each dATP, dCTP, dGTP and dTTP ([^{3}H] dTTP 2000 cpm/pmole), 0.1 mM each CTP, UTP and GTP, 20 µg/ml DNA and extract. Incubation was for 20 min at 30°. 100% corresponds to 80 pmol dTMP incorporated.

In order to further characterize the specificity of the in vitro N4 replication reaction, the in vitro products were hybridized to nitrocellulose filters containing N4 DNA fragments generated by HpaI endonuclease. As seen in Figure 1, initiation of replication is not random; in fact, it is very specific. Initiation occurs in both terminal HpaI fragments G (3.8 Kb) and K (2.1 Kb), occurring more frequently in HpaI G. DNA synthesis then proceeds toward the middle of the genome, rightward from HpaI K and leftward from HpaI G.

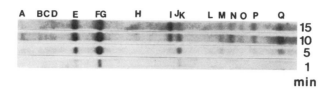

FIGURE 1. Origin and direction of in vitro N4 DNA replication. Top. HpaI restriction map of N4 DNA (1). Arrows indicate origin and direction of replication. Bottom. N4 DNA was incubated with extract as described in Table 2 except that α^{32}PdCTP (Amersham, 2000 Ci/mmole, 50 μCi/ml) was used as label. At the indicated incubation times aliquots were phenol extracted, digested with HpaI endonuclease, hybridized to nitrocellulose blots containing size-fractionated, HpaI-restricted N4 DNA and autoradiographed as described in ref. 1.

N4-coded Proteins Required in the In Vitro System.

N4 replication activity was measured in extracts made from cells infected with N4 mutants in genes required for in vivo DNA synthesis. Table 3 shows that extracts from N4 am12 infected cells have wild type levels of replicating activity demonstrating that the function of the N4 dns gene product is not required in the in vitro system. In contrast, extracts from N4 am25, N4 33am 7 or N4 amD11 infections do not support in vitro replication, indicating that the wild type products of these genes are required for in vivo as well as for in vitro replication.

Replication activity of an inactive mutant-infected extract can be restored by a wild type protein fraction containing the missing component. We have used this complementation assay to purify the protein component missing in 33 am7 infections and found it to be an N4-coded single-stranded DNA binding protein (Table 3). We have also purified the N4-coded DNA polymerase and an N4 coded 5' → 3'

TABLE 3

REPLICATION ACTIVITY OF N4 MUTANT INFECTED CELL EXTRACTS[a]

Infecting phage	Supplemented with	Activity, pmoles dTMP incorporated
--	--	0.8
N4+	--	23
dnp am25	--	0
dnp am25	N4 DNA polymerase	30
exo amD11	--	1.5
exo amD11	N4 exonuclease	23
dbp 33am7	--	1.3
dbp 33am7	N4 DNA binding protein	19
dns am12	--	24

[a]Extracts were prepared and activity was measured as described in Table 2. Where indicated, 0.4 μg N4 DNA binding protein, 0.19 units N4 DNA polymerase (1 unit incorporates 10 nmoles total nucleotide into acid insoluble material at 30°C in 15 min), or 1.5 units N4 exonuclease (1 unit produces 1 nmol acid soluble material from N4 DNA in 15 min at 37°C) were added.

exonuclease which also restore activity to wild type levels when added to their respective mutant extracts (Table 3).

DISCUSSION

Coliphage N4's unusual genome structure, unique transcriptional program among DNA containing phages and the involvement of the virion RNA polymerase in DNA synthesis prompted our study of N4 DNA replication.

We have found that N4 codes for most functions required for replication of its genome depending only on the E. coli ligase, gyrase and DNA polymerase I 5' → 3' exonuclease activities. Five N4 genes required for replication have been identified genetically. These are

dnp, dbp, exo, vrp and dns which code respectively for a
DNA polymerase, a DNA binding protein, a 5' → 3' exonuclease,
a virion associated DNA-dependent RNA polymerase and an
unknown function.

Two of the N4-coded gene products required for replication: the N4 DNA polymerase and the N4 exonuclease have
been purified by conventional methods (J. K. Rist, D.
Guinta, unpublished). The development of a cell-free
system for N4 DNA replication has allowed us to identify a
third gene product since the activity of a N4 33am7 infected cell extract can be restored to wild type levels by
addition of wild type protein fractions containing the
missing factor. After purification to apparent homogeneity
using this complementation assay, we have identified the
product missing in N4 33am7 infected cell extracts as an
N4 coded DNA binding protein.

The in vitro system does not require the product of
the N4 dns gene. This gene product may be required for a
post-initiation event or a reaction which is not occurring
in vitro. We have not yet attempted to develop an in
vitro system dependent upon the N4 virion associated RNA
polymerase because mutants in this function have pleio-
tropic effects upon gene expression. The eventual reconstitution of the replication system using purified
components requires the identification of additional genes
required for replication and the biochemical fractionation of the system.

The in vitro N4 replication system shows specificity
for initiation (Fig. 1), although its relationship to the
in vivo origin remains to be demonstrated. To define the
in vivo origin, we have isolated replicative intermediates
(J. K. Rist, unpublished). Labelled replicating N4 DNA,
centrifuged to equilibrium in neutral CsCl gradients has a
higher buoyant density than virion N4 DNA suggesting the
presence of single-stranded DNA. Analysis of these
molecules by electronmicroscopy reveals linear duplex
molecules with single stranded branches and/or single-
stranded tails. These observations suggest that bacteriophage N4 DNA may replicate by a strand displacement
mechanism.

Bacillus subtilis phage φ29 (11-13), adenovirus
(14,15) and parvoviruses (16,17) replicate their DNA by
strand-displacement. Both φ29 and adenovirus DNAs contain
proteins covalently bound to the 5' ends of their DNAs
(18-20) and initiation of replication requires newly
synthesized terminal protein which is used as a primer

(21-24). Helper-dependent parvoviruses initiate replication using the 3' end of a terminal hairpin as a primer (17). In contrast, N4 virion DNA does not have a covalently bound terminal protein nor a terminal hairpin. Further experiments are underway to determine the mechanism of initiation of N4 DNA replication.

REFERENCES

1. Zivin R, Malone C, Rothman-Denes LB (1980). Physical map of coliphage N4 DNA. Virology 104:205.
2. Zivin R, Zehring W, Rothman-Denes LB (1981). Transcriptional map of bacteriophage N4: location and polarity of N4 RNAs. J Mol Biol 152:335.
3. Falco SC, Vander Laan K, Rothman-Denes LB (1977). Virion-associated RNA polymerase required for bacteriophage N4 development. Proc Natl Acad Sci USA 74:520.
4. Falco SC, Rothman-Denes LB (1979). Bacteriophage N4-induced transcribing activities in Escherichia coli. I. Detection and characterization in cell extracts. Virology 95:454.
5. Falco SC, Zehring W, Rothman-Denes LB (1980). DNA-dependent RNA polymerase from bacteriophage N4 virions. Purification and characterization. J Biol Chem 255:4339.
6. Falco, SC, Zivin R, Rothman-Denes LB (1978). Novel template requirements of N4 virion RNA polymerase. Proc Natl Acad Sci USA 75:3220.
7. Zehring W, Falco SC, Malone C, Rothman-Denes LB (1983). Bacteriophage N4-induced transcribing activities in E. coli. III. A third cistron required for N4 RNA polymerase II activity. Virology, in press.
8. Zehring W, Rothman-Denes LB (1983). Purification and characterization of coliphage N4 RNA polymerase II activity from infected cell extracts. J Biol Chem, in press.
9. Rothman-Denes LB, Schito GC (1974). Novel transcribing activities in N4-infected Escherischia coli. Virology 60:65.
10. Guinta D (1983). Ph.D. Thesis, The University of Chicago.
11. Harding, NE, Ito J. (1980). DNA replication of bacteriophage φ29: characterization of the intermediates and location of the termini of replication. Virology 104:323.

12. Inciarte MR, Salas M, Sogo JM (1980). The structure of replicating DNA molecules of Bacillus subtilis phage φ29. J Virol 34:187.
13. Sogo JM, Garcia JA, Penalva MA, Salas M (1982). Structure of protein-containing replicative intermediates of Bacillus subtilis phage φ29. Virology 116:1.
14. Lechner RL, Kelly Jr. TJ (1977). The structure of replicating adenovirus 2 DNA molecules. Cell 12:1007.
15. Winnacker EL (1978). Adenovirus DNA: structure and function of a novel replicon. Cell 14:761.
16. Hauswirth WW, Berns KI (1977). Origin and termination of adeno-associated virus DNA replication. Virology 78:488.
17. Berns KI, Hauswirth WW (1982). Organization and replication of parvovirus DNA. In Kaplan AS (ed): "Organization and Replication of Viral DNA," CRC press, p. 3.
18. Salas M, Mellado RF, Vinuela E, Sogo JM (1978). Characterization of a protein covalently linked to the 5' termini of the DNA of Bacillus subtilis phage φ29. J Mol Biol 119:269.
19. Ito J (1978). Bacteriophage φ29 terminal protein - its association with the 5' termini of the φ29 genome. J Virol 28:895.
20. Rekosh DM, Russell, WC, Bellet AJD (1977). Identification of a protein linked to the ends of adenovirus DNA. Cell 11:283.
21. Penalva MA, Salas M (1982). Initiation of phage φ29 DNA replication in vitro: formation of a covalent complex between the terminal protein, p3 and 5'-dAMP. Proc Natl Acad Sci USA 79:5522.
22. Lichy, JH, Horwitz, MS, and Hurwitz J (1981). Formation of a covalent complex between the 80,000-dalton adenovirus terminal protein and 5'-dCMP in vitro. Proc Natl Acad Sci USA 78:2678.
23. Challberg MD, Ostrove JM, Kelly Jr. TJ (1982). Initiation of adenovirus DNA replication: detection of covalent complexes between Nucleotide and the 80.000 kilodalton terminal protein. J Virol 41:265.
24. Tamanoi, F, Stillman, BW (1982). Function of the adenovirus terminal protein in the initiation of DNA replication. Proc Natl Acad Sci USA 79:2221.

IV. REPLICATION OF *ESCHERICHIA COLI* CHROMOSOME AND PLASMIDS

Mechanisms of DNA Replication and Recombination, pages 257–273
© **1983 Alan R. Liss, Inc., 150 Fifth Avenue, New York, NY 10011**

THE Escherichia coli ORIGIN OF REPLICATION : ESSENTIAL[1]
STRUCTURE FOR BIDIRECTIONAL REPLICATION

Mituru Takanami, Satoshi Tabata, Atsuhiro Oka,
Kazunori Sugimoto, Hitoshi Sasaki

Institute for Chemical Research, Kyoto University
Uji, Kyoto 611, Japan

Seiichi Yasuda, Yukinori Hirota

National Institute of Genetics
Mishima 411, Japan

ABSTRACT: The 245 base-pair oriC sequence of the
E.coli K-12 chromosome was subjected to localized
mutagenesis in vitro which uses sodium bisulfite,
and a large number of mutants carrying a single to
multiple GC to AT changes were isolated. The
replicating ability of these mutants was assayed by
introducing into cells, and the correlation between
the position of mutation and replicating function
was analysed. Combining the results obtained from
this and previous studies together, we concluded
that oriC provides multiple interaction sites
precisely separated by spacer sequences, and
initiation factors cooperatively interact on these
sites to form an initiation complex.
   The functional role of the 245 bp oriC sequence
was investigated by using an in vitro replication
system and various Ori + and Ori − plasmids. The
result of analysis clearly indicated that DNA
replication started at a region near, but outside,
the 245 bp oriC sequence, and proceeded bidirectionally. The replicating function was reduced

1. This work was supported by Grants from the
Ministry of Education, Science and Culture of Japan.

by mutations introduced only in oriC, but not by
sequence replacements in the flanking regions. The
result indicates that the 245 bp oriC sequence
contains information enough for directing
bidirectional replication in the vicinity of oriC.
It was also noted that initiation of DNA synthesis
at the specific region required the dnaA-
complementing fraction from cells harboring a dnaA-
carrying plasmid. Based on the above observations, a
possible model for bidirectional replication
directed by oriC is discussed.

INTRODUCTION

The replication origin of the E.coli K-12 chromosome has
been isolated as autonomously replicating molecules (oriC
plasmid)(1-5). We have cloned it on the pBR322 vector, as
replicating function of oriC can be assayed by introducing
it into PolA⁻ cells(5,6). By using this assay procedure (Ori
assay), deletion was introduced from both sides of the oriC
carrying segment, and finally the DNA region essential for
replicating function was localized to a sequence of minimum
232 base-pairs(bp) and maximum 245 bp in length(6). As the
next step, we introduced mutations at three restriction
sites which located within oriC (BamHI, AvaII and HindIII
sites) and found that Ori function was destroyed by either
insertion or deletion of short sequences at these sites, but
not by base-substitution(7,8). On the other hand, Ori⁻
mutants directed by base subtitions were isolated from
other parts of oriC(9).These results led to a concept that
oriC contains two categories of sequences, one specifying
interaction with initiation factors(recognition sequences)
and the other spacing the recognition sequences in
appropriate distances(spacer sequences)(8).
In this paper, a large number of base-substitution
mutants were isolated by localized mutagenesis in vitro, and
the correlation between the Ori phenotype and the position
of substitution was systematically investigated. As a
consequence, we could deduce more precise features on the
structural organization of oriC.
Recently, an in vitro replication system of oriC
plasmids was constructed by Fuller et al.(10). This has
enabled us to examine the function of oriC directly. By
using this system, Kaguni et al.(11) have shown by electron
microscopy that oriC plasmids replicate bidirectionally
starting around oriC. We also investigated the functional

role of the 245 bp oriC sequence by using a similar in vitro
system and various Ori⁺ and Ori⁻ mutants previously
constructed. As a result, we could define the functional
role of the oriC sequence more precisely.

MATERIALS AND METHODS

oriC plasmids

OriC⁺ plasmids used were pTSO182(6), pTSO196(6), pTSO290(6),
pTSO293(6), pTSO236(6), pKA22(8) and pMY129 (constructed by
M.Yamada). The chromosomal moieties carried by these
plasmids are shown in Fig.1. OriC⁻ mutants which showed
typical Ori⁻ phenotype and used for in vitro replication
analysis were pTSO279(7 bp deletion at the BglII site)(6),
pTSO202(4 bp insertion at the BamHI site)(6), pTSO207(15 bp
deletion at the BamHI site)(6), pTSO190(4 bp insertion at
the HindIII site)(6), pTSO209(5 bp deletion at the HindIII
site) (6), pKA32 (3 bp insertion at the AvaII site)(8) and

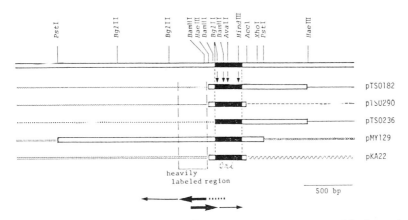

Fig.1 Restriction maps of oriC plasmids. Filled and open
boxes represent the oriC sequence and remaining chromosomal
moieties, respectively, and other lines indicate pBR322
moieties. pTSO196 and pTSO293 are identical to pTSO182 and
pTSO290, respectively, except that restriction sites in the
outside of oriC had been modified(6). Differences in pBR322
sequences adjacent to chromosomal moieties were shown by
changing the shapes of lines. Relevant restriction sites in
the vicinity of oriC were indicated, and sites at which
deletion or insertion had been introduced were shown by
vertical arrows. Horizontal arrows show the approximate
region where bidirectional replication starts.

pKA62(3 bp deletion at the AvaII site)(8)(see Fig.1 for
restriction sites). Other plasmids used for construction of
base-substitution mutants were pAO120(12), pAO097(12),
pAO117(12), and M13mp7.

Localized mutagenesis by sodium bisulfite

For isolation of Ori⁻ mutants, the oriC area was divided
into three parts: the left BglII to BamHI sites(positions 22
to 92), the BamHI to HindIII sites(positions 92 to 243), and
the HindIII to AccI sites(positions 243 to 290), and
heteroduplex molecules carrying these regions as
single-stranded gaps were constructed, as described
previously(13). After mutagenesis by sodium bisulfite under
relatively weak conditions, Ori⁻ clones were isolated.
Alternatively, the area encompassing oriC (positions 41 to
422) was first cloned on M13mp7 vector. Phage DNAs carrying
respective strands were hybridized with vector RF DNA, and
the resulting heteroduplexes were treated with sodium
bisulfite under relatively strong conditions, and after
repairing, transformation was carried out. RF DNA was
prepared as a mixture, from which the oriC area(-41 to 290)
was cut out, and re-inserted into pBR322 vectors for Ori
assay. Ori⁺ plasmids were isolated, and the positions of GC
to AT substitutions were determined. The detail of the
conditions will be published elsewhere(12).

Ori assay

When Ori phenotypes were examined, a oriC plasmid was
introduced into both PolA⁻(W3110 polA) and PolA⁺ cells
(C600) by transformation. Under the conditions, Ori⁺
plasmids gave Apʳ(or Kmʳ) transformants of polA⁻ cells as
well as of polA⁺ cells. In contrast, typical Ori⁻ plasmids
previously constructed and pBR322 vector yielded only
transformants of polA⁺ cells. Most of mutants obtained by
base substitution, however, showed intermediate phenotypes:
upon transformation into polA⁻ cells, colonies were formed
at reduced rates, indicating occurrence of some defect in
Ori function. Since the colony sizes were different with
mutants, we devided mutants into five classes from strong
Ori⁻ (-5) to weak Ori⁻(-1) for convenience, depending on the
growing rate.

Enzyme fractions and reaction conditions for in vitro
replication

The enzyme fractions used for in vitro replication analysis
were essentially identical to those described by Fuller et
al.(10), except for their sources. Fraction II was prepared
either from E.coli HMS83 (F⁻ polA polB rha lys thyA lacZ
str)(13) or JE107251 (HfrP4X8 dnaA725 thy str)(14), and
fraction III(a dnaA complementing fraction) was prepared

from E.coli JE6087(F⁻ Δ proB lac) which harbors a
dnaA-carrying plasmid pSY405(14). The chromosomal gene
carried by pSY405 is assumed to be only dnaA, as additional
chromosomal sequences in pSY405 were about 400 bp at the
5'-side and a part of the dnaN gene at the 3'-side of the
dnaA gene.

The reaction mixture(25 μl/tube) contained 40 mM Hepes
buffer(pH 7.6), 11 mM Mg acetate, 2 mM ATP, 500 μM each of
CTP, GTP and UTP, 21 mM creatine phosphate, 2.5 μg creatine
kinase, 50 μM each of dATP, dGTP, and dCTP, 20 μM dTTP
containing 5-methyl(^{3}H)dTTP(220 cpm per pmol of total dTTP),
1.25 μg bovine serum albumin, and 5% polyvinyl alcohol(W/V).
0.2 μg of oriC plasmids, 0.12 mg of fraction II from either
HMS83(polA polB) or JE107251(dnaA725) and 0.05 mg of
fraction III were added. When fraction II from the dnaA
strain was used, dnaA protein was inactivated by incubation
for 10 min at 37°C in the presence of 5 mM ATP, 50 mM
creatine phosphate and 0.2 mg/ml creatine kinase. Incubation
was carried out at 30°C for indicated periods. Reaction was
terminated by 5% TCA, precipitates were mounted on glass
filters, and radioactivity was determined.

Analysis of in vitro reaction products

dNTPs in the above reaction mixtures were replaced by 10 uM
each of four (α^{32}P)dNTPs, and 1 μM to 20 μM ddTTP were added.
Incubation was made for 10 min at 30°C, unless otherwise
noted. Reaction was terminated by adding SDS to 1%, and
after shaking with 80% phenol, aqueous layers were passed
through Agarose A5m columns(0.6 cm x 20 cm). The DNA
fraction was collected, and digested with restriction
enzymes. Resulting fragments were resolved by 5%
polyacrylamide gel electrophoresis, and identified by
autoradiography.

Separation of labelled strands

The area from the PstI site in the pBR322 moiety and the
PstI site at the left of oriC in pTSO182 was cloned on the
PstI site of M13mp7. The resulting phage DNAs carrying
leftward(L) and righward(R) strands, respectively, were
fixed on either nitrocellulose or DBM paper (about 0.5
μg/cm). Synthesized DNA on pTSO182 in the presence of
ddTTP(ddTTP/dTTP = 1 or 1.5) was heat-denatured and added
onto the filters, and hybridization was carried out either
in the presence of 50% formamide-0.5% SDS-X5 SSC (42°C) or
0.5% SDS-X5 SSC(65°C). After washing filters, hybridized DNA
was dissociated by incubation for 2 hr at 50°C in the
presence of 70% formamide--1 mM EDTA. To the dissociated DNAs
was added M13mp7 DNA carrying either L-strand or R strand
and After annealing, products were digested with HaeIII,
SalI and BglII, and electrophoresed.

RESULTS AND DISCUSSION

Analysis of Sequence Organization by construction of base substitution mutants

The entire region from positions -41 to +290 encompassing the 245 bp oriC sequence was subjected to in vitro mutagenesis by sodium bisulfite under relatively strong conditions. Ori$^+$ mutants were isolated, and positions of GC to AT replacements were determined. Alternatively, oriC was divided into three areas, and each area was subjected to localized mutagenesis by the same mutagen under relatively weak conditions, and Ori$^-$ mutants carrying a single to a few GC to AT changes were isolated, as in a previous paper(9). The results are summarized in Fig.2, in which number indicated below the sequence are frequencies of GC to AT replacements at each position in Ori$^-$ mutants independently isolated. In the outside of oriC, replacements at 34 positions out of 37 GC pairs were obtained, indicating the frequency of replacement is 92%. Within oriC, however, replacements at 70 positions out of 101 GC pairs were obtained. Assuming that the frequency of replacement is identical in the outside and inside of oriC, GC pairs at 23 positions is expected to be unreplaceable positions among the remaining 30 positions. Experimentally, GC pairs at 17 positions were identified to be unreplaceable, and replacement at these positions results in some defect of Ori function. These positions are shown in Fig.2 by thick arrows.

Although these unreplaceable GC pairs were found in many different parts of oriC, distribution was not completely random and two or three unreplaceable GC pairs were often lying close. The areas among unreplaceable GC pairs were filled with replaceable GC pairs, so that replaceable GC pairs appeared to be clustering. The BamHI, AvaII and HindIII sites in which insertion or deletion of short sequences destroy Ori function(8) were just located within these clusters of replaceable GC pairs. It should be mentioned that all these unreplaceable GC pairs were found in the "consensus" sequence, defined by Ziskind et al.(15), except the one at position 196.

A characteristic of defective mutants caused by base-substitution was that their phenotype was generally much weaker than that of the typical Ori$^-$ mutants, previously constructed by introducing deletion or insertion, and upon transformation onto PolA$^-$ cells, the formation of small colonies was observed. It was also significant that

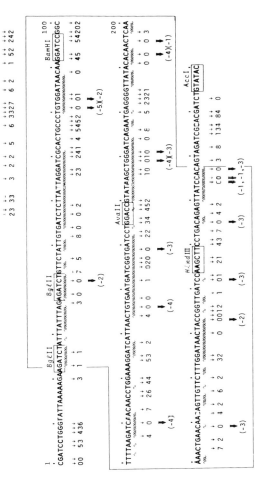

Fig.2 Effect of GC to AT substitutions on Ori phenotypes. The sequence from positions -41 to 290 is shown, and the 245 bp oriC sequence was boxed. GATC at positions -3 to 1 is an artificial insert, used for destruction of the BamHI site. Number indicated by arrows below the sequence are frequencies of GC to AT substitutions detected at each position in independently isolated Ori⁻ mutants. Zero means no replacement was identified. Among the 30 GC positions at which substitution was not isolated, replacement at each of 17 positions was found to cause some defect in Ori function. These essential GC positions are shown by thick arrows, and effect on Ori phenotypes was indicated by dividing from strong Ori⁻ (-5) to weak Ori⁻ (-1).

the addition of two or more changes each showing a weak Ori⁻
phenotype results in more strong Ori⁻ phenotypes. For
instance, the Ori⁻ phenotype of mutants at positions 264 and
265 was ranked as (-1), respectively, but that of the mutant
carrying two changes at these sites was ranked as (-4).

The result can be interpreted by the view previously
proposed(8): oriC contains two categories of sequences, one
specifying interaction with initiation factors (recognition
sequence), and the other spacing the recognition sequences
in fixed distances(spacer sequence). In each recognition
sequence, several residues may participate in interaction
with an initiation factor, so that defectiveness becomes
more apparent by substitution of not a single but multiple
residues. In the spacer regions, on the contrary, not
base-substitution, but insertion or deletion of short
sequences results in more serious effect, as it alters the
distances between the recognition sequences. So far little
is known about how many factors are actually involved in the
initiation reaction on oriC, but oriC is large enough to
provide interaction sites for several components. Our base
substitution data also suggest that several recognition
sequences are present in oriC. We assume that multiple
factors cooperatively interact on oriC to form an active
"initiation complex".

Analysis of function by an in vitro replication system

The extent and time course of DNA synthesis directed by
various oriC plasmids are shown in Fig.3. As has been
observed by Fuller et al.(10), incorporation by oriC
plasmids steeply increased after a few minutes lag, and
reached to a plateau of which the level was about 50-70% of
the template added(Fig.3 A). In contrast, the extent of DNA
synthesis with mutant plasmids was much lower than that of
the wild type, although their levels depended on the type
and positions of mutation(Fig.3 B). The result is consistent
with that of the in vivo Ori assay(6-8). By omitting
fraction III, the extent of DNA synthesis directed by Ori⁺
and Ori⁻ plasmids was markedly reduced, respectively.

To identify the site and direction of replication,
incorporation was limited by adding a chain terminator ddTTP
together with (α^{32}P)dNTPs. The products were digested with
appropriate restriction enzymes, and labelled fragments were
identified by polyacrylamide gel electrophoresis, followed
by autoradiography. Restriction patterns generated from
pTS0182 at increasing levels of ddTTP are shown in Fig.4 A
a-e. At higher ratios of ddTTP/dTTP, only a few fragments

were heavily labelled. The specific activities of fragments estimated from the label and length were plotted along the sequence of pTSO182, as shown in Fig.4 B. It is clear that the replication starts from the left side of oriC and proceeds symmetrically in both directions. Essentially identical labelling patterns were obtained without ddTTP, but by a brief incubation at reduced substrate concentrations. The bidirectional replication at the left side of oriC was also observed with other OriC+ plasmids. As an example, labelling patterns obtained from pMY129 are shown in Fig.5. Since these Ori+ plasmids carry different sequences in the flanking regions of oriC (see Fig.1) and the extent of DNA synthesis was reduced only by mutations

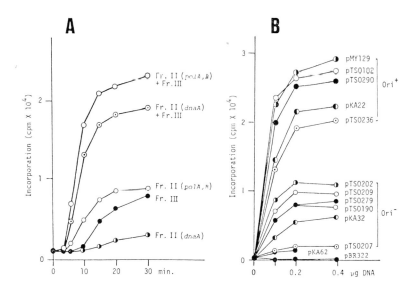

Fig.3 Replication of oriC plasmids in the in vitro system.
A: Kinetics of DNA synthesis. pTSO182(Ori+)(0.2 ug each) and indicated fractions were added to the reaction mixtures containing (³H)dTTP, described in the Method section, and incubated at 30°C at indicated periods.
B: The extent of DNA synthesis directed by Ori+ and Ori⁻ plasmids. The reaction mixtures were identical to those in (A), except that indicated amounts of various plasmids, 0.12 mg of fraction II from HMS83(polA polB) and 0.05 mg of fraction III were added. Incubation was carried out for 30 min. at 30°C.

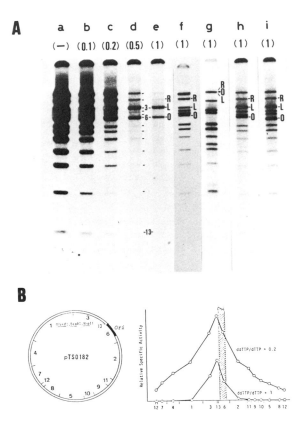

Fig.4 (A) DNA synthesis directed by oriC plasmids in the presence of ddTTP. The composition of reaction mixtures and conditions for analysis were given in the Method section. (a–e): EcoRI–XhoI–HinfI digests of DNA formed on pTS0182. (f): EcoRI–XhoI–HinfI digest of DNA formed on pTS0182 without fraction III. (g): SalI–XhoI–HinfI–PstI digest of pTS0279. (h): EcoRI–XhoI–HinfI digest of pTS0202. (i): EcoRI–XhoI–HinfI digest of pTS0209. The ddTTP/dTTP ratios in reaction mixtures were indicated above the columns. Labels R,L and 0 by the side of columns indicate fragments generated from the right and left of oriC, and those containing oriC.

(B) Relative specific activities of fragments in (A) d & e were estimated from label and chain length, and plotted along the linear map of pTS0182. Numbers correspond to those on the circular map. Shaded boxes represent oriC.

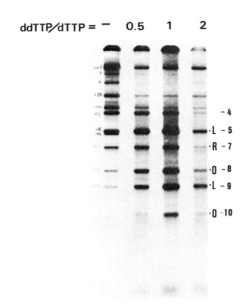

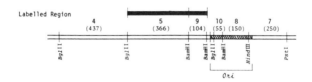

<u>Fig.5</u> The starting site of DNA replication on pMY129. DNA synthesized on pMY129 at the indicated ratios of ddTTP/dTTP were digested with <u>Pst</u>I, <u>Hind</u>III, <u>Bgl</u>II and <u>Bam</u>HI. Labelled fragments were shown above the restriction map. The conditions for labelling and restriction analysis of labelled fragments were identical to those given in the legend to <u>Fig.4</u>. Aliquots from each sample have been analysed to adjust the exposure time. Number by the side of columns correspond to those on the restriction map. Labels R, L and O by the side of column indicate fragments generated from the right and left of <u>oriC</u> and those containing <u>oriC</u>.

introduced in oriC, we concluded that the oriC sequence contained information enough to induce bidirectional replication at its adjacent region.

To investigate the initial incorporation site more precisely, the products synthesized with pTSO182 at higher ddTTP concentrations were digested into more small pieces. Typical autoradiograms are shown in Fig.6. Labels were mainly identified in five bands at ddTTP/dTTP=1, and in two bands at ddTTP/dTTP=2. The labelled regions are shown on the restriction map(Fig.6). The labelling patterns were not changed by prolonged incubation, and identical results were obtained by using fraction II from either the HMS83(polA polB) or dnaA strain. At ddTTP/dTTP=2, the relative specific activities of fragments roughly estimated were highest in fragment 16 and decreased toward left. Little radioactivity was detected in fragment 18. It is therefore likely that

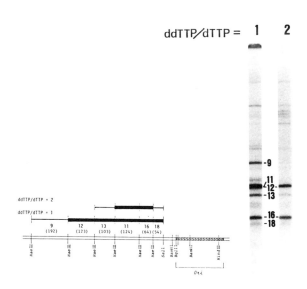

Fig.6 The starting site of DNA replication on pTSO182. DNA formed on pTSO182 at ddTTP=dTTP ratios of 1 and 2 were digested by HaeIII-SalI. Labelled fragments were indicated above the restriction map. Number by the side of columns correspond to those on the restriction map.

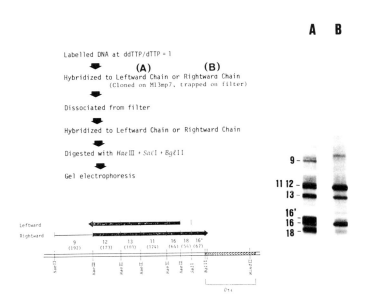

Fig.7 Strand separation of DNA formed on pTSO182 at ddTTP/dTTP ratio = 1.

A: Synthesized DNA on pTSO182 at ddTTP/dTTP = 1 was hybridized to a filter trapping the leftward(L) strand, and then dissociated. Phage DNA carrying the L-strand was added and after annealing, the product was digested with HaeIII, SalI and BglII, and resolved by gel electrophoresis. Under the conditions, rightward chains synthesized only yield labelled restriction fragments. As a control, phage DNA carrying rightward(R) strand was added to the dissociated fragment and digested with the same enzymes, but no signficant restriction bands were yielded.

B: Labelled DNA was hybridized to a filter trapping the R chain. After dissociating hybridized fragments, phage DNA carrying the R chain was added. The mixture was digested with HaeIII, SalI and BglII, and electrophoresed. Under the conditions, leftward chains formed only yield labelled restriction fragments. Labelled restriction fragments were numbered, and indicated on the restriction map.

the initial DNA incorporation predominantly begins from the
right of fragment 16, possibly in the fragment 18 area(67 to
122 bp outside of the oriC sequence). This region is not on
the E.coli chromosomal moiety, but on the pBR322 moiety.
Note that in the above analysis, newly formed chains are
identified as a proper band only when traversed a certain
restriction fragment area, and those initiating or
terminating in the restriction fragment area are not
identified. Therefore, the initiation site is localized more
precisely by digesting into smaller pieces.

In order to determine the polarity of initially
synthesized chains, the region covering from the left of
oriC to oriC was cloned on M13mp7, and labelled fragments
were fractionated by hybridization to immobilized phage DNA
carrying each strand. The fractionated chains were
dissociated, re-anealled to phage DNA carrying each strand,
and digested with restriction enzymes. The resulting
fragments were resolved by gel electrophoresis. As shown in
Fig.7, the chain in the leftward direction yielded mainly
three bands, corresponding to fragments 11+12, 13 and 16. On
the contrary, the chain in the rightward direction yielded
five bands corresponding to fragments 11+12, 13, 16, 16' and
18. The regions covered by these fragments are shown in the
lower part of Fig.7. The result implies that the leftward
DNA synthesis is predominantly initiated at the fragment 18
area, whereas the rightward DNA synthesis is initiated
mainly in the fragment 9 area and proceeds toward oriC.

Fraction II alone from the HMS83 strain has a
considerable activity to generate DNA synthesis(see Fig.3
A). When the products were analysed, however, a striking
difference was found in the labelling pattern. As shown in
Fig.4 A-f, all the regions of plasmids had been labelled
rather uniformly even at higher ddTTP concentrations. Under
the conditions, the level of incorporation by Ori⁻ plasmids
decreased further and pBR322 vector does not promote any
incorporation, indicating that the labelling of DNA by
fraction II alone depends on the oriC sequence. It is
therefore evident that DNA initiation is restricted to a
specific region by the component(s) supplemented by fraction
III. Since the addition of fraction III increases the level
of DNA synthesis, we must assume that the component(s)
contained in fraction III has at least two functions; one
stimulating the overall initiation of DNA replication and
the other restricting the DNA initiation site. The most
likely component supplemented by fraction III is dnaA
protein. If this is the case, the latter function appears to
require a relatively large quantity of dnaA protein, for

fraction II of the HMS83 strain contains a certain level of dnaA protein.

As three OriC⁻ plasmids, pTSO202, pTSO209, and pTSO279, each containing mutation at different part of oriC, directed significant levels of DNA synthesis(see Fig.2 b), synthesized products were analysed as above. The labelling patterns were rather similar to those generated by OriC⁺ plasmids but without fraction III, although the labels in fragments generated from the left of oriC to oriC area appeared to be slightly heavy(Fig.4 A-g,h,i). The result implies that the oriC sequence is associated with the function of the component which restricts the DNA initiation site.

A replication model of oriC plasmids

Participation of rifampicin-sensitive RNA polymerase in the replication of oriC plasmids has been demonstrated(11). It is therefore reasonable to assume that the leftward DNA synthesis is preceded by primer RNA synthesis initiated at an upstream region(at the 5'-side), most likely within oriC, and switched to DNA at the outside of oriC by the action of the component supplemented by fraction III. As mentioned previously, fraction II alone from the HMS83 strain had the ability to promote oriC-dependent DNA synthesis, but lacked the activity initiating DNA synthesis at the specific region. At a low level of the supplemented component, the RNA-DNA transition may take place randomly.

Our model for the bidirectional replication directed by oriC is as follows. An initiation complex, named ORISOME, is constructed from multiple components under the direction of the oriC sequence, and initiates leftward transcription probably within oriC. The resulting primer RNA is switched to DNA at an adjacent region of oriC. The formation of such a leading strand would result in opening of the duplex DNA and induce synthesis of a lagging strand in the reverse direction, possibly by the mechanism demonstrated with the ϕx174 genome(16,17). The resulting lagging strand may become the leading strand for the rightward replication. This model is supported by an experiment, in which products labeled at high ddTTP concentrations were fractionated by using separated strands (Fig.7). We assume that long specific sequences are not involved in the RNA to DNA transition and initiation of the lagging strand, as the similar labelling patterns were obtained with three Ori⁺ plasmids differing in the sequences adjacent to oriC.

If primer RNA is indeed initiated within oriC by rifampicin-sensitive RNA polymerase, promoter must be present in oriC. Our attempts, however, to identify promoter within oriC by using a promoter-cloning vector(18) and by an in vitro transcription system were unsuccessful, in contrast to the report by Lother & Messer(19) who assigned two promoters in the opposite directions in oriC by in vitro transcription experiments. The initiation of primer RNA synthesis in oriC is assumed to be highly regulated, so that such a promoter could be identified only in a very specified system. According to the result in this paper, the RNA-DNA transition occurs at the outside of oriC. Two RNA-DNA transition sites, however, had been assigned within oriC by Okazaki et al.(20) based on analysis of in vivo products. It would be necessary to re-investigate the transition sites in the in vitro system.

REFERENCES

1. Hiraga,S. (1976) Proc. natn. Acad. Sci. U.S.A., 73, 198-202.
2. Yasuda, S. & Hirota, Y. (1977) Proc. natn. Acad. Sci. U.S.A., 74, 5458-5462.
3. von Meyerburg,K., Hansen,F.G., Nielsen,L.D. & Riise,E. (1978) Molec. gen. Genet., 160, 287-295.
4. Miki,T., Hiraga,S., Nagata,T.& Yura,T.(1978) Proc. natn. Acad. Sci. U.S.A., 75, 5099-5103.
5. Sugimoto,K., Oka,A.,Sugisaki,H., Takanami,M., Nishimura,A., Yasuda,S. & Hirota,Y. (1979) Proc. natn Acad. Sci. U.S.A., 76, 575-579.
6. Oka,A., Sugimoto,K., Takanami,M. & Hirota,Y. (1980) Molec. gen. Genet., 178, 9-20.
7. Hirota,Y., Oka,A., Sugimoto,K., Asada,K., Sasaki,H. & Takanami,M.(1981) ICN-UCLA symp. Mol. Cell. Biol., vol.20, p1-12.
8. Asada,K., Sugimoto,K., Oka,A.,Takanami,M.& Hirota,Y. (1982) Nucleic Acids Res., 10, 3745-3754.
9. Oka,A., Sugimoto,K., Sasaki,H. & Takanami,M. (1982) Gene 19, 59-69.
10. Fuller,R.S., Kaguni,J.M. & Kornberg,A.(1981) Proc. natn. Acad. Sci. U.S.A., 78, 7370-7374.
11. Kaguni,J.M., Fuller,R.S. & Kornberg,A. (1982) Nature 296, 623-627.

12. Oka,A., Sugimoto,K., Sasaki,H. &. Takanami,M. Manuscript in preparation.
13. Campbell,J.L., Soll,L.& Richardson,C.C.(1972) Proc.natn. Acad. Sci. U.S.A., 69, 2090-2094.
14. Yasuda,S., Takagi,T. & Hirota,Y. (1981) Annual Report Natn. Inst. Genetics, Japan, No.32.
15. Arai,K. & Kornberg,A. (1981) Proc. natn. Acad. Sci. U.S.A., 78, 69-73.
16. Zyskind,J.W., Harding,N.E., Takeda,Y., Cleary,J.M. & Smith,D.W., ICN-UCLA Symp. Mol. Cell. Biol., 22, 13-25.
17. Arai,K., Low,R.L. & Kornberg,A. (1981). Proc. natn. Acad. Sci. U.S.A., 78, 707-711.
18. Morita, M., Sugimoto,K, Oka,A., Takanami,M., Hirota,Y. (1981) ICN UCLA Symp. Mol. Cell. Biol., vol.20, p29.
19. Lother,H. & Messer,W. (1981) Nature 294, 376-378.
20. Okazaki,T., Hirose,S., Fujiyama,A. & Kohara,Y. (1980) ICN-UCLA Symp. Mol. Cell. Biol., vol.19, p429.

Mechanisms of DNA Replication and Recombination, pages 275–288
© 1983 Alan R. Liss, Inc., 150 Fifth Avenue, New York, NY 10011

ENZYMES IN THE INITIATION OF REPLICATION
AT THE E. COLI CHROMOSOMAL ORIGIN[1]

Robert S. Fuller, LeRoy L. Bertsch, Nicholas E. Dixon,
James E. Flynn, Jr., Jon M. Kaguni, Robert L. Low,[2]
Tohru Ogawa and Arthur Kornberg

Department of Biochemistry, Stanford University
School of Medicine, Stanford, California 94305

ABSTRACT   Efficient replication of exogenously added
plasmids containing the E. coli origin of replication,
oriC, occurs in vitro when an ammonium sulfate fraction of soluble proteins is supplemented with substrates, other small molecules, and polyvinyl alcohol.
This system faithfully reflects characteristics of in
vivo initiation at oriC including requirements for RNA
polymerase, DNA gyrase, products of genes for initiation and propagation of chromosomal replication (including dnaA protein), and the oriC sequence; replication is bidirectional from oriC. This in vitro system
has provided a functional assay for purification of
dnaA protein to near homogeneity from a plasmid-containing, overproducing strain. dnaA protein binds
specifically to oriC as measured by a Millipore filter
binding assay and by protection of the unique HindIII
site within oriC. A partially reconstituted system
contains purified proteins (RNA polymerase, DNA gyrase
A and B subunits, DNA polymerase III holoenzyme, SSB,
primse, the dnaA, dnaB and dnaC proteins and proteins
i and n') and additional fractions that include: (i)
a "specificity factor" identical to topoisomerase I,
required to inhibit dnaA-independent, oriC-independent

[1]This work was supported by grants from the National
Institutes of Health and the National Science Foundation.
[2]Present address: Department of Pathology, Washington
University School of Medicine, St. Louis, Missouri 63110.

DNA synthesis; (ii) protein HU, a small, basic histone-like protein required both to act positively and
to enhance the specificity activity of topoisomerase
I; and (iii) at least two other factors partially resolved by chromatography on Amicon Red A agarose.
Initial kinetics of oriC replication in the reconstituted system tentatively identifies stages in which
these various components participate.

INTRODUCTION

Replication in vitro of the single-stranded DNA bacteriophages M13, G4 and φX174 has illuminated the mechanisms of priming of nascent chains in E. coli (1) as well
as the steps by which RNA primers are extended by the DNA
polymerase III holoenzyme (2). A model of the replication
fork in the E. coli chromosome incorporates features of
these phage replication systems, in particular the highly
processive movement of a multisubunit primosome on the
lagging strand and of DNA polymerase III holoenzyme on the
leading strand (3). The mechanism by which replication
forks arise in bidirectional replication of the chromosome
demands study of enzymology of initial events at the chromosomal origin of replication, oriC.
Isolation of oriC in a functional form as an autonomously replicating sequence in plasmids and bacteriophages
(4,5), definition of its functional domains by deletion and
mutation (6,7), and compilation of an enterobacterial oriC
consensus sequence (8) have provided convenient, functionally characterized templates for analyzing oriC replication in vitro. The replication of oriC-containing plasmids by a crude protein fraction from E. coli cells has
been described (9). Key features of this system are its
dependence on a protein fraction precipitated over a narrow
range of ammonium sulfate from a crude lysate, an absolute
dependence on 6 to 8% polyvinyl alcohol (PVA) or similar
polymers, and a substantial stimulation by inclusion of an
ATP-regenerating system of creatine kinase and creatine
phosphate (9). The reaction reflects characteristics of in
vivo chromosomal replication: (i) replication requires the
oriC sequence, begins at or near oriC, and proceeds bidirectionally; (ii) initiation requires RNA polymerase as
judged by sensitivity to rifampicin; and (iii) the reaction
depends on the dnaA, dnaB and dnaC proteins, SSB (single-

strand binding protein) and DNA gyrase A and B subunits (9,10, and unpublished observations).

In vitro oriC replication provides an assay for purification of novel factors required for replication of oriC plasmid, but not, for example, φX174. Dependence on the dnaA gene product has provided an in vitro mutant-complementation assay for purification of dnaA protein by functional criteria (11). Partial reconstitution of the reaction using a combination of purified replication proteins and crude fractions has provided an assay for other positively-acting factors as well as for factors required for specificity. Mechanistic details of origin recognition, priming of bidirectional leading strands, and genesis and propagation of replication forks can be directly probed in the reconstituted system by identification, isolation, and characterization of intermediates and products of the reaction.

RESULTS AND DISCUSSION

dnaA Protein

The product of the dnaA gene plays a key role at an early stage in initiation of a cycle of chromosome replication (12); with the exception of plasmid pSC101 (13,14), the dnaA requirement is a unique feature of chromosomal replication. A functional interaction with RNA polymerase, suggested by genetic and physiological experiments, makes its mechanism of action especially intriguing (15-17). Despite cloning (18,19) and sequencing (20) of the gene and identification of the gene product (18,19,21), little progress had been made in defining its function biochemically. Recently, however, development of the oriC plasmid in vitro replication system provided a direct assay for dnaA protein function (9). Without a functional assay, purification of dnaA protein yielded a preparation with demonstrable oriC-specific binding activity (22).

Assay and overproduction. An ammonium sulfate fraction (fraction II) of dnaA mutant cells is inactive for oriC replication unless a source of wild-type dnaA protein is provided (9). Addition of fraction II from the strain HB101(pBF101) containing a multicopy dnaA$^+$ plasmid stimulates the dnaA mutant fraction 10- to 15-fold more than does a fraction from the parental strain (9). Overproduction up to 300-fold was achieved by placing the dnaA gene

under inducible control of the λp_L promoter in plasmid
pBF110 (11).

Purification of the dnaA protein. dnaA protein was
purified from heat-induced cells containing pBF110. Purification involved lysozyme-heat lysis in 20 mM spermidine,
ammonium sulfate fractionation, chromatography on BioRex 70
(BioRad), and High Performance Liquid Chromatography (HPLC).
HPLC on an Altex 3000SW gel-filtration column separated the
complementing activity into two peaks: an excluded peak (2
to 4 x 10^5 daltons) of aggregated dnaA protein and an
included peak of higher specific activity which eluted
after bovine serum albumin and corresponds to monomeric
protein (Table 1). As measured by polyacrylamide gel
electrophoresis, purified monomeric dnaA protein had a
molecular weight of 52,000 ± 1000 daltons and was about 95%
pure.

Purified dnaA protein binds specifically to oriC.
dnaA protein purified without a functional assay specifically retained restriction fragments containing oriC in a
nitrocellulose filter binding assay (22). In the in vitro
oriC replication system, DNA gyrase is required and linearized oriC plasmid is inactive as template (unpublished
data), suggesting interaction of dnaA protein and oriC
within a covalently closed circular, supercoiled DNA molecule. Initially, the specificity of binding of purified
dnaA protein was assessed by comparing binding of ^{3}H-
labeled M13oriC26 RFI and M13oriC26Δ221 RFI, the latter
having a deletion of oriC entailing removal of approximately 330 base pairs (23). In the absence of competing
DNA, dnaA protein retained ^{3}H-M13oriC26 about 3-fold better
than ^{3}H-M13oriC26Δ221. In the presence of a 10-fold excess
of unlabeled ColE1 DNA, the discrimination by dnaA protein
increased to 30-fold. On HPLC, the dnaA polypeptide copurified with both oriC replication activity and oriC-specific
binding (Table 1). When ^{3}H-M13oriC26 DNA was linearized by
cleavage with EcoR1, which recognizes a single site outside
oriC, binding by dnaA protein was reduced 3-fold, indicating a preference for a supercoiled molecule. dnaA protein
bound linearized oriC plasmid about 2-fold better than
linearized deletion plasmid, indicating that some degree of
dnaA-specificity persists in the case of linearized oriC
DNA. In addition, dnaA protein specifically protected
against cleavage by HindIII at the HindIII site within the
minimal origin sequence (6). Dependence on dnaA protein
concentration was virtually identical for HindIII protection and for Millipore filter binding. Protection of

TABLE 1

ACTIVITIES OF PURIFIED dnaA PROTEIN

Fraction	oriC Replication[a] u/mg x 10^{-5}	oriC Binding[b] u/mg x 10^{-4}
HPLC aggregate	5	7
HPLC monomer	15	28

[a]Activity in complementation assay. One unit is 1 pmol/min of nucleotide incorporated in a 20-min assay with 600 pmol (as nucleotide) of M13oriC26 RFI (9,11).
[b]Activity in Millipore filter binding assay. One unit is 1 fmol oriC plasmid retained in an assay containing 25 fmol of ^{3}H-M13oriC26 RFI and 1 µg unlabeled ColE1 DNA (11).

the HindIII site is particularly interesting, for this site, within a relatively nonconserved region of oriC, is flanked by inverted repetitions of a nine-base pair sequence (5'-TTAT$_A^C$CACA-3') which is highly conserved in four positions within oriC (8) and which also appears in or near the origin of replication of the dnaA-dependent replicon pSC101 (K. Armstrong and E. Ohtsubo, personal communication) and between the dual promoters of the dnaA gene (20). Whether this sequence is required for the binding of dnaA protein is clearly of interest.

Fractionation and Reconstitution of oriC Plasmid Replication.

A partially reconstituted (recon) system for oriC plasmid replication was another approach to probing initiation at oriC and provided assays for purification of novel factors. The crude fraction (fraction II) was supplemented with purified proteins known or expected to be required for the reaction (Table 2). Dependence of the recon reaction on fraction II provides an assay for novel components in this crude fraction. Fraction II from a dnaA mutant strain (WM433) (9) made possible the use of both dnaA-protein dependence and oriC specificity as "guideposts" for the authenticity of the assays.

TABLE 2
COMPONENTS OF RECON SYSTEM

Components required for φX174 SS→RF

    dnaB protein
    dnaC protein
    Primase
    DNA polymerase III holoenzyme
    SSB
    Proteins i and n'

Additional factors required for oriC replication

    DNA gyrase A and B subunits
    dnaA protein
    RNA polymerase
    Additional components in fraction II

Topoisomerase I is a specificity factor for oriC replication. When fraction II was fractionated on DEAE cellulose (Figure 1), a bound fraction eluted by 0.2 M KCl stimulated DNA synthesis in the recon assay, but this synthesis was neither oriC-dependent nor dnaA-dependent; the recon system supplemented with the DEAE-bound fraction supported DNA synthesis dependent on φXRFI DNA to the same extent as with the oriC template M13oriC26 RFI. This dnaA-independent, oriC-independent reaction required 7% PVA, DNA gyrase, RNA polymerase and dnaB and dnaC proteins. Addition of the DEAE flow-through fraction restored specificity (Figure 1); the reaction dependent on φXRFI was inhibited and dnaA-dependence was restored on M13oriC26 RFI. Using inhibition of φXRFI-dependent DNA synthesis as an assay, "specificity factor" was purified 2500-fold to yield a homogeneous 100,000-dalton polypeptide. Identity of this specificity factor with E. coli topoisomerase I (ω-protein) (24) is based on several criteria: (i) specific activities in the oriC specificity assay and in relaxation of supercoiled DNA; (ii) molecular weights; (iii) inhibition by antibody prepared against topoisomerase I; (iv) deficiency of specificity factor in a topA deletion strain (24); and (v) cleavage specificities on single-

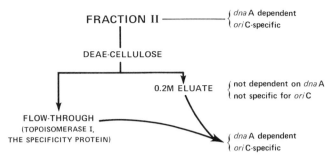

FIGURE 1. Chromatography of fraction II on DEAE cellulose resolves a required fraction (0.2 M eluate) from specificity protein (topoisomerase I in flow-through) as measured by activity in oriC recon and specificity assays.

stranded DNA (J. Kaguni, F. Dean and N. Cozzarelli, unpublished data). Topoisomerase I is not, however, the only factor required to confer oriC-specificity on the oriC replication reaction; additional factors, including protein HU (25) (see below), appear to be required.

Protein HU is required for oriC replication. Under conditions in which replication was dnaA protein-dependent and oriC-specific (i.e., in the presence of the DEAE-flow-through and 0.2 M KCl eluate), a fraction eluted with slightly higher salt (0.25 M KCl) was stimulatory by 3- to 5-fold. Purification of the stimulatory activity yielded a protein with a subunit molecular weight of about 10,000 daltons. It is identical to protein HU provided by J. Yaniv (25) by these criteria: (i) comigration in SDS-poly-acrylamide gel electrophoresis; (ii) specific activity of HU in the recon assay; and (iii) inhibition of oriC replication by protein HU antibody (provided by J. Yaniv) and neutralization of this inhibition by the purified factor. Inhibition of oriC replication by anti-HU antibody to background levels and reversal of this inhibition by HU or the purified factor establish that oriC replication is completely dependent on protein HU. Optimal levels of protein HU are reached at a weight ratio of protein to DNA of about 1:10.

Protein HU also acts as a specificity factor in oriC replication when added together with topoisomerase I. Much higher levels of protein HU are required and the effect is maximal at a weight ratio to template DNA of about 1:1. At this ratio, protein HU can convert SV40 form I DNA into a chromatin-like beaded form in the presence of a type I topoisomerase (26). Although multiple mechanisms may suppress the dnaA- and oriC- independent reactions in vitro, the most physiologically relevant mechanism may be the assembly of the template into a particular nucleoprotein structure.

Requirements for replication in the recon system. At least two additional factors are needed; these are distinguished by Amicon Red A agarose chromatography of fraction II. When flow-through and 0.5 M KCl eluate fractions from this column are combined with the purified components in the recon system, efficient DNA synthesis is observed (Table 3) equivalent to nearly complete replication of the template in 30 minutes. This reaction is completely dependent on dnaA, dnaB and dnaC proteins, primase, DNA polymerase III holoenzyme, and RNA polymerase. A requirement for gyrase is suggested by sensitivity of the reaction to the gyrase inhibitors: nalidixate, oxolinate, and novobiocin. A more profound dependence on protein HU is indicated by inhibition by anti-HU antibody. Dependence on the crude Red A agarose fractions is only 2.5- to 3-fold. The high backgrounds observed in the absence of these fractions may be due to contamination of some of the purified proteins with these factors. A high level of synthesis in the absence of topoisomerase I is not dependent on dnaA protein.

Initial kinetics and a model for the early phase of oriC replication. OriC plasmid replication with fraction II as sole source of enzymatic activity, or with the better defined recon system, begins after a lag of 3.5 to 5 min. With the recon system, the lag can be abolished by preincubation of a mixture of all reaction components in the absence of deoxynucleoside triphosphates (dNTPs) for 5 to 10 min; upon addition of dNTPs, DNA synthesis begins immediately. Omission of gyrase, RNA polymerase, and the Red A agarose bound fraction from the preincubation results in the full lag of 3.5 minutes. Omission of dnaA protein or dnaB and dnaC proteins results in a lag of only about 1.5 min. When primase, holoenzyme and the Red A agarose flow-through fraction are omitted from the preincubation, no lag occurs (Table 4).

TABLE 3
REQUIREMENTS FOR oriC REPLICATION IN RECON SYSTEM

Omission	DNA synthesis[a] (pmol/30 min)
Experiment 1	
None	590
φX SS→RF components[b]	25
dnaA protein	8
RNA polymerase	7
Gyrase A and B subunits	355
Protein HU	216
Red A agarose bound fraction	183
Red A agarose flow-through fraction	243
Both Red A agarose fractions	126
Experiment 2	
None	488
dnaB protein	8
dnaC protein	6
DNA pol III holoenzyme	4
Primase	46

[a] Input M13oriC26 RFI template, 600 pmol as nucleotide.
[b] These are dnaB and dnaC protein, primase, pol III holoenzyme, SSB, and proteins i and n'.

These data suggest action at different stages for these various factors. Early action by gyrase and topoisomerase I may be required for preparation of the template in an active form, perhaps involving assembly of protein HU with template in a nucleoprotein complex resistant to the dnaA-independent reaction. Subsequent transcription by RNA polymerase may result in priming or some other form of activation of the origin. RNase H might be involved in processing an RNA polymerase transcript into a primer in a fashion analogous to its action in ColE1 replication (27).

TABLE 4
LATENCIES OF ACTION OF oriC REACTION COMPONENTS

Component omitted[a] from preincubation	Observed lag time (minutes)
None	0
DNA gyrase A + B	3.5
RNA polymerase	3.5
Red A agarose bound	3.5
dnaA protein	1.5
dnaB + dnaC proteins	1.5
Primase + holoenzyme	0
Red A agarose flow-through	0

[a]Deoxynucleoside triphosphates were omitted from the preincubation (5 min at 30°C) in each case. The individual components, if omitted, were added along with deoxynucleoside triphosphates.

It is unclear how dnaA protein interacts with RNA polymerase (15,16). Possibly, dnaA protein, targeted to oriC by the specificity of its binding, interacts with RNA polymerase at some step after promoter binding and initiation of transcription; a notion supported but not proven by the kinetic data above. In any event, the time of action of dnaB and dnaC proteins cannot be distinguished kinetically from the time of action of dnaA protein. From their roles in the replication of φX174 single-stranded DNA (28,29), it is likely that dnaB and dnaC proteins are involved in primosome formation (30) and the initiation of lagging strand replication. A role for dnaA protein in such events has not been ruled out. Nor has a role for primase in priming leading strand synthesis been eliminated, but the kinetic data suggest a relatively late involvement of both primase and the DNA polymerase III holoenzyme in initiation.

Future studies of oriC plasmid replication in vitro. A major goal is purification of the remaining factors required for oriC replication. With a reaction reconstituted from well-defined and purified components, several issues should be clarified: (i) whether RNA polymerase recognizes promoters within oriC and how its transcripts are utilized (e.g., as primers or not); (ii) how dnaA protein and RNA polymerase interact; (iii) whether DNA gyrase and topoisomerase I establish a topological equilibrium important both for specificity and positive function or whether the function of one might be targeted directly to oriC; (iv) whether nucleoprotein structure of the template, involving protein HU and perhaps other proteins as well, is crucial for correct and efficient replication of oriC; and finally, (v) how closely the bidirectionally elongating replication forks established in oriC replication in vitro (9,10) resemble the model proposed on the basis of the enzymology of single-stranded bacteriophage DNA replication (3).

REFERENCES

1. Stayton MM, Bertsch L, Biswas S, Burgers P, Dixon N, Flynn JE Jr, Fuller R, Kaguni J, Kobori J, Kodaira M, Low R, Kornberg A (1982). Enzymatic recognition of DNA replication origins. Cold Spring Harbor Symp Quant Biol 47, in press.
2. Burgers PMJ, Kornberg A (1982). ATP activation of DNA polymerase III holoenzyme from Escherichia coli. II. Initiation complex: stoichiometry and reactivity. J Biol Chem 257:11474.
3. Dixon NE, Bertsch LL, Biswas SB, Burgers PMJ, Flynn JE, Fuller RS, Kaguni JM, Kodaira M, Stayton MM, Kornberg A (1983). Single-stranded phages as probes of replication mechanisms. In Cozzarelli N (ed): "Mechanisms of DNA Replication and Recombination - UCLA Symposia on Molecular and Cellular Biology," New Series, Vol 10, New York, NY: Alan R Liss, Inc (this volume).
4. Yasuda S, Hirota Y (1977). Cloning and mapping of the replication origin of Escherichia coli. Proc Natl Acad Sci USA 74:5458.
5. von Meyenburg Y, Hansen FG, Nielsen LD, Riise E (1978). Origin of replication, oriC, of the Escherichia coli

chromosome on specialized transducing phages λasn.
Molec Gen Genet 160:287.

6. Oka A, Sugimoto K, Takanami M, Hirota Y (1980).
Replication origin of the Escherichia coli K-12 chromosome: the size and structure of the minimum DNA
segment carrying the information for autonomous replication. Molec Gen Genet 178:9.

7. Hirota Y, Oka A, Sugimoto K, Asada K, Sasaki H,
Takanami M (1981). Escherichia coli origin of replication: structural organization of the region essential
for autonomous replication and the recognition frame
model. In Ray DS (ed): "Structure and DNA-Protein
Interactions of Replication Origins, ICN-UCLA Symposia
on Moecular and Cellular Biology," Vol 22, New York:
Academic Press, p 1.

8. Zyskind JW, Cleary JM, Brusilow WSA, Harding NE, Smith
DW (1983). Chromosomal replication origin from the
marine bacterium Vibrio harveyi functions in Escherichia coli: oriC consensus sequence. Proc Natl Acad
Sci USA 80:1164.

9. Fuller RS, Kaguni JM, Kornberg A (1981). Enzymatic
replication of the origin of the Escherichia coli
chromosome. Proc Natl Acad Sci USA 78:7370.

10. Kaguni JM, Fuller RS, Kornberg A (1982). Enzymatic
replication of E. coli chromosomal origin is bidirectional. Nature 296:623.

11. Fuller RS, Kornberg A (1983). Purified dnaA protein
in initiation of replication at the E. coli chromosomal origin of replication. Proc Natl Acad Sci USA,
in press.

12. Hirota Y, Mordoh J, Jacob F (1970). On the process of
cellular division in E. coli. III. Thermosensitive
mutants of Escherichia coli altered in the process of
DNA initiation. J Mol Biol 53:369.

13. Hasunuma K, Sekiguchi M (1977). Replication of plasmid pSC101 in Escherichia coli K12: requirement for
dnaA function. Molec Gen Genet 154:225.

14. Frey J, Chandler M, Caro L (1979). The effects of an
Escherichia coli dnaAts mutation on the replication of
the plasmids ColE1, pSC101, R100.1 and RTF-TC. Molec
Gen Genet 174:117.

15. Bagdasarian MM, Izakowska M, Bagdasarian M (1977).
Suppression of the dnaA phenotype by mutations in the
rpoB cistron of ribonucleic acid polymerase in
Salmonella typhimurium and Escherichia coli. J
Bacteriol 130:577.

16. Atlung T (1981). Analysis of seven dnaA suppressor loci in Escherichia coli. In Ray DS (ed): "Structure and DNA-Protein Interactions of Replication Origins, ICN-UCLA Symposia on Molecular and Cellular Biology," Vol 22, New York: Academic Press, p 297.

17. Zyskind JW, Deen LT, Smith DW (1977). Temporal sequence of events during the initiation process in Escherichia coli deoxyribonucleic acid replication: roles of the dnaA and dnaC gene products and ribonucleic acid polymerase. J Bacteriol 129:1466.

18. Hansen FG, von Meyenberg K (1979). Characterization of the dnaA, gyrB and other genes in the dnaA region of the Escherichia coli chromosome on specialized transducing phages λtna. Molec Gen Genet 175:135.

19. Yuasa S, Sakakibara Y (1980). Identification of the dnaA and dnaN gene products of Escherichia coli. Molec Gen Genet 180:267.

20. Hansen EB, Hansen FG, von Meyenburg K (1982). The nucleotide sequence of the dnaA gene and the first part of the dnaN gene of Escherichia coli K-12. Nucleic Acids Res 10:7373.

21. Kimura M, Yura T, Nagata T (1980). Isolation and characterization of Escherichia coli dnaA amber mutants. J Bacteriol 144:649.

22. Chakrabority T, Yoshinaga K, Lother H, Messer W (1982). Purification of the E. coli dnaA gene product. EMBO Journ 1:1545.

23. Kaguni LS, Kaguni JM, Ray DS (1981). Replication of M13oriC bacteriophages in Escherichia coli rep mutant is dependent on the cloned Escherichia coli replication origin. J Bacteriol 145:974.

24. Liu LF, Wang JC (1979). Interaction between DNA and Escherichia coli DNA topoisomerase I. J Biol Chem 254:11082.

25. Rovière-Yaniv J, Gros F (1975). Characterization of a novel, low-molecular-weight DNA-binding protein from Escherichia coli. Proc Natl Acad Sci USA 72:3428.

26. Rovière-Yaniv J, Yaniv M, Germond J-E (1979). E. coli DNA binding protein HU forms nucleosome-like structure with circular double-stranded DNA. Cell 17:265.

27. Itoh T, Tomizawa J (1980). Formation of a primer for initiation of replication of colE1 DNA by ribonuclease H. Proc Natl Acad Sci USA 77:2450.

28. Arai K, Low R, Kobori J, Shlomai J, Kornberg A (1981). Mechanism of dnaB protein action. V. Association of

dnaB protein, protein n', and other prepriming proteins in the primosome of DNA replication. J Biol Chem 256:5273.

29. Kobori JA, Kornberg A (1982). The Escherichia coli dnaC gene product. III. Properties of the dnaB-dnaC protein complex. J Biol Chem 257:13770.

30. Arai K, Low RL, Kornberg A (1981). Movement and site selection for priming by the primosome in phage φX174 DNA replication. Proc Natl Acad Sci USA 78:707.

Mechanisms of DNA Replication and Recombination, pages 289–301
© **1983 Alan R. Liss, Inc., 150 Fifth Avenue, New York, NY 10011**

INTERACTION OF PROTEINS WITH ORIC,
THE REPLICATION ORIGIN OF E. COLI

Heinz Lother,[1] Rudi Lurz,[1] Elisha Orr,[2]
and Walter Messer[1]

[1]Max-Planck-Institut f. molekulare Genetik,
Berlin 33 (Dahlem), Germany
[2]Department of Genetics, University of Leicester,
Leicester LE1 7RH, England

ABSTRACT Regulatory sites within the E. coli
replication origin, oriC, were defined by their
interactions with initiation proteins: RNA
polymerase, dnaA protein, DNA gyrase, protein
fraction B'.

INTRODUCTION

The control of replication in bacteria is exerted at
the level of the initiation of a new cycle of chromosome
replication. Therefore, the initiation event is a key
process in the bacterial cell cycle.
The isolation of minichromosomes (1-4), plasmids
which contain and use the unique replication origin of E.
coli, oriC, opened the possibility for an analysis of the
initiation of DNA replication in bacteria at the molecular
level.
The replication origin has been located within a
stretch of 422 bp (5). Its nucleotide sequence has been
determined (6, 7). Subsequently, a segment of 245 bp,
cloned in plasmid pBR322, has been defined as the smallest
DNA fragment which is able to promote the replication of
adjacent DNA: the minimal replication origin (8).
However, sequences to the right of the minimal replication
origin were shown to be required for efficient
bidirectional replication in vivo (9) but apparently not,
in a joint replicon, in vitro (10).

A number of proteins were found to be required for the initiation of replication. They were defined by mutations in initiation genes (13) and by the analysis of in vitro initiation systems using as templates the DNA of small single stranded phages (14) or, recently, oriC containing DNA (15).

Here we present data which demonstrate that the origin of replication is one of the most complicated regulatory regions within the bacterial chromosome. We have defined some of the elements of this structure by analyzing the interactions of several proteins with oriC DNA.

RESULTS AND DISCUSSION

Promoters and Transcripts in oriC.

Early experiments suggested a requirement for RNA polymerase in the initiation process, initiation could be blocked with rifampicin (16-18). Since DNA polymerases are incapable to start new DNA chains, the synthesis of a primer RNA, presumably by RNA polymerase, is required. Therefore, we searched for promoters within and ext to oriC with two independent techniques. First we mapped RNA polymerase binding sites using electron microscopy (19), then we analyzed transcripts syntnesized in vitro using purified RNA polymerase and various restriction fragments. Determination of the size and 5'-termini of the transcripts by conventional RNA sequencing techniques revealed two promoters located within oriC (20).

One of these promoters (Pori-l) is completely contained within the minimal replication origin (Fig. 1). The transcript starts with nucleotide 166, transcription is leftward. The second promoter (Pori-r) is located immediately to the right of the rightward boundary of the minimal origin (Fig. 1). Rightward transcription started predominantly with nucleotide 323. Some transcripts started with nucleotide 313 .

Using proper conditions in the in vitro transcription assay, transcripts were found which were terminated within the DNA fragments used. For both promoters, Pori-l and Pori-r, these terminated transcripts had a length of approximately 110 nucleotides (Fig. 2). Starting at Pori-l three terminated transcripts of very similar size were found (Fig. 2). All of them were terminated within the

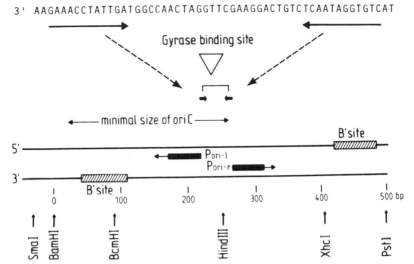

FIGURE 1. Protein recognition sites within oriC.
The positions of binding sites for gyrase, the membrane
derived protein fraction B',and of origin promoters Pori-1
and Pori-r are shown. Transcripts start at the base of the
arrows. Two of the inverted repeats, close to the HindIII
site, are indicated.

minimal origin, close to a position where RNA-DNA
junctions had been located (21). All terminated
transcripts were quite resistant to the attack of several
RNases, suggesting a high proportion of secondary
structure. They could result from true termination or
could be due to strong pause signals present in oriC
(Lother, manuscript in preparation).

The geometric arrangement of the two origin promoters
suggests that they are responsible for the synthesis of an
RNA primer. However, this has not been demonstrated and a
regulatory role of the transcripts, in a way similar to
the small RNA transcribed from the origin region of
plasmid ColE1 (22), has not been excluded.

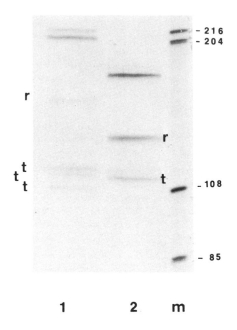

FIGURE 2. Transcripts obtained from restriction
fragments from oriC.
Autoradiogram of a urea-polyacrylamide gel of transcripts
obtained using purified RNA polymerase and DNA fragments
from oriC:
1. BglII (38) - HindIII (244)
2. HindIII (244) - XhoI (417)
Conditions were as described (20). Heparin was used to
inhibit reinitiation.
m = marker, r = runoff transcripts, t = terminated
transcripts.

Binding of the dnaA protein to oriC DNA

The dnaA protein is a key protein in the initiation
process. It is the only known protein specific for
initiation of oriC. Phages and plasmids, with the

exception of pSC101, replicate independently of the presence of functional dnaA.

The dnaA protein was purified from cells infected with a vector in which the dnaA gene had been cloned. The purified product bound specifically to DNA fragments containing oriC and only a few base pairs of regions outside oriC (23). Specific binding of the dnaA protein to oriC was resistant to competition by a 200 fold excess of calf thymus DNA, whereas competition showed normal kinetics with unspecific DNA binding proteins used as a control (manuscript in preparation).

Interaction of DNA Gyrase with oriC.

Replication forks proceed bidirectionally from oriC around the negatively supercoiled covalently closed circle of chromosomal or minichromosomal DNA. This topological status is actively maintained by topoisomerases. Thermal inactivation of DNA gyrase, a type II topoisomerase, in conditional lethal mutants defective in either the gyrA or gyrB gene products suggested that initiation at oriC is inhibited by the loss of gyrase activity (24,25). We, therefore, analyzed binding of purified gyrase to minichromosome DNA by electron microscopy.

DNA gyrase was incubated with pOC51 (12) DNA. The complexes were fixed, and the DNA was cleaved with either SmaI or XhoI, enzymes which cleave the minichromosome only once, in order to allow a positioning of the complexes.

Orientation of measured molecules, determination of peak positions and plotting of histograms were done using a computer program. The histograms in Fig. 3 show identical data, but using different factors (0.5 - 3.0) for computation. These factors are values used by the program to determine the peak positions and to orient the individual molecules. They are sigma values (in per cent of the total length of the molecule) and determine which individual measurements are grouped into one peak. Computing the results with different factors allows to discriminate between significant peaks which are present in all computed histograms and peaks which occur at varying positions. As shown in Fig. 3, there is one specific peak of interaction of gyrase with the DNA within oriC at position 235 ± 23. In addition unspecific binding at many or random sites in the rest of the molecule was observed.

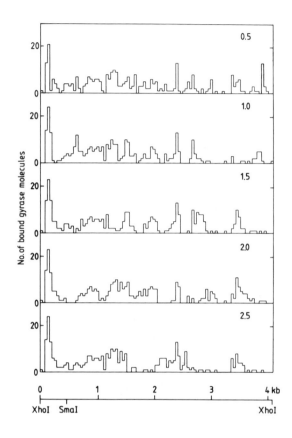

FIGURE 3. Effect of various peak widths used in computation on the positioning of gyrase complexes.
pOC51 ccc DNA (20µg/ml) was incubated with 30µg/ml of each gyrase subunit A and subunit B in: 25mM HEPES pH8.0, 80mM KCl, 4mM Mg-acetate, 2mM ATP, 5mM dithiothreitol for 10 min at 30 C. Glutaraldehyde (0.2%) was added and incubation continued for 15 min. Glutaraldehyde and unbound enzyme were separated from the complexes by gel filtration on Sepharose 4B. The DNA was cleaved with XhoI and again purified on Sepharose 4B. Complexes were adsorbed to mica, washed with distilled water for 2 hours, stained with 2% uranylacetate for 5 min, dehydrated with ethanol and shadowed with Pt/Ir. Primary magnification was 6000x. The positions of DNA gyrase molecules were measured using a digitizer and computed using different sigma values for the broadness of the peaks (see text).

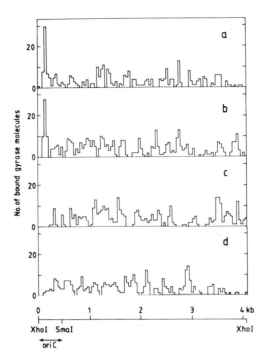

FIGURE 4. Positions of DNA gyrase on pOC51 DNA.
Binding conditions, electron microscope preparation and
computation of peak positions were as described in Fig. 3.
a.) +ATP; b.) -ATP; c.) ATP replaced by ß-γ-imido ATP; d.)
B. subtilis gyrase instead of E. coli gyrase.

As shown elsewhere (H. Lother, R. Lurz, E. Orr,
submitted), the A subunit is the subunit which mediates
the binding. Specific comp exes of gyrase with oriC were
formed with (Fig. 4a) and without (Fig. 4b) metabolic
energy. However, addition of the ATP analogue ß-γ -imido
ATP (Fig. 4c), as well as the addition of oxolinic acid
(data not shown) abolished specific binding. Gyrase
isolated from the Gram positive bacterium Bacillus
subtilis showed unspecific binding but no specific peak at
oriC (Fig. 4d). A similar replicon specificity was
observed in an experiment in which an enzyme made up from
B. subtilis A subunit and E. coli B subunit was inactive

in supercoiling ColE1 DNA while the reciprocal combination was fully active (26).

The replicon specificity in binding of DNA gyrase implies that supercoiling may not only be a critical step in the initiation of replication but also a selecting mechanism for the replication of different extra-chromosomal elements in a given host.

Interaction of oriC DNA with a Membrane Derived Protein Fraction.

A DNA-binding protein derived from the membrane of E. coli was found to bind specifically to oriC DNA if it is in single stranded form (27,28). This protein(s) recognizes two regions in oriC: Binding occurred to the strand reading 3' - 5' in the orientation of the E. coli genetic map between positions 38 and 165, i.e. within the minimal origin. Binding to the strand reading 5' - 3' occurred to the right of the minimal origin between positions 417 and 488 (Fig. 1).

Features of oriC Structure Relevant to its Interaction with Protein.

A comparison of the nucleotide sequences of the replication origins of different Gram negative bacteria showed that the majority of nucleotide differences between these origins occurred in clusters, interspersed with regions which were very highly conserved (29,30). Induced nucleotide changes in the variable regions were mostly permissible whereas insertions and deletions normally abolished origin function (8). This led to the hypothesis that regions of interaction with proteins (the conserved regions) have to be precisely spaced for efficient binding of replication proteins (31).

We reached the same conclusion using a different approach. Within oriC invertedly repeated sequence blocks are arranged such that they form two centers of coordinated dyad symmetry (12). The sequence blocks which are members of the two symmetrical structures are to a large extend identical to the conserved sequences described by Zyskind et al. (29,30). Since on one hand the conservation of these blocks in the different origins indicates their importance for origin function, on the

other hand none of these blocks is unique to the origin
(12), it must be their spacing as exemplified by their
symmetric arrangement which is important for a functional
origin. One of these blocks, 9 bp long, is paricularly
noteworthy. It occurs four times within oriC, two of
these positions are shown in Fig. 1, the two other
positions are at bp 186-194 and 80-88, respectively.

A Model for the Interaction of Initiation Proteins with
oriC.

This section must necessarily be speculative.
However, the results obtained so far suggest a tentative
model whose predictions can be tested. There is indirect
evidence for an interaction of three of the proteins
mentioned above. Suppressor mutations for dnaA mutants are
found in rpoB, the gene for the ß' subunit of RNA
polymerase (32). Similar genetic experiments suggest an
interaction between dnaA and gyrase (33).
Two of the 9 bp repeats mentioned above are located
between the two origin promoters in such a way that they
may be sites for positive regulation of these promoters
(20). A likely protein for such positive control is the
dnaA protein which acts early and positively in
initiation.
Within the limits of detection, gyrase interacted
with oriC at the same position. We could imagine that the
two proteins bind and act sequentially at the same or a
similar site. Gyrase might bind first to the 9 bp repeats
shown in Fig. 1. In the presence of ATP it might cause a
translocation using the identical sites to the left,
resulting in a loop out of the region around the HindIII
site (Fig. 1). This in turn might present an "activated"
form of the origin for the interaction with dnaA, RNA
polymerase and other initiation proteins. This mechanism
would also explain the inhibition of specific binding of
gyrase to oriC by ß-γ-imido-ATP although specific binding
itself occurred in the absence of ATP, since ATP would
induce the translocation - or the attempted translocation
in the case of the analogue. This proposed formation of an
"activated" loop is very similar to the mechanism
suggested for the interaction of T4 topoisomerase and the
T4 replication origin (34).
The role of the membrane protein B' is at present
unclear. Its binding to DNA strands of opposite polarity

make it a good candidate for a protein which promotes the segregation of daughter chromosomes via attachment to the membrane. Alternatively it might be responsible for correct inception by binding to the non-coding strand when transcription from origin promoters occurs.

ACKNOWLEDGEMENTS

The able and dedicated technical assistance of B. Rueckert and M. Hearne are gratefully acknowledged. This work was supported in part by grant Me659 of the Deutsche Forschungsgemeinschaft, grant No. G80/0084/0CB from the Medical Research Council (U.K.) and by a short term EMBO award (E. O.).

REFERENCES

1. Yasuda, S, Hirota, Y (1977). Cloning and mapping of the replication origin of Escherichia coli. Proc. Natl. Acad. Sci. USA 74:5458.
2. Messer, W, Bergmans, HEN, Meijer, M, Womack, JE, Hansen, FG, Meyenburg, Kv (1978). Minichromosomes: plasmids which carry the E. coli replication origin. Mol. Gen. Genet. 162:269.
3. Miki, T, Hiraga, S, Nagata, T, Yura, T (1978). Bacteriophage lambda carrying the E. coli chromosomal region of the replication origin. Proc. Natl. Acad. Sci. USA 75:5099.
4. Leonard, AC, Weinberger, M, Munson, BR, Helmstetter, CE (1980). The effects of oriC containing plasmids on host cell growth. ICN-UCLA Symp. Mol. Cell. Biol. 19:171.
5. Messer, W, Meijer, M, Bergmans, HEN, Hansen, FG, Meyenburg, Kv, Beck, E, Schaller, H (1979). Origin of replication, oriC, of the Escherichia coli K-12 chromosome: nucleotide sequence. Cold Spring Harb. Symp. Quant. Biol. 43:139.
6. Meijer, M, Beck, E, Hansen, FG, Bergmans, HEN, Messer, W, Meyenburg, Kv, Schaller, H (1979). Nucleotide sequence of the origin of replication of the E. coli K-12 chromosome. Proc. Natl. Acad. Sci. USA 76:580.
7. Sugimoto, K, Oka, A, Sugisaki, H, Takanami, M,

Nishimura, A, Yasuda, Y, Hirota, Y (1979). Nucleotide sequence of Escherichia coli replication origin. Proc. Natl. Acad. Sci. USA 76:575.

8. Oka, A, Sugimoto, K, Takanami, M, Hirota, Y (1980). Replication origin of theEscherichia coli K-12 chromosome: the size and structure of the minimum DNA segment carrying the information for autonomous replication. Mol. Gen. Genet. 178:9.

9. Meijer, M, Messer, W (1980). Functional analysis of minichromosome replication: bidirectional and unidirectional replication from the Escherichia coli replication origin, oriC. J. Bact. 143:1049.

10. Kaguni, JM, Fuller, RS, Kornberg, A (1982). Enzymatic replication of E. coli chromosomal origin is bidirectional. Nature 296:623.

11. Lother, H, Buhk, HJ, Morelli, G, Heimann, B, Chakraborty, T, Messer, W (1981). Genes, transcriptional units and functional sites in and around the E. coli replication origin. ICN-UCLA Symp. Mol. Cell. Biol. 21:57.

12. Buhk, HJ, Messer, W (1983). Replication origin region of Escherichia coli: Nucleotide sequence and functional units. Gene: in press.

13. Wechsler, JA (1978). The genetics of E. coli DNA replication. In: Molineux I, Kohiyama M (eds), "DNA Synthesis, Present and Future," Plenum Publishing Corp., New York, p 49.

14. Kornberg, A (1980). "DNA Replication." W. H. Freeman and Company, San Francisco, USA.

15. Fuller, RS, Kaguni, JM, Kornberg, A (1981). Enzymatic replication of the origin of the E. coli chromosome. Proc. Natl. Acad. Sci. USA 78:7370.

16. Lark, KG (1972). Evidence for direct involvement of RNA in the initiation of DNA replication in E. coli 15T. J. Mol. Biol. 64:47.

17. Messer, W (1972). Initiation of DNA replication in E. coli B/r. Chronology of events and transcriptional control of initiation. J. Bact. 112:7.

18. Zyskind, JW, Deen, LT, Smith, DW (1977). Temporal sequence of events during the initiation process in E. coli deoxyribonucleic acid replication: roles of the dnaA and dnaC gene products and ribonucleic acid polymerase. J. Bact. 129:1466.

19. Morelli, G, Buhk, HJ, Fisseau, C, Lother, H, Yoshinaga, K, Messer, W (1981). Promoters in the region of the E. coli replication origin. Mol.

Gen. Genet. 184:255.

20. Lother, H, Messer, W (1981). Promoters in the E. coli replication origin. Nature 294:376.

21. Okazaki, T, Hirose, S, Fujiyama, A, Kohara, Y (1980). Mapping of Initiation sites of DNA replication on prokaryotic genomes. ICN-UCLA Symp. Mol. Cell. Biol. 19:429.

22. Tomizawa, J, Itoh, T (1981). Plasmid ColE1 incompatibility determined by interaction of RNA I with primer transcript. Proc. Natl. Acad. Sci. USA 78:6096.

23. Chakraborty, T, Yoshinaga, K, Lother, H, Messer, W (1982). Purification of the E. coli dnaA gene product. EMBO J. 1:1545.

24. Orr, E, Fairweather, NF, Holland, IB, Pritchard, RH (1979). Isolation and characterization of a strain carrying a conditional lethal mutation in the cou gene of E. coli K-12. Mol. Gen. Genet. 177:103.

25. Kreuzer, KN, Cozzarelli, NR (1979). E. coli mutants thermosensitive for DNA gyrase subunit A: effects on DNA replication, transcription and bacteriophage growth. J. Bact. 140:424.

26. Orr, E, Staudenbauer, WL (1982). Bacillus subtilis DNA gyrase: purification of subunits and reconstitution of supercoiling activity. J. Bact. 151:524.

27. Jacq, A, Lother, H, Messer, W, Kohiyama, M (1980). Isolation of a membrane protein having an affinity to the replication origin of E. coli. ICN-UCLA Symp. Mol.Cell. Biol. 19:189.

28. Jacq, A, Kohiyama, M, Lother, H, Messer, W (1983). Recognition sites for a membrane-derived DNA binding protein preparation in the E. coli replication origin. Mol. Gen. Genet.: in press.

29. Zyskind, JW, Harding, NE, Takeda, Y, Cleary, JM, Smith, D. W. (1981) The DNA replication origin region of the enterobacteriaceae. ICN-UCLA Symp. Mol. Cell. Biol. 22:13.

30. Zyskind, JW, Cleary, JM, Brusilow, WSA, Harding, NE, Smith, D. W.:(1983) Chromosomal replication origin from the marine bacterium Vibrio harveyi functions in Escherichia coli: oriC consensus sequence. Proc. Natl. Acad. Sci. USA 80:1164.

31. Asada, K, Sugimoto, K, Oka, A, Takanami, M, Hirota, Y (1982). Structure of replication origin of the Escherichia coli K-12 chromosome: the presence of spacer sequence in the ori region carrying

information for autonomous replication. Nucl. Acids
Res. 10:3745.

32. Bagdasarian, MM, Izakowska, M, Bagdasarian, M (1977).
Suppression of the dnaA phenotype by mutations in the
rpoB cistron of RNA polymerase in S. typhimurium
and E. coli. J. Bact. 130:577.

33. Filutowicz, M, Jonczyk, P (1981). Essential role of
the gyrB gene product in the transcriptional event
coupled to dnaA-dependent initiation of E. coli
chromosome replication. Mol. Gen. Genet. 183:134.

34. Liu, LF, Chung-Cheng, L, Alberts, BM (1979). T4 DNA
topoisomerase: a new ATP-dependent enzyme essential
for initiation of T4 bacteriophage DNA replication.
Nature 281:456.

Mechanisms of DNA Replication and Recombination, pages 303–315
© 1983 Alan R. Liss, Inc., 150 Fifth Avenue, New York, NY 10011

ANALYSIS OF RECESSIVE HIGH COPY NUMBER MUTANTS OF COLE1[1]

Randolph C. Elble, Jane C. Schneider, Joe Tamm,
Mark A. Muesing, and Barry Polisky

Program in Molecular, Cellular, and Developmental Biology
Department of Biology, Indiana University
Bloomington, Indiana 47405

ABSTRACT The copy number of the E. coli plasmid ColE1
is controlled by a 108 nucleotide plasmid-encoded RNA
known as RNA1. RNA1 acts in trans to inhibit formation
of a functional RNA primer for replication. Using a
novel drug selection procedure we have isolated several
high copy number plasmid mutants of a ColE1 derivative
plasmid. In incompatibility tests in vivo these mutants
are recessive, indicating a defective RNA1 function.
The sequence location of the mutants has been
determined. They occur 5' to the initiating nucleotide
of RNA1 but do not appear to affect the activity of the
RNA1 promoter. We have examined the synthesis and
stability of RNA1 produced in vivo by wild type and copy
number mutant plasmids. The copy number mutants
overproduce an RNA1 transcript which is 4-6 nucleotides
shorter at the 5' terminus than RNA1 synthesized in
vitro from a wild-type template. In addition, the
shorter RNA1 transcript produced by the mutants is
metabolically stable in vivo in contrast to RNA1
produced by cop^+ plasmids, which is highly unstable.

INTRODUCTION

The multicopy plasmids, best exemplified by ColE1 and
its relatives, have served as useful models for understanding
how the initiation of DNA replication is controlled. The

[1]This work was supported by grant GM24212 from the
National Institutes of Health.

nature of the regulatory activity that establishes and maintains plasmid copy number at a constant level has been approached at both the genetic and biochemical levels. Genetic studies have shown that ColE1 encodes information that acts negatively to control copy number and has permitted identification of the target sites for these regulators. The biochemical roles of these effectors have been elucidated by the development of an in vitro system capable of carrying out correct initiation of DNA synthesis using purified host enzymes (1).

Two regions of the ColE1 genome carry information involved in the control of initiation of DNA replication and plasmid copy number. The first is located in a symmetrically transcribed region approximately 400 nucleotides upstream from the replication origin. This region codes on one strand for part of the primer precursor and on the other strand for a small RNA of 108 nucleotides, known as RNA1. The larger RNA is processed by RNaseH to yield a molecule of 555 nucleotides, which serves as a primer for the initiation of DNA replication in vitro (1). The smaller RNA has been shown to inhibit DNA replication by interfering with the processing of the primer by interacting with the complementary structure in the primer precursor (2).

The second trans-acting regulatory element was mapped by Twigg and Sherratt (3) to a region bounded by two HaeII restriction sites 806 and 184 nucleotides downstream from the replication origin of pMB1. This region is located between the replication origin and the mobilization genes of ColE1, is not essential for plasmid replication, and encodes a small polypeptide known as the rop protein (4). The rop protein is a 63aa polypeptide that is entirely conserved between pMB1 and ColE1 derivatives. The rop gene product has been shown to specifically repress transcription initiated at the primer promoter in vivo (4).

In this paper we describe several new copy number mutations with unusual properties. These mutants are shown to map outside the RNA1 coding sequence and do not affect the RNA1 promoter. Nonetheless, these mutants behave genetically as recessive mutants. We have compared RNA1 synthesized in vivo by these cop^- mutants with that made by cop^+ ColE1 plasmids. RNA1 made by cop^+ plasmids is unstable in vivo. However, the cop^- mutants appear to produce an aberrantly stable species of RNA1.

MATERIALS AND METHODS

Bacterial Strains and Plasmids

E. coli strain DG75 (5) was the host strain in all experiments except for incompatibility experiments in which HB101 was the host. All cop⁻ mutants were derived from pNOP42, a 4.2 kilobase-pair cop⁺ derivative of ColE1.

Extraction and Analysis of DNA

DNA was isolated as described by Maniatis et al. (6) and analyzed on agarose gels.

RNA Gels

RNA was analyzed on either 7% or 12% acrylamide gels containing 50% urea. Samples were suspended in 80% deionized formamide, 2X sequencing buffer, 0.5% bromphenol blue, 0.05% xylene cyanol and boiled prior to loading.

RNA Blots

RNA was isolated from mid-log phase cells by hot SDS-phenol extraction and treated with pancreatic DNase I. Fifty micrograms of RNA was applied to each lane of a 7% acrylamide-urea gel and electrophoresed. The RNA was then transferred electrophoretically (7) to diazotized paper (8).

RNAI Probe Preparation and Hybridization

A probe specific for detecting RNA1 was isolated by cloning a 150 base-pair HaeIII-FnuDII restriction fragment containing the RNAI coding sequence into mp8, a derivative of the single-stranded DNA phage, M13 (9). Preparation of the ³²P-labeled probe and hybridization conditions were essentially as described by Hu and Messing (9) except that a synthetic oligonucleotide was used to prime the DNA synthesis reaction. After hybridization and washing, the blots were exposed using an intensifying screen.

RESULTS

Isolation of ColE1 Copy Number Mutants

We have developed a simple technique to select spontaneous high copy number mutants of ColE1 derivatives that carry the bla gene. The technique is based on the fact that the levels of β-lactamase encoded by plasmids is proportional to gene dosage (10); consequently, high copy number mutants for some plasmids can be selected simply on plates containing elevated drug levels. However, the presence of feeder colonies can often reduce the efficiency of this selection procedure due to the limited number of cells (10^3-10^4 per plate) that can be conveniently tested. We found that methicillin, a drug that binds tightly to the plasmid-encoded β-lactamase but is not cleaved by this enzyme, greatly potentiates the effects of ampicillin added as a selective agent. We tested a variety of concentrations of these drugs and found that 4 mg/ml methicillin and .4 mg/ml ampicillin in L-plates permitted us to plate 10^9 cop$^+$ plasmid-containing cells per plate with very little background growth (11). About 70% of the cells containing a ColE1 plasmid capable of growth on these plates contain a high copy number plasmid mutant.

Using this procedure, we isolated a large number of mutants derived from the cop$^+$ ColE1 derivative pNOP42 (11). Colonies were picked at random and the plasmid contained in them was characterized by agarose gel electrophoresis of cleared lysates. The plasmids we studied further have been designated cop3, cop4, and cop7. By densitometric measurements, we have estimated the copy number of cop7 to be about 500 copies per chromosome in mid-log phase cultures compared to about 100 for cop3 and 4 and about 20 copies per chromosome for the wild-type parent plasmid, pNOP42 (Figure 1). In late stationary phase, copy numbers per Klett unit of culture of all plasmids increases but the relative copy number levels are not changed compared to mid-log phase cultures (Figure 1). Each of the mutants is 4.2 kb and shows no gross sequence differences from pNOP42 when probed with a variety of restriction enzymes. Thus, these mutants appeared to contain small sequence alterations, possibly point mutations.

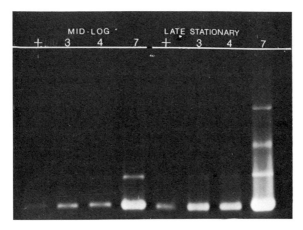

FIGURE 1. Cleared lysates from mid-log and late stationary phase cultures run on an 0.8% agarose gel showing relative copy numbers of cop^+ and cop^- plasmids. DNA from the same number of mid-log or late stationary Klett units was loaded onto each lane. The lowest band represents supercoiled monomer. Higher bands represent supercoiled dimer, trimer, and tetramer, respectively.

We determined the nucleotide sequence alteration in these mutants. Our sequencing efforts were guided by the fact that most previously described copy number mutants characterized by ourselves (11, 12), and others (2, 13, 14) mapped in the overlap region of the coding sequences for RNA1 and the 5'-terminal region of the replication primer. Consequently, we isolated and sequenced restriction fragments in this region. The sequence alterations in these plasmids are shown in Figure 2. None of these mutations is located in the overlap region of RNA1 and the primer. Each is located upstream of the RNA1 gene; thus they are located exclusively in the sequence encoding the primer. cop4 and cop7 are single base-pair alterations at positions 131 and 128 of the primer RNA. Each is an AT → GC transition. The location of these mutations is particularly intriguing in light of our previous observation that a single bp change at position -21 is known to create a temperature-sensitive plasmid copy number phenotype (11). cop4 and cop7, however, have elevated copy number at both 30° and 42°C (data not shown). Cop3 is a 13 bp deletion between positions 152 and 165 of the primer RNA (Figure 2).

```
                           COP 3
        -80              -60    _____  -40              -20         pppACA⟶.
(5')  CCGGATTAGCAGAGCGATGATGGCACAAACGGTGCTACAGAGTTCTTGAAGTAGTGGCCCGACTACGGCTACACTAGAAGGACAGTATTTGG
(3')  GGCCTAATCGTCTCGCTACTACCGTGTTTGCCACGATGTCTCAAGAACTTCATCACCGGGCTGATGCCGATGTGATCTTCCTGTCATAAACC
                                                    COP 4  G  C  COP 7        ▲RNA1start
                                                           C  G
                                                           C  G
```

FIGURE 2. DNA sequence of the region preceding the start site of RNAI and containing the cop3, 4, and 7 mutations. The numbers refer to the start position of RNA1 transcription. The boxes centered around −35 and −10 are the highly conserved sequences characteristic of procaryotic promoters. The replication primer is transcribed from the top strand. Position 1 of RNA1 corresponds to position 111 of the primer.

From their location between the −10 and −35 region of the RNA1 promoter, it was conceivable that the high copy number phenotypes of cop4 and cop7 were due to their effect on the strength of the RNA1 promoter. If these changes reduced the efficiency of the promoter, less RNA1 would be synthesized and copy number might be increased. The possibility that these mutations affected the RNA1 promoter seemed unlikely since they map in a region that, in general, is not highly conserved among promoters (15). Alternatively, the mutations might alter the secondary structure of the primer precursor such that it did not respond to RNA1 levels in vivo (2).

To test whether the RNA1 promoter was affected by these mutations, restriction fragments containing the promoter were isolated from pNOP42, cop3, cop4, and cop7 DNAs. These fragments were cloned upstream of the E. coli galactokinase structural gene contained in the promoter probe plasmid pKO1 (16). In this plasmid, expression of the galk gene is dependent upon upstream insertion of a functional promoter. Extracts from cells containing these promoters driving galk synthesis were assayed for galk activity. The results (not shown) demonstrated no detectable difference in the RNA1 promoter strength between the mutants and wild-type sequences. These results ruled out the possibility that the mutants were altered in RNA1 synthesis due to a defective promoter.

Incompatibility Behavior of cop⁻ Mutants

To gain insight into the nature of these mutations, we carried out an in vivo analysis of their incompatibility properties. We expected that if the mutation led to some defect in RNA1 structure or stability, the mutant would behave as a recessive copy number mutant; i.e., it would be displaced by a wild-type ColE1 plasmid in co-resident tests (17). Alternatively, if the lesion affected the secondary structure of the primer and had no effect on RNA1 synthesis such that primer precursor processing was no longer modulated by RNA1, then the mutants would behave as cis-dominant copy number mutants; i.e., they would displace a resident ColE1 plasmid. In previous studies of this type, cis-dominant mutants have been localized to the overlap region of primer-RNA1; specifically in each of the three potential loops near the 5' terminus of the primer (2, 14). A recessive mutation in ColE1 has been localized to the transcription termination stem of RNA1 (12). This mutation weakens the termination stem and results in readthrough transcription of RNA1 in vitro. The results of cis-trans tests using cop3, cop4, and cop7 are shown in Table 1. To our surprise, all of these mutants behaved as recessive mutants despite the fact that they do not map within the RNA1 structural gene. These results suggested that these mutants were defective in some aspect of RNA1 synthesis or function.

TABLE 1
INCOMPATIBILITY BEHAVIOR OF cop⁻ MUTANTS

Plasmid		Plasmid Retained After		
Incoming	Resident	25 Generations		
A	B	A only	A + B	B only
pNOP42	pNopt	11	53	32
cop3	pNopt	0	0	96
cop4	pNopt	15	10	75
cop7	pNopt	2	0	93
pOP42	pNopt	2	2	84

Approximately 400 ng of plasmid A were transformed into calcium chloride treated HB101 cells carrying a cop⁺ tet^R resident plasmid (Nopt). An aliquot of the transformation mixture was spread on plates containing ampicillin (100 µg/ml) and tetracycline (10 µg/ml). After overnight incubation at 30°C one colony was resuspended in 1 ml saline. A 10^{-4} dilution was used to inoculate 5 ml of 2XYT broth. Cells were grown for 24 hrs at 37°C without antibiotics to allow segregation, then plated on 2XYT plates. About 50 colonies from each transformation were transferred onto plates containing either no antibiotics, a single antibiotic or both. The number of colonies that grew on the double antibiotic plate was subtracted from the number of colonies on either of the single antibiotic plates to obtain the number of single plasmid cells. The resulting value was divided by the number that grew on the plate without antibiotics, and multiplied by 100. pNopt is a 3.7 kb cop⁺ ColE1 plasmid carrying a tet^R marker.

Analysis of Steady-state Levels of RNAI

In order to measure the steady-state concentration of RNAI, total cell RNA was extracted from mid-log cells, run on

7% acrylamide gels, and then blotted electrophoretically onto diazotized paper. The blots were hybridized with a [^{32}P]-labeled RNAI-specific probe derived from a 150 bp HaeIII-Fnu DII restriction fragment. The autoradiogram of the hybridized blot (Figure 3b) shows two bands of equal intensity in cop$^+$ RNA. The upper band comigrates with RNAI synthesized in vitro (data not shown). The lower band (RNAI*) occurs only in vivo and is of much greater intensity in the cop7, cop3, and cop4 patterns than RNAI. The RNAI* transcript comprises about 4% of total cellular RNA in cop3, 4, and 7 extracts and is visible by ethidium-bromide staining (Figure 3a). No other cop$^-$ mutant thus far investigated shows this high level of RNAI* on Northern blots. The half-lives of RNAI and RNAI* have been measured in cop$^+$- and cop7-containing cells by pulse-chase experiments (data not shown). RNAI and RNAI* both have a half-life of about one minute in cop$^+$ cells. In cop7-containing cells, however, RNAI has a half-life of about three minutes, and RNAI* has a half-life of about 30 minutes. Thus the RNAI* transcript in cop7-containing cells is aberrantly stable. The RNAI* band from cop7 RNA has been eluted from acrylamide gels, labeled at the 5' end using γ-^{32}P ATP and T4 polynucleotide kinase, and sequenced. The results show that 4-6 nucleotides that occur at the 5' end of RNAI in vitro are missing from the stable in vivo RNAI* band (results not shown). It is not yet known whether RNAI* represents a different transcriptional start from RNAI or endonucleolytic processing of the longer transcript.

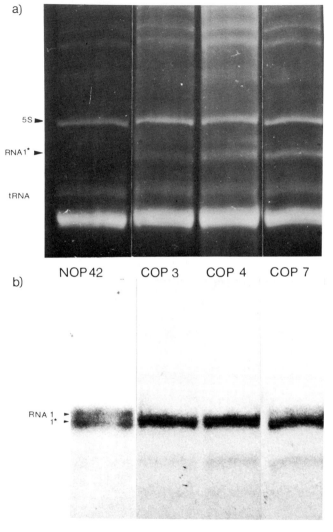

a)

5S ▶

RNA1* ▶

tRNA

NOP 42   COP 3   COP 4   COP 7

b)

RNA 1 ▶
1* ▶

FIGURE 3. Analysis of steady-state levels of RNA1 in vivo. Total cell RNA was isolated from cells containing pNOP42 (cop⁺) or the cop⁻ plasmids cop4, cop4, and cop7. Part a shows the RNA analyzed on a 7% acrylamide gel and stained with ethidium bromide. A prominent band labeled RNA1* is present in extracts from cop3, 4, and 7 that is not present in the extract from pNOP42 or plasmid-free cells (not shown). The RNA in these lanes was transferred to diazotized paper and probed with an RNA1-specific probe as described in

Methods (part b). The lane containing RNA from pNOP42 has been exposed about 10-fold longer than the remaining three lanes to generate a visible signal. The position of the prominent labeled band in part b corresponds in position to the band labeled RNA1* in part a.

Discussion

We have described the isolation and characterization of several cop⁻ mutants of ColE1 isolated by a novel drug selection procedure. These mutations map upstream from the replication origin in a region which encodes the replication primer RNA. Although these mutations lie immediately upstream of the 5'-terminus of RNA1, they appear not to affect the promoter of RNA1. This conclusion stems from observations that RNA1 is made by these mutants in vivo and in vitro. Nevertheless, when tested in co-residence with a cop⁺ ColE1 plasmid in vivo, these cop⁻ mutants display sensitivity to wild-type incompatibility determinants. These observations create a paradox since RNA1 is believed to be the major, if not the sole, molecular component of ColE1 incompatibility. A simple scheme for the relationship between replication control and incompatibility on the one hand, and RNA1 and the primer on the other, would predict that high copy number mutants could arise by mutation in RNA1 or primer. This situation is complicated by the fact that these components overlap such that mutations in one also alter the other. Nevertheless, since the primer is thought not to be diffusible, high copy number mutations altering RNA1 synthesis or conformation could have a variety of incompatibility phenotypes. Mutations altering critical aspects of RNA1 secondary structure could conceivably be recessive, neutral, or dominant with respect to their activity to wild-type primer in trans. Certain mutations could also create new incompatibility groups (2, 14). A mutation mapping outside RNA1 but within the primer would be expected to be recessive only if it altered the promoter activity of RNA1 or altered the initiation position of RNA1. We do not know if the RNA1 produced in vivo by the cop7, cop3, and cop4 plasmids initiates at new sites. The smaller RNA1-related species produced by these plasmids can also be detected at low levels in cop⁺ plasmid-containing cells. In contrast to this cop⁺ transcript, however, RNA1* is extremely stable in vivo. The basis for this stability is not known, nor is it apparent whether RNA1* can interact with the primer.

The region of the plasmid genome in which these mutations occur plays a critical role in the regulation of copy number. We have isolated a number of mutations in the region between the -10 and -35 components of the RNA1 promoter. This region is unusually GC rich (75%). In addition to cop4 and cop7, two temperature-sensitive replication mutations, pOM1 and pEW27, also map in this region (11). We have suggested previously that this region of the primer RNA is involved in a secondary structural configuration which is essential for proper RNA1 interaction. An understanding of this configuration will require detailed biochemical information about the structure of the primer.

ACKNOWLEDGEMENTS

M.A.M. was a predoctoral trainee supported by a training grant in Molecular Biology from the National Institutes of Health. B.P. was a recipient of a Research Career Development Award from the National Institute of Allergy and Infectious Diseases.

REFERENCES

1. Itoh T. and Tomizawa J. (1980). Proc. Natl. Acad. Sci. (USA) 77:2450.
2. Tomizawa J. and Itoh T. (1981). Proc. Natl. Acad. Sci. (USA) 78:6096.
3. Twigg A.J. and Sherratt D. (1980). Nature 283:216.
4. Cesareni G., Muesing M.A., and Polisky B. (1982). Proc. Natl. Acad. Sci. (USA) 79:6313.
5. O'Farrell P.H., Polisky B., and Gelfand D.H. (1978). J. Bacteriol. 134:645.
6. Maniatis T., Fritsch E.F., and Sambrook J. (1982). "Molecular Cloning: A Laboratory Manual," New York: Cold Spring Harbor, pp. 368-369.
7. Stellwag E.J. and Dahlberg A.E. (1980). Nucleic Acids Research 8(2):299.
8. Maniatis et al., op. cit., pp. 342-343.
9. Hu N. and Messing J. (1982). Gene 17:271.
10. Uhlin B. and Nordstrom K. (1977). Plasmid 1:1.

11. Wong E.M., Muesing M.A., and Polisky B. (1982). Proc. Natl. Acad. Sci. (USA) 79:3570.
12. Muesing M., Tamm J., Shepard H.M., and Polisky B. (1981). Cell 24:235.
13. Conrad S.E. and Campbell J.L. (1979). Cell 18:61.
14. Lacatena R.M. and Cesareni G. (1981). Nature 294:623.
15. Rosenberg M. and Court D. (1979). Ann. Rev. Genet. 13:319.
16. McKinney K., Shimatake H., Court D., Schmeissner U., Brady C., and Rosenberg M. (1981). In "Gene Amplification and Analysis, J.G. Chirikjian, T.S. Papas (eds.), Elsevier/North-Nolland, Amsterdam, Vol. 2, pp. 384-417.
17. Shepard H.M., Gelfand D.H., and Polisky B. (1979). Cell 18:267.

Mechanisms of DNA Replication and Recombination, pages 317–326
© 1983 Alan R. Liss, Inc., 150 Fifth Avenue, New York, NY 10011

PROPERTIES OF THE PURIFIED dnaK GENE
PRODUCT OF ESCHERICHIA COLI[1]

Maceij Zylicz and Costa Georgopoulos

Department of Cellular, Viral and Molecular Biology
University of Utah Medical Center
Salt Lake City, Utah 84132

ABSTRACT   The dnaK gene of Escherichia coli was
originally discovered because mutations in it, like
dnaK756, did not allow bacteriophage λ DNA replication.
The dnaK gene product (gpdnaK[2]) has also been shown
to be essential for E. coli viability, since bacteria
carrying the dnaK756 mutation do not form colonies
at 42°C. gpdnaK has been identified as protein B66.0,
a member of the heat-shock group of E. coli proteins.
Purified gpdnaK has been shown to possess both DNA-
independent ATPase and autophosphorylating activities.
Purified gpdnaK has also been shown to interact in
vitro with the gpO and gpP replication functions of
bacteriophage λ. The interaction with gpP confirms
the results of earlier genetic experiments. gpdnaK
has been shown to be essential for an in vitro DNA
replication system. Purified gpdnaK has retained
this biological activity.

INTRODUCTION

Previous work from our laboratory (1,2) and that of
others (3,4) has uncovered the existence of two closely
linked genes, dnaK and dnaJ, which map at 0.3 minutes of
the E. coli genetic map. The products of these genes (previously referred to as groP) are required for λ DNA

[1]This work was supported by NIH grant GM23917.
[2]gp refers to gene product e.g. gpdnaK refers to the
gene product of the dnaK⁺ gene.

replication (1,2,3,4) and for E. coli viability at high temperature (2,3,4,5). The two genes form an operon with the genetic order of the loci being thr...promoter dnaK dnaJ...leu (5,6). Mutations in the dnaK and dnaJ genes have been shown to inhibit both bacterial DNA and RNA syntheses at the non-permissive temperature (3,5,6,8). Mutations in gene P enable λ to bypass the block exerted by bacterial hosts carrying mutations in the dnaK and dnaJ genes (1,2,3,4) suggesting a potential protein-protein interaction between gpP and gpdnaK and gpdnaJ. In this report we explore the properties of purified gpdnaK and show that some of the biological predictions of genetic experiments are fulfilled.

METHODS AND MATERIALS

All the bacterial and phage strains used in this work have been previously described (2,6,9,10). The purification of gpdnaK and the methods of procedure have also been described (10).

RESULTS

Figure 1 shows the proof that gpdnaK is protein B66.0 of E. coli. Panel A shows the position of wild type gpdnaK. Panel B shows that gpdnaK756 codes for a protein with a more acidic isoelectric point. Panel C shows the positions of both gpdnaK and gpdnaK756 in mixed extracts. Panel D shows the position of gpdnaK Ts$^+$7, coded by a temperature resistant revertant at 42°C of dnaK756 (9). Panel E shows the position of gpdnaK Ts$^+$ 26, coded by another temperature resistant revertant. gpdnaK Ts$^+$26 codes for a protein with more basic isoelectric point than wild type gpdnaK. Panel F shows the positions of both gpdnaK and gpdnaK Ts$^+$26 in mixed extracts. These results clearly demonstrate that gpdnaK is the protein located in two-dimensional gels in the indicated position (the B66.0 protein).

The purification of gpdnaK was aided by the cloning of the dnaK$^+$ gene onto the "runaway" plasmid pMOB45 (10,11).

A   B   C

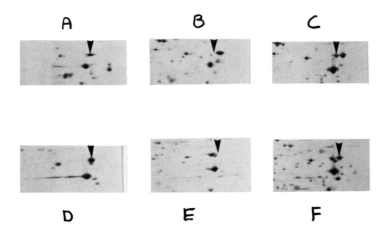

D   E   F

Figure 1. Identification of gpdnaK on two-dimensional gels. The $[^{35}S]$ met-labeled extracts in the various panels (A-F) are described in the text. Only the regions of the gels corresponding to a pH gradient of 5.8 to 5.0 (in the horizontal dimension) and an M_r of 55,000 to 80,000 (in the vertical dimension) are shown. The arrow points to the position of wild type gpdnaK.

Bacteria carrying the pMOB45 dnaK$^+$ plasmid were shown to overproduce gpdnaK about fifty-fold. Starting with such strains we purified gpdnaK to homogeneity. The purified protein has been shown to possess an ATPase activity which breaks down ATP to ADP and P_i (10). The K_m of the enzyme has been estimated to be 0.2mM ATP. The rate of hydrolysis is 10nmoles of ATP per mg of purified protein per minute at 30°C. Surprisingly, the ATPase activity of gpdnaK is both DNA independent and resistant to heating at 95°C for ten minutes. The proof that the ATPase activity is a property of gpdnaK and not due to another contaminating protein relies on the following data. (i) As Table 1 shows the levels of the heat resistant ATPase activity in crude extracts of various E. coli strains varies. Bacteria carrying the dnaK756 mutation consistently showed five- to ten- fold lower levels of dnaK-specified ATPase activity.

TABLE 1
ATPase ACTIVITY OF VARIOUS E. COLI STRAINS

Source of Extract	Total ATPase*	Heat Resistant ATPase*
dnaK$^+$	328	0.2
dnaK756	324	0.03
dnaK$^+$(pMOB45)	330	0.9
dnaK$^+$(pMOB45dnaK$^+$	324	11.2

*The total ATPase activity refers to nmoles of $[\alpha-^{32}P]$ ATP molecules broken per mg protein; the heating resistant ATPase activity was assayed after incubating the crude extracts at 95°C for five minutes. The dnaK$^+$ and dnaK756 extracts were assayed after immunoprecipitation with anti-gpdnaK IgG antibodies in the presence of N-ethylmaleimide, nalidixic acid and rifampicin (10).

Bacteria which overproduced gpdnaK, as judged by SDS-PAGE, showed a corresponding increase in the heat-resistant ATPase activity. (ii) As Figure 2 shows, crude extracts derived from dnaK756 bacteria possess a relatively more thermolabile ATPase activity than those from dnaK$^+$. (3) The heat-resistant ATPase activity of dnaK756 bacteria co-migrates in native isoelectrofocusing gels with the altered gpdnaK756 protein (9,10). All these data taken together prove that gpdnaK codes for the heat-resistant ATPase activity that we described.

When purified gpdnaK is incubated in vitro with $[\gamma-^{32}P]$ ATP a certain fraction of it becomes phosphorylated (Figure 3). Phosphoamino acid anslysis indicated that gpdnaK is phosphorylated at a threonine residue. We believe this to be an autophorphorylating reaction because (i) similarly to the ATPase activity, it is resistant to heating at 95°C for ten minutes and (ii) gpdnaK is phosphorylated in immune complexes with purified anti-gpdnaK IgG antibodies. The kinase activity of gpdnaK is

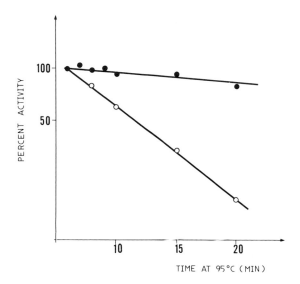

Figure 2. gp<u>dnaK</u> is more resistant to heat inactivation than gp<u>dnaK</u>756. —●–●–●– <u>dnaK</u>+ extracts;
–0–0–0– <u>dnaK</u>756 extracts.

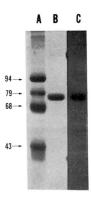

Figure 3. <u>In vitro</u> phosphorylation with $[\gamma\text{-}^{32}P]$
ATP of purified gp<u>dnaK</u>. (A) Molecular weight standards.
(B) Coomassie blue stained gp<u>dnaK</u>. (C) Autoradiogram
of the gel shown in lane B.

very specific since added bovine serum albumin, casein and lysozyme are not phosphorylated. Phosphorylated gpdnaK possesses a more acidic isoelectric point as judged by two-dimensional electrophoresis. By labeling dnaK+ bacteria with $^{32}P_i$ it can be demonstrated that a certain fraction of gpdnaK exists in vivo in a phosphorylated form (Figure 4). The in vivo and in vitro phosphorylated forms of gpdnaK co-migrate in two-dimensional gels.

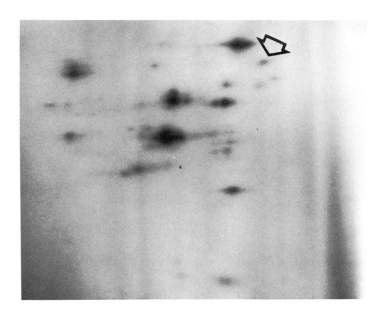

Figure 4. gpdnaK is phosphorylated in vivo. Auto-radiogram of two-dimensional gel electrophoresis of ^{32}P-labeled E. coli extracts. The arrow points to the position of phosphorylated gpdnaK.

In collaboration with Jonathan Lebowitz and Roger McMacken we have been able to demonstrate in vitro that our purified gpdnaK has retained biological activity (10). The assay system for in vitro DNA replication of λdv is that described by Wold et al. (12). The replication of λdv is carried out in a certain fraction of E. coli extracts and requires exogenously added gpP and gpO (Table 2).

TABLE 2
gpdnaK IS REQUIRED FOR λdvDNA
REPLICATION IN VITRO*

Extract		pmoles dNTP
(A)	Complete system*	42.0
(B)	As in (A) - gpP	3.0
(C)	As in (A) + antigpdnaK IgG	4.0
(D)	As in (C) + 2μg gpdnaK	39.0
(E)	As in (D) - gpP	3.5

* The DNA replication system is identical to that
described in Wold et al. (12). This experiment
was carried out in collaboration with Jonathan
Lebowitz and Roger McMacken.

Addition of purified IgG from sera directed against gpdnaK
inhibits this in vitro DNA replication, showing that
gpdnaK is necessary in the system. This inhibition,
however, can be reversed by the addition of excess gpdnaK
(Table 2). This proves that our purified gpdnaK
preparation has retained biological activity.

DISCUSSION

The purified protein of the dnaK[+] gene of E. coli
has been shown to possess multiple, interesting biological
properties. It possesses a DNA-independent, heat-
resistant ATPase activity. The ATPase activity is due
to gpdnaK because gpdnaK756 exhibits a more heat-labile
ATPase activity, with a more acidic isoelectrofocusing
point. The ATPase activity is retained in immune
complexes with purified anti-gpdnaK IgG. Another interesting biological property of purified gpdnaK is its in vitro
phosphorylation at a threonine residue. A fraction of
gpdnaK can be shown to be phosphorylated in vivo as well.
It is not known whether the in vivo phosphorylated amino

acid is also threonine. The in vitro phosphorylation is probably due to an autophosphorylation reaction because (i) the activity, similar to the ATPase, is heat-resistant and (ii) it occurs in immune complexes with purified anti-gpdnaK IgG. The phosphorylating activity of gpdnaK is very specific since IgG, casein, bovine serum albumin, and lysozyme are not phosphorylated. We have shown recently that purified gpdnaK is active in an in vitro DNA replication system which is dependent on the λ replication functions gpO and gpP (10). Here we have shown that purified gpdnaK is also active in a λdv in vitro replication system. The exact role that gpdnaK plays in these DNA replication systems has not been established. The most likely step is at the pre-priming level of primer RNA synthesis.

It has been shown recently that gpdnaK belongs to the heat-shock group of E. coli proteins (9,13). These proteins are transiently overproduced upon a temperature shift (or during "stress" conditions). We have shown that gpdnaK is the inhibitor of this transient over-production (K. Tilly, M.Z. and C.G., unpublished results). A remarkable observation made is that a protein similar in size and amino acid composition (up to fifty percent homology) to gpdnaK, is also found in eukaryotes (14). It is called hsp70 and is a member of the eukaryotic heat-shock proteins. Like gpdnaK, hsp70 has been shown to be a modulator of the heat-shock response in Drosophila (15). As of now, no enzymatic activities have been shown to be associated with hsp70. It would be interesting to see whether the structural and regulatory similarities between hsp70 and gpdnaK have been conserved at the enzymatic levels as well.

REFERENCES

1. Georgopoulos CP, Herskowitz I (1971). Escherichia coli mutants blocked in lambda DNA synthesis. In Hershey AD (ed): "Bacteriophage λ," New York: Cold Spring Harbor Laboratory, p. 553.
2. Georgopoulos CP (1977). A new bacterial gene (groPC) which affects λ DNA replication. Mol Gen Genet 151:35.

3. Saito H, Uchida H (1977). Initiation of the DNA replication of bacteriophage lambda in Escherichia coli K12. J Mol Biol 113:1.

4. Sunshine M, Feiss M, Stuart J, Yochem J (1977). A new host gene (groPC) necessary for lambda DNA replication. Mol Gen Genet 151:27.

5. Saito H, Uchida H (1978). Organization and expression of the dnaJ and dnaK genes of Escherichia coli K12. Mol Gen Genet 164:1.

6. Yochem J, Uchida H, Sunshine M, Saito H, Georgopoulos CP, Feiss M (1978). Genetic analysis of two genes dnaJ and dnaK necessary for Escherichia coli and bacteriophage lambda DNA replication. Mol Gen Genet 164:9.

7. Itikawa H, Ryu J-I (1979). Isolation and character-ization of a temperature-sensitive dnaK mutant of Escherichia coli B. J Bacteriol 138:339.

8. Wada M, Kadokami Y, Itikawa H (1982). Thermosensitive synthesis of DNA and RNA in dnaJ mutants of Escherichia coli K12. Jpn J Genet 57:407.

9. Georgopoulos CP, Tilly K, Drahos D, Hendrix R (1982). The B66.0 protein of Escherichia coli is the product of the dnaK$^+$ gene. J. Bacteriol 149:1175.

10. Zylicz M, LeBowitz J, McMacken R, Georgopoulos CP (1983). The dnaK protein of Escherichia coli possesses an ATPase activity and autophosphorylating activity and is essential in an in vitro DNA replication system. Submitted to Proc Ntl Ac Sci USA.

11. Bittner M, Vapnek D (1981). Versatile cloning vectors derived from the runaway replication plasmid pKN402. Gene 15:319.

12. Wold MS, Mallory JB, Roberts JD, LeBowitz JH, McMacken R (1982). Initiation of bacteriophage λ DNA replication in vitro with purified λ replication proteins. Proc Ntl Ac Sci USA 79:6176.

13. Yamamori T, Ito K, Nakamura Y, Yura T (1978). Transient regulation of protein synthesis in Escherichia coli upon shift-up of growth temperature.

14. Craig E, Ingolia T, Slater M, Manseau L, Bardwell J (1982). Drosophila, Yeast and E. coli genes related to the Drosophila heat-shock genes. In Schlessinger MJ, Ashburner M, Tissieres A (eds): "Heat shock from bacteria to man," New York: Cold Spring Harbor Laboratory, p. 11.

15. DiDomenico BJ, Bugaisky GE, Lindquist S (1982). The heat shock response is self-regulated at both the transcriptional and post-transcriptional levels, Cell 31:593.

Mechanisms of DNA Replication and Recombination, pages 327–336
© **1983 Alan R. Liss, Inc., 150 Fifth Avenue, New York, NY 10011**

PURIFIED ROP PROTEIN INHIBITS
COLE1 REPLICATION <u>IN VITRO</u>

Rosa M. Lacatena, David W. Banner,
and Gianni Cesareni

European Molecular Biology Laboratory,
Postfach 10.2209, D-6900 Heidelberg

ABSTRACT   We show that a peptide of 63 amino-
acids (P63) which is encoded in the region
previously identified as the <u>rop</u> gene (6) is a
negative regulator of ColE1 plasmid replication.
Purified P63, when added to a crude extract
active in plasmid replication,   specifically
inhibits   replication   of   ColE1   and   pMB1
derivatives but not RSF1030 derivatives. We have
not been able to demonstrate any phenotype of
purified   P63   on   a   replication   system
reconstituted from purified enzymes (9).

INTRODUCTION

Plasmids   of   the   ColE1   family   regulate   their
replication at the level of primer formation. Two regions
near the replication origin have been shown to be involved
in the control of plasmid copy number (1, 2). The first is
located approximately 500 nucleotides upstream from the
replication origin and encodes an RNA molecule of 108
nucleotides (RNA1) which prevents primer processing by base
pairing with part of the complementary structure in the
primer precursor (3, 4, 5).   The second regulation
mechanism is mediated by an inhibitor (Rop) which acts <u>in</u>
<u>trans</u> by interfering with transcription starting from the
primer promoter (6).   The inhibitor is encoded in a region
500 nucleotides downstream from the replication origin
which directs the synthesis of a 63 amino-acid (aa) peptide

(P63). Here we report the purification of the 63 aa protein and its identification as the <u>rop</u> gene product.

RESULTS

Construction of a Rop Overproducer

In order to increase and control the level of production of the 63 amino-acid protein we have constructed a plasmid in which the pBR322 <u>Sau3a</u> fragment which codes for P63 has been inserted in front of the λpL promoter in the unique BamHI site of plasmid pLC28 (7). The construction of the overproducer plasmid, pL6, is illustrated in Figure 1.

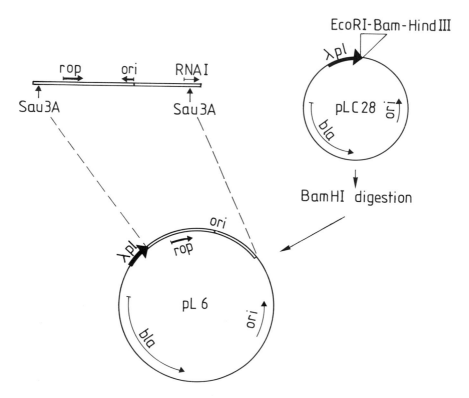

FIGURE 1. Construction of the P63 overproducer plasmid pL6

Plasmid pL6 was transformed into a minicell producing strain which contains a λ prophage and the protein products synthesized by purified minicells were analyzed on SDS PAGE gels (Figure 2). pL6 directs the synthesis of a new peptide, not synthesized by the original vector pLC28 (lanes 5 and 6 in Figure 2), which comigrates with P63. This protein product is at least 20 times more abundant than the corresponding P63 product synthesized by pBR322 when the β-lactamase bands are taken as standards (compare lanes 3 and 6 in Figure 2).

Figure 3 shows the induction of P63 synthesis directed by λpL under the control of a thermosensitive repressor in

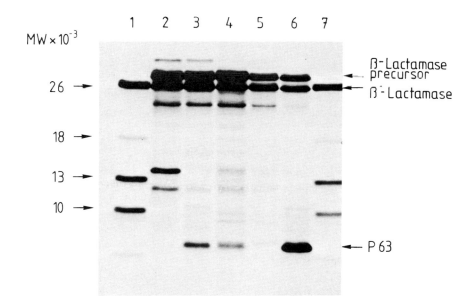

FIGURE 2. SDS PAGE (20%) of the protein products of E. coli minicells. Purified minicells were incubated for 30' at 37°C in the presence of 5 μcuries of ^{35}S methionine. Slots 1 and 7 have been loaded with tritiated concanavalin A (Amersham) as a molecular weight marker. Corresponding molecular weights are shown on the left. Slots 2-6 have been loaded with extracts of minicells containing (slot 2) pD1180, a deletion derivative of pBR322, (slot 3) pBR 322, (slot 4) pUC8-HpaII309, (slot 5) pLC28, (slot 6) the P63 overproducing plasmid pL6.

a λ prophage. No band can be identified as P63 in total cell extracts of strains which harbour pL6 or pLC28 and have been labelled with ^{35}S methionine for 10' at 30°C. When the temperature is raised to 42°C, however, a protein product which comigrates with P63 appears in extracts derived from the strain containing pL6 but not from the strain containing pLC28. After 45' incubation at 42°C the P63 band contains approximately 1/100 of the total radioactivity.

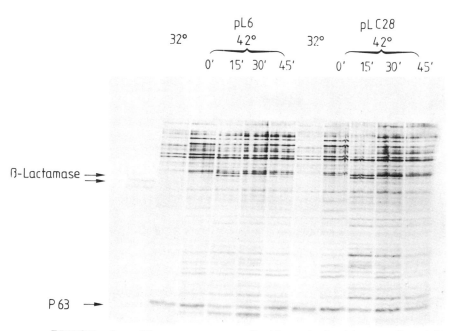

FIGURE 3. Time course of the induction of P63 synthesis in the overproducer plasmid pL6 under the control of a thermosensitive repressor. Bacterial strains containing a λ prophage with a thermosensitive repressor (cI857) and either the overproducer plasmid pL6 or the vector pLC28 were grown in minimal M9 medium at 30°C up to O.D. 0.3 at 600 nm. At this time pL transcription was derepressed by changing the temperature to 42°C. 200 μl samples were withdrawn at different times after induction and incubated for 10' at 42°C in the presence of 5 μcuries of ^{35}S methionine. Samples were electrophoresed on a 20% SDS polyacrylamide gel.

Large Scale Purification of P63

Cells were grown at 32°C in 100 l of culture medium (1000 g Bacto Tryptone; 500 g NaCl; 500 g Yeast Extract) to an O.D.$_{600}$ of 0.6. After a rapid temperature change to 42°C growth was allowed to continue for 3 hrs to a final O.D.$_{600}$ of 0.9. 110 g wet weight of cells were harvested using a continuous flow centrifuge, frozen in liquid nitrogen and stored at -70°C. All further operations were carried out at $4-7^{\circ}$C. All buffers contained 50 mM sodium phosphate, pH 7.4, 1 mM EDTA, 0.02% sodium azide, 1 mM DTT and sodium chloride as indicated. 50 g cells were suspended in 100 ml buffer made 0.45 M in sodium chloride and 0.1 mM in PMSF. After passage twice through a French pressure cell operated at 1000 p.s.i., the solution was cleared by centrifugation for 10' at 10,000 r.p.m. and the supernatant further cleared by centrifugation at 60,000 r.p.m. for 2 hrs.

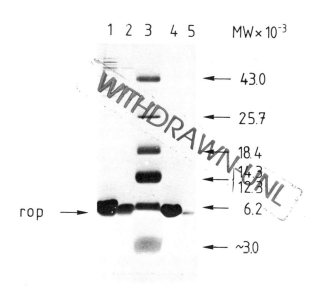

FIGURE 4. Purification of P63. Samples were electrophoresed on a 20% SDS PAGE and after fixation protein bands were visualized by Coomassie staining. Lanes 1 and 2 peak fraction (10 and 2 μg) of the gradient elution from a DEAE column. Lane 3 prestained molecular weight standards (Bethesda Research Laboratories, Inc.). Lane 4 and 5 redissolved crystals (10 and 2 μg).

This solution was diluted with buffer to a salt concentration of 0.15 M, passed twice rapidly through 50 ml of phosphocellulose (Whatman) and loaded on a 200 ml DEAE cellulose column. This was washed with 1 l of starting buffer and the protein eluted using a 2 l linear gradient of salt from 0.15 M to 0.45 M. As assessed by SDS PAGE the rop protein elutes as a single sharp peak at about 0.275 molar salt and is, at best, about 90% pure (Figure 4, slots 1 and 2). Gel filtration on Ultrogel ACA54 (LKB) in 0.1 M salt removes high molecular weight contaminants and gives an apparent molecular weight of 17-18,000. This is to be contrasted with the apparent molecular weight of 6,000 from SDS PAGE. Further purification is obtained by repeated crystallization using, for example, polyethylene glycol (PEG) containing solutions buffered to pH 6 with citrate. Small crystals grow readily on the addition of magnesium sulphate to about 0.1 M. They dissolve at pH 7.5-8.0 to give pure protein (Figure 4, slots 4 and 5). The identity of the protein has been confirmed by amino-acid analysis and partial sequencing (R. v.d. Broek and A. Tsugita, in preparation). About 30 mg of protein are obtained from 50 g cells.

Purified P63 Inhibits Plasmid Replication in vitro

In order to prove that P63 is the product of the rop gene we have tested its activity on plasmid replication in a crude extract. An extract able to support ColE1 type replication was prepared from a culture of C600 cells containing the plasmid pUC8 (8). The incorporation of tritiated dTTP when the extract is incubated at 30°C for 60' in different conditions is shown in Table 1.

TABLE 1
INHIBITION OF IN VITRO REPLICATION BY PURIFIED P63

Plasmid	dNTP Incorporation (pmoles)[a] P63	
	−	+
None	1	1
pBR322	21	2
RSF1030	19	18

[a]The experiments were performed in the conditions already described for fraction 1(1)

While dTTP incorporation in the absence of exogenously added plasmid is negligible, up to 20 picomoles of dNTP are incorporated in acid insoluble material when 1 μg of pBR322 or of RSF1030 is added to the extract. pBR322 replication, however, is reduced to control levels in the presence of 1 μg of purified P63 protein. This P63 dependent inhibition of plasmid replication is specific for replicons derived from pMB1 or ColE1, while RSF1030 replication, for instance, is not affected by the addition of purified P63. These results prove that P63 is an inhibitor of plasmid replication and, since P63 is encoded the region which contains the rop gene (6), strongly suggests that P63 is the product of the rop gene. Genetic data had previously suggested that the product of the rop gene might regulate plasmid replication by affecting the transcription of the primer precursor (6). We have therefore tested whether this activity could be shown in an in vitro system constituted by purified RNA polymerase and P63. Figure 5 panel A (lane 2) shows the gel pattern of transcripts obtained in vitro from plasmid pBR322 with purified RNA polymerase. Transcripts which initiate from the primer promoter were identified by digestion with the enzyme RNaseH (lane 3). However no qualitative or quantitative difference was reproducibly observed in the band pattern of the transcripts obtained in the presence of 0.5 μg of purified P63 (lane 4 and 5). These data suggest that the inhibitory activity of the rop gene product on primer transcription is somewhat different from classical repression of initiation of transcription. Furthermore the footprinting experiments shown in Figure 5 panel B shows that Rop does not bind (lane 2 and 3) to a fragment of DNA containing the primer promoter and 40 bases downstream in conditions in which RNA polymerase binding is clearly observed (lane 4).

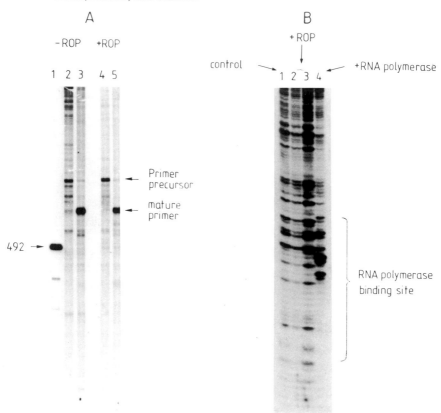

FIGURE 5. Panel A: <u>in vitro</u> transcription of plasmid pBR322. Transcription and RNaseH processing was carried out according to Itoh and Tomizawa (9). Transcripts were analyzed on a polyacrylamide (3%) urea gel. Incubation was for 30' at 30°C in presence of RNA polymerase (lane 2) RNA polymerase + RNaseH (lane 3) RNA polymerase + Rop (0.5 μg) (lane 4) RNA polymerase +RNaseH + Rop (0.5 μg) (lane 5). Lane 1 molecular weight markers. Panel B: Footprinting of Rop and RNA polymerase on the primer promoter. The <u>HpaII</u> site at position 3046 in the pBR322 map was end labelled by filling in with DNA polymerase Klenow subunit and the <u>HpaII-TaqI</u> fragment containing the primer promoter was used in the footprinting experiment. Approximately 0.1 μg of DNA was digested for 15' at 37°C with 0.2 ng of DNase I. The degradation products were analyzed on a sequencing (6%) gel (lane 1). The same experiment was repeated in the presence of 4 μg (lane 2) and 16 μg (lane 3) of Rop or 1.5 μg of RNA polymerase (lane 4).

DISCUSSION

ColE1 is the only replicon whose regulation of initiation of replication has been reproduced in a purified *in vitro* system. Tomizawa and Itoh (4) have shown that addition of RNA1 to an *in vitro* system constituted by purified RNA polymerase, DNA polymerase I and RNaseH inhibits initiation of plasmid replication by preventing the maturation of the primer.

A second regulatory mechanism participates in the control of plasmid replication (2). This mechanism has been recently shown *in vivo* to be dependent on the product of the *rop* gene which regulates plasmid replication by interfering with transcription starting from the promoter of the primer (6). Here we report the purification of the *rop* gene product and its identification as a slightly acidic protein of 63 amino-acids.

The purified Rop protein can specifically inhibit ColE1 plasmid replication in crude extracts. However we have not been able to show any activity of the purified protein in an *in vitro* transcrition system. These results, taken together with the *in vivo* data (6), indicate that some factor present in the crude extract, but absent from the purified system, is essential for Rop inhibitory activity. Preliminary data from our laboratory suggest that this factor is plasmid encoded.

ACKNOWLEDGMENTS

Some of the P63 used in this work was purified from a strain containing a different plasmid (pMAM7) which is approximately 5 times more efficient than our pL6 in the overproduction of the peptide. We are grateful to B. Polisky and M. Muesing for providing the plasmid before publication.

We would also like to thank W.L. Standenbauer and G. Hillenbrand for advices concerning the *in vitro* system and I. Benner for typing the manuscript.

REFERENCES

1. Conrad SE, Campbell JL (1979). Characterization of an improved *in vitro* DNA replication system for *Eschrichia coli* plasmids. Nucleic Acids Res 6:3289-3303.

2. Twigg AJ, Sherratt D (1980). Trans complementable copy number mutants of plasmid ColE1. Nature 283:216–218.
3. Muesing M, Tam J, Shepard HM, Polisky B (1981). A single base-pair alteration is responsible for the DNA overproduction phenotype of a plasmid copy number mutant. Cell 24:235–242.
4. Tomizawa J, Itoh T (1981). Plasmid ColE1 incompatibility determined by interaction of RNAI with primer transcript. Proc Natl Acad Sci USA 78:6096–6100.
5. Lacatena RM, Cesareni G (1981). Base pairing of RNAI with its complementary sequence in the primer precursor inhibits ColE1 replication. Nature 294:623–626.
6. Cesareni G, Muesing MA, Polisky B (1982). Control of ColE1 DNA replication: The rop gene product negatively affects transcription from the replication primer promoter. Proc Natl Acad Sci USA 79:6313–6317.
7. Remant E, Straussens P, Fiers W (1981). Plasmid vectors for high-efficiency expression controlled by the pL promoter of coliphage lambda. Gene 15:81–93.
8. Viciza J, Messing J (1982). The pUC plasmids, an M13mp7-derived system for insertion mutagenesis and sequencing with synthetic universal primers. Gene 19:259–268.
9. Itoh T, Tomizawa J (1980). Formation of an RNA primer for initiation of replication of ColE1 DNAA by ribonuclease H. Proc Natl Acad Sci USA 77:2450–2454.

Mechanisms of DNA Replication and Recombination, pages 337–349
© 1983 Alan R. Liss, Inc., 150 Fifth Avenue, New York, NY 10011

polA[+]-INDEPENDENT REPLICATION OF CONCATEMERIC pBR322
IN sdrA MUTANTS OF Escherichia coli K-12.

Tokio Kogoma[*†] and Nelda L. Subia[*]

Department of Biology[*]
and
Department of Cell Biology,
Cancer Research and Treatment Center[†]
University of New Mexico
Albuquerque, NM 87131, USA

ABSTRACT sdrA mutants of E. coli capable of
continued DNA replication in the absence of
protein synthesis can dispense with the origin
of replication, oriC, and the dnaA protein.
Plasmid pBR322 was found to replicate, in these
sdrA mutants, in concatemeric form. The
replication occurred in the absence of DNA
polymerase I activity. The formation of
concatemeric plasmid molecules in polA mutants
was dependent on recA[+] and sdrA[-]. The sequence
of pBR322 containing ori and Pori (promoter for
transcription of the primer RNA) appeared to be
required for the replication.

INTRODUCTION

Stable DNA replication (SDR) mutants of E. coli are
capable of continued DNA replication in the absence of
protein synthesis (1,2). The mutations map at two distinct
loci; sdrA, located between metD and proA (3), and sdrT,
located at or near dnaT (2). Whereas the sdrT mutant is
dependent on recA[+] for both normal DNA replication and SDR,
in the sdrA mutants recA[+] is essential only for SDR but not
for normal DNA replication during the cell cycle in the
presence of protein synthesis.
Extragenic suppressor mutations (rin) which
specifically suppress the defect of recA mutations in SDR of

sdrA stains (but not defects in the recombinational or proteolytic function of recA) have been isolated and mapped (4). It has been proposed (3) that E. coli has an alternative initiation pathway distinct from the normal oriC$^+$ dnaA$^+$-dependent initiation mechanism. Since sdrA mutations apparently allow for constitutive expression of the alternative mechanism (1), the product of the sdrA$^+$ gene may normally repress it. This alternative mechanism may involve a recombinational activity requiring recA$^+$ protein. Furthermore, the induced SDR in sdrA$^+$ cells may be due to a recA$^+$ dependent inactivation of the sdrA$^+$ gene product (5).

The above proposal has been strongly supported by the recent observation (6) that the oriC site and the dnaA protein are dispensable in sdrA mutants: oriC can be deleted from the sdrA mutant chromosome, and the dnaA gene can be inactivated by the insertion of a transposon (Tn10) without loss of viability of sdrA mutant cells. The results imply that DNA replication in sdrA mutants is initiated at site(s) other than oriC and suggest that the initiation may involve recA$^+$protein. In this report we describe our recent observation that plasmid pBR322 replicates, in sdrA mutants, in concatemeric form and that replication occurs in the absence of DNA polymerase I activity despite the fact that the plasmid normally requires pol I for initiation of its replication (7). The formation of concatemeric plasmid molecules in sdrA and sdrA polA mutants is dependent on recA$^+$.

RESULTS

During the course of the experiments to examine replication of various types of plasmids in sdrA mutants, we noticed that cleared lysates of sdrA224 mutant cells transformed with monomer pBR322 contained, in addition to normal forms of the plasmid DNA, other DNA species whose mobilities in agarose gel electrophoresis were slower than that of the contaminating chromosomal DNA (Figure 1). These slowly migrating DNA species were consistently observed with sdrA mutants including another allele, sdrA102, but seldom with sdrA$^+$ strains. pBR322 contains single sites for EcoRI, HindIII and PstI (8). The digestion of the DNA samples with one of those restriction enzymes prior to electrophoresis resulted in the appearance of a single band the mobility of which was identical to that of linearized pBR322 plasmid DNA (data not shown). The results indicated to us that the

TABLE 1

E. coli STRAINS

Strain	Relevant genotype	Source, reference, construction
JC10240	recA56 srl300::Tn10	(23)
JC10284	srlR::Tn10Δ(srlR-recA)306	(24)
CM5389	polA12	N. Grindley
AQ634 (1)[*]	sdrA$^+$	Derived from LS534 (6)
AQ666 (1)	sdrA224	sdrA isolate of AQ634
AQ685 (2)[*]	sdrA$^+$ metD88 proA3	(6)
AQ699 (2)	sdrA224 metD88	(6)
AQ738 (2)	sdrA102 metD88	This work
AQ991 (2)	sdrA$^+$ (pBR322)	This work
AQ993 (2)	sdrA224 (pBR322)	This work
AQ997 (2)	sdrA102 (pBR322)	This work
AQ1558 (2)	sdrA224 Δ(srlR-recA)306	P1 JC10284 x AQ699, transf. with pBR322
AQ1647 (2)	sdrA224 recA56 (pBR322)	P1 JC10240 x AQ685, transf. with pBR322
AQ1648 (2)	sdrA$^+$ recA56 (pBR322)	P1 JC10240 x AQ699, transf. with pBR322
AQ1523 (1)	sdrA$^+$ polA12	P1 CM5389 x AQ634
AQ1525 (1)	sdrA224 polA12	P1 CM5389 x AQ666
AQ1635 (1)	sdrA224 polA$^+$ (pTK51)	This work
AQ1636 (1)	sdrA$^+$ polA12 (pTK51)	This work
AQ1637 (1)	sdrA224 polA12 (pTK51)	This work
AQ1639 (1)	sdrA224 polA$^+$ (pTK51) (F'128)	This work
AQ1640 (1)	sdrA$^+$ polA12 (pTK51) (F'128)	This work
AQ1641 (1)	sdrA224 polA12 (pTK51) (F'128)	This work

[*]The remaining genotype: (1)F$^-$ argH ilv metB his-29 trpA9605 pro thyA deoB(or C); (2)F$^-$ argH metB1 his-29 trpA9605 proA3 thyA deoB(or C) rpoB

slowly migrating species were either pBR322 in concatemers (i.e., circular oligomers), or catenates of monomers and/or multimers. To examine if the DNA species contained concatemers, a derivative (pTK27) of pBR322 (Figure 4) was introduced into sdrA224 mutant cells, and the plasmid purified by two cycles of dye-CsCl equilibrium centrifuga-

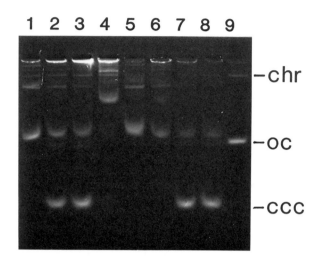

FIGURE 1. Agarose gel electrophoresis of plasmid DNA from transformants of sdrA224 and sdrA102 mutants. AQ699 (sdrA224) and AQ738 (sdrA102) were transformed with monomer pBR322. Four independent transformants of each mutant were grown in M9 glucose medium to 3×10^8 cells/ml and incubated in chloramphenical (250 µg/ml) for 12 hours for plasmid amplification. Cleared lysates were prepared and electrophoresed in a 1% agarose slab gel(0.5 µg/ml ethidium bromide) at 50 V for about four hours.

tion. The purified plasmid DNA exhibited, upon electrophoresis, discrete bands which appeared to represent supercoiled CCC forms of multimers (i.e., monomers, dimers, trimers, etc.) (Figure 2a). Digestion of the DNA with a single site restriction enzyme yielded a single species of linear DNA of the expected molecular size (2.25 Kb) (Figure 2a). The result shown in Figure 2b indicates that plasmid DNA partially digested with HindIII contained linearized molecules of molecular sizes equivalent to multiples of the unit length. Preliminary examination by electron microscopy has indicated that the DNA samples indeed contained

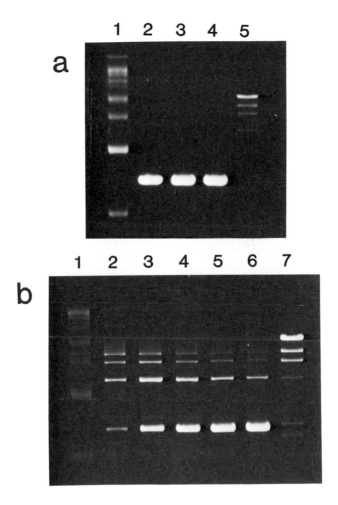

FIGURE 2. Restriction endonuclease digestion of pTK27. pTK27 plasmid DNA was isolated from an sdrA224 strain and purified through two cycles of dye-CsCl equilibrium centrifugation. a). 0.8 μg of pTK27 DNA was treated with no enzyme (lane 1), EcoRI (lane 2), PstI (lane 3), or HindIII (lane 4) at 37°C for 60 min. Lane 5, λDNA digested with HindIII. b). pTK27 DNA (0.8 μg) was treated with HindIII at 37°C for 60 min. at concentrations of: 0 (lane 1), 0.01 (lane 2), 0.02 (lane 3), 0.03 (lane 4), 0.04 (lane 5) or 0.05 units/μl (lane 6). Lane 7, λDNA digested with HindIII. All samples were electrophoresed in a 1.2% agarose slab gel as in Figure 1.

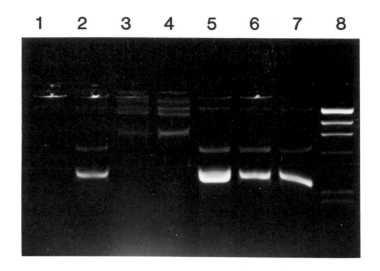

FIGURE 3. The effects of <u>recA</u> mutations on the
formation of pBR322 concatemers. Cells from 1.5 ml of an
overnight culture were lysed by boiling for 40 seconds with
lysozyme (72 µg/ml) in 100 µl of sucrose (8%), Triton X-100
(5%), Tris (50 mM, pH 8.0)-EDTA (50 mM). After lysis
samples were centrifuged in an Eppendorf microfuge for 15
min. Pellets were removed and samples were precipitated by
adding 10 µl of 3M Na-acetate and 100 µl of isopropanol.
After 30 min. at -80°C samples were centrifuged for 15 min.
Pellets were washed with 200 µl of 80% ethanol and then
dried under vacuum for one hour. Dried pellets were
suspended in 50 µl of 0.3 M Na-acetate and 150 µl of 95%
ethanol added. After 30 min. at -80°C samples were
centrifuged and pellets were washed and dried as before.
Dried pellets were suspended in 20 µl of Tris (10 mM, pH
8.0)-EDTA (1 mM). 8 µl of samples were electrophoresed in
a 1% agarose gel at 20 V for 6 hours. Lanes 1, sdrA224
(AQ699); 2, sdrA$^+$/pBR322 (AQ991); 3, sdrA224/pBR322 (AQ993);
4, sdrA102/pBR322 (AQ997); 5, sdrA224 Δ(srlR-recA)306/pBR322
(AQ1558); 6, sdrA224 recA56/pBR322 (AQ1647); 7, sdrA$^+$ recA56
/pBR322 (AQ1648); 8,λDNA digested with HindIII.

TABLE 2

TRANSFORMATION FREQUENCY OF polA AND sdrA MUTANTS
WITH pBR322, pTK27 OR pSC138

Exp.	Strain		No. of Ap^r transformants with:		
			pBR322	pTK27	pSC138
I.	AQ1523	30°C	46	49	--
		42°C	0	0	--
	AQ1525	30°C	2,700	3,300	--
		42°C	3,600	3,600	--
II.	AQ1523	37°C	0	15	545
	AQ1525	37°C	2,205	1,198	775

pSC138 (mini F) consists of the EcoRI f5 fragment of F
factor and an Ap^r fragment.

supercoiled CCC DNA with molecular sizes larger than the
unit length of the plasmid (unpublished). We conclude that
pBR322 and its derivatives exist in sdrA mutants as
concatemeric form consisting of unit size molecules joined
in a head-to-tail fashion. The above results do not,
however, exclude the possibility that the plasmid also
exists as catenates.

The results shown in Figure 3 indicate that the
formation of concatemers in sdrA mutants requires the
recA$^+$ phenotype. Thus, the introduction of the recA56 or
recAΔ306 mutation into sdrA mutants resulted in the specific
disappearance of concatemeric plasmid DNA without affecting
monomeric forms.

Initiation of the replication of ColE1-type plasmids
(including pBR322) has been shown, in vivo (7,9) and in
vitro (10), to require active DNA polymerase I (polI).
Therefore, we constructed polA12 sdrA224 double mutants to
examine the requirement of pBR322 replication in sdrA
mutants for pol I activity. polA12 is a missense mutation
giving rise to a temperature-sensitive pol I phenotype (71).
The double mutant was transformed with pBR322, or its
derivative pTK27, selecting for Ap^r. The results summarized
in Table 2 show that pBR322 and pTK27 could transform the
polA12 sdrA224 double mutant even at the restrictive
temperature for the mutant pol I. An analysis of plasmid
DNA from such a transformant grown at 42°C showed specific
disappearance of monomeric plasmid DNA without affecting

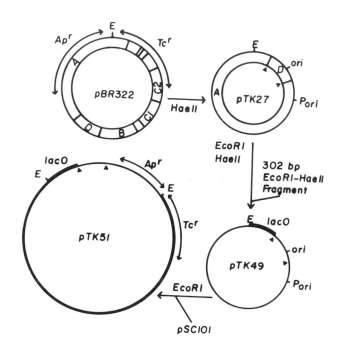

FIGURE 4. Construction of plasmids. pTK27 was one of the smallest plasmids found after complete digestion of pBR322 with HaeII and subsequent ligation (see Text). pTK49 was derived by replacing the EcoRI-HaeII fragment of pTK27 with the 302-bp EcoRI-HaeII fragment of mWB2342 (12) containing the lacO lacP sequences. pTK51 was a chimeric plasmid between pTK49 and pSC101 recombined at the EcoRI site.

the concatemeric form (data not shown). This result contrasts with the above observation with sdrA recA double mutants which indicated that recA mutations affected only the concatemeric form (Figure 3). Taken together, these results indicated to us that the sdrA mutation allows pBR322 plasmid to replicate in the absence of pol I and that the plasmid replication becomes dependent on the recA$^+$ gene product, yielding the concatemeric forms.

What sequence(s) of pBR322 are necessary for replication in the absence of pol I in sdrA mutants? To

TABLE 3

β-GALACTOSIDASE ACTIVITY IN sdrA AND polA MUTANTS
HARBORING pTK51 PLASMID

Strain	Relevant genotype	β-galactosidase activity[a]
AQ1523	sdrA$^+$ polA12	0.33
AQ1525	sdrA224 polA12	0.49
AQ1635	sdrA224 polA$^+$ (pTK51)	23.31
AQ1636	sdrA$^+$ polA12 (pTK51)	3.65
AQ1637	sdrA224 polA12 (pTK51)	11.95
AQ1639	sdrA224 polA$^+$ (pTK51) (F'128)	12.45
AQ1640	sdrA$^+$ polA12 (pTK51) (F'128)	0.71
AQ1641	sdrA224 polA12 (pTK51) (F'128)	0.65

[a]Enzyme units/O.D. (at 600 nm) − 0.5 (about 4 x
10^8 cells/ml). The β-galactosidase was assayed
according to (14).

obtain a minimum sequence required for replication in polA12
sdrA224 double mutants, pBR322 plasmid DNA was digested with
multi-site restriction enzymes such as HaeII and HinfI,
ligated with T4 DNA ligase, and used to transform Ap^S polA12
sdrA224 cells to Ap^r. The smallest plasmid obtained, which
still replicated in polA sdrA double mutants, was pTK27
consisting of the A and D HaeII fragments (Figure 4).
Digestion with HinfI indicated that only the small HinfI
fragment downstream from ori could be removed from the pTK27
sequence without losing the ability to replicate. Repeated
attempts to find a plasmid with the D fragment in the
reversed orientation failed although a plasmid containing
two D fragments in tandem in the correct orientation was
found to be viable. Since the HaeII site lies between Pori
(promoter for transcription of the primer RNA) and ori (8),
the above results strongly suggest the requirement of the
ori site as well as the intact sequence between ori and Pori
for replication in sdrA polA double mutants. The
mini-plasmid (pTK27) was found to also replicate in
sdrA$^+$ polA$^+$ strains.
        To demonstrate the dependence of pTK27 replication in
polA mutants on the sdrA mutation, a chimeric plasmid

(pTK51) was constructed between pTK27 and pSC101 (Figure 4). The chimera also had a 302 bp HaeII-EcoRI fragment containing the lac operator sequence (12). pSC101 is a low copy number plasmid as opposed to pTK27 (pBR322) which replicates at a high copy number. A chimera between a high copy and a low copy number plasmid replicates at a high copy number (13). Only when the replication from the origin of the high copy number plasmid is prevented, does the chimera replicate as a low copy number plasmid (13). pTK51 was found to replicate in sdrA polA double mutants as a high copy number plasmid, rending the lac$^+$ cells to express the Lac$^+$ phenotype in the absence of inducers. This Lac$^+$ phenotype without induction results from sequestering of lac repressor molecules by the lac operator sequence which exists at a high copy number, resulting in constitutive expression of the lac operon. Thus, an sdrA polA double mutant containing pTK51 (AQ1637) formed blue colonies on X-gal plates (14). When F'128 carrying sdrA$^+$ (1) was introduced into the cells, colonies were white on X-gal plates (Lac$^-$). This is consistent with our earlier observation (1) that sdrA mutations are recessive to sdrA$^+$ and demonstrates the dependence of pBR322-type plasmid replication on sdrA mutations in the absence of pol I. This was quantified by measuring β-galactosidase activity of cultures grown at 42°C. The results are summarized in Table 3.

DISCUSSION

Several plasmids have been observed to form concatemers (circular oligomers) in wild-type E. coli strains (15,16,17). Both formation and resolution (interconversion) of the concatemers are dependent on recA$^+$. Some plasmids are known to contain a cis-acting recombinogenic element (18). It has been reported that pBR322 carries such a recombinogenic sequence near the BamHI site in the tet gene (19). Since the effects of recA and recF mutations on the formation and resolution of concatemers appears to be identical to the effects on the intermolecular homologous recombination of plasmids, it is postulated that concatemers are generated during the process of genetic recombination of plasmids (20).

We have shown that a strikingly large portion of pBR322 and its derivatives are in concatemeric form (with perhaps, some catenates) when replicated in sdrA mutants (Figure 1).

The formation of concatemers is $recA^+$ dependent (Figure 3).
It is possible that the sdrA mutations may bring about a
condition in which intermolecular recombination of plasmids
is drastically stimulated and/or the resolution of
concatemers is severely retarded. For example, plasmid
molecules may accumulate single-strand gaps which are known
to be highly recombinogenic. Alternatively, the concatemers
may arise as a necessary consequence of plasmid replication
in sdrA mutants. A recombination event between
intramolecular homologous sequences may provide a site at
which plasmid replication is initiated by a novel way other
than by the normal $polA^+$ dependent manner. This view is
consistent with the observation that (1) plasmid replication
in concatemeric forms can occur in the absence of pol I
activity (Table 2) and that (2) it is dependent on
$recA^+$ (Figure 3) and $sdrA^-$ (Table 3). This hypothesis
predicts the inability of pBR322 to replicate in sdrA
mutants in the absence of both $polA^+$ and $recA^+$ functions.
   The fact that sdrA mutant cells can survive the
deletion of the oriC sequence and the complete inactivation
of the dnaA gene without deleterious consequences (6)
implies that rounds of chromosome replication in sdrA
mutants are initiated at another site(s), completed and
processed for cell division. Although the mechanism of the
initiation process is not known at present, it most likely
involves a recombinational event, perhaps similar to a
mechanism proposed for DNA replication in later stages of T4
phage infection (21). This recombinational initiation
mechanism in sdrA mutants has been postulated to represent a
back-up system alternative to the normal $oric^+$ $dnaA^+$
dependent initiation pathway (3,4). This notion has been
strongly supported by the finding (22) that dasF mutations
isolated as suppressor mutations of $dnaA^{ts}$ are most likely
allelic to sdrA mutations.
   Crucial to the understanding of the significance of the
postulated initiation pathway in E. coli is to determine
whether or not DNA replication is initiated at a fixed
orgin(s) on the chromosome and, if so, to determine the
structure of the site(s). The results described in this
report (Figure 4) indicate that replication of pBR322 in
polA sdrA double mutants requires the intact Pori-ori region
but not the sequence identified as a recombinogenic site
(19) since pTK27 is devoid of the sequence (Figure 4). If
$polA^+$ independent replication of concatemeric pBR322 occurs
by way of the recombinational initiation pathway as
suggested above, pBR322 may contain a sequence identical or

similar to one of the alternative origins of chromosome replication in E. coli sdrA mutants.

ACKNOWLEDGEMENTS

We thank Drs. B. Bachmann, N. Grindley and D. Natvig for bacterial strains and plasmids; Dr. D. Bear for help in electron microscopy. The excellent technical assistance of T. Torrey and J. Malone was greatly appreciated. This work was supported by grants from the National Institutes of Health (GM22092 and MBS Grant PRO8139).

REFERENCES

1. Kogoma, T (1978). J. Mol. Biol. 121:55.
2. Lark, KG, Lark, CA, Meenen, EA (1981). In Ray, DS (ed.), The Initiation of DNA Replication, ICN-UCLA Symposia on Molecular and Cellular Biology, Vol. 22, Academic Press, NY, p 337.
3. Kogoma, T, Torrey, TA, Subia, NL, Pickett, GG (1981). In Ray, DS (ed.), The Initiation of DNA Replication, ICN-UCLA Symposia on Molecular and Cellular Biology, Vol. 22, Academic Press, NY, p 361.
4. Torrey, TA, Kogoma, T (1982). Mol. Gen. Genet. 187:225.
5. Kogoma, T, Torrey, TA, Connaughton, MJ (1979). Mol. Gen. Genet. 176:1.
6. Kogoma, T, von Meyenburg, K (1983). The EMBO Journ. 2, 463.
7. Backman, K, Betlach, M, Boyer, HW, Yanofsky, S (1978). Cold Spr. Harbor Symp. Quant. Biol. 43:69.
8. Sutcliffe, JG (1978). Cold Spr. Harbor Symp. Quant. Biol. 43:77.
9. Kingsbury, DT, Helinski, DR (1973). J. Bacteriol. 114:116.
10. Sakakibara, Y, Tomizawa, J (1974). Proc. Natl. Acad. Sci. USA 71:802.
11. Kelley, WS (1980). Genetics 95:15.
12. Barnes, WM, Beran, M (1983). Nuc. Acid Res. 11:349.
13. Cabello, F, Timmis, K, Cohen, SN (1976). Nature 259:285.
14. Miller, JH (1972). Exp. in Mol. Genet. Cold Spr. Harbor Laboratory, Cold Spr. Harbor, NY.
15. Hobom, G, Hogness, D (1974). J. Mol. Biol. 88:65.

16. Bedbrook, JR, Ausubel, FM (1976). Cell 9:707.
17. Potter, H, Dressler, D (1977). Proc. Natl. Acad. Sci. USA 77:4168.
18. Kolondner, R (1980). Proc. Natl. Acad. Sci. USA 77:4847.
19. James, AA, Kolondner, R (1983). UCLA Symposia on Molecular and Cellular Biology, J. Cell. Biochem. Suppl. 7B, Abstract 888.
20. James, AA, Morrison, PT, Kolondner, R (1982). J. Mol. Biol. 160:411.
21. Mosig, G, Luder, A, Rowen, L, MacDonald, P, Bock, S (1981). In Ray, DS (ed.), The Initiation of DNA Replication, ICN-UCLA Symposia on Molecular and Cellular Biology, Vol. 22, Academic Press, NY, p 277.
22. Kogoma, T, Torrey, TA, Atlung, T (1983). UCLA Sypmposia on Molecular and Cellular Biology, J. Cell. Biochem. Suppl. 7B, Abstract 750.
23. Csonka, LN, Clark, AJ (1980). J. Bacteriol. 143:529.
24. Csonka, LN, Clark, AJ (1979). Genetics 93:321.

V. EUKARYOTIC DNA REPLICATION

Mechanisms of DNA Replication and Recombination, pages 353 –366
© **1983 Alan R. Liss, Inc., 150 Fifth Avenue, New York, NY 10011**

SYMMETRY AND SELF-REPAIR IN ADENO-ASSOCIATED VIRUS
DNA REPLICATION[1]

K.I. Berns, R.J. Samulski[2], A. Srivastava[3]
and N. Muzyczka

Department of Immunology and Medical Microbiology,
University of Florida, College of Medicine,
Gainesville, Florida 32610

ABSTRACT   The adeno-associated virus (AAV) genome
contains an inverted terminal repetition of 145
bases. The first 125 bases are palindromic and hairpin to serve as a primer for DNA replication. We
have recently cloned the intact AAV genome into
pBR322. Upon transfection of human cells in the
presence of helper adenovirus infection, the AAV DNA
is rescued from the recombinant molecule and replicated. In addition to the clone containing the
intact genome, several clones were isolated that contain deletions restricted to one or both terminal
repetitions. Those clones with deletions in either
the right or left terminal repeats can also be rescued upon transfection into human cells. The virus
produced contain full length DNA in which the deletions in the terminal repetition have been repaired
and the original sequence restored. Three of four
clones with deletions in both terminal repeats could
not be rescued. The fourth, which could be rescued,

[1]This work was supported by grants from the Public
Health Service (AI 16326) and the American Cancer Society
(NP353).
        [2]Present address: Department of Microbiology,
State University of New York, Stony Brook, New York 11794.
        [3]Present address: Department of Medicine,
University of Arkansas Medical School, Little Rock,
Arkansas 72205.

contained a 113 base deletion on one end and a
deletion of 9 nucleotides on the other end. From
these data a model for self repair via a circular
intermediate in DNA replication has been developed.
This system has also permitted us to begin the study
of the requirement for specific terminal sequences in
DNA replication.

INTRODUCTION

Adeno-associated virus (AAV) is a defective
parvovirus which requires coinfection with an adenovirus
or herpes simplex virus in order to replicate (1-4). The
viral genome is a linear, single stranded DNA 4675 bases
long (5-7). Recently the entire nucleotide sequence of
the AAV 2 genome has been determined (8). A striking
feature of the sequence is an inverted terminal repetition
of 145 bases (9,10). The first 125 bases of the repeat
form an overall palindromic sequence in which bases 1-41
are complementary to bases 85-125. In between there are
two shorter palindromes from 42-62 and from 64-84 (10).
When the sequence is folded so as to permit the maximum
amount of potential base pairing, a T-shaped structure is
formed (Fig. 1). In this structure only 7 of the first
125 bases are unpaired and all are A's or T's. Six of the
unpaired bases are required to permit the two short
internal palindromes to hairpin and the seventh separates
the two internal palindromes. Interestingly, all but one
of the potential base pairs in the short arms of the
T-shaped structure are GC base pairs.

Two types of heterogeneity in the sequence of the
terminal repetition have been observed. First, 65% of the
5' termini of virion DNA are missing either one or both
terminal T's (11). Second, between nucleotides 42-84
there are two possible sequences that occur with equal
frequency in virion DNA (10,12). The two sequences are
consistent with an inversion of the first 125 bases.
Because the sequence from 1-41 is perfectly complementary
to the sequence from 125-85, an inversion would not alter
these sequences.

A model for AAV DNA replication which is consistent
both with the specific properties of the terminal repetition and the properties of replicative intermediates

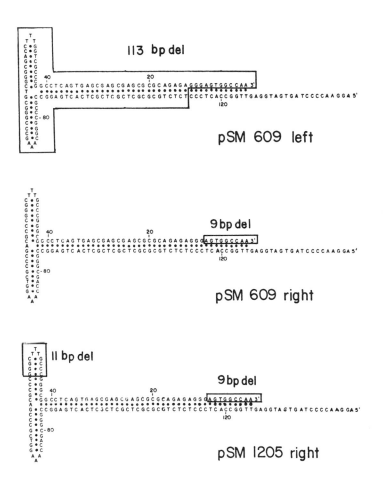

FIGURE 1. Nucleotide sequences of the inverted terminal repeats of pSM609 and pSM1205. The terminal AAV sequences are represented as a single-stranded sequence in a form which illustrates the position and size of the three palindromes within the terminal repeat. The remaining sequences are not palindromic but are part of the total inverted terminal repetition. The boxes indicate the positions of the deletions in the left and right termini of pSM609 and the right terminus of pSM1205. The DNA sequence of pSM1205 is identical to that of pSM609, its parent, except for an additional 11 bp deletion in one

of the palindromes at the right end of pSM1205. By
convention (8) the right ends of both pSM609 and 1205 are
in the flop orientation. Because of the magnitude of the
deletion in the left end of pSM609 and 1205, no orientation can be assigned. For comparison, the sequence of the
pSM609 left end is written as if it were in the flip
orientation.

isolated from infected cells is illustrated in Fig. 2.
The 3' terminal repetition is presumed to fold over to
form a hairpinned T-shaped structure that can function as
a primer for DNA synthesis. Subsequently, the parental
strand is cleaved between bases 125 and 126 at a point
opposite the original 3' base (now on the progeny strand).
The cleavage dissolves the hairpin and leaves the parental
strand with a shortened 3' end that can serve as a primer
for a repair type of DNA synthesis to fill the gap. As a
consequence of hairpin formation and the subsequent
cleavage the 3' terminal 125 bases of the parental strand
are transferred to the 5' end of the progeny strand and
the sequence is inverted. A possible first intermediate
in replication could be a single stranded circle
stabilized by hydrogen bonding between the complementary
bases in the inverted terminal repeats. An intermediate
of this type is an attractive possibility because the ends
of the DNA would be in a duplex structure and thus
topologically equivalent to the putative duplex
replicative intermediates that initiate subsequent rounds
of replication.

The model utilizes the self base pairing properties
of the inverted terminal repetition and predicts the
observed inversion. It further predicts the existence of
replicative intermediates in which the complementary DNA
strands are covalently crosslinked at the termini. Such
intermediates have been isolated from AAV-infected cells
(13,14). Finally, the model predicts that DNA replication
should begin near the 5' end of the progeny strand and
terminate at the 3' end. In vivo pulse-labeling experiments have confirmed this as well (15).

Recently, the AAV genome has been cloned intact into
the bacterial plasmid pBR322. When this recombinant
plasmid was transfected into human cells in culture in the
presence of a concomitant helper adenovirus infection, the

AAV genome was rescued, replicated, and infectious virus were produced (16). The existence of this biologically active clone has permitted further detailed studies on the role of the terminal repetition in AAV DNA replication.

RESULTS

The recombinant clone of AAV DNA inserted into pBR322 offered the first opportunity to directly test the hypothesis that the inverted sequences in the terminal repetition result from the mode of DNA replication. Obviously, each end of the AAV genome in the recombinant was present as a unique sequence (i.e., in only one orientation). Fortuitously, both ends of the insert were in the same orientation. When DNA from the virions produced by rescue of the genome from the recombinant clone was analyzed, both orientations of the terminal repetition were found at both ends of the DNA (16) in equal proportions. Thus, the inversion is clearly a consequence of DNA replication.

As noted above, 65% of the ends of AAV DNA are missing one or both 5' terminal T's; 15% are missing both. In pSM620 both termini of the insert were missing both T's, yet the vast majority of the progeny virion DNA molecules had T as the 5' terminal nucleotide (16,18). Thus, the second type of sequence heterogeneity noted in AAV virion DNA is also clearly a dynamic consequence of the process of DNA replication. The results suggest that the process of nicking the hairpin structure in the replicative intermediate is not absolutely specific for the site between bases 125 and 126; it may also occur after base 123 or after base 124.

The biological activity of the clone also allowed a direct test of the effects of alterations in the sequence of the terminal repeat. The clone was constructed by tailing the 3' ends of the duplex form of AAV virion DNA (formed by annealing the isolated plus and minus strands originally encapsidated in separate virions, 5, 6, 17) with poly dC and by tailing the 3' ends of pBR322 DNA after digestion with the restriction enzyme PstI with poly dG. The modified DNAs were joined by annealing and transfected into E. coli. Covalent linkage between AAV and pBR322 sequences thus occurred in vivo. During this

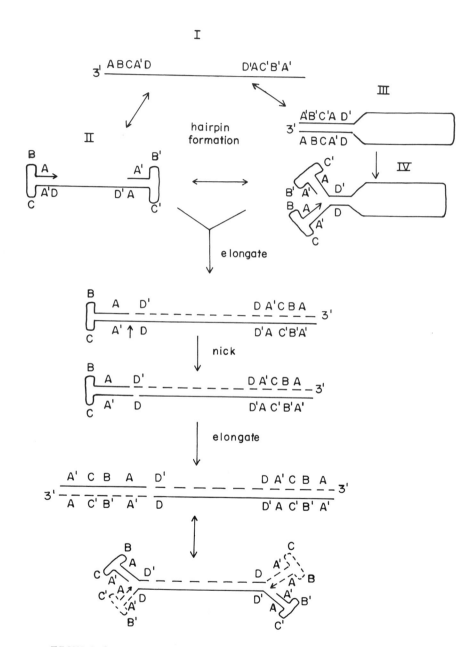

FIGURE 2. Model for AAV DNA replication. The
terminal sequence domains represented by primed letters
are the complement of those represented by unprimed

letters. B and C are the two smaller palindromes (nucleotides 42-62 and 64-94) which are contained within the
larger palindrome AA' (nucleotides 1-41 and 85-125). D
sequences (nucleotides 125-145) are not part of the
terminal palindrome but are part of the terminal inverted
repetition. Details of the sequence domains and
replicative steps are discussed in the text.

process most of the recombinant plasmids suffered
deletions in one or both terminal repetitions. As a
result, the initial cloning process provided a pool of
altered molecules to test for changes in biological
activity.

The plasmid pSM609 had 113 bases deleted from the
left terminal repeat and 9 bases deleted from the right
(Fig. 1). Surprisingly, when this plasmid was transfected
into human cells in the presence of a concomitant
adenovirus infection, the AAV genome was rescued,
replicated, and infectious virus were produced (18). Even
more unexpected was the finding that the DNA isolated from
the virions had the deletion repaired and the original
sequence restored. It seems likely that the biological
activity and repair processes observed in the case of
pSM609 were a direct consequence of the structure of the
inverted terminal repetition. The deletion at the right
end of pSM609 removed only the terminal 9 bases. Thus,
the 3' terminus at this end of the genome would still have
been able to hairpin and serve as a primer for DNA
synthesis. The first 9 bases inserted would have repaired
the deletion. Evidence to support this notion was
presented above in that absence of the two terminal T's in
pSM620 still permitted rescue and restoration of the T's
at the termini of the DNA in progeny virions.

The mechanism of repair of the longer deletion (113
bases) at the left end of pSM609 is more problematic. A
possible mechanism of repair is illustrated in Fig. 3. A
nick at the right end of the insert in pSM609 would permit
initiation of DNA synthesis from the 3' hairpin that could
then form. Elongation of the progeny strand would
displace a parental strand which could then form a single-
stranded circular structure of the type shown in Fig. 3.
The structure would be stabilized by base pairing between
the right terminal repetition and the 32 bases remaining

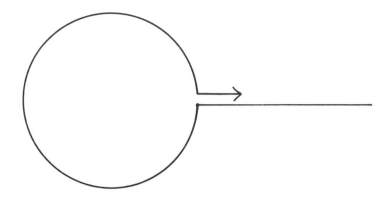

FIGURE 3. A potential intermediate for the repair of the large terminal deletion (113 bp) at the left end of pSM609. See text for details.

(114-145) in the left terminal repeat. The 3' end of the displaced strand would be the shorter strand in the terminal duplex structure and could serve as a primer for a repair type of synthesis. This would result in a circular structure equivalent to the putative first replicative intermediate pictured in Fig. 2 (III or IV); once it was formed, normal DNA replication could ensue.

The implications of the biological activity of pSM609 bear strongly on the consequences of the symmetry of the inverted terminal repetition in AAV DNA. Because of this sequence arrangement, all of the sequences in the palindromic portion (the first 125 bases) are repeated four times (twice at each end of the DNA). The results with pSM609 suggest that the presence of just one set of these sequences in the right configuration is sufficient to regenerate the entire wild type sequence during the process of DNA replication. By the same token, deletion of even a short sequence completely from the terminal repetition should prevent the regeneration of the wild type sequence and presumably have a serious effect on DNA replication. To test this prediction pSM1205 was created by deletion of the 11 bases from 47-57 on the right end of pSM609 (Fig. 1). The deletion removed a short symmetrical sequence from one of the small internal palindromic sequences that had already been deleted from the left end

of pSM609. Thus, this sequence was totally missing in
pSM1205. Upon transfection of this plasmid into human
cells with an adenovirus coinfection, the AAV insert did
appear to be rescued, but no mature monomer length DNA nor
infectious virions could be detected and only a small
amount of presumed concatemeric replicative intermediates
could be observed. Therefore, complete removal of a very
short (11 base) sequence from the terminal repetition at
both ends of the genome led to a complete cessation of
mature genome production. The deletion shortened one of
the cross arms of the T-shaped folded structure (Fig. 1)
by about 50%. A possible consequence, as judged by the
experimental results, might have been that the specific
nicking of the hairpin structure invoked in the model of
AAV DNA replication (Fig. 2) was inhibited; hence, no
production of mature monomer length genomic DNA occurred.

The conclusions drawn from the results obtained for
pSM609 and pSM1205 have been supported by the results
obtained with a series of recombinant plasmids containing
terminal deletions in the AAV genome (Table 1). The

TABLE 1
PROPERTIES OF RECOMBINANT PLASMIDS

| Plasmid | Deletion (bp) | | Virus |
	Left	Right	
pSM620	2[a]	2[a]	+
pSM621	80	0	+
pSM802	0	100	+
pSM703	100	95	−
pSM704	116	90	−
pSM803	111	83	−
pSM609	113[a]	9[a]	+
pSM1205	113[a]	20[a]	−
pSM209	0	600	−

[a]Determined by DNA sequencing; all other sizes
are estimates from restriction enzyme digestions.
The 2 base deletion in pSM620 represents the
missing T's at both 5' termini (see text for
details).

results may be summarized as follows. 1) A deletion of
any part of the palindromic sequence in only one end is
viable. 2) A deletion that extends beyond the terminal
repetition is not viable. 3) Deletions that remove any
sequence totally from both ends are not viable.
4) Deletions within both terminal repetitions are viable
if one of the deletions is short enough to allow stable
hairpin formation (the limit is probably less than 40
bases but greater than 9 bases).

DISCUSSION

Linear DNA genomes have a specific problem in the
replication of their termini. All DNA polymerases that
have been characterized require, in addition to a
template, a primer with an available 3' OH group to join
with the first 5' deoxynucleoside triphosphate. In order
to maintain the 5' terminal sequence, the primer, if RNA
or protein, must be eventually removed and replaced with a
suitable DNA sequence. If, as in the case of the
parvoviruses, the primer is DNA, specific enzymatic
reactions of repair type are required to maintain a
constant 5' sequence. Beyond the problem involved in
replication, linear DNA molecules are also susceptible to
exonucleolytic degradation. Presumably, as a response to
both of these problems, linear DNA genomes have evolved
specialized terminal nucleotide sequences. Those
characterized to date include natural terminal repetitions
in many bacteriophage DNAs (19) and herpes simplex virus
DNA (20); inverted terminal repetitions in the DNAs of
adenovirus (21,22), AAV (9,10), herpes simplex virus (20),
and vaccinia virus (23,24); and palindromic terminal
sequences in the DNAs of parvoviruses (10) and poxviruses
(25). Specialized palindromic sequences that covalently
crosslink complementary strands have been reported for
poxviruses (26), paramecium mitochondrial DNA (27), and
yeast chromosomal DNA (28). By virtue of its palindromic
inverted terminal repetition AAV DNA is an excellent
example of such specialized terminal nucleotide sequences.
Clearly this structure can function as a DNA primer for
replication that is identical for both complementary
strands. The construction of the biologically active
recombinant plasmid of AAV and pBR322 has now permitted a
clear demonstration of the ability of this type of

structure to also maintain the terminal sequence by an active repair process during DNA replication.

It is not clear whether the rescue of the AAV insert from the recombinant plasmid involves excision followed by replication or selective replication out of the AAV genome. It is possible to follow the fate of the parental plasmid. Under these conditions pBR322 DNA free of AAV sequences cannot be detected, thus the latter form of rescue might seem more likely. However, it is possible that as little as 1% of the input DNA is involved in the rescue process and it is doubtful that this small a proportion could have been detected.

Another pertinent question involves the specificity of the rescue process. AAV causes latent infections in the absence of a helper virus coinfection by integration into host cell DNA (29,30). The integrated genome can be efficiently rescued by subsequent superinfection with the helper virus. Within the limits of detection afforded by restriction enzyme analysis no specificity in the cell sequences flanking the AAV insert could be detected (31). The ability of pSM609 to be rescued implies that any viral sequence specificity does not reside at the very ends of the AAV genome (at least within the terminal 9 bases). These facts suggest two possibilities; either there is sequence specificity in the terminal repetition but the signal sequence is internally located, or the critical feature is the self complementarity of the overall sequence, enough of which must be retained to allow potential formation of a cruciform structure. The latter possibility is especially intriguing in light of the similarity of cruciform structures to putative recombination intermediates. Resolution of this problem awaits further experiments in which part or all of the AAV terminal repetition is replaced with substitute sequences that retain the ability to form the cruciform structure.

In many respects, therefore, the recombinant clone of AAV DNA in pBR322 offers an attractive model to study the structural requirements for AAV DNA replication, recombination, and repair.

REFERENCES

1. Atchison RW, Casto BC, Hammon, WMcD (1965). Adenovirus-associated defective virus particles. Science 194:754.
2. Hoggan MD, Blacklow NR, Rowe WP (1966). Studies of small DNA viruses found in various adenovirus preparations: Physical, biologcal, and immunological characteristics. Proc Natl Acad Sci USA 55:1457.
3. Parks WP, Melnick JL, Rongey R, Mayor HD (1967). Physical assay and growth cycle studies of a defective adeno-satellite virus. J Virol 1:171.
4. Buller RML, Janik JE, Sebring ED, Rose JA (1981). Herpes simplex virus types 1 and 2 help adeno-associated virus replication. J Virol 40:241.
5. Rose JA, Berns KI, Hoggan MD, Koczot FJ (1969). Evidence for a single-stranded adenovirus-associated virus genome: formation of a DNA density hybrid on release of viral DNA. Proc Natl Acad Sci USA 64:863.
6. Mayor HD, Torikai K, Melnick J, Mandel M (1969). Plus and minus single-stranded DNA separately encapsidated in adeno-associated satellite virions. Science 166:1280.
7. Gerry HW, Kelly TJ, Jr, Berns KI (1973). Arrangement of nucleotide sequences in adeno-associated virus DNA. J Mol Biol 79:207.
8. Srivastava A, Lusby EW, Berns KI (1983). Nucleotide sequence and organization of the adeno-associated virus 2 genome. J Virol 45:555.
9. Kozcot FJ, Carter BJ, Garon CF, Rose JA (1973). Self-complementarity of terminal sequences within plus or minus strands of adenovirus-associated virus DNA. Proc Natl Acad Sci USA 70:215.
10. Lusby E, Fife KH, Berns KI (1980). Nucleotide sequence of the inverted terminal repetition in adeno-associated virus DNA. J Virol 34:402.
11. Fife KH, Berns KI, Murray K (1977). Structure and nucleotide sequence of the terminal regions of adeno-associated virus DNA. Virology 78:475.
12. Spear IS, Fife KH, Hauswirth WW, Jones CJ, Berns KI (1977). Evidence for two nucleotide sequence orientations within the terminal repetition of adeno-associated virus DNA. J Virol 24:627.

13. Straus SE, Sebring E, Rose JA (1976). Concatemers of alternating plus and minus strands are intermediates in adenovirus-associated virus DNA synthesis. Proc Natl Acad Sci USA 73:742.

14. Hauswirth WW, Berns KI (1979). Adeno-associated virus DNA replication: non unit-length molecules. Virology 93:57.

15. Hauswirth WW, Berns KI (1977). Origin and termination of adeno-associated virus DNA replication. Virology 79:488.

16. Samulski RJ, Berns KI, Tan M, Muzyczka N (1982). Cloning of adeno-associated virus into pBR322: rescue of intact virus from the recombinant plasmid in human cells. Proc Natl Acad Sci USA 79:2077.

17. Berns KI, Adler S (1972). Separation of two types of adeno-associated virus particles containing complementary polynucleotide chains. J Virol 9:394.

18. Samulski RJ, Srivastava A, Berns KI, Muzyczka N (1983). Rescue of adeno-associated virus from recombinant plasmids: gene correction within the terminal repeats of AAV. Cell, in press.

19. MacHattie LA, Ritchie DA, Thomas CA, Richardson CL (1967). Terminal repetition in permuted T2 bacteriophage DNA molecules. J Mol Biol 23:355.

20. Roizman B (1980). Herpes simplex viruses. In Tooze J (ed): "Molecular Biology of Tumor Viruses," 2nd ed, Vol 2, New York: Cold Spring Harbor Laboratory, p 615.

21. Garon CF, Berry KN, Rose JA (1972). A unique form of termininal redundancy in adenovirus DNA molecules. Proc Natl Acad Sci USA 69:2391.

22. Wolfson J, Dressler D (1972). Adenovirus-2 DNA contains an inverted terminal repetition. Proc Natl Acad Sci USA 69:3054.

23. Witlek R, Menna A, Muller KH, Schumperli D, Bosely PG, Wyler R (1978). Inverted terminal repeats in rabbit poxvirus and vaccinia virus DNA. J Virol 28:171.

24. Garon CF, Barbosa E, Moss B (1978). Visualization of an inverted terminal repetition in vaccinia virus DNA. Proc Natl Acad Sci USA 75:4863.

25. Baroudy BM, Venkatesan S, Moss B (1982). Incompletely base-paired flip-flop terminal loops link the two DNA strands of the vaccinia virus genome into one uninterrupted polynucleotide chain. Cell 28:315.

26. Geshelin P, Berns KI (1974). Characterization and localization of the naturally occurring crosslinks in vaccinia virus DNA. J Mol Biol 88:785.

27. Cummings DJ, Pritchard AE (1982). Replication mechanism of mitochondrial DNA from Paramecium aurelia: sequence of the cross-linked origin. In Slonimski P, Borst P, Attardi G (eds): "Mitochondrial Genes," New York: Cold Spring Harbor Laboratory, p 441.

28. Forte MA, Fangman WL (1979). Yeast chromosomal DNA molecules have strands which are cross-linked at their termini. Chromosoma 72:131.

29. Hoggan MD, Thomas GF, Johnson FB (1972). Continuous "carriage" of the adenovirus-associated virus genome in cell cultures in the absence of helper adenovirus. Proc 4th Lepetit Colloq, North-Holland, Amsterdam, 243.

30. Cheung AK-M, Hoggan MD, Hauswirth WW, Berns KI (1980). Integration of the adeno-associated virus genome into cellular DNA in latently infected human Detroit 6 cells. J Virol 33:738.

31. Berns KI, Cheung AK-M, Ostrove JM, Lewis M (1982). Adeno-associated virus latent infection. In Mahy BW, Minson AC, Darby GK (eds): "Virus Persistence," Cambridge (U.K.): Cambridge University Press, p 249.

Mechanisms of DNA Replication and Recombination, pages 367–379
© 1983 Alan R. Liss, Inc., 150 Fifth Avenue, New York, NY 10011

PARVOVIRUS DNA REPLICATION[1]

M. Goulian, R. Kollek[2], D. Revie, W. Burhans
C. Carton and B. Tseng

Department of Medicine
University of California San Diego
La Jolla, California 92093

Autonomous parvoviruses are small DNA viruses that infect a wide range of animal cells (1,2,3). The genome is a 5 kb single strand (-) with 1-200 bp foldbacks at both termini. It is known that the infecting viral single strand is converted in the cell to a duplex replicative form (RF) but much about the subsequent steps in replication remains uncertain. A portion of the bulk RF in the cell is hinged (terminally cross-linked) at the "left" end (viral 3′ terminus) and, in addition to the monomer duplex molecules, there is a smaller amount of left end-to-left end dimer duplex molecules.

Studies on parvovirus H-1 synthesis several years ago showed double-stranded Y-shaped molecules by electron microscopy with the sequence of oriented molecules indicating progression of the fork from right to left. This, together with results from a limited form of the Danna-Nathans experiment, analyzing specific activities in restriction fragments of pulse labeled RF, suggested the familiar semi-discontinuous mechanism,

[1]This work was supported by NIH grants CA 11705 (M.G.) and GM 29091 (B.T.) (and training support from HL 07107 to D.R. and from Deutsche Forchungs-gemeinschaft to R.K.). B.T. is a Leukemia Society of America Scholar
[2]Present address: Heinrich-Pette Institute, Universität Hamburg, Hamburg, Germany

with the replication fork moving from right to left (see Rhode, S in ref. 1). However, attempts to find the Okazaki pieces (or RNA primers) implied by this mechanism have failed (4). A recent confirmation of that result is illustrated here (Fig. 1).

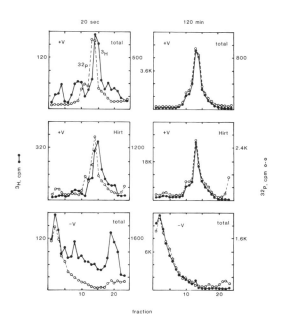

FIGURE 1. Alkaline velocity sedimentation of extracts from H-1 parvovirus infected (+V) and uninfected (-V) NB cells. All cells were prelabeled (5h) with $^{32}P_i$ and at 18 h post-infection pulse-labeled with ^{3}H-dThd for either 20 sec (left side) or 2 h (right side). The top and bottom figures were total cell detergent/protein lysates ("total"); the middle figures were Hirt extracts. The extracts were analyzed by alkaline sucrose gradient centrifugation. Direction of sedimentation was to the left; the major DNA peak (^{32}P) in the infected cells (fractions 13-14) is the 16 S monomer duplex RF.

The 5-6S peak of nascent Okazaki fragments is prominent in the 20 sec labeled uninfected cells (Fig. 1). It is not seen in the corresponding uninfected cells, neither in the Hirt fraction nor the whole cell lysate, which show instead the peak of nascent DNA (^{3}H) to be sedimenting slightly more slowly than monomer duplex RF (^{32}P).

For the helper-dependent parvovirus, adeno-associated virus (AAV), there is substantial evidence for a mechanism, similar to that of adenovirus, of continuous replication of both strands, in the 5´-3´ direction by strand displacement, with initiation at either (or both) termini (2) (although the protein priming mechanism of adenovirus does not apply to AAV).

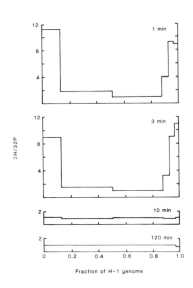

FIGURE 2. Relative ratios of pulse label to uniform label in restriction fragments of H-1 RF. H-1 RF (16 S monomer duplex) uniformly prelabeled with ^{32}P$_i$ and pulse labeled with ^{3}H (see Fig. 1) was isolated on high salt neutral sucrose gradients and then purified further as monomer length strands on alkaline sucrose gradients. The single strands were renatured to form duplex monomer RF, which was then digested with restriction endonucleases Hinc II and Hind III, and the fragments separated by gel electrophoresis.

Because of similarities between autonomous and helper dependent forms of parvovirus, and the apparent absence of short nascent pieces, we favor a form of continuous (as opposed to semi-discontinuous) strand displacement mechanism for the autonomous parvovirus, as well. In support of this kind of mechanism may be cited the results of our own attempts to show the direction of replication in RF (Fig. 2).

With the shorter labeling times (Fig. 2; 1 and 3 min) the highest specific activities were at <u>both</u> termini of monomer duplex RF, indicating that replication terminates at both right and left ends, consistent with a continuous displacement synthesis mechanism, as with adenovirus and AAV.

We are uncertain how to account for the earlier electron microscopic results showing double-stranded Y-shaped molecules but, recently, Ross Inman (in a collaborative study with our laboratory) has observed single strands bridging the fork of such molecules, suggesting that they could be generated by a strand-switching event, which Berns and associates postulated several years ago as one of the mechanisms by which defective molecules of AAV are generated (2).

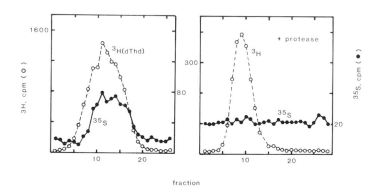

fraction

FIGURE 3. Alkaline CsCl equilibrium centrifugation of ^{35}S-methionine labeled H-1 RF. Virus infected cells were labeled with ^{35}S-methionine and ^{3}H-dThd and RF isolated and centrifuged to equilibrium in alkaline CsCl. The sample on right was treated with protease; the one on left was not. Direction of increasing density is to the left.

The autonomous parvoviruses have an additional feature in common with adenovirus, a protein covalently associated with both 5´ termini of the RF. This was originally shown by anomalously low density of the DNA even in strong alkali, which was corrected by protease treatment (5).

The terminal protein can be labeled with ^{35}S-methionine, which remains associated with the DNA after treatment with 6 M guanidinium HCl and sedimentation in 4 M guanidinium HCl sucrose gradients (not shown), or banding in alkaline CsCl (Fig. 3).

The terminal protein can also be detected by anomalously slow migration of terminal restriction fragments of RF in SDS gels, again corrected by protease (Fig. 4).

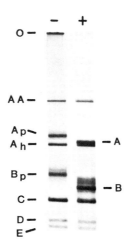

FIGURE 4. Agarose gel (1.8%) electrophoresis on H-1 RF digested with Hae III, without (-) and with (+) treatment with pronase. Samples were heated to 60° in 0.5% SDS before loading, and gel buffer contained 0.5% SDS. A and B are left and right terminal fragments; Ap and Bp refer to terminal DNA-protein complexes. A_h is "hinged" form of left hand terminus; A-A is central fragment from dimer duplex RF. Heterogeneity of the right end of the virus has been observed previously (6).

Initially, the parvoviral terminal protein seemed a candidate for a DNA chain priming function, as was then postulated and is now demonstrated, for adenovirus (7), but an alternative function has been proposed by Ward and colleagues, based on the sequence data (3).

The problems of preservation of terminal sequences, and initiation of new strands, which are solved for adenovirus by the terminal protein, appear to have been solved for the helper dependent parvovirus AAV by a terminal "flip-flop" mechanism and initiation of the new strand by the terminal hairpin (2). The autonomous parvovirus also has terminal palindromes that would allow self initiation but a flip-flop mechanism has been shown only for the right end (3,8).

One of the proposed schemes for the replication of autonomous parvovirus that would account for some of the known features is shown here in simplified form (Fig. 5). (This omits details e.g. of the flip-flop mechanism).

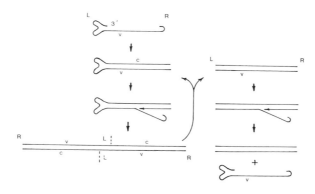

FIGURE 5. Proposed mechanism for replication of autonomous parvovirus.

It shows:
1) Self-primed synthesis of parental RF from the infecting viral single strand. The evidence supporting this (1) showed hinged molecules derived from the labeled viral strands.
2) Continuous replication by strand displacement (see above).

3) An obligatory dimer intermediate. This is supported by demonstration in Ward's laboratory (1) (and confirmed by ourselves), that a short pulse label appears in both dimer and monomer RF. The mechanism of formation of the dimer, by unfolding of the hairpin, would account for an "origin" at the left end without flip-flop, and would be consistent with our result showing termination of replication at both ends (Fig. 2, above). However, in contrast to the results with AAV, our results (and also those of Ward [1]) are against the the dimer being simply a precursor of monomer, since the ratio of label in the two forms remains approximately the same with longer labeling or chase.

4) It is usually assumed that initiation of replication on duplex RF as well as the conversion of displaced strands to RF utilizes self priming at a terminal fold-back. The demonstration of flip-flop supports this for initiation at the right end of RF.

5) An alternative route for the displaced single-strand (aside from being re-cycled back to RF), is to be packaged as virions.

In an attempt to identify some of the requirements for parvoviral DNA synthesis we initially studied the effects of inhibitors on the in vitro synthesis of viral DNA in nuclei from infected cells (9).

There was inhibition of viral DNA synthesis by either aphidicolin or ddTTP (in contrast to their effects on nuclei from uninfected cells, in which aphidicolin inhibited more strongly whereas ddTTP had virtually no effect). The effects were more pronounced when the early and late periods of incubation were examined separately by pulse labeling. Aphidicolin caused strong inhibition initially but this diminished to essentially no effect later in the incubations; whereas ddTTP caused inhibition throughout the incubation period, particularly in the latter part.

The results indicate that both DNA polymerase (pol) α and a ddTTP-inhibitable enzyme, probably pol γ, participate in H-1 DNA synthesis. Gel analysis of the products of these reactions suggested that ddTTP inhibited primarily the size fraction that would contain displaced single strands that were being converted back to monomer duplex RF. This is of interest in view of what will be discussed next.

In our initial attempts to isolate and characterize
the enzymes that carry out H-1 viral DNA replication we
looked for an activity in uninfected cells that would
convert viral single strands to duplex RF, a reaction
known to be carried out by several prokaryotic (and
tumor viral) enzymes, but which had not been demonstrated for purified mammalian DNA polymerases.

Crude extracts from uninfected cells were able to
carry out extensive strand synthesis, with a small
amount of the product reaching the size of RF. The
amount of synthesis was reduced partially by either
aphidicolin or ddTTP but length of the product was
affected much more by ddTTP, suggesting the greater
importance of pol γ in this reaction (10). When the
extract was fractionated on DEAE cellulose, the activity
of pol γ and the activity on H-1 viral DNA template
paralleled each other whereas pol α separated.

Using human placenta as a source, the procedure was
scaled up and, in addition to a DEAE step, it was
fractionated on phosphocellulose, DNA cellulose and
glycerol gradient achieving a 10,000-fold purification
by the DNA cellulose step (10). The activities on H-1
DNA, and on poly rA-oligo dT, remained associated
through these steps (and also hydroxylapatite), with a
constant ratio of the two activities. The enzyme from
these fractions showed the ability to efficiently
synthesize RF-like molecules from viral single strands
(Fig. 6).

RF

ss

FIGURE 6. 1% neutral agarose gel of ^{32}P-labeled
product of purified pol γ on H-1 viral single strand
template. Positions of duplex monomer RF (RF) and viral
single strand (ss) are indicated.

The characteristic sensitivity of pol γ to ddTTP and N-ethylmaleimide and resistance to aphidicolin and salt were observed with the purified enzyme (10).

Recently, we have returned to the cell extract for other possible activities that participate in this reaction. A crude extract, prepared in a somewhat different way from the previous carries out fairly efficient conversion of viral single strand template to duplex molecules (Fig. 7).

RF —

ss —

FIGURE 7. Activity of crude cell extract on H-1 single stranded DNA template. Ammonium sulfate concentrate of 100,000xg supernatant of Dounce homo-genate of uninfected NB cells (S-100) was incubated with H-1 viral single strand for 2 h. The product was analyzed in a 1% agarose gel. "RF" and "ss" refer to the positions of monomer duplex and single strand, respectively.

An ammonium sulfate fraction, which retains most of the pol α activity and little of the pol γ, shows similar ability to synthesize full length strand. There is some inhibition by ddTTP but marked depression by aphidicolin, indicating that pol α is required for the reaction. An additional requirement for this reaction is ATP.

The activity for synthesis of full length duplex product on H-1 single strands has been further frac-tionated into a still relatively crude pol α-containing fraction and a fraction with very little polymerase activity; the two together reproduce the reaction but do not when separate.

A highly purified preparation of pol α has also been examined for its ability to carry out conversion of H-1 viral single strands to RF. It was found that the purified pol α did carry out the reaction to a small extent in the presence of ATP but not without ATP (Fig. 8).

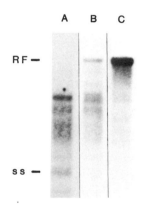

FIGURE 8.   Product of pol α on H-1 viral single stranded template; effects of ATP and primase.   Pol α (purified from mouse cells and free of detectable nuclease or primase activity) was incubated with H-1 single strands for 2 h, without (A) and with (B) ATP, and with both ATP and purified mouse primase (11,12) (C).

This ability of pol α has been reported recently by Faust and Rankin (13), who also found stimulation by ATP.   Preparations of pol α that utilize primed single stranded templates have also been described by Earl Baril and Melvin DePamphilis and their colleagues (14,15); these preparations are not stimulated by ATP.

Because of the recently observed tendency for pol α and primase to be associated during purification (16,17), and the availability of purified primase (11,12), in a collaborative study (M.G. and B.T.) we have looked into the possibility that the presence of primase in preparations of pol α may enhance the synthesis of full length product on H-1 viral single strands and may account for some of the variability in results between different preparations of pol α.

In the presence of primase (+ ATP) pol α was much more efficient in synthesizing full length duplex product than pol α (+ ATP) alone (Fig. 8). No effect of primase on this reaction was observed without ATP, and the efficiency was improved even further when the other 3 rNTPs were included.

It should be pointed out that the differences in total amount of synthesis between these various conditions were relatively small compared to the effect on the size of the full length product.

Analysis of the products of these reactions is still in progress but it does appear that most of the product in all cases is self initiated, i.e. is covalently associated with the template. The stimulatory effect of ATP on pol α alone is not yet understood but no evidence for re-initiation has been seen. With primase present (+ ATP, with or without the other 3 rNTPs), re-initiation does take place, but chain extension appears to be facilitated, in addition. The relationship of this latter effect to the stimulation of pol α by ATP remains to be determined.

It is likely that the effects we observe with the S-100 and the ammonium sulfate fraction are also related, at least in part, to what we observe with the purified enzymes. However, interestingly, the DNA product from the ammonium sulfate fraction is essentially all in one piece by alkaline gel analysis. The form of pol α in this case may still have associated with it the accessory factor(s) described by Baril and DePamphilis; however, this by itself would not explain the requirement for ATP.

In conclusion, both DNA polymerases α and γ are capable of carrying out the conversion of the H-1 viral single stranded DNA to a duplex RF-like structure. Our results with incubations of nuclei were consistent with pol γ carrying out this reaction for displaced single strands. There is recent evidence, provided by Robert Bates (18), that pol α is required for the initial synthesis of parental RF from the infecting viral single strands. Systems capable of reproducing displacement synthesis in vitro are required to pursue the additional enzymatic requirements for viral RF replication.

REFERENCES

1. Ward DC, Tattersall P (eds.) (1978). "Replication of Mammalian Parvovirus," Cold Spring Harbor Laboratory.
2. Berns KI, Hauswirth NW (1982). Organization and replication of parvovirus DNA. In Kaplan AS (ed): "Organization and Replication of Viral DNA," CRC Press, p. 3.
3. Astell CR, Thomson M, Chow MB, Ward DC (In press). Structure and replication of minute virus of mice DNA. Cold Spring Harbor Laboratory.
4. Tseng BY, Grafstrom RH, Revie D, Oertel W, Goulian M (1979). Studies on early intermediates in the synthesis of DNA in animal cells. Cold Spring Harbor Symposia Quant Biol, Vol XLIII.
5. Revie D, Tseng BY, Grafstrom RH, Goulian M (1979). Covalent association of protein with replicative form DNA of parvovirus H-1. Proc Natl Acad Sci 76: 5539.
6. Rhode SL, III (1977). Replication process of the parvovirus H-1. IX. Physical mapping studies of the H-1 genome. J Virol 22:446.
7. Kelly TJ, Jr. (1982). Organization and replication of adenovirus DNA. In Kaplan AS (ed): "Organization and Replication of Viral DNA," CRC Press, p. 115.
8. Rhode SL, III, Klaassen, B (1982). DNA sequence of the 5′ terminus containing the replication origin of parvovirus replication form DNA. J Virol 41: 990.
9. Kollek R, Tseng BY, Goulian M (1982). DNA polymerase requirements for parvovirus H-1 DNA replication in vitro. J Virol 41:982.
10. Kollek R, Goulian M (1981). Synthesis of parvovirus H-1 replicative form from viral DNA by DNA polymerase γ. Proc Natl Acad Sci 78:6206.
11. Tseng BY, Ahlem CN (1982). DNA primase activity from human lymphocytes; synthesis of oligoribonucleotides that prime DNA synthesis. J Biol Chem 257:7280.
12. Tseng BY, Ahlem CN (In press). A DNA primase from mouse cells, purification and partial characterization. J Biol Chem.

13. Faust EA, Rankin CD (1982). In vitro conversion of MVM parvovirus single-stranded DNA to the replicative form by DNA polymerase α from Ehrlich ascites tumor cells. Nucl Acid Res 10:4181.

14. Lamothe P, Baril B, Chi A, Lee L, Baril E (1981). Accessory proteins for DNA polymerase α activity with single-strand DNA templates. Proc Natl Acad Sci 78:4723.

15. Pritchard CG, DePamphilis ML (In press). Preparation of DNA polymerase α $[C_1C_2]$ by reconstituting DNA polymerase α with its specific stimulatory co-factors, C_1C_2. J Biol Chem.

16. Conaway RC, Lehman IR (1982). A DNA primase activity associated with DNA polymerase α from Drosophila melanogaster embryos. Proc Natl Acad Sci 79:2523.

17. Yagura T, Koza T, Seno T (1982). Mouse DNA replicase; DNA replicase associated with a novel DNA polymerase activity to synthesize initiator RNA of strict size. J Biol Chem 25/:11121.

18. Robertson AT, Stout ER, Briggs LL, Bates RC (1983). Aphidicolin inhibition of bovine parvovirus (BPV) DNA replication in vivo. Fed Proc 42:1983.

Mechanisms of DNA Replication and Recombination, pages 381–393
© **1983 Alan R. Liss, Inc., 150 Fifth Avenue, New York, NY 10011**

THE REPLICATION OF ADENOVIRUS DNA

Bruce W. Stillman

Cold Spring Harbor Laboratory
Cold Spring Harbor, New York 11724

ABSTRACT Analysis of virus mutants that are
defective for DNA replication in vivo and in vitro
has enabled the characterization of adenovirus
encoded proteins that are required for DNA synthesis.
A DNA binding protein and a DNA polymerase have
previously been described. Two other mutants have
been characterized. First, the onset of DNA
replication in vivo for the host range mutant,
Ad5hr7, is delayed by approximately 10hr in HeLa
cells and the total amount of intracellular virus DNA
at late times is reduced when compared with wild type
infected cells. Second, nuclear extracts prepared
from cells infected with the temperature sensitive
mutant, Ad2ts111, are defective for DNA synthesis in
vitro. The defective extracts can be complemented by
addition of purified adenovirus DNA-binding protein,
even though the mutation maps outside the coding
region for this protein.

INTRODUCTION

The use of conditionally defective mutants in
prokaryotic DNA replication systems has aided the identification of proteins required for DNA synthesis in vitro.
For the human adenoviruses, genetic analysis has also
contributed to the identification of proteins required
for lytic growth of the virus (1). Furthermore, with the
development of a cell free system for the replication of
adenovirus DNA (2), the characterization of virus mutants
that are defective for DNA replication in vivo and in
vitro has helped elucidate the functions of virus coded
proteins required for virus DNA synthesis.
Adenovirus virions contain a linear duplex DNA of

approximately 36 kilobase pairs which has a 55,000 dalton
(55K) protein covalently attached to each 5' end,
(terminal protein, TP; refs 3,4). However the TP is
synthesized during lytic infection as an 80K precursor
protein (pTP, refs 5,6) which becomes covalently attached
to all nascent replicating strands (6-10). The pTP is
cleaved to the 55K TP during virion morphogenesis (9).
The pTP is required for the initiation of DNA replication, which occurs at origins of replication located at
each end of the DNA. The pTP covalently binds a dCMP
residue (11) and this complex acts as a primer for
subsequent DNA synthesis. This priming reaction also
requires a virus coded DNA polymerase (12-14), a protein
encoded by the host cell (15), and specific DNA sequences
located near the termini of adenovirus DNA (16). By
creating site specific mutations in the origin DNA, and
assaying the products for template activity, the DNA
sequences required for DNA replication initiation have
been identified (43). A virus coded single strand DNA
binding protein (72K DBP) is primarily required for the
elongation of DNA synthesis (15, 17-20), but may also
enhance the initiation reaction.

Both the 72K DBP and the DNA polymerase were identified by the characterization of temperature sensitive
mutants Ad5ts125 and Ad5ts149 respectively, that were
both defective for DNA synthesis in vivo and in vitro
(14, 17-21). We have characterized two other conditionally lethal mutants of adenovirus, Ad5hr7 and
Ad2ts111. Ad2ts111 virus is defective for DNA replication in vivo, and is able to complement the replication
defects of Ad5ts125 and Ad5ts36 (another mutant in DNA
polymerase) in vivo (22-24). The map location of the
Ad2ts111 mutation is within the left 30% of the virus
genome (24). The Ad5hr7 mutant is a member of the complementation group II mutants which maps in the E1b region
of the genome (25-27) and it fails to produce the E1b 57K
tumor antigen (28,29).

MATERIALS AND METHODS

Cells and Virus. HeLa cells were grown in monolayer and
spinner cultures and 293 cells were grown as monolayer
cultures. Ad5hr7 was obtained from J. Williams and was
titred by plaque assay and grown on 293 cells. Ad2ts111

was obtained from J. Sussenbach and Ad5ts125 was obtained
from H. Ginsberg. Both were grown and titred in HeLa
cells at 32.5°C; the titer of these viruses at the non-
permissive temperature (38.5°C) was 10^3 to 10^4 fold lower
that the titer at 32.5°C.

Isolation of intracellular DNA. DNA was isolated by the
"Hirt" procedure which has been described previously
(44), except that pronase (1mg/ml) was added to the cell
lysis solution. Extrachromosomal DNA was then purified
by extraction with phenol, chloroform and isoamylalcohol
and then ethanol precipitation.

DNA replication in vitro. The isolation of nuclear
extracts and conditions for DNA replication reactions
have been described previously (2,5,14). Ad2ts111 and
Ad5ts125 nuclear extracts were prepared from cells
infected at 39.5°C, and harvested at 21hr post infection.
DNA binding protein (72K) was purified from Ad2wt
infected HeLa cells by the procedure of Schechter et al.
(45).

RESULTS

The map locations of some conditionally defective
mutants for adenovirus DNA replication are shown in Fig.
1, together with the maps for the E1 and E2 early transcription regions. The E2 region encodes the three virus
proteins required for the replication of adenovirus DNA
in vitro (5, 12-15, 17-20). We have recently identified
a DNA polymerase that is encoded by the E2B region of the
virus genome (14). This DNA polymerase was first identified as being required for formation of the pTP-dCMP
complex and is tightly bound to pTP (12, 13). The
purified DNA polymerase complements in vitro defective
nuclear extracts prepared from Ad5ts149 infected cells
and the mutation maps within a region that is contained
within a large open reading frame in the virus DNA
sequence (30-32). The 140K protein can be translated
from mRNA derived from the E2B region of the genome (14).
The Ad5ts125 mutation lies within the E2A 72K DBP coding
region and this mutation also affects the replication of
adenovirus DNA in vitro (19-21). Defective nuclear
extracts prepared from Ad5ts125 infected cells can be

complemented _in vitro_ by purified 72K DBP (20,21,Fig. 3).

Delayed Replication of Adenovirus DNA.

We have analysed two other mutants that are conditionally defective for DNA synthesis _in vivo_. First, the Ad5hr7 mutant (26), which maps within the E1B region (25,27, Fig. 1), is an early mutant that is defective for growth on HeLa cells, but can replicate on 293 cells, which are human embryo kidney cells that contain and express early regions E1A and E1B. The mutant fails to synthesize the E1B 57K tumor antigen in HeLa cells (28,29) and has been reported to synthesize virus DNA in these cells (28).

As part of an analysis of the effects of early mutants on the synthesis of virus DNA _in vivo_, it was

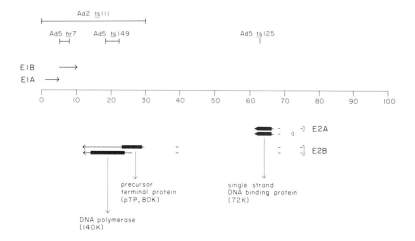

Figure 1. Map of adenovirus DNA showing the location of early regions 1 and 2, and the map location of some adenovirus mutants that are defective for DNA replication. The multiple mRNA species transcribed in regions E1A and E1B are not shown. The mRNA and protein products from E2A and E2B have been described previously (1,5,14). The map location of adenovirus mutants have also been described previously (1,24,27).

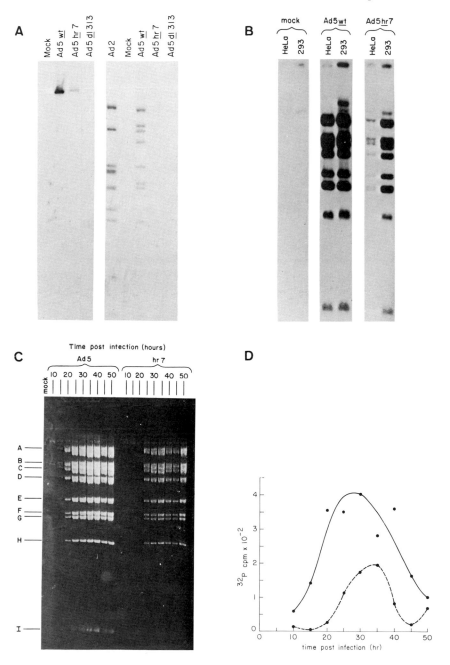

Figure 2. See legend at foot of page 386.

confirmed that Ad5hr7 did indeed synthesize virus DNA,
but the amount of virus DNA synthesized was consistently
lower than that found in Ad5wt infected cells labeled
from 17 to 22 hours post infection (Fig. 2A). As a
control, the host range mutant Ad5dl313, which contains a
deletion encompassing the entire E1B region (30), did not
replicate its DNA in HeLa cells. In order to determine
the relative levels of Ad5wt and Ad5hr7 DNA in the
permissive (293) and non- permissive (HeLa) cells, virus
DNA labeled with ^{32}P from 37 to 40hr post infection, was
isolated, digested with HindIII and electrophoresed on an
agarose gel. DNA was detected by autoradiography (Fig.
2B) and the amount of ^{32}P label reflected the total
amount of DNA detected by EtBr staining of the gel (data
not shown). The amounts of DNA in Ad5wt infected HeLa
and 293 cells are similar; however, the total amount of
Ad5hr7 DNA in HeLa cells is less than in 293 cells. This
suggests that DNA synthesis is either reduced or delayed
in Ad5hr7 infected HeLa cells.

The time course of both the accumulation and rate of
synthesis of either wild type or mutant virus DNA was
compared in HeLa cells (Fig. 2C and D). Figure 2C shows
the accumulation of virus DNA from infected cells at
various times post infection, and demonstrates that

Figure 2. Detection of adenovirus DNA in the Hirt
supernatants from mutant infected cells. A. DNA from
HeLa cells infected at a m.o.i. of 10 pfu/cell and
labeled with ^{32}P (20uCi/ml) from 17 to 22 hr. post
infection. DNA was subjected to elecrophoresis prior to
(left) or after (right) digestion with HindIII. B. DNA
from HeLa or 293 cells infected with Ad5wt, Ad5hr7 or
mock infection was labeled with ^{32}P (20uCi/ml) from 37 to
40hr post infection and then isolated, digested with
HindIII and subjected to agarose gel electrophoresis
followed by autoradiography. C. Hirt supernatent DNA
isolated from infected HeLa cells at various times post
infection. DNA was labeled with ^{32}P for 1hr prior to
isolation and digested with HindIII before
electrophoresis. The gel shows EtBr staining of DNA. D.
The ^{32}P labeled DNA in C was excised from the gel and
counted in a scintillation counter. Graph shows ^{32}P
label incorporated into virus DNA per 1hr for each time
point for Ad5wt and Ad5hr7.

Ad5hr7 begins DNA synthesis much later than Ad5wt DNA synthesis and that the total amount of intracellular DNA is less than wild type. The DNA shown in Fig. 2C was also labeled with ^{32}P for 1hr prior to harvesting the cells. The total amount of radioactivity in each restriction fragment was determined and the total amount of radioactivity in virus DNA expressed as a function of time post infection (Fig. 2D). It can be seen that the initial rate of synthesis of virus DNA is similar in both wild type and infected cells, but that the onset of DNA synthesis is delayed by approximately 10hr when compared with wild type infected cells. These results suggest that the Ad5hr7 mutation does not prevent DNA synthesis in HeLa cells, but that the onset of DNA synthesis is delayed. This phenotype is recessive since Ad5hr7 makes normal amounts of DNA in infected 293 cells.

DNA Replication _in_ _vitro_ with extracts from mutant infected cells.

Another mutant that is defective for adenovirus DNA synthesis _in_ _vivo_ is the Ad2ts111 temperature sensitive mutant (22,23). We have prepared extracts from Ad2wt, Ad2ts111 and Ad5ts125 infected cells grown at the non-permissive temperature (39.5°C) and assayed for DNA synthesis _in_ _vitro_ using HindIII digested Ad2 DNA-protein complex as template DNA. Under these conditions, origin specific DNA synthesis should occur only on the origin containing HindIII G and K terminal restriction fragments (33). Figure 3 demonstrates that nuclear extracts from Ad2wt infected cells support DNA synthesis. Similarly, extracts from Ad2wt infected cells that had been passed through DEAE cellulose (equilibrated in 0.2M NaCl containing buffer), still supported DNA synthesis; however DEAE cellulose extracts prepared from either Ad2ts111 or Ad5ts125 infected cells were inactive. The DEAE extracts were prepared so as to remove the large amount of degraded cell DNA present in the Ad2ts111 nuclear extracts (unpublished observations). Both Ad5ts125 (34,35) and Ad2ts111 (unpublished observations) were capable of formation of the pTP-dCMP initiation complex _in_ _vitro_, which suggests that both extracts contain functional pTP and 140K DNA polymerase. When purified DBP prepared from Ad2 infected HeLa cells was added to each of the DEAE nuclear extracts, specific

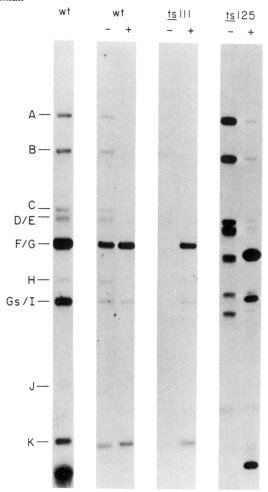

Figure 3. Replication _in_ _vitro_ with extracts from
Ad2wt, Ad2ts111 and Ad5ts125 infected cells. Nuclear
extracts were prepared from infected cells grown at
39.5°C and fractionated by chromatography on DEAE
cellulose. Replication was assayed on HindIII digested
Ad2 DNA protein complex in the absence (-) or presence
(+) of purified Ad2wt DBP. After replication, the DNA
was treated with pronase and subjected to agarose gel
elctrophoresis and detected by autoradiography. The
left track shows replication with unfractionated
extracts from wild type infected cells.

DNA replication of the HindIII G and K restriction fragments was observed (Fig. 3). An additional fragment (Gs) was also observed in each of the lanes where specific DNA synthesis occurred and and corresponds to the single strand nascent DNA chains produced by multiple rounds of initiation in vitro (33) (the Ks fragment migrated off the end of the gel). Complementation by DBP of defective Ad5ts125 nuclear extracts has been reported previously (20,24,34), and is expected as the mutation in Ad5ts125 DBP renders that protein thermolabile. Conversely, complementation by DBP of Ad2ts111 nuclear extracts was unexpected since the mutation lies outside of the gene for DBP, within the left 30% of the adenovirus genome (Fig. 1).

DISCUSSION

Two mutants of adenovirus have been studied with respect to their effects of virus DNA replication in vivo and in vitro. The Ad5hr7 mutation, which maps within the E1B regions, causes a delay in the onset of DNA synthesis when compared with the time course of a wild type virus infection. The final yield in the HeLa cell of both virus (26) and DNA (Fig. 2) is reduced when compared with Ad5hr7 infection in 293 cells, which suggests that an E1B gene product functions early in the lytic cycle to permit normal DNA synthesis. HeLa cells infected with Ad5hr7 do not contain the E1B 57K tumor antigen (28,29), but contain proteins from the E1A (11K), E1B (15K), E2 (72K), E3 (19K) and E4 (10K) regions (29) (the E4 10K protein (36) was originally reported to be an E1A gene product by Ross et al. (ref. 29)). Thus, it is possible that the E1B 57K protein functions to render the cells more permissive for virus DNA replication, presumably by interacting with a cellular protein. Such a cell protein may be present in some embryonic cells, since Harrison et al. (26) reported that Ad5hr7 will grow efficiently in human embryo kidney cells (HEK). Interestingly, Ad5wt, which grows efficiently in HeLa cells, produced 3.3 fold more virus after infection in HEK cells. Since 293 cells were derived from HEK cells, it is possible that the complementation of Ad5hr7 observed in these cells for virus yield (26) and amount of intracellular DNA (Fig. 2B) is due to a cellular factor and not the E1B 57K protein that is synthesized in these transformed cells.

The Ad5hr7 mutant fails to transform rat embryo cells
(37). An analogous situation occurs with the transformation defective host range mutants of polyoma virus
(38-40). These mutants are able to grow on mouse embryo
cells, but not on mouse 3T3 cells. These results suggest
that adenovirus, and possibly other DNA tumor viruses,
produce a protein or proteins that interacts with
proteins to increase the efficiency with which virus DNA
is replicated in cells. Some cells derived from
embryonic tissue express the required cellular factor(s)
and thereby alleviate the need for the virus protein.

A second virus mutant, Ad2ts111, was defective for
the replication of adenovirus DNA in vitro and the
replication activity could be restored by addition of
purified DBP from wild type infected cells. Since the
mutation in Ad2ts111 lies outside the coding regions for
the DBP (24), the protein altered in Ad2ts111 infected
cells must affect either the synthesis or the activity of
this protein. These possibilities are currently under
investigation. The precise map location of the Ad2ts111
mutation is not known, however, physical mapping studies
of D'Halluin et al (24) suggest that the virus contains a
single mutation which lies within the left 30% of the
virus genome.

The Ad2ts111 mutant is defective for replication of
virus DNA at the non-permissive temperature in vivo and
it has been suggested that this is the result of an
alteration in an early protein expressed just after lytic
infection (23). Ad2ts111 infection of cells at the
non-permissive temperature also leads to the degradation
of a portion of cell chromosomal DNA (23). Other host
range mutants of adenovirus type 12 and type 5 also have
this DNA degradation phenotype and the mutations have
been mapped by complementation studies to the E1B region
of the genome (41,42). Whether or not the Ad2ts111
mutation maps within the E1B region is currently under
investigation. Further investigation of this mutant and
the function of the E1B gene product in stimulating DNA
synthesis in vivo should provide valuable information as
to the nature of gene products involved in both DNA
synthesis and cell transformation.

ACKNOWLEDGEMENTS

I wish to thank E. Woodruff and P. Lalik for excellent technical assistance and F. Tamanoi for helpful discussions. The work was supported by a National Cancer Institute Center grant and the author is a Rita Allen Foundation Scholar.

REFERENCES

1. Flint, S.J. and Broker, T.R. (1980). Lytic infection by adenoviruses. In Tooze, J. (ed.) "DNA Tumor Viruses." Cold Spring Harbor Laboratory, New York. p 443.
2. Challberg, M.D. and Kelly T.J. Jr. (1979). Proc. Natl. Acad. Sci. USA 76: 655.
3. Robinson, A.J., Younghusband, H.B. and Bellett A.J.D. (1973). Virology 56: 54.
4. Rekosh, D.M.K., Russell, W.C., Bellett A.J.D. and Robinson, A.J. (1977). Cell 11: 283.
5. Stillman, B.W., Lewis, J.B., Chow, L.T., Mathews, M.B. and Smart, J.E. (1981). Cell 23: 497.
6. Challberg, M.D., Desiderio, S.V. and Kelly T.J. Jr. (1980). Proc. Natl. Acad. Sci. USA 77: 5105.
7. Stillman, B.W. and Bellett A.J.D. (1979). Virology 93: 69.
8. van Wielink, P.S., Naaktgeboren, N. and Sussenbach, J.S. (1979). Biochim. Biophys. Acta 563: 89.
9. Challberg, M.D. and Kelly, T.J. Jr. (1981). J. Virol. 38: 272.
10. Stillman, B.W. (1981). J. Virol. 37: 139.
11. Lichy, J.H., Horwitz, M.S. and Hurwitz, J. (1981). Proc. Natl. Acad. Sci. USA 78: 2678.
12. Enomoto, T., Lichy, J.H., Ikeda, J-E. and Hurwitz, J. (1981). Proc. Natl. Acad. Sci. USA 78: 6779.
13. Lichy, J.H., Field, J., Horwitz, M.S. and Hurwitz, J. (1982). Proc. Natl. Acad. Sci. USA 79: 5225.
14. Stillman, B.W., Tamanoi, F. and Mathews, M.B. (1982). Cell 31: 613.
15. Nagata, K., Guggenheimer, R.A., Enomoto, T., Lichy, J.H. and Hurwitz, J. (1982). Proc. Natl. Acad. Sci. USA 79: 6438.
16. Tamanoi, F. and Stillman, B.W. (1982). Proc. Natl. Acad. Sci. USA 79: 2221.

17. van der Vleit, P.C. and Sussenbach, J.S. (1975). Virology 67: 415.
18. van der Vleit, P.C., Levine, A.J., Ensinger, M.J. and Ginsberg, H.S. (1975). J. Virol. 15: 348.
19. Horwitz, M.S. (1978). Proc. Natl. Acad. Sci. USA 75: 4291.
20. Kaplan, L.M., Ariga, H., Hurwitz, H. and Horwitz, M.S. (1979). Proc. Natl. Acad. Sci. USA 76: 5534.
21. Ostrove, J.M., Rosenfeld, P., Williams, J. and Kelly, T.J. Jr. (1983). Proc. Natl. Acad. Sci. USA 80: 935.
22. Martin, G.R., Warocquier, R., Cousin, C., D'Halluin, J.C. and Boulanger, P.A. (1978). J. Gen. Virol. 41: 303.
23. D'Halluin, J.C., Allart, C., Cousin, C., Boulanger, P.A. and Martin, G. (1979). J. Virol. 32: 61.
24. D'Halluin, J.C., Cousin, C. and Boulanger P. (1982). J. Virol. 41: 401.
25. Frost, E. and Williams, J. (1978). Virology 91: 39.
26. Harrison, T., Graham, F. and Williams, J. (1977). Virology 77: 319.
27. Galos, R.S., Williams, J., Shenk, T. and Jones, N. (1980). Virology 104: 510.
28. Lassam, N.J., Bayley, S.T. and Graham, F.L. (1978). Virology 87: 463.
29. Ross, S.R., Levine, A.J., Galos, R.S., Williams, J. and Shenk, T. (1980). Virology 103: 475.
30. Galos, R.S., Williams, J., Binger, M-H and Flint, S.J. (1979). Cell 17: 945.
31. Aleström, P., Akusjärvi, G., Pettersson, M. and Pettersson, U. (1982). J. Biol. Chem. 257: 13492.
32. Gingeras, T.R., Sciaky, D., Gelinas, R.E., Bing-Dong, J., Yen, C.E., Kelly, M.M., Bullock, P.A., Parsons, B.L., O'Neill, K.E. and Roberts, R.J. (1982). J. Biol. Chem. 257: 13475.
33. Horwitz, M.S. and Ariga, H. (1981). Proc. Natl. Acad. Sci. USA 78: 1476.
34. Challberg, M.D., Ostrove, J.M. and Kelly T.J. Jr. (1982). J. Virol. 41: 265.
35. Friefeld, B.R., Krevolin, M.D. and Horwitz, M.S. (1983). Virology 124: 380.
36. Sarnow, P., Hearing, P., Anderson, C.W., Reich, N. and Levine, A.J. (1982). J. Mol. Biol. 162: 565.
37. Graham, F.L., Harrison, T. and Williams, J. (1978). Virology 86: 10.

38. Benjamin, T.L. (1970). Proc. Natl. Acad. Sci. USA
    67: 394.
39. Goldman, E. and Benjamin, T.L. (1975). Virology
    66: 372.
40. Feunteun, J., Sompayrac, L., Fluck, M. and
    Benjamin, T. (1976). Proc. Natl. Acad. Sci. USA
    73: 4169.
41. Lai Fatt, R.B. and Mak, S. (1982). J. Virol. 42:
    969.
42. Ezoe, H., Lai Fatt, R.B. and Mak, S. (1981). J.
    Virol. 40: 20.
43. Stillman, B.W. and Tamanoi, F. (1983). Cold Spring
    Harbor Symp. Quant. Biol. 47: (in press).
44. Hirt, B. (1967). J. Mol. Biol. 26: 365.
45. Schechter, N.M., Davies, W. and Anderson, C.W.
    (1980). Biochemistry 19: 2802.

Mechanisms of DNA Replication and Recombination, pages 395–421
© 1983 Alan R. Liss, Inc., 150 Fifth Avenue, New York, NY 10011

IN VITRO SYNTHESIS OF FULL LENGTH ADENOVIRAL DNA

Ronald A. Guggenheimer,[*] Kyosuke Nagata,[*] Jeffrey Field,[*]
Jeff Lindenbaum,[*] Richard M. Gronostajski,[*]
Marshall S. Horwitz[+] and Jerard Hurwitz[*]

Departments of Developmental Biology and Cancer,[*]
Microbiology and Immunology and Cell Biology[+]
Albert Einstein College of Medicine
Bronx, New York 10461

ABSTRACT   We have shown that five protein fractions are required for the reconstitution of Adenovirus (Ad) DNA synthesis when Ad DNA-prot is used as a template. Three of these proteins are viral coded; the 80 kd pre-terminal protein (pTP), the 140 kd Ad polymerase (Ad Pol) and the 72 kd DNA binding protein (Ad DBP). The two remaining proteins, purified from uninfected nuclear extracts of HeLa cells, have been designated nuclear factors I and II.

Four discrete assays have been devised to assess the role of each of the proteins during Ad DNA replication. Here we show that nuclear factor I is involved in the initiation of Ad DNA replication and subsequent elongation of nascent chains to 25-30% the length of full-sized Ad DNA and that nuclear factor II is required for the production of full length 34S Ad DNA-prot.

In addition, we describe an in vitro system that replicates plasmid DNA molecules containing the origin of Ad DNA replication but no terminal protein. Replication of such plasmid DNA molecules proceeds by a similar pTP-primed mechanism as is used by Ad DNA-prot. The results of these studies suggest a requirement for a specific DNA sequence and/or structure for the initiation of viral DNA synthesis.

INTRODUCTION

The Adenovirus (Ad) genome is a linear duplex DNA molecule of approximately 36,000 base pairs (bp). Ad serotypes 2 and 5, the types most commonly used for in vitro DNA replication studies, contain inverted terminal repeats of 102 and 103 bp, respectively (1-3). The 5' terminus of each viral DNA strand is covalently linked to a 55 kd protein, which has been termed the terminal protein (TP) (4-6).

Replication of the DNA-protein complex (Ad DNA-prot) can be initiated at either end of the double stranded parental molecule. Elongation of the nascent strand occurs in the 5' → 3' direction with concomitant displacement of the non-template strand (type I replication). Due to the presence of terminal repetitive sequences, the 5' and 3' ends of the displaced strand can anneal to form a duplex region identical to the termini of the duplex parental DNA. Replication of the displaced strand (type II replication) then initiates at its 3' terminus with elongation proceeding until a full length double stranded progeny molecule has been formed (7-9).

Due to its simple mechanism of replication and the availability of an in vitro system dependent on added DNA, Adenovirus has proved to be an especially useful model system for studies on eukaryotic replication mechanisms. The original in vitro system utilized nuclear extracts prepared from Ad infected HeLa cells to which hydroxyurea had been added at 2 h. post infection (10,11). The addition of hydroxyurea blocked DNA replication, but permitted the accumulation of viral proteins required for replication. DNA synthesis in this system required Ad DNA-prot, $MgCl_2$, ATP, and the infected nuclear extract. No synthesis occurred with deproteinized Ad DNA or with extracts prepared from uninfected cells. Replication in vitro mimicked the in vivo mechanism in that initiation occurred at either terminus of the template and the nascent chain was elongated continuously until a full genome length progeny strand had been formed. DNA synthesis in vitro retained the sensitivity to dideoxy TTP and aphidicolin characteristic of Ad DNA replication in vivo.

Our work has been directed toward reconstituting Ad DNA replication in vitro with purified proteins. We have isolated five highly purified protein fractions which when combined, catalyze the synthesis of full-length Ad DNA-prot. Three of these proteins are viral coded and are purified from cytoplasmic extracts of Ad infected HeLa cells. They include the 80 kd precursor (pTP) of the 55 kd terminal protein (12-14), a 140 kd DNA dependent-DNA polymerase (Ad Pol) (12) and the 72 kd Ad DNA binding protein

(Ad DBP) (15-17). Nucleotide sequence analysis of the gene encoding the Ad DBP predicted a protein consisting of 529 amino acid residues with a molecular weight of 59 kd (18). This is in contrast with an apparent native molecular weight of 72 kd as determined by SDS-polyacrylamide gel electrophoresis. The prevelance of proline residues in the Ad DBP or a high degree of phosphorylation of the Ad DBP (19) may explain the overestimation of the molecular weight of Ad DBP by SDS-polyacrylamide gel electrophoresis. The two remaining factors have been isolated from nuclear extracts of uninfected HeLa cells. These are nuclear factor I, a 47 kd polypeptide which is involved in the initiation and early elongation of DNA synthesis in vitro (20,21) and nuclear factor II, a protein of approximately 30 kd which is required for the elongation of replicating intermediates to full length Ad DNA-prot (21,22).

In order to assess the role of each of the proteins in Ad DNA replication in vitro we have devised four assays which permit the examination of different partial reactions involved in viral DNA replication. These include:

1) The initiation reaction; pTP-dCMP complex formation - A number of studies using both crude extracts and purified protein fractions have provided strong evidence for a novel mechanism of initiation of Ad DNA replication in which the pre-terminal protein is used as a primer (14,21,23-26) . As first proposed by Rekosh (5), this mechanism involves the covalent linkage of a dCMP residue, the first nucleotide of the nascent chain, to the 80 kd precursor of the terminal protein. Subsequent elongation of the pTP-dCMP complex occurs using the 3'OH of the dCMP residue as a more conventional primer.

2) pTP-26 mer formation - Limited elongation of the pTP-dCMP initiation complex to a pTP-26 mer can occur in the presence of dCTP, dATP, dTTP and dideoxy GTP (25). Formation of this complex is the result of termination of DNA chain synthesis at the 26th nucleotide of the template strand, the first site requiring dGMP incorporation.

3) Replication of restriction fragments - Replication of various segments of the Ad genome may be monitored by treating the Ad DNA-prot template with restriction endonucleases prior to in vitro DNA synthesis reactions. Fragments derived from internal regions of the Ad DNA-prot are devoid of the terminal protein while those derived from the ends of the molecule contain the 55 kd terminal protein. Fragments containing the terminal protein at their ends are replicated specifically (27).

4) Synthesis of full length Ad DNA-prot - The size of products formed in in vitro DNA replication reactions may be monitored by alkaline sucrose gradient centrifugation. As stated above, synthesis of full length 36 kbp Ad DNA-prot requires the presence of Ad DBP, pTP, Ad Pol, and nuclear factors I and II.

The work presented in this paper summarizes the role of the host protein factors in the replication of Ad DNA-prot in vitro. The purification, physical properties and functions of each of the three viral coded polypeptides have been described elsewhere and will be only briefly mentioned here. In addition, results of preliminary studies on the replication of plasmid DNA molecules containing the left terminus (0-9.4 map units) of Ad5 DNA will be presented.

MATERIALS AND METHODS

Nucleic acids and proteins - Ad2 DNA-prot was isolated as previously described (17). Plasmid pLAI was a gift of Dr. B. Stillman. pLAl was propagated in E. coli strain SK2267 and isolated by cesium chloride and neutral sucrose gradient centrifugation.

Purification of the pTP-Ad Pol complex (12), Ad DBP (12), nuclear factor I (20) and nuclear factor II (22) were as described. Partially purified nuclear fractions from uninfected HeLa cells, used for studies with EcoRI digested pLAI DNA, were purified through DEAE-cellulose, phosphocellulose and denatured DNA-cellulose column chromatography. Phosphocellulose and denatured DNA-cellulose fractions were free of detectable DNA polymerase activities. Details of the purification procedure will be described elsewhere (28). HeLa and calf thymus type I topoisomerases were a gift of Dr. L. Liu. E. coli type I topoisomerase was a gift of Dr. J. Wang.

Assay Conditions: Conditions for Ad DNA complementing assays, pTP-dCMP complex formation assays, pTP-26 mer assays and replication of restriction fragments have been previously described (20,22). Analysis of the products of the aforementioned assays and conditions for alkaline sucrose density gradient centrifugation have also been described. Conditions for assays using Eco RI digested pLAI DNA were identical to those using Ad DNA-prot as template except that EcoRI digested pLAI DNA and AdDBP were present at 0.2 µg and 1.0 µg, respectively.

RESULTS

Requirements for Ad DNA synthesis with purified proteins and Ad DNA-prot as template

DNA synthesis using the complete system containing the pTP-Ad Pol fraction, Ad DBP, nuclear factors I and II, and Ad DNA-prot was totally dependent on the presence of DNA, pTP-Ad Pol and Ad DBP (Table 1). In the presence of both the pTP-Ad Pol fraction and Ad DBP, synthesis was stimulated five-fold by the addition of nuclear factor I, but not by the addition of nuclear factor II. The simultaneous addition of nuclear factors I and II stimulated DNA synthesis 15-fold. Stimulation of DNA synthesis by nuclear factor II was completely dependent on the presence of nuclear factor I. Maximal replication of Ad DNA-prot with purified proteins required ATP and was inhibited by aphidicolin to the same extent found with crude extracts which supported in vitro Ad DNA replication.

TABLE 1

Requirements for Ad DNA-prot replication
with purified proteins

Conditions	DNA synthesis (%)	
	A	B
Complete	100	
- pTP-Ad Pol fraction or -Ad DBP or -Ad DNA-prot	< 1	
- nuclear factor I and - nuclear factor II	6	
- nuclear factor II	30	
- nuclear factor I	9	
Complete		100
+ aphidicolin (10 ⎍M)		45
+ aphidicolin (100 ⎍M)		10
- ATP		31

100% synthesis in A and B were 6.13 and 5.9 pmol of dTMP incorporation in 60 min at 30°, respectively.

Partial reactions of Ad DNA replication in vitro

A) Initiation of DNA replication: pTP-dCMP complex formation

Ad DNA synthesis commences with the covalent linkage of dCMP to the pTP. Maximal pTP-dCMP complex formation required the pTP-Ad Pol complex, nuclear factor I, Ad DBP and ATP (Table 2). In the presence of Ad DBP, the initiation reaction was dependent on nuclear factor I; the omission of the pTP-Ad Pol complex or omission of either the pTP or Ad Pol (29) reduced pTP-dCMP complex formation below detectable levels. The pTP-dCMP complex was detected in reactions containing the pTP-Ad Pol complex alone and this reaction was stimulated by ATP (maximally two-three fold). Nuclear factor I stimulated this reaction two-fold and again, ATP stimulated this reaction three-fold. Nuclear factor II had no effect on pTP-dCMP complex formation (21).

TABLE 2

Requirements for pTP-dCMP Complex Formation

pTP-AdPol	Factor I	AdDBP	ATP	pTP-dCMP formed (%)
+	+	+	+	100
+	+	+	-	46
+	-	+	± or	< 5
-	+	+	+	< 5
+	+	-	+	35
+	-	-	+	16
+	+	-	-	14
+	-	-	-	8

100% synthesis was 0.93 fmol pTP-dCMP complex formed in 60 min at 30°.

Previous studies have shown that pTP-dCMP complex could be formed with single-stranded DNA. This reaction required the pTP-Ad Pol complex and was inhibited by the addition of ATP or Ad DBP or E. coli DNA binding protein. Nuclear factor I had no effect on pTP-dCMP complex formation on single-stranded DNA catalyzed by the pTP-Ad Pol fraction; the inhibition of complex formation on single-stranded DNA by either ATP or Ad DBP was also unaffected by nuclear factor I (20,30).

B) Limited Elongation in the presence of Ad DNA-prot

The elongation reaction directed by the Ad DNA-prot can be examined in three different assays. (i) In the presence of dCTP, dATP, dTTP and ddGTP, a complex of 26 deoxynucleotides covalently linked to the pTP is formed. Elongation proceeds to this point due to the fact that no dGMP incorporation is directed by the template Ad DNA until the 26th nucleotide (25). Under the conditions used, the product (pTP-26 mer) migrates in SDS-polyacrylamide gels with an apparent molecular weight of 88 kd. In such reactions, we have always detected the accumulation of the pTP-dCMP complex migrating as an 80 kd product as well.

The effect of nuclear factor I on the formation of the 80 kd and 88 kd product using Ad DNA-prot as template was examined in the presence of the purified pTP-AdPol fraction alone or with the addition of Ad DBP (Fig. lA). Both the 80 kd and the 88 kd complex were formed by the pTP-Ad Pol fraction alone. In this case, the predominant product was the 80 kd complex. Formation of the 80 and 88 kd complexes was stimulated four to five-fold by the addition of nuclear factor I. In the presence of Ad DBP, the 88 kd complex form predominated, and again, nuclear factor I stimulated the production of both the 80 kd and 88 kd complexes.

(ii) The effect of nuclear factor I on the rate and extent of DNA synthesis with Ad DNA-prot as a template was examined. In these experiments, the products of in vitro reactions were subjected to KpnI digestion and analyzed by agarose gel electrophoresis (Fig. lB). In the presence of the pTP-Ad Pol fraction and Ad DBP, labeling of the terminal fragments G and F was detected. The addition of nuclear factor I stimulated the rate at which these two fragments (2088 and 2340 bp long, respectively) were labeled and resulted in the furthur extension of the DNA synthesized to fragments A and B. These results suggest that nuclear factor I acts in concert with the pTP, Ad Pol and Ad DBP in the early stages of the elongation reaction. Products synthesized by these proteins are elongated to approximately 30% of the length of full-sized Ad DNA-prot. The further addition of Ad DNA-prot, pTP-Ad Pol, Ad DBP, nuclear factor I, ATP, the dNTPs or extended incubation times, did not alter the final length of the products. However, the addition of nuclear factor II resulted in the production of full-length Ad DNA.

(iii) Digestion of Ad DNA-prot with restriction enzyme XbaI yields two fragments (E and C) from each end which are covalently linked to protein. The E fragment (1368 bp) and the C fragment (4212 bp) are replicated specifically if XbaI digested Ad DNA-prot is used as template. Terminal fragment synthesis by the pTP-Ad Pol and Ad DBP was shown to be stimulated by nuclear factor I (21).

C) Synthesis of full length Ad DNA-prot

The size of products synthesized in vitro by the pTP-Ad Pol complex, Ad DBP and nuclear factor I in the presence and absence of nuclear factor II was determined by alkaline sucrose density gradient centrifugation (Fig. 2). Omission of nuclear factor II resulted in synthesis of DNA chains of 20S which were approximately 25% the length of full sized Ad DNA. Synthesis of 20S DNA in the absence of nuclear factor II was hardly affected by the addition of 100 μM aphidicolin. Addition of nuclear factor II resulted in the synthesis of full sized 34S Ad DNA. Synthesis of full length Ad DNA was first detected after 30-60 min of incubation at 30°C (Fig. 2, insert) and was inhibited by 100 μM aphidicolin.

Properties of Nuclear Factor II

Highly purified preparations of nuclear factor II contain a DNA topoisomerase activity. The topoisomerase activity copurified with nuclear factor II complementing activity through filtration on an Ultragel AcA54 column (Fig. 3). Both nuclear factor II complementing activity and the topoisomerase activity present in nuclear factor II preparations were similarly affected by N-ethylmaleimide and heat (Fig. 4). The topoisomerase activity present in nuclear factor II preparations did not require ATP for the relaxation of ØX 174 RFI DNA. The addition of Mg^{++} stimulated relaxation of ØX 174 RFI DNA by the topoisomerase but was not absolutely required. NaCl at 0.2M inhibited the topoisomerase activity approximately 80% (Fig.5).

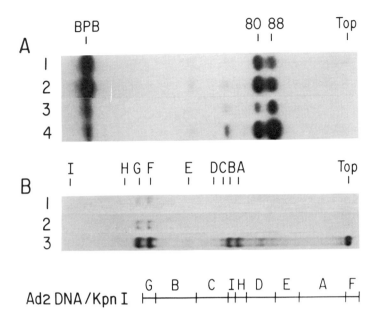

Fig.1. Effects of nuclear factor I on partial elongation reactions

(A) Effect of nuclear factor I on the formation of the partially elongated protein-nucleic acid complexes. Assay conditions were as described (22). Lane 1, pTP-Ad Pol fraction; lane 2, pTP-Ad-Pol fraction and nuclear factor I; lane 3, pTP-Ad Pol fraction and Ad DBP; lane 4, pTP-Ad Pol fraction, nuclear factor I and Ad DBP. In the absence of the pTP-Ad Pol fraction no complex was detected. The numbers 80 and 88 refer to the size in kilodaltons of the complex. BPB, bromophenol blue.

(B) Restriction endonuclease analysis of Ad DNA synthesized <u>in</u> <u>vitro</u>. Reaction mixtures (10 μl) contained 25mM Hepes/NaOH (pH 7.5), 5mM MgCl$_2$, 4mM dithiothreitol, 3mM ATP, 40 μM each dATP, dGTP, and dTTP, 4 μM α-[^{32}P]dCTP (23.5 Ci/mmole), 2 μg bovine serum albumin, 0.02 μg Ad DNA-prot, pTP-Ad Pol fraction and Ad DBP in the absence (lane 1) or presence (lanes 2 and 3) of nuclear factor I. Reaction were incubated at 30°C for 30 min (lane 2) or 120 min (lanes 1 and 3). After terminating the reactions, restriction enzyme Kpn I was added and the mixtures were incubated for 60 min at 37°C. [^{32}P]-dCMP labeled products were isolated and subjected to neutral agarose gel electrophoresis as described (22). Ad2-DNA fragments (A-I) generated by Kpn I digestion are shown in the map under panel B.

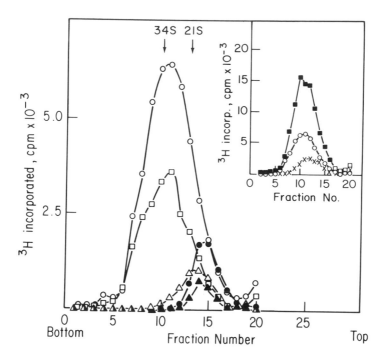

Fig. 2. Alkaline sucrose gradient sedimentation of Ad DNA synthesized in vitro. Products were formed with pTP-Ad Pol fraction, Ad DBP and nuclear factor I in the absence (△) or in the presence of nuclear factor II (O) or HeLa type I DNA topoisomerase (☐). Similar reactions were carried out in the presence of 100 µM aphidicolin and are represented by the filled in symbols (▲,◉). Reactions were carried out at 30°C for 120 min. Products were processed as described (22). Arrows indicate the positions of the 34S [^{14}C] Ad-DNA single strands as an internal marker and of the 21S [^{32}P] single stranded DNA as an external marker in a parallel gradient. Sedimentation was from right to left. The insert shows profiles of products formed with pTP-Ad Pol fraction, AdDBP, nuclear factor I and nuclear factor II at 30°C for 30 min (x), 120 (o) or 360 min (■).

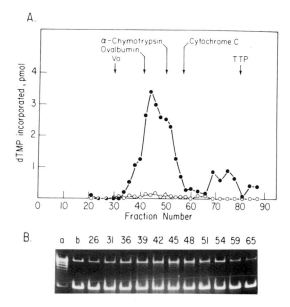

Fig. 3. Ultrogel AcA54 filtration.

(A)   A 10 ml column of Ultrogel AcA54 was calibrated using Ovalbumin, α-chymotrypsin and cytochrome c. Vo and Vi were determined using [^{3}H]-labeled ØX174 RFI DNA and [^{3}H]-dTTP, respectively.   Nuclear factor II was filtered separately under identical conditions; aliquots of each fraction were assayed for complementing activities in the absence (O) or in the presence (●) of nuclear factor I as described (22).

(B)   Aliquots of fractions from the Ultragel column were assayed for DNA topoisomerase activity as described (22).   Lane a, fraction applied to the column; b, DNA alone with no incubation; the fractions (26 to 65) obtained from the Ultrogel column were assayed for topoisomerase activity.

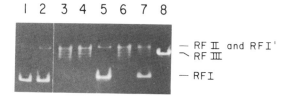

Fig. 4. Effect of N-ethylmaleimide and heat on the DNA topoisomerase activity associated with nuclear factor II. The DNA topoisomerase activity associated with nuclear factor II was assayed as described (22). Numbers (see below) in parentheses indicate percent DNA synthesis after various treatments of nuclear factor II before addition to the Ad DNA-prot replication system in the presence of pTP-Ad Pol fraction, Ad DBP and nuclear factor I. In the absence of nuclear factor II (41%). Lane 1, minus nuclear factor II and no incubation; lane 2, minus nuclear factor II with incubation; lane 3, nuclear factor II (100%); lane 4, nuclear factor II plus a mixture of 0.2 mM N-ethylmaleimide and 0.4 mM dithiothreitol (103%); lane 5, nuclear factor II treated with 10 mM N-ethylmaleimide for 30 min at 0°C and then neutralized with 20 mM dithiothreitol prior to the reaction (51%); lane 6, nuclear factor II heated at 45°C for 15 min prior to the reaction (83%); lane 7, nuclear factor II heated at 90°C for 2 min prior to the reaction (66%); lane 8, ØX RFI DNA digested with XhoI as a marker of RFIII DNA.

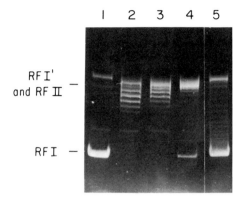

RF I'
and RF II

RF I

Fig. 5.   Requirements for Nuclear Factor II Associated Topoisomerase Activity.   Topoisomerase activity was assayed as described (22). Lane 1, omit nuclear factor II; lane 2, add nuclear factor II; lane 3, nuclear factor II in the presence of 3 mM ATP; lane 4, nuclear factor II omitting $MgCl_2$; lane 5, nuclear factor II in the presence of 0.2 M NaCl.

These results suggest that the topoisomerase activity present in nuclear factor II preparations is a type I topoisomerase.   In addition, eukaryotic type I topoisomerase isolated from Hela cells or calf thymus substituted for nuclear factor II in the in vitro Ad DNA replication system (Fig.6) while the E. coli type I topoisomerase did not.   The eukaryotic type I topoisomerases showed the same dependency on nuclear factor I for stimulation of Ad DNA replication as did nuclear factor II and supported the synthesis of full length DNA when used in lieu of nuclear factor II (Fig. 2).

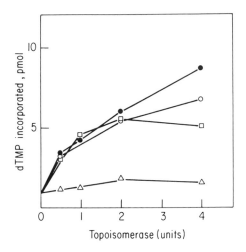

Fig. 6. **Replacement of nuclear factor II by DNA topoisomerases.** Reaction mixtures were as previously described (22) and contained either nuclear factor II (●), type I topoisomerase purified from HeLa cells (O), type I topoisomerase purified from calf thymus (□) or <u>E. coli</u> type I topoisomerase (△).

Nuclear Factor I Contains a DNA Binding Activity

Incubation of highly purified nuclear factor I with $5'[^{32}P]$labelled DNA fragments, generated by an Eco RI, Pvu II double digestion of plasmid pLAI DNA (see Fig 7), led to the retention of radioactivity on nitrocellulose filters (Fig.8A). Retention of labelled DNA on nitrocellulose by nuclear factor I did not require Mg^{++} or ATP (data not shown). Elution of the bound DNA from nitrocellulose, and analysis by polyacrylamide gel electrophoresis (Fig.8B) indicated that nuclear factor I bound selectively to a 451 bp fragment. The 451 bp fragment contained adenoviral sequences extending from the Eco RI site that linked the left end of the Ad genome to vector sequences up to the Pvu II site, 451 bp from the left end of the genome. Selective binding of nuclear factor I to the 451 bp sequence did not require the sequence to be located at the termini of the DNA fragments in which it was located. In addition, nuclear factor I binding was unaffected by the distance of the 451 bp sequence from the termini of DNA fragments in which it was contained (31).

Evidence for an interaction between the Ad Pol and Ad DBP

Evidence suggesting a specific interaction between the 140 kd Ad Pol and Ad DBP has been obtained using synthetic homopolymers as the primer-template for deoxynucleotide incorporation. In the presence of poly dT: oligo dA, dAMP incorporation was completely dependent on the presence of the pTP-Ad Pol fraction and Ad DBP (Table 3A). In the absence of the pTP-Ad Pol or Ad DBP no incorporation was observed. DNA polymerase α did not replace the pTP-Ad Pol fraction and E. coli DNA binding protein (SSB) failed to substitute for the Ad DBP. The effect of the Ad DBP was shown to be specific for the 140 kd Ad Pol subunit of the pTP-Ad Pol fraction (Table 3B). Using separated 80 kd and 140 kd polypeptides (29), dAMP incorporation was observed only when the Ad Pol and Ad DBP were combined. The addition of the 80 kd pTP inhibited the synthesis of poly dA. Products synthesized in the presence of the pTP-Ad Pol fraction, Ad DBP and poly dT: oligo dA were greater than 10 kb in length (data not shown).

TABLE 3

Poly dT:oligo dA primed DNA synthesis using
the Ad DNA polymerase

Additions	dAMP incorp., (pmol)	
	A	B
pTP-Ad Pol + Ad DBP	18.1	
omit pTP-Ad Pol	0.13	
omit Ad DBP	0.21	
omit pTP-Ad Pol/add polymerase α	0.14	
omit Ad DBP/add E. coli ssb	0.21	
pTP + Ad Pol + Ad DBP		10.2
omit Ad Pol		0.13
omit pTP		28.3
omit Ad DBP		0.1
omit Ad DBP, omit Ad Pol		0.1
omit Ad DBP, omit pTP		0.1

Reaction mixtures (50 μl) contained 50 mM Tris-HCl (pH 7.5), 10 mM MgCl₂, 4.5 mM dithiothreitol, 10 μg bovine serum albumin, 0.4 μg poly dT, 0.4 μg oligo dA, 8 μM [³H] -dATP (~2500 cpm/pmol), pTP-Ad Pol or isolated subunits, and 0.8 μg AdDBP or E. coli ssb.

DNA synthesis using a plasmid containing the origin of Ad DNA replication

pLAI is a 7.5 kb plasmid containing the 3.3 kb Bgl II E fragment derived from the left hand terminus (0-9.4 map units) of Ad 5 DNA (Fig.8). The construction of this plasmid had been described elsewhere (32). Digestion of pLAI with Eco RI restriction endonuclease yields a linear duplex DNA in which the cloned DNA sequence is present at one end of the molecule. Plasmid DNA so digested, functioned as a template for DNA synthesis in an in vitro system containing the pTP-Ad Pol, Ad DBP and a partially purified fraction isolated from nuclear extracts of uninfected Hela cells (Fig.9). pLAI digested with Xba I or Hpa I endonucleases did not support DNA synthesis; in these cases the adenovirus sequence is neither present at the terminus nor does it remain as an intact colinear structure. Removal, with nuclease S1, of the 4 base 5' extension generated by Eco RI digestion of plasmid pLAI, had no effect on the ability of the plasmid DNA to function as a template for DNA synthesis or to direct pTP-dCMP complex formation in an in vitro system containing the pTP-Ad Pol fraction, Ad DBP and a partially purified nuclear fraction (data not shown). pBR322 DNA digested with Eco RI did not support DNA synthesis.

DNA synthesis using Eco RI digested pLAI DNA was shown to have the following properties:

(i) pTP-dCMP complex formation occurred only if the adenovirus derived sequence in pLAI was intact and present at the terminus of the linear duplex DNA molecule (Fig. 10A). This suggests that Xba I and Hpa I digested pLAI DNA fail to support DNA synthesis due to their inability to direct pTP-dCMP complex formation. These results indicate that a specific DNA sequence, structure or both must be present at the termini of DNA molecules for the initiation reaction to occur. pTP-dCMP complex formation using Eco RI digested pLAI DNA was 100-fold less efficent than Ad DNA-prot, demonstrating, quantitatively, the important role of the 55 kd terminal protein in the initiation of adenoviral DNA replication. pTP-dCMP complex formation using Eco RI digested pLAI required the presence of the pTP-Ad Pol and a partially purified nuclear fraction (Fig.10B). Ad DBP inhibited non-specific synthesis which occurred in the presence of the pTP-Ad Pol and partially purified nuclear fraction.

(ii) Limited elongation of the pTP-dCMP complex was shown to occur in the presence of dATP, dTTP, dCTP and ddGTP (Fig.10C). As observed with pTP-dCMP complex formation, the limited elongation reaction using Eco RI digested pLAI DNA required the pTP-Ad Pol and partially purified nuclear fraction. Addition of Ad DBP inhibited non-specific synthesis and stimulated the elongation of the pTP-dCMP complex (three-fold). It should be noted that the elongated product migrated as a doublet in SDS-polyacrylamide gels, suggesting that the initiation reaction on Eco RI digested pLAI DNA did not occur at a unique site at the terminus of the DNA.

(iii) Analysis by benzoyl-naphthoyl DEAE-cellulose chromatography (16) demonstrated that 90% of the acid insoluble radioactivity synthesized in reactions primed with Eco RI digested pLAI DNA was protein linked. Alkaline agarose gel electrophoresis demonstrated that the size of these protein linked products was 7.5 kb (data not shown). These results demonstrated that full-sized, protein linked DNA chains were formed in vitro when Eco RI digested pLAI DNA was used as a template in conjunction with the pTP-Ad Pol fraction, Ad DBP and partially purified nuclear fraction.

(iv) DNA synthesis using Eco RI digested pLAI DNA did not require ATP and was inhibited by aphidicolin to a similar extent as reactions containing Ad DNA-prot as template (data not shown). Aphidicolin did not inhibit the initiation and partial elongation reactions (Figs.10B and 10C) using Eco RI digested pLAI DNA, suggesting that the inhibitory action of aphidicolin in Ad Pol catalyzed reactions requires extensive elongation. This appeared to be the case when either Eco RI digested pLAI DNA or Ad DNA-prot were used as templates.

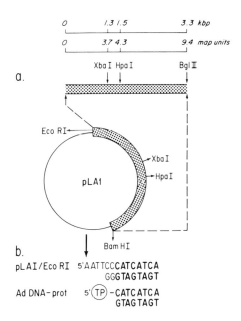

Fig. 7. Structure of plasmid pLAI. pLAI DNA contains the BglII E fragment of Ad 5 DNA (▨). Construction of the plasmid was as described (32). The sequences that appear at the end of pLAI after digestion with EcoRI and Ad DNA-prot are shown below the structure.

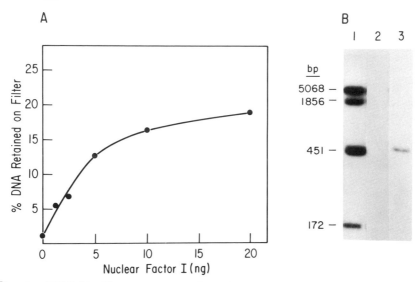

Fig. 8. DNA binding property of nuclear factor I.

(A) Nitrocellulose filter binding assay. DNA binding activity of nuclear factor I (glycerol gradient fraction) was assayed as a function of the amount of protein added. pLAl DNA was digested with <u>EcoRI</u> and <u>PvuII</u> and treated with <u>E. coli</u> alkaline phosphatase. The 5' ends of DNA fragments were labeled with γ-[^{32}P] ATP and T4 polynucleotide kinase. Nuclear factor I was incubated in a reaction mixture (50 μl) containing 25 mM Hepes/NaOH (pH 7.5), 5 mM MgCl$_2$, 4 mM dithiothreitol, 3 mM ATP, 10 μg BSA, 150 mM NaCl and 40 ng 5'- [^{32}P] labeled DNA (1.6 x 10^2 cpm/ng). After 20 min at 30°C the mixture was filtered through a nitrocellulose membrane (Millipore HAWP). The filter was washed with 0.5 ml of a buffer containing 25 mM Hepes/NaOH (pH 7.5), 5 mM MgCl$_2$, 4 mM dithiothreitol and 150 mM NaCl five times, dried and the radioactivity retained was counted by Cerenkov radiation.

(B). Analysis of DNA retained on nitrocellulose filters. DNA fragments retained on the filter under the above condition were eluted with 10 mM Tris-HCl (pH 7.5), 5 mM EDTA-0.2% SDS. <u>E. coli</u> tRNA (30 μg) was added to each eluate and the DNA was precipitated with ethanol. DNA fragments in each sample were analyzed by polyacrylamide slab gel electrophoresis (3.5%; 10 cm x 15 cm x 0.1 cm) with Tris-borate buffer (89 mM Tris-base/89 mM boric acid/2 mM EDTA) and autoradiographed. Lane 1, control sample; lane 2, no protein added; lane 3, nuclear factor I (10 ng).

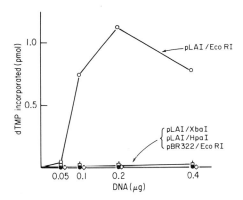

Fig. 9.   DNA requirement for synthesis using plasmid pLAI.
Reaction mixtures (50 µl) contained 25 mM Hepes-NaOH (pH 7.5),
5 mM MgCl$_2$, 4 mM dithiothreitol, 3 mM ATP, 40 µM each of
dATP, dCTP, and dGTP, 4 µM [^{3}H]-dTTP (4000 cpm/pmol), 10 µg
bovine serum albumin, the pTP-Ad Pol (glycerol gradient fraction,
0.002 unit), 0.3 µg of the partially purified nuclear fraction
(denatured DNA-cellulose fraction), and a five-fold excess
(weight:weight) of Ad DBP to DNA. Reactions were carried out
at 30°C for 60 min.

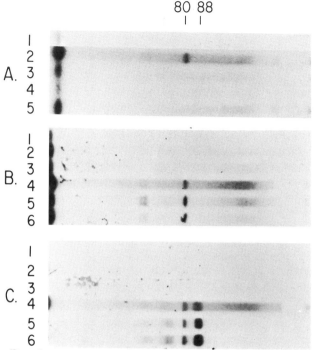

Fig. 10. Partial reactions using pLAI DNA as template.

(A) Template DNA requirement for pTP-dCMP complex formation. Reaction conditions were as described in the legend to Fig. 9 except that dATP, dGTP, and dTTP were omitted and α-[^{32}P]-dCTP (410 Ci/mmol) was present at 0.5 µM. Lane 1, no DNA present; lane 2, EcoRI digested pLAI DNA; lane 3, HpaI digested pLAI DNA; lane 4, XbaI digested pLAI DNA and lane 5, EcoRI digested pBR322.

(B) Protein Requirements for pTP-dCMP complex formation using EcoRI digested pLAI DNA. Lane 1, no proteins added; lane 2, the pTP-Ad Pol complex alone; lane 3, the pTP-Ad Pol complex plus Ad DBP; lane 4, the pTP-Ad Pol complex plus the partially purified nuclear fraction; lane 5, the pTP-Ad Pol fraction, Ad DBP and the partially purified nuclear fraction (complete); lane 6, complete plus 100 µM aphidicolin.

(C) Protein requirements for the partial elongation reaction using EcoRI digested pLAI DNA. Reactions were as described in (10A) except that dATP, dTTP and dideoxy GTP were present at 40 µM each. Lanes 1-6 were as above in B.

DISCUSSION

In vitro synthesis of full length 34S Ad DNA has been accomplished with five purified protein fractions, three of which are viral coded. These include the pTP, Ad Pol and Ad DBP. The two remaining proteins, nuclear factors I and II, have been purified from uninfected nuclear extracts of HeLa cells and were free of detectable DNA polymerase activities (20,22). This indicates that the 140 kd Ad Pol is the only polymerase required for in vitro synthesis of full-length Ad DNA. The possibility that Ad DNA replication in vivo proceeds by a more complex mechanism involving one or more cellular DNA polymerases cannot be excluded. However, addition of highly purified DNA polymerases α, β or γ, to the reconstituted in vitro system was without effect (21).

Ad DNA replication commences with the transfer of dCMP to the pTP. This reaction is catalyzed by the 140 kd Ad Pol and in the presence of the Ad DBP is completely dependent on nuclear factor I. pTP-dCMP complex formation is also observed when ØX174 single- stranded DNA is used as template. Nuclear factor I had no effect on the priming reaction with single-stranded DNA, while ATP and the Ad DBP or E. coli SSB were inhibitory (20,30). These characteristics distinguish priming on single-stranded DNA templates from specific priming on Ad DNA-prot. The different protein requirements for pTP-dCMP complex formation with ØX174 single-stranded DNA and Ad DNA-prot suggest that nuclear factor I, Ad DBP and ATP, may stimulate the initiation reaction on Ad DNA-prot by exposing a single stranded DNA region near the terminus of the duplex molecule. Such a single-stranded region may contain a specific DNA sequence or structure particularly suited for supporting pTP-dCMP synthesis. Evidence for a specific sequence and/or structural requirement for the initiation of Ad DNA replication in vitro has been obtained by studying the replication of cloned Ad DNA fragments which lack the 55 kd parental terminal protein (this work,32,33). Nuclear factor I has been shown to bind specifically to a 451 bp DNA fragment derived from the origin of Ad DNA replication. This interaction may aid in the separation of the strands of the template Ad DNA-prot. Strand separation would expose single stranded DNA regions thereby facilitating entry into and recognition of such regions by the pTP and Ad Pol. Alternatively, nuclear factor I may direct the pTP and Ad

Pol to the origin of Ad DNA replication by protein protein interaction.

After initiation of daughter strand synthesis, elongation of nascent DNA chains ensues utilizing the 3' OH end of the initiating dCMP residue as a primer. It has been postulated that as elongation proceeds, dissociation of the Ad Pol from the pTP occurs (21). The postulated dissociation of subunits during the course of elongation remains to be demonstrated. In the presence of nuclear factor I, Ad DBP, ATP, the pTP and Ad Pol, DNA chains of approximately 9 kb are formed. It is not known whether nuclear factor I remains associated with the replication fork or remains near the origin of replication during nascent DNA chain elongation. Polymerization of dNTPs on synthetic homopolymer template primers by the Ad Pol has been shown to be a highly processive process in the presence of the Ad DBP (unpublished data). Ad DBP may increase the processivity of the Ad Pol by binding to single-stranded DNA regions generated by fork movement and/or through direct interaction with the Ad Pol. The processive action of the Ad Pol with synthetic polynucleotides was unaffected by either nuclear factor I or nuclear factor II.

Synthesis of full length Ad DNA requires nuclear factor II in addition to the pTP, Ad Pol, Ad DBP and nuclear factor I. Purified nuclear factor II preparations contain a DNA topoisomerase activity. Furthermore, type I DNA topoisomerases from other eukaryotic sources substitute for nuclear factor II in the in vitro Ad DNA replication system. A possible role for the topoisomerase activity of nuclear factor II in the replication of Ad DNA is the relaxation of positive superhelical turns generated by fork movement on a template with constrained termini. The exact relationship of nuclear factor II and the HeLa type I topoisomerase has yet to be discerned. Nuclear factor II has an apparent native molecular weight of 30 kd while the HeLa type I topoisomerase has been shown to be a single polypeptide of approximately 100 kd (34). The discripancy in size may reflect extensive proteolysis of nuclear factor II during purification or the presence of a different topoisomerase.

The rate of in vitro Ad DNA synthesis in the reconstituted system described above is approximately 20 nucleotides/sec at 30°C. In addition, the rate of poly dA synthesis in reactions containing poly dT: oligo dA, Ad DBP and the Ad Pol was approximately 20 nucleotides/sec at 37°. These values are close to the reported in vivo rate of Ad DNA replication (35) as well as host DNA synthesis (36).

Ad DNA replication in vitro is sensitive to aphidicolin even in the reconstituted system free of DNA polymerase α, its presumed target in eukaryotic cells. Since the Ad Pol is insensitive to the drug (21,29), aphidicolin may act at some site other than the Ad Pol. The existance of another aphidicolin sensitive host replication factor in addition to DNA polymerase α could explain why host DNA replication is more sensitive to aphidicolin than is the activity of the purified DNA polymerase α (37,38). Other investigators have also proposed a second site of aphidicolin action, based on observations that Ad DNA replication in vivo is less sensitive to aphidicolin than host DNA replication, and that the effect of aphidicolin on DNA polymerase α but not on Ad DNA replication is competitive with dCTP (38,39). Alternatively, the Ad Pol may be sensitive to aphidicolin but this sensitivity only becomes manifest with the synthesis of long DNA chains. This is in keeping with our findings that only the extensive elongation of nascent DNA chains by the Ad Pol using either Ad DNA-prot or EcoRl digested pLAl DNA as template is inhibited by aphidicolin.

The host coded factors identified using the in vitro Ad DNA replication system are of interest with regard to their possible role in host DNA replication. It is our hope that these factors might form the starting point for the reconstitution of other replication systems which more closely resemble cellular DNA replication.

REFERENCES

1. Steenbergh, P.H., Maat, J., van Ormondt, H. and Sussenbach, J.S. Nucleic Acid Res. 4, 4371-4390 (1977).
2. Arrand, J.R. and Roberts, R.J. J. Mol. Biol. 128, 577-594 (1979).
3. Shinagawa, M. and Padmanabhan, R. Biochem. Biophys. Res. Comm. 87, 679-685 (1979).
4. Robinson, A.J., Younghusband, H.B. and Bellett, A.J.D. Virology 56, 54-69 (1973).
5. Rekosh, D.M.K., Russell, W.C., Bellett, A.J.D. and Robinson, A.J. Cell 11, 283-295 (1977).
6. Carusi, E.A. Virology 76, 380-394 (1977).
7. Horwitz, M.S. J. Virology 8, 675-683 (1971).

8. Horwitz, M.S. J. Virology 18, 307-315 (1976).
9. Lechner, R.L. and Kelly, T.J. Jr. Cell 12, 1007-1020 (1977).
10. Challberg, M.D. and Kelly, T.J. Jr. PNAS 76, 655-659 (1979).
11. Challberg, M.D. and Kelly, T.J. Jr. J. Mol. Biol. 135, 999-1012 (1979).
12. Enomoto, T., Lichy, J.H., Ikeda, J.-E. and Hurwitz, J. PNAS 78, 6779-6783 (1981).
13. Ikeda, J.-E., Enomoto, T. and Hurwitz, J. PNAS 78, 884-888 (1981).
14. Stillman, B.W., Lewis, J.B., Chow, L.T., Mathews, M.B. and Smart, J.E. Cell 23, 467-508 (1981).
15. Van der Vliet, P.C. and Sussenbach, J.S. Virology 67, 415-426 (1975).
16. Horwitz, M.S., Kaplan, L.M., Abboud, M., Maritato, J., Chow, L.T. and Broker, T.R. CSHSOB 43, (1979).
17. Kaplan, L.M., Ariga, H., Hurwitz, J. and Horwitz, M.S. PNAS 76, 5534-5538 (1979).
18. Kruijer, W., van Schaik, F.M.A. and Sussenbach, J.S. Nucleic Acid Res. 9, 4439-4457 (1981).
19. Linne, T. and Philipson, L. Eur. J. Biochem. 103, 259-270 (1980).
20. Nagata, K., Guggenheimer, R.A., Enomoto, T., Lichy, J.H. and Hurwitz, J. PNAS 79, 6438-6442 (1982).
21. Lichy, J.H., Nagata, K., Friefeld, B.R., Enomoto, T., Field, J., Guggenheimer, R.A., Ikeda, J.-E., Horwitz, M.S. and Hurwitz, J. CSHSOB 47, (1982).
22. Nagata, K., Guggenheimer, R.A. and Hurwitz, J. PNAS (in press).
23. Challberg, M.D., Desiderio, S.V. and Kelly, T.J. Jr. PNAS 77, 5105-5109 (1980).
24. Challberg, M.D. and Kelly, T.J. Jr. J. Virology 38, 272-277 (1981).
25. Lichy, J.H., Horwitz, M.S. and Hurwitz, J. PNAS 78, 2678-2682 (1981).
26. Challberg, M.D., Ostrove, J.M. and Kelly, T.J. Jr. J. Virology 41, 265-270 (1982).
27. Horwitz, M.S. and Ariga, H. PNAS 78, 1476-1480 (1981).
28. Guggenheimer, R.A., Nagata, K., Lindenbaum, J. and Hurwitz, J. (in preparation).
29. Lichy, J.H., Field, J., Horwitz, M.S. and Hurwitz, J. PNAS 79, 5225-5229 (1982).
30. Ikeda, J.-E., Enomoto, T. and Hurwitz, J. PNAS 79, 2442-2446 (1982).
31. Nagata, K., Guggenheimer, R.A. and Hurwitz, J. (in preparation).

32. Tamanoi, F. and Stillman, B.W. PNAS 79, 2221-2225 (1982).
33. van Bergen, B.G.M., van der Ley, P.A., van Driel, W., van Mansfield, A.D.M. and van der Vliet, P.C. Nucleic Acid Res. 11, 1975-1989 (1983).
34. Liu, L.F. and Miller, K.G. PNAS 78, 3487-3491 (1981).
35. Bodner, J.W. and Pearson, G.D., Virology 100, 208-211 (1980).
36. Cairns, J. J. Mol. Biol. 15, 372-373 (1966).
37. Longiaru, M., Ikeda, J.-E., Jarkovsky, Z., Horwitz, S.B. and Horwitz, M.S. Nucleic Acid Res. 6, 3369-3386 (1979).
38. Kwant, M.M. and van der Vliet, P.C. Nucleic Acid Res. 8, 3993-4007 (1980).
39. Pincus, S., Robertson, W. and Rekosh, D. Nucleic Acid Res. 9, 4919-4938 (1981).

Mechanisms of DNA Replication and Recombination, pages 423–447
© 1983 Alan R. Liss, Inc., 150 Fifth Avenue, New York, NY 10011

PAPOVAVIRUS CHROMOSOMES AS A MODEL FOR MAMMALIAN DNA REPLICATION[1]

M.L. DePamphilis, L.E. Chalifour, M.F. Charette, M.E. Cusick[2]
R.T. Hay[3], E.A. Hendrickson, C.G. Pritchard[4], L.C. Tack[5],

P.M. Wassarman, D.T. Weaver, and D.O. Wirak

Department of Biological Chemistry, Harvard Medical School
Boston, MA 02115, USA

INTRODUCTION

Our objective is to understand, at the molecular level, how mammalian chromosomes are replicated, and eventually, how this process is related to the control of cell proliferation. For this purpose, simian virus 40 (SV40) and polyoma virus (PyV) provide relatively simple, but appropriate, model systems. Both viruses replicate in the nuclei of their respective hosts, monkey and mouse cells, as small circular chromosomes whose histone composition and nucleosome structure are indistinguishable from those of its host. Both genomes have been completely sequenced and contain 5243 and 5292 bp respectively. With the exception of initiation of viral DNA replication, which requires viral encoded T-antigen (T-Ag), all subsequent steps in viral DNA replication and chromatin assembly appear to be carried out by the host cell. Furthermore, the final stages in replicon maturation

[1] Parts of this work were supported by the National Cancer Institute, the National Science Foundation, the American Cancer Society, and the American Heart Association.
[2] Present address: Biology Department, California Institute of Technology, Pasadena, CA.
[3] Present address: MRC Virology Unit, Glasgow, Scotland.
[4] Present address: Syva Company, Palo Alto, CA.
[5] Present address: Department of Biochemistry, Scripps Clinic and Research Foundation, La Jolla, CA.

appear to be the same for both virus and cell since the topological problems in separating two sibling viral chromosomes
are analogous to the merging of adjacent replicons; the ability
to rotate one DNA strand about the other is as restricted in
an "infinitely" long linear molecule as in a small covalently-
closed circular one. Although SV40 and PyV differ in their
DNA sequences, genetic organization and host ranges, their
mechanisms for DNA replication appear quite similar. The
past 10 years has evolved a detailed picture of the intermediates in viral DNA replication, the sequence of events at
their replication forks and the structure of their replicating
and mature chromosomes. These data have recently been reviewed
in depth (1) and compared with those from other eukaryotic
chromosomes (2). Here we attempt to summarize our current
view of papovavirus DNA replication and highlight some of our
recent results.

THE LYTIC CYCLE

Of the 11 forms of papovavirus DNA identified in a lytic
infection (1), only 5 appear to act as major intermediates
or products in viral DNA replication (Fig. 1). T-Ag binds
to the DNA region containing the unique ori sequence and
initiates bidirectional replication. The resulting replicating
intermediates (RI) contain two forks traveling at similar but
not identical rates (3,4) Although T-Ag is tightly bound
to only about 8% of the total viral chromosomes, most of the
replicating chromosomes, which comprise 1-2% of the population,
contain T-Ag (5,6). As replication continues past 70% completic
T-Ag is rapidly lost from the replicating chromosomes (Fig. 2B;
ref. 5) and RI begin to accumulate until late replicating
intermediates about 90% completed (RI*) are about 3-fold more
abundant than RI at earlier stages in replication (Fig 2A;
refs. 5,7,8). The presence of supercoils in the unreplicated
portion of RI demonstrates that parental strands remain
covalently-closed.

Although termination of replication does not require a
unique DNA sequence, replication forks are arrested at several
specific DNA sites within the termination region (4). Most
forks are arrested when replication is 91% completed, and the
two forks are separated by about 470 bp of unreplicated DNA
centered at the expected termination site. Perhaps attenuation
of replication prior to merging of two oncoming forks promotes
separation of sibling molecules by allowing helicase and topoisomerase activities to unwind the prefork DNA and separate

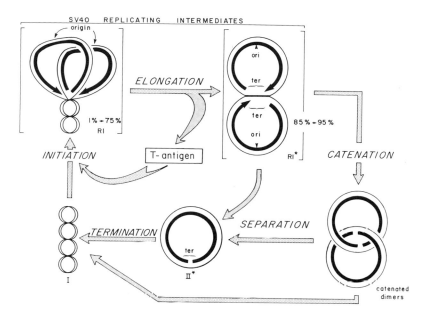

FIGURE 1.   Major forms of viral DNA in a lytic infection.

the resulting single-stranded regions to release circular
monomers containing a gap in the nascent strand within the
termination region (II*).   Separation is not inhibited by
the presence of gaps in the termination region (8), and
topoisomerase I can operate on single-stranded DNA (9).
Should this mechanism fail and the two forks merge before
separation occurs, then the resulting catenated dimers could
be separated by topoisomerase II.   Under optimal in vivo
conditions for SV40 DNA replication (0.8 - 1.0X Dulbecco's
Modified Eagle's Medium), II* is the predominant product of
separation (Fig. 3; ref. 8).   However, if the osmolarity of
the culture medium is increased slightly, then separation is
dramatically inhibited, DNA replication is retarded and
catenated dimers are rapidly accumulated (Fig. 3; ref. 10).
Since the catenated dimers produced at higher osmolarities
contain from 1 to 20 intertwines, unwinding of prefork parental
DNA strands must be inhibited under these conditions (10).
Therefore, instead of the intertwines being reduced to zero
before the forks meet (resulting in II*), catenated duplex
DNA molecules arise with one intertwine for every double-helix
turn that was not removed from the parental DNA templates
prior to replication.   Never-the-less, returning to normal
osmolarity allows separation of catenated dimers into monomers

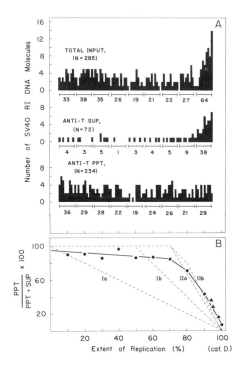

FIGURE 2. The fraction of replicating SV40 chromosomes associated with T-Ag as a function of the extent of their replication. DNA was purified from the 90S pool of viral chromosomes before and after immunoprecipitation with either anti-T serum or anti-T monoclonal antibody (5). Panel A: The number of SV40(RI) DNA at each 1% increment in the extent of their replication was determined by electron microscopy (vertical bars) prior to immunoprecipitation (total input), for the immunoprecipitate (anti-T ppt), and for the supernatant fraction (anti-T sup). The number of molecules in each 10% increment is indicated under the horizontal bars. Total number scored is N. Panel B: The fraction of replicating molecules containing T-Ag (●) was calculated for each 10% increment. Bgl I cut molecules were used to score RI between 85 and 99% replicated (▲). Catenated dimers were defined as molecules 100% replicated. About 75% of the total RI in the 90S pool contained T-Ag. The dashed lines were calculated by assuming 1 mole of T-Ag per RI with T-Ag loss beginning immediately after initiation (Ia), or after 70% replication (IIa), or by assuming 2 moles of T-Ag per RI with T-Ag loss beginning after initiation (Ib) or after 50% replication.

Relative Osmolality
0.4 0.6 0.8 1.0 1.2 1.4 1.7

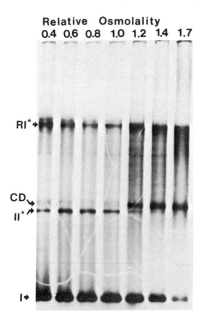

FIGURE 3.   Relative amounts of SV40 DNA as form I, II*,
catenated dimers containing 2 superhelical DNA molecules (CD),
and RI* following a 20 min labeling period carried out at
different osmolarities.   SV40 infected CV-1 cells were
incubated for 36 hrs after infection with 1X Dulbecco's
Modified Eagle's Medium (DMEM) at 37°C.   The medium was then
changed to that concentration of DMEM indicated plus [³H]Thd
and incubated for an additional 20 min at 37°C.   DNA was
extracted by the method of Hirt, purified and then fractionated by agarose gel electrophoresis and the [³H]DNA detected
by fluorography.   Methods and identification of dimeric DNA
forms were as described by Sundin and Varshavsky (10).   In
addition, standards of SV40 I, II and RI DNA were analyzed.

(10).   Such a fail-safe mechanism for mammalian chromosomes
would prevent extensive intertwining of two sister chromatids,
which could promote recombination and mitotic dysjunction in
each cell cycle, while still avoiding a dangerous requirement
for unique termination signals that might be lost during
chromosome rearrangements.   Once DNA replication terminates,
the superhelical turns found in form I DNA are automatically
accounted for by the presence of an average of 24 nucleosomes
which were assembled during the elongation phase.

THE EVENTS AT DNA REPLICATION FORKS

The data available for SV40 and PyV DNA replication forks can be interpreted on the basis of a simple model (Fig. 4) which has previously been discussed in detail (1,2, 11,25). DNA synthesis is predominantly, if not exclusively, continuous on forward arms and discontinuous on retrograde arms where the direction of synthesis must be opposite to the direction of fork movement. Thus, as forks advance, the forward arm is maintained as duplex DNA while a single-stranded DNA region is exposed on the retrograde arm that acts as an "initiation zone" for the synthesis of Okazaki fragments By some stochastic process, primase selects one of several preferred DNA sites to initiate synthesis of a short RNA primer on whose 3'-OH end DNA polymerase alpha[C1C2] rapidly begins DNA synthesis, continuing until a gap of about 15 bases remains to be completed. One or more protein cofactors are then required for alpha[C1C2]-pol to fill the gap and allow DNA ligase to join the 3'-OH end of the Okazaki fragment to the 5'-PO4 end of the long nascent DNA strand. During this process, RNA primers are excised in two steps: removal of the bulk of the primer does not require concomitant DNA synthesis whereas removal of the RNA-DNA junction is facilatated by DNA synthesis The repeated initiation of Okazaki fragments an average of once every 135 bases and the limitation of their mature length to 290 bases can be accounted for by the periodic arrangement of nucleosomes in front of the fork and the assumption that nucleosome "disassembly" to allow DNA unwinding is the rate limiting step. This would limit the size of the initiation zone to 220 ± 74 bases, the average distance from one near-randomly spaced nucleosome core to the next (Fig. 14). The broad size-range (40 to 290 bases) for mature Okazaki fragments, and the heterogeneous base composition of their RNA primers reflects the stochastic choice among different initiation sites located at varying distances from the 5'-ends of long nascent DNA strands.

DNA Polymerase Alpha[C_1C_2].

The evidence is compllelling that DNA polymerase alpha is solely responsible for DNA synthesis at replication forks (1,2). Alpha-pol copurifies with replicating chromosomes. Aphidicolin, a drug that specifically inhibits alpha-pol, blocks all steps in DNA synthesis at forks. ddTTP, a nucleotide that can inhibit beta and gamma-pol without inhibiting alpha-pol, does not inhibit any step in DNA replication.

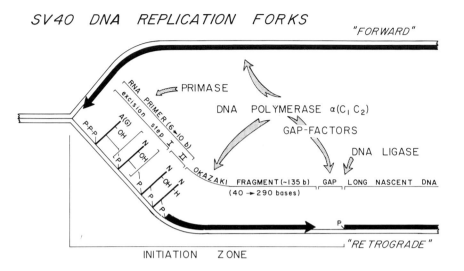

FIGURE 4. The events at an SV40 DNA replication fork.

Finally, a subcellular system inactivated with N-ethylmaleimide (NEM) requires addition of alpha-pol, but not of
beta or gamma-pol, for reactivation (12-14). However, as
originally shown with HeLa cells (15), essentially all of
the alpha-pol activity can be extracted from CV-1 monkey cells
as a complex with at least two proteins, C1 and C2, that act
as stimulatory cofactors (16). The native alpha[C1C2]-pol
(Fig. 5) is at least 30-fold more active on extensively
single-stranded DNA (ssDNA) substrates (e.g. denatured DNA)
than purified alpha-pol. The C1C2 complex is readily separated from alpha-pol either by addition of Triton X-100 or by
chromatography on phosphocellulose, and the two components
independently purified free of dsDNA and ssDNA endo- and exonucleases, RNA polymerase, DNA ligase and ATPase activities;
C1C2 is also free of DNA polymerase activity. The "holo"-
enzyme is rapidly reconstituted under normal assay conditions,
and can be purified on DEAE-Biogel (Fig. 6). The native and
reconstituted "holo"-enzymes are so far indistinguishable
from each other but are easily distinguished from "core"-
enzyme by three criteria: (i) physical separation on DEAE-
Biogel, (ii) ssDNA inhibition of alpha-pol but not of alpha
[C1C2]-pol, and (iii) C1C2 stimulation of alpha-pol but not
of alpha[C1C2]-pol activity on substrates with a high ratio

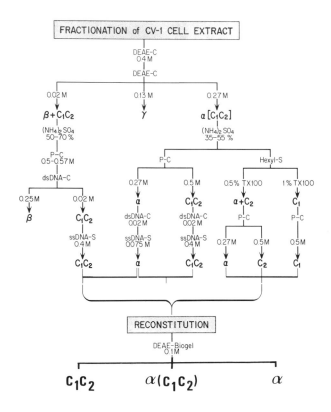

FIGURE 5. Preparation of DNA polymerase alpha[C1C2] by reconstituting DNA polymerase alpha with its specific stimulatory cofactors C1C2. The KCl concentration required to elute the designated protein is given. The concentration under DEAE-Biogel refers to alpha[C1C2]-pol. Beta and gamma refer to those DNA polymerases. DEAE-C is DEAE cellulose. P-C is phosphocellulose. Hexyl-S is hexylsepharose. dsDNA-C is double-stranded DNA cellulose. ssDNA-S is single-stranded DNA sepharose.

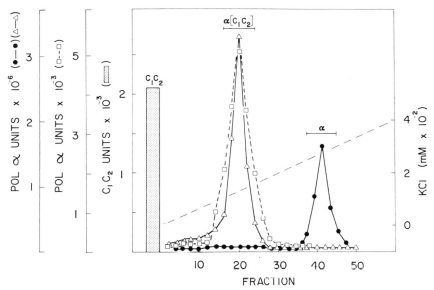

FIGURE 6. Chromatographic elution profiles for C1C2, DNA polymerase alpha[C1C2], and DNA polymerase alpha on DEAE-Biogel. Elution was with a linear gradient of KCl (----) in 50 mM KPO4 (pH 7.2), 1 mM EDTA, 1 mM dithiothreitol, and 10% glycerol at 4°C. Native alpha[C1C2]-pol (▲), reconstituted alpha[C1C2]-pol (■), and alpha-pol (●).

TABLE I

C_1C_2 STIMULATES DNA POLYMERASE α ON DNA SUBSTRATES WITH HIGH TEMPLATE:PRIMER RATIOS

SUBSTRATE	BASES TEMPLATE PRIMER	DNA SYNTHESIS (pmoles/15')		$\dfrac{\alpha[C_1C_2]}{\alpha}$
		α	$\alpha[C_1C_2]$	
DENATURED DNA	"high"	0.5	17	34
PARVOVIRUS DNA	5000	0.3	33	110
M13 ssDNA + Hha I DNA PRIMERS	3000	0.3	32	107
poly(dT):oligo(dA)$_{10}$	1000	0.4	16	40
poly(dT):oligo(rA)$_{10}$	1000	1.5	49	33
DNase I act. DNA	50	6.1	5.6	0.9
LAMBDA DNA ENDS	12	2.2	1.8	0.8
Eco RI DNA SITES	4	0.02	0.02	1.0

of ssDNA template per DNA or RNA primer (Table I). Stimulation
by C1C2 is specific for alpha-pol from its own cell type;
monkey C1C2 does not stimulate HeLa or calf thymus alpha-pol,
and HeLa C1C2 does not stimulate CV-1 alpha-pol (17). Hydrophobic chromatography on hexylsepharose separates the stimulatory complex into two proteins which, in the case of HeLa
cells, have been purified to homogeneity (15). It is important
to note that the response of alpha[C1C2]-pol to aphidicolin,
ddTTP, and NEM is the same as that of alpha-pol alone (Fig. 7),
and therefore our conclusion that alpha-pol is the replicative
polymerase remains unchanged.

The C1C2 complex functions as primer recognition proteins
(17). When the DNA substrate concentration is low [v=V(S)/K],
C1C2 stimulates alpha-pol from 180 to 1800 fold, depending
on the substrate and enzyme source, by reducing the Km for
the DNA substrate and, more specifically, the primer itself
(Table II). Both "core" and "holo"-enzyme requires a minimum
amount of template per primer and exhibit the same activity
on substrates with low template per primer ratios (Table I).
However, stimulation by C1C2 requires more ssDNA per primer
than needed for optimal activity of the "core"-enzyme (Table I).
Optimal stimulation occurs at about 1000 bases per primer.
The additional template is not needed to allow extensive
elongation since stimulation is observed with only the first
dNTP incorporated. Furthermore, C1C2 has no effect on the
Km values for dNTP substrates, the frequency or intensity of
DNA sequence signals that arrest alpha-pol (18), or the
processivity of alpha-pol (9-11 bases). Therefore, C1C2
specifically increases the ability of alpha-pol to find a
primer and insert the first nucleotide. The fact that extensive ssDNA primer-templates are better substrates for C1C2
than optimally activated DNase I treated DNA is for alpha
demonstrates that ssDNA participates in the reaction, presumably
by allowing alpha[C1C2]-pol to identify the template, bind
and then slide along the template until it finds a primer
(Fig. 8). This mechanism would also account for the ability
of alpha[C1C2]-pol to find and use very short primers before
they dissociate from the template. In contrast, ssDNA
inhibits alpha-pol, demonstrating that binding is nonproductive,
and the enzyme must continually rebind to locate the primer.

The properties of alpha[C1C2]-pol make it an excellent
candidate for carrying out DNA synthesis on the extensively
single-stranded replicating intermediates of parvovirus and
adenovirus DNA. With mammalian DNA, alpha[C1C2]-pol is designed
to use the low concentration of replication forks (about 0.04%
of total DNA) present during S phase 10^2-10^3 fold more effectivel

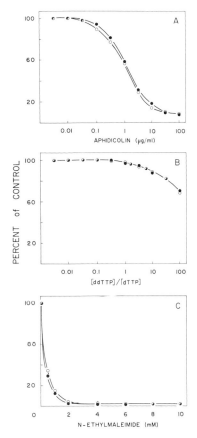

FIGURE 7. Effects of inhibitors on alpha (●) and alpha[C1C2] (○) polymerases.

TABLE II

DNA POLYMERASE α COFACTORS C_1C_2 FUNCTION AS PRIMER RECOGNITION PROTEINS

C_1C_2:	(1) decreases the relatively high Km values for DNA substrates with high template:primer ratios to values below those for DNA substrates with low template:primer ratios.
	(2) decreases the Km for the primer itself.
	(3) stimulates incorporation of the first dNTP.
	(4) stimulates the use of primers as short as dinucleotides.
	(5) has NO effect on the Km values for dNTPs.
	(6) has NO effect on DNA synthesis arrest signals in the template.
	(7) has NO effect on enzyme processivity.

than alpha-pol alone, an advantage that remains even if the
polymerase brings along its own primase.

The arrest of replication forks in the termination region
may reflect the inherent ability of replicative DNA polymerases
such as alpha and phage T4 to dissociate from the template
at specific DNA sequences (11,18). Further examination of
this phenomenon has revealed at least two types of arrest
sites: (i) cruciform structures in the template, and (ii)
some non-cruciform GC rich sequences. Both types of arrest
sites can be equally potent. With cruciforms, DNA polymerase
stops at the base of the stem regardless of the direction
from which it approaches. With non-cruciform sites, DNA
polymerase is arrested only when approaching from one direction;
no stops are observed with the complementary template. Neither
direct nor inverted repeats per se appear to act as arrest
sites. Non-cruciform arrest sites cannot be accurately predicted. Direct comparison of arrest sites in SV40 DNA in
vivo and in vitro revealed similar phenomena but the frequency
and intensity of sites expressed on bare DNA with purified
enzymes were generally greater. Although arrest of replication
forks occurred throughout the genome, the strongest sites were
found in the termination region. Specific stimulatory proteins,
such as C1C2 for alpha-pol and the products of genes 45, 44/62
and 32 for T4 polymerase had no effect on arrest sites. C1C2
stimulated initiation of DNA synthesis, and the T4 accessory
proteins stimulated processivity at least 10 fold. The
inherent ability of replicative DNA polymerases to recognize
specific DNA sequences may allow cells to moderate and
coordinate events in replication (11,18).

OKAZAKI FRAGMENTS AND THE ORIGIN OF DNA REPLICATION

To determine the relationship between the initiation of
Okazaki fragments at replication forks throughout the genome
and the initiation of DNA synthesis at the origin of replication,
we have mapped, at single nucleotide resolution, the 5'-ends
of nascent DNA chains and examined their structure. Three
regions of the genome were selected and cloned into phage M13
to provide purified single-stranded copies of each strand
(Fig. 9; ref. 19). One region of 311 bp contained the 65 bp
genetically-defined origin of replication (ori). A second
region of similar size contained the Msp I site (about 300
bp from ori), and the third region contained the Eco RI site
(about 1700 bp from ori). 5'-End labeled nascent DNA chains
were then prepared from purified SV40(RI) DNA (Figs. 9,10),

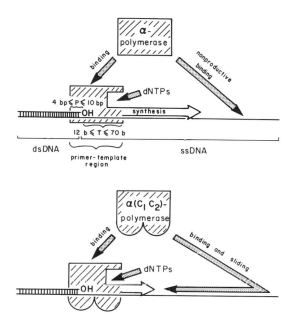

FIGURE 8. Model for the action of DNA polymerase alpha with and without C1C2 attached (17).

fractionated by gel electrophoresis inorder to isolate [5'-^{32}P] DNA chains 100-200 bases long, and then hybridized to each of the six M13 cloned SV40 DNA templates. Since the genomic location and orientation of each template were known from DNA sequence analysis, the position and orientation of the annealed DNA strands were unambiguously defined. The hybrids were then purified from free [5'-^{32}P]DNA by gel filtration, cut at a unique restriction site, and fractionated by gel electrophoresis. Their 5'-terminal nucleotides were identified and mapped by direct comparison with the [3'-^{32}P]DNA fragments released during Maxam-Gilbert sequencing of the appropriate DNA strand labeled at the same restriction site used to cut the nascent DNA.

Nascent DNA chains were identified by the presence of short oligoribonucleotides covalently attached to their 5'-ends. This was accomplished with three different experiments. First, the locations of rN-p-dN junctions were identified by phosphorylating the 5'-ends of all polynucleotide chains using unlabeled ATP and then "unmasking" those DNA ends attached to RNA by removing the RNA with alkali (Fig. 10). The resulting 5'-OH ends on the DNA were then labeled with ^{32}P so that only those DNA chains that were previously linked to RNA at their 5'-

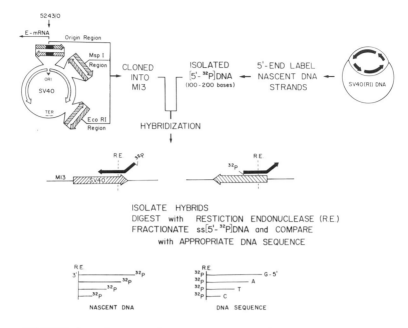

FIGURE 9. Protocol for mapping 5'-ends of nascent DNA on the SV40 genome.

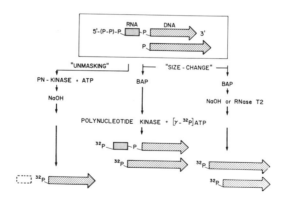

FIGURE 10. Two procedures for radiolabeling the 5'-ends of nascent DNA strands.

ends were now radiolabeled. The second method measured the change in size of polynucleotide chains that were 5'-end labeled before treatment with either alkali or RNase T2 with the same chains labeled after digesting the RNA. In this way, both the size of the putative RNA primer and the genomic position of its 5'-end could be determined. The third method measured the average length of all RNA primers that annealed to a particular DNA template by digesting the nascent [5'-^{32}P] DNA strands from the size-change experiment with T4 DNA polymerase 3'-5' exonuclease (19). With 5'-labeled DNA, the products are mono-, di- and trinucleotides. With 5'-labeled RNA-DNA, the products include the undigested RNA primers whose sizes can then be determined by gel electrophoresis (19).

Examples of these three procedures using nascent SV40 DNA from the ori region have been published (19), and they are typical of the data obtained from both the Msp I and Eco RI regions. Figure 11 shows densitometer tracings of the "unmasking" and "size-change" experiments with 5'-end labeled chains that annealed to M13 cloned SV40 DNA strands representing either the retrograde or forward sides of SV40 replication forks in the Msp I region. As previously observed in the ori region (19), RNA primed initiation events were not found on forward templates or with M13 alone. In contrast, RNA-p-DNA covalent linkages were found at specific nucleotide locations on the retrograde template with some sites used more frequently than others. When the RNA was digested prior to labeling 5'-ends, a pattern of [5'-^{32}P]DNA bands was observed that is remarkably similar to those representing the RNA-p-DNA linkages, indicating the absence of specific DNA breaks. When the 5'-ends of all polynucleotide chains were labeled, regardless of the presence or absence of RNA, again a similar set of bands were observed in the gel but these were displaced an average of 8-10 bases towards longer lengths, indicating that all of the DNA chains in the major bands contained an RNA primer at their 5'-ends. Since an end-group analysis of the total population of nascent DNA revealed that only 50% had RNA primers, the 5'-ends of nascent DNA chains beginning with RNA must be located at specific DNA sites on the template while the 5'-ends of nascent chains without RNA must be distributed throughout the template sequence. Therefore, assuming that an RNase H activity removes the bulk of the primer, the second step in excision which removes the RNA-p-DNA junction must involve a nuclease activity similar to E. coli DNA polymerase I 5'-3' exonuclease or phage T7

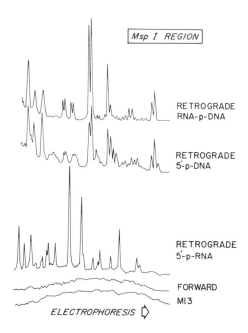

FIGURE 11. Mapping the 5'-ends of nascent DNA chains from the Msp I region in SV40 genomes.

exonuclease (5'-3') that degrades both RNA and DNA, resulting in a variable loss of dNMPs. Furthermore, analysis of the average size of RNA primers in all three regions revealed oligoribonucleotides 6 to 10 bases long covalently attached to the 5'-ends of about 40% of the DNA chains, and 1 or 2 ribonucleotides attached to the 5'-ends of about 10% of the DNA. These data are consistent with a two-step excision of RNA primers which, under steady-state conditions in vivo, leaves about half the nascent DNA chains free of RNA. Thus, the mapping data reveals 5'-ends of RNA at frequencies of 1/6 to 1/15 bases throughout the genome with RNA-p-DNA junctions consistently appearing a few bases downstream, while the overall end analysis reveals that these 5'-RNA termini result from RNA primers attached to the 5'-ends of DNA. A consensus sequence was therefore possible to obtain by lining up all the 5'-ends of RNA primers from all three genomic regions and making a histogram of the initiation site sequences (Fig. 12). Initiation of Okazaki fragment synthesis is promoted at 5'-Pu-p-T-3' sites in the template that tend to be preceded by purines and followed by Py-p-Pu. About 80% of the "PuT" sites in the 400 bases of template analyzed were used to initiate RNA primer synthesis. Secondary sites

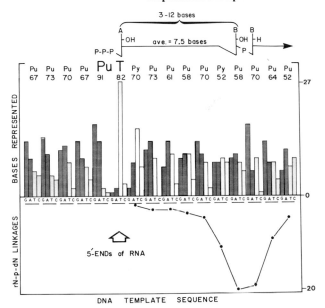

FIGURE 12. Histogram of the initiation sites that promote synthesis of RNA primed Okazaki fragments.

appear to be "PuC", 25% of which were used at lower frequencies than PuT sites. The histogram also reveals that the transition from RNA to DNA synthesis occurs an average of 8.5 bases downstream, and that RNA primers are NOT of unique length, but vary from at least 6 to 11 bases. Thus it appears that initiation of Okazaki fragments occurs throughout the genome by synthesis of RNA primers at specific sites on retrograde templates.

A map of the RNA-p-DNA linkages in the ori region (Fig. 13) suggests a simple relationship between the initiation of Okazaki fragment synthesis and initiation of replication at ori (Fig. 14; ref. 19). T-Ag together with cellular proteins binds to the ori region and generates a replication bubble of ssDNA in ori. Then by some stochastic process, one of several possible Okazaki fragment initiation sites is chosen on the DNA strand that also serves as template for E-mRNA synthesis, and the same mechanism used to initiate RNA-primed Okazaki fragment synthesis at replication forks throughout the genome is used to initiate synthesis of the "first Okazaki fragment" in ori. This DNA synthesis, however, continues and becomes the forward arm of the fork in the early gene region. As soon as a new initiation zone is exposed on the retrograde arm of this fork, initiation of a second

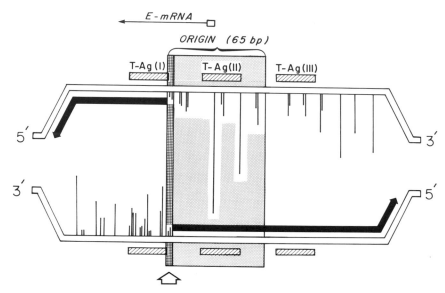

FIGURE 13. Initiation of DNA synthesis in the SV40
ori region. Indicated are the 3 major T-Ag DNA binding
sites, the 65 bp ori sequence containing the major initiation
site for E-mRNA, the direction of E-mRNA synthesis, the
positions and relative frequencies of rN-p-dN junctions
(vertical bars), the direction and location of continuous
DNA synthesis (wide solid arrows), and the transition site
where discontinuous DNA synthesis (presence of rN-p-dN
junctions) changes to continuous DNA synthesis (absence of
rN-p-dN junctions). This transition site, which is by
definition the origin of bidirectional DNA replication,
occurs within a 2 bp border between the strongest T-Ag site
and the genetically required ori sequence.

Okazaki fragment occurs in the opposite direction by the same
stochastic site selection process and becomes the forward
arm of the replication fork traveling towards the late gene
region. This process is repeated an average of 38 times
per genome. In order to establish the bidirectional transition point, something must physically block initiation
events on the late mRNA side of ori because the sequence
initiation signals are clearly there. A 27 bp palindrome
exsists within ori, and PuT sites exsist on both strands.
A stochastic site selection process occurs at ori because

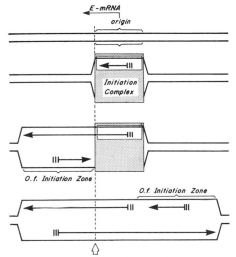

FIGURE 14. Model for the initiation of SV40 DNA replication. Arrows with 3 vertical bars are RNA-primed nascent DNA chains. "O.f." is Okazaki fragment.

if all the sites were used in every genome, the resulting DNA fragments would have been much shorter than 100 bases and therefore excluded from our mapping experiments. If transcription is involved in initiation of DNA replication, it could be either to "open" the ori region, or to provide a primer for one or two unique start sites. We are now mapping the locations of 5'-pppRNA-DNA chains in an effort to clarify these points.

CONTROL OF INITIATION OF DNA REPLICATION

We are examining the ability of embryonic mouse cells to replicate and express SV40 and PyV genomes after micro-injection of viral DNA (20). Mouse somatic cells permit PyV DNA but not SV40 DNA replication. Viral DNA injected into oocyte nuclei accurately expresses both early (T, mT and t-antigens) and late (VP1) genes in accord with the fact that oocytes are actively transcribing cell DNA. In contrast, 1-cell embryos exhibit little transcriptional activity, relying instead on stored mRNA, and thus do not express viral genes either. However, following development to the 2-cell and later stages, injected viral genomes are expressed along

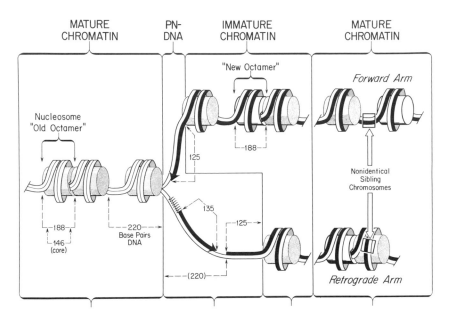

FIGURE 15. Nucleosome organization at SV40 replication forks (21). Nascent DNA (solid wide arrows) with an RNA primer is shown. Numbers indicate DNA lengths in nucleotides.

with the normal onset of mouse transcriptional activity. PyV, but not SV40, DNA appears to replicate after injection into the nuclei of 2-cell embryos; neither DNA replicates in 1-cell embryos. Thus, mouse embryonic cells do not appear promiscuous towards DNA replication, but like somatic cells, appear to require specific interactions between DNA sequences and replication proteins.

CHROMATIN STRUCTURE AT DNA REPLICATION FORKS

Replicating SV40 chromosomes consist of at least four components (Fig. 15): (i) Mature chromatin exists in front of forks as it is in nonreplicating mature SV40 chromosomes (1). (ii) The actual sites of DNA synthesis are not organized into nucleosomes but exist in a region of prenucleosomal DNA (PN-DNA). The nascent DNA in this region can be excised with exonucleases, released as fragments by MNase prior to digestion, and the Okazaki fragments can be released as dsDNA by single-strand specific endonucleases (21) (iii) Newly replicated DNA on both arms of the fork is rapidly assembled

into nucleosomes but the newly assembled chromatin is hypersensitive to nonspecific endonucleases even in the absence
of PN-DNA (22). (iv) The structure of mature chromatin
containing nascent DNA is not the same in two sibling molecules. Unique restriction sites can be accessible in one
sibling chromosome and not in the other, and the frequency
at which this occurs is consistent with a nearly random
distribution of nucleosomes along the DNA sequence of both
arms of replication forks (23). Therefore, the structure
of chromatin on one arm does not appear to direct assembly
of chromatin on the other arm, as might be expected if the
relationship between chromatin structure and gene activity
is to be maintained during replication.

To further examine the relationship between chromatin
structure, chromatin assembly and DNA replication, the distribution of "old histone octamers" in front of SV40
replication forks (Fig. 15) to the two arms of forks has
been followed in the presence of cycloheximide to prevent
assembly of nucleosomes from newly synthesized histones.
The four protocols are outlined in figure 17. SV40 infected
cells were treated with sufficient cycloheximide to inhibit
protein synthesis 95% and then replicating viral DNA was
radiolabeled either in vivo or in vitro. Continued DNA
replication in the absence of protein synthesis produced
nucleosome deficient chromosomes as judged by four criteria:
(i) increased sensitivity to micrococcal nuclease (MNase),
(ii) decreased sedimentation rate, (iii) increased amounts
of PN-DNA and internucleosomal DNA, and (iv) decreased superhelical density in the DNA (Fig. 16) which is a direct
measurement of the number of nucleosomes per genome. Nascent
nucleosomal DNA was then isolated following extensive MNase
digestion and the DNA hybridized to separated strands of
individual DNA restriction fragments (Fig. 17). Radiolabeled nascent nucleosomal DNA from each protocol annealed
equally well to both strands of each fragment whereas labeled
Okazaki fragments annealed only to the strand representing
the retrograde template. Therefore, since old histone octamers
are stable and do not exchange histones during replication
(2,24), segregation of prefork octamers must occur dispersively
to both arms of forks rather than conservatively to only one
arm. Seidman et al. (24) previously reported in similar
experiments using SV40 tsB11, a mutant that fails to assemble
virus at the restrictive temperature, that nucleosome segregation
was conservative to the forward arm. They postulated that
virus assembly may artifactually protect nascent DNA from
MNase. However, we obtained the same results with tsB11 at

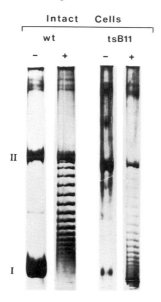

Intact   Cells

wt   tsB11

FIGURE 16. Gel electrophoresis of SV40 [³H]DNA from infected cells before (-) and after (+) treatment with cycloheximide. Positions of SV40(I) and (II) DNA are indicated. wt is wild type virus and tsB11 is a mutant in virus assembly.

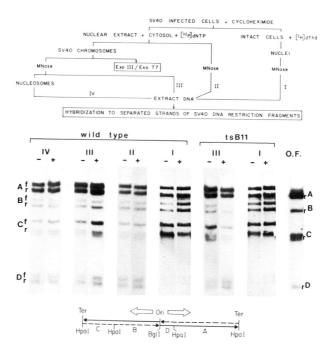

FIGURE 17. Hybridization of nascent nucleosomal DNA to the forward (f) and retrograde (r) strands of restriction fragments.

its restrictive temperature as we did with wild type at the permissive temperature.

In a second experiment we measured the ratio of labeled nascent DNA released from replicating chromosomes by E. coli Exo III (3'-5') to that released by T7 Exo (5'-3'). If nucleosome segregation in the presence of cycloheximide is dispersive the ratio should not change, but if segregation is conservative, then more nascent DNA will be excised from the arm deficient in nucleosomes. We found that the ratio after cycloheximide treatment was the same as before, consistent with dispersive nucleosome segregation.

The data for SV40 supports the following model for chromatin assembly. Nucleosomes are assembled on both arms of the fork as rapidly as sufficient duplex DNA becomes available. Both prefork histone octamers and newly synthesized histone octamers are utilized on both sides of the fork with assembly on one sibling molecule occuring independently of assembly on the other. Chromatin maturation then occurs on both sides of the fork, perhaps by addition of histone H1.

REFERENCES

1. DePamphilis ML, Wassarman PM (1982). Organization and replication of papovavirus DNA. In Kaplan AS (ed): "Organization and Replication of Viral DNA", CRC Press, p. 37.
2. DePamphilis ML, Wassarman PM (1980) Replication of eukaryotic chromosomes: a close-up of the replication fork. Ann. Rev. Biochem. 49:627.
3. Tapper DP, Anderson S, DePamphilis ML (1979). Maturation of replicating SV40 DNA molecules in isolated nuclei by continued bidirectional replication to the normal termination region. Biochim. Biophys. Acta 565:84.
4. Tapper DP, DePamphilis ML (1980). Preferred DNA sites are involved in the arrest and initiation of DNA synthesis during replication of SV40 DNA. Cell 22:97.
5. Tack LC, DePamphilis ML (1983). Analysis of SV40 chromosome:T-antigen complexes: T-antigen is preferentially associated with early replicating DNA intermediates. J. Virology, in press.
6. Segawa M, Sugano S, Yamaguchi N (1980). Association of SV40 T-antigen with replicating nucleoprotein complexes of SV40. J. Virology 35:320.
7. Tapper DP, DePamphilis ML (1978). Discontinuous DNA replication: accumulation of SV40 DNA at specific stages

in its replication. J. Mol. Biol. 120:401.

8. Tapper DP, Anderson S, DePamphilis ML (1982). Distribution of replicating DNA in intact cells and its maturation inisolated nuclei. J. Virology 41:877.

9. Liu LF, Depew RE, Wang JC (1976). Knotted ssDNA rings: a novel topological isomer of circular ssDNA formed by treatment with E. coli omega protein. J. Mol. Biol. 106:439. Champoux JJ (1977). Renaturation of complementary ssDNA circles: complete rewinding facilitated by the DNA untwisting enzyme. Proc. Natl. Acad. Sci. USA 74:5328.

10. Sundin O, Varshavsky A (1981). Arrest of segregation leads to accumulation of highly intertwinded catenated dimers: dissection of the final stages of SV40 DNA replication. Cell 25:659.

11. DePamphilis ML, Cusick ME, Hay RT, Pritchard C, Tack LC, Wassarman PM, Weaver DT (1983). Chromatin structure, DNA sequence and replication proteins: searching for the principles of eukaryotic chromosome replication. In deRecondo AM (ed): "New Approaches in Eukaryotic DNA Replication", Plenum Press, p. 203.
DePamphiils ML, Anderson S, Cusick M, Hay R, Herman T, Krokan H, Shelton E, Tack L, Tapper D, Weaver D, Wassarman PM (1980). The interdependence of DNA replication, DNA sequence and chromatin structure in SV40 chromosomes. In Alberts B (ed): "Mechanistic Studies of DNA Replication and Genetic Recombination", ICN-UCLA Symposia on Molecular and Cellular Biology, vol. 19, Academic Press, p. 55.

12. Edenberg HJ, Anderson S, DePamphilis ML (1978). Involvement of DNA polymerase alpha in SV40 DNA replication. J. Biol. Chem. 253:3273.

13. DePamphilis ML, Anderson S, Bar-Shavit R, Collins E, Edenberg H, Herman T, Karas B, Kaufmann G, Krokan H, Shelton E, Su R, Tapper D, Wassarman PM (1979). Replication and structure of SV40 chromosomes. Cold Spring Harbor Symp. Quant. Biol. 43:679.

14. Krokan H, Schaffer P, DePamphilis ML (1979). The involvement of eukaryotic DNA polymerases alpha and gamma in the replication of cellular and viral DNA. Biochemistry 18:4431.

15. Lamothe P, Baril B, Chi A, Lee L, Baril E (1980). Proc. Natl. Acad. Sci. USA 78:4723.

16. Pritchard CG, DePamphilis ML (1983). Preparation of DNA polymerase alpha with its specific stimulatory cofactors, C1C2. J. Biol. Chem., in press.

17. Pritchard CG, Weaver DT, Baril EF, DePamphilis ML (1983). DNA polymerase alpha cofactors C1C2 function as primer recognition proteins. J. Biol. Chem., in press.
18. Weaver DT, DePamphilis ML (1982) Specific sequences in native DNA that arrest the progress of DNA polymerase alpha. J. Biol. Chem. 257:2075.
19. Hay RT, DePamphilis ML (1982). Initiation of SV40 DNA replication in vivo: location and structure of 5'ends of DNA synthesized in the ori region. Cell 28:767.
20. Wirak DO, Chalifour LE, Wassarman PM, DePamphilis ML (1983). Replication and expression of SV40 and polyoma DNA during early mammalian development. J. Cellular Biochem., supplement 7B, Abs. 829, p. 129.
21. Cusick ME, Herman TM, DePamphilis ML, Wassarman PM (1981). Structure of chromatin at DNA replication forks: prenucleosomal DNA is rapidly excised from replicating SV40 chromosomes by micrococcal nuclease. Biochemistry 20:6648.
22. Cusick ME, Lee K-S, DePamphilis ML, Wassarman PM (1983). Structure of chromatin at DNA replication forks: nuclease hypersensitivity results from both prenucleosomal DNA and an immature chromatin structure. Biochemistry, in press.
23. Tack LC, Wassarman PM, DePamphilis ML (1981). Chromatin assembly: the relationship of chromatin structure to DNA sequence during SV40 replication. J. Biol. Chem. 256:8821.
24. Seidman MM, Levine AJ, Weintraub H (1979). The asymmetric segregation of parental nucleosomes during chromosome replication. Cell 18:439.
25. Anderson S, DePamphilis ML (1979). Metabolism of Okazaki fragments during SV40 DNA replication. J. Biol. Chem. 254:11495.

Mechanisms of DNA Replication and Recombination, pages 449–461
© 1983 Alan R. Liss, Inc., 150 Fifth Avenue, New York, NY 10011

REPLICATION OF VACCINIA VIRUS

Bernard Moss, Elaine Winters and Elaine V. Jones

Laboratory of Biology of Viruses, National Institute
of Allergy and Infectious Diseases, Bethesda, MD 20205

ABSTRACT The vaccinia virus genome consists of a
single uninterrupted polydeoxyribonucleotide chain
folded into a linear base-paired structure with hairpin loops at each end. The hairpin loops are A+T-rich,
incompletely base-paired, and exist in isomeric forms
that are inverted and complementary in sequence (flip-
flopped). Examination of replicating DNA molecules
revealed the presence of palindromes formed from the
hairpin loop structures. These palindromes were found
in DNA that sedimented faster than unit length molecules in neutral sucrose density gradients. Replication models that account for flip-flop sequence inversions and concatemeric structures were considered.
Replication of vaccinia virus DNA in the cytoplasm of
infected cells appears to be carried out largely by
viral enzymes. The location of the DNA polymerase
gene within a 2,000 bp segment of the 187,000 bp
genome was determined by rescue of the phosphonoacetate resistance marker of mutant virus.

INTRODUCTION

Poxviruses are distinguished by their large size,
complex morphology, and ability to replicate within the
cytoplasm of infected cells (1). Infectious particles
contain a lipoprotein membrane and a core structure. DNA
and enzymes needed for transcription of early genes are
located within the core. Vaccinia virus, the most
intensively studied member of this family, has a linear
double-stranded DNA genome of about 187,000 bp with several
distinctive features (Fig. 1). An inability to separate
the two DNA strands upon alkaline sucrose gradient

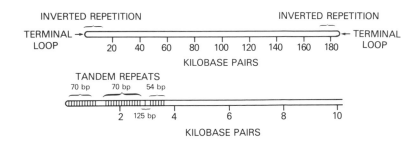

FIGURE 1. Structural features of the vaccinia virus genome. Representations of the entire genome and an expansion of the 10,000 bp inverted terminal repetition are shown.

centrifugation led Berns and co-workers (2) to suggest that the genome contains cross-links. Visualization of approximately twice genomic length single-stranded DNA circles by electron microscopy of denatured molecules established that the cross-links are near the ends of the linear duplex. Similar evidence for covalently linked DNA strands has been obtained for other members of the poxvirus family indicating that this is a characteristic feature. The precise nature of the link between the two DNA strands was established by nucleotide sequencing (3). An A+T-rich incompletely base-paired hairpin loop was found at each end of the DNA molecule (Fig. 2). These loops exist in two isomeric forms that are inverted and complementary in sequence (flip-flopped). Thus, the genome of vaccinia virus and presumably other poxviruses consists of a single uninterrupted polynucleotide chain folded into a linear duplex structure.

A second feature of interest is a 10,000 bp inverted terminal repetition (Fig. 1). The presence of identical structures at each end of the vaccinia virus genome was suggested by restriction endonuclease mapping (4) and visualized by electron microscopy (5). Short, tandemly repeated sequences (Fig. 1) as well as at least three early genes are located within the inverted repetition (6, 7). The predominant repeating unit is 70 bp long and contains single HinfI, TaqI, and MboII sites. In vaccinia virus, there are two sets of 70 bp repeats. The first set of 13 repeats starts 87 bp from the proximal end of the

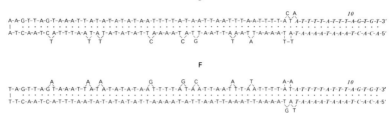

FIGURE 2. Nucleotide sequences of the terminal loops of vaccinia virus DNA. The S and F forms are inverted and complementary in sequences (flip-flopped). Reprinted with permission (3).

terminal loop and is separated by 325 bp from a second set of 18 tandem 70 bp repeats. Nucleotide sequencing indicated that the most proximal 70 bp repeat is partially overlapped by two successive 125 bp repeats which are followed by 8 tandem 54 bp repeats that lack restriction endonuclease sites (Fig. 1). The three different length repeats have extensive sequence homology: the 70 bp and 125 bp repeats contain duplications and triplications, respectively, of a segment of the 54 bp repeat. Apparently, these repeats have evolved by unequal cross-over. Some variation in the number of 70 bp tandem repeats has been noted with individual plaque isolates. A closely related virus, rabbitpox, contains only about 8 copies of the terminal set of 70 bp repeats and even fewer in the proximal set but has the same number of 125 and 54 bp repeats as vaccinia virus (our unpublished observations). Nucleotide sequencing indicates that the vaccinia virus and rabbitpox repeats are identical. The repeats of cowpox are very similar to those of vaccinia virus but show some sequence divergence (8).

Replication of poxvirus DNA occurs within the cytoplasm of infected cells. Autoradiographic studies indicate that DNA synthesis occurs in localized regions termed factory areas (9). Although the nucleus is required for formation of mature virus particles, vaccinia virus DNA replication can occur in enucleated cells (10, 11). Accordingly, it seems possible that a majority of the enzymes and protein factors are virus coded. In this regard, several new enzymes including a DNA polymerase (12, 13), DNA ligase (14), DNA unwinding enzyme (15), several DNases (16), and a thymidne kinase (17) have been found in infected cells or virus particles. Genetic evidence of viral origin, however, is lacking except in the cases of thymidine kinase (17-19) and DNA polymerase (20). The DNA polymerase consists of a single polypeptide chain with a molecular weight of about 110,000 that can be separated chromatographically from host DNA polymerases (12, 13). The most highly purified preparations contain 5'-3' exonuclease activity suggesting that it may be an integral part of the polymerase. The thymidine kinase has a subunit molecular weight of 19,000 (19, 21) and the amino acid composition has been inferred from the DNA sequence (22).

Studies with infected tissue culture cells indicated that replication starts near both ends of the vaccinia virus genome and involves discontinuous DNA synthesis and RNA primers (23-27). Self-priming and de novo synthesis models have been considered to explain the role of the terminal loop in replicating the ends of the genome (3). Additional structural, genetic and biochemical studies are needed to understand the basic mechanisms of poxvirus DNA replication. In this communication, we describe the structures at the ends of replicating DNA molecules and the genetic locus of the vaccinia DNA polymerase.

RESULTS

Concatemeric Ends of Replicating DNA.

Previous studies of Moyer and Graves (28) suggested that the terminal sequences of rabbitpox virus DNA are present as palindromes in replicating molecules. Similar evidence also has been obtained with vaccinia virus (29). Vaccinia virus replicates within the cytoplasm starting 1 to 2 h post-infection (30, 31). To characterize the ends of replicating molecules, DNA was isolated from the cytoplasm at various times after virus inoculation and digested with appropriate restriction endonucleases. Generally, BstEII or SalI, which cut about 1.3 kb and 3.6 kb from the ends of the genome, respectively, were used. After agarose gel electrophoresis, the DNA fragments were transferred to nitrocellulose and probed by hybridization with ^{32}P-labeled vaccinia virus DNA. In the

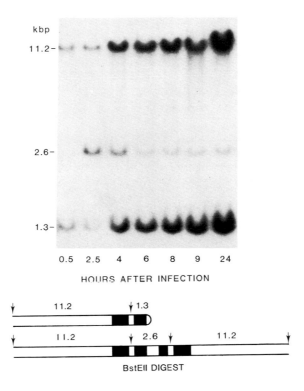

FIGURE 3. Detection of terminal sequences in replicating
vaccinia virus DNA. Cytoplasmic DNA, isolated at the
indicated times after infection, was digested with BstEII
subjected to agarose gel electrophoresis, transferred to a
nitrocellulose membrane, and hybridized to ^{32}P-labeled
70 bp terminal repeat fragment. An autoradiograph is
shown. Below, a representation of the end of mature
vaccinia virus DNA and postulated concatemeric structures
of replicating DNA with 70 bp tandem repeats shaded.
Reprinted with permission (29).

experiments shown in Fig. 3, the ^{32}P-labeled 70 bp repeat
was hybridized to BstEII digested cytoplasmic DNA fragments.
Two labeled bands of 1.3 kb and about 11.2 kb, representing
the terminal and adjacent DNA fragments respectively, were
detected when mature virion DNA was examined in this
manner. The same two bands were found using DNA isolated
from the cytoplasm at 0.5 h after infection. By 2.5 h,

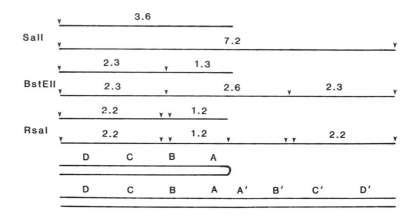

FIGURE 4. Restriction endonuclease maps of mature and concatemeric end sequences. For each restriction enzyme, arrowheads indicate cleavage sites and numbers indicate the size in kb of fragments detected by Southern blotting. Results obtained with mature and replicating DNA are shown. Palindromic model for replicating ends is shown at bottom.

however, an additional band of 2.6 kb was detected. When similar experiments were carried out with SalI digested DNA, the terminal fragment of 3.6 kb as well as a new fragment of 7.2 kb were seen. In both cases, the fragment unique to replicating DNA was precisely twice the size of the terminal fragment of mature DNA.

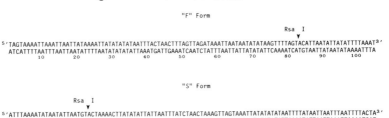

FIGURE 5. Predicted nucleotide sequences of palindromes formed from terminal hairpins. RsaI restriction endonuclease sites are shown.

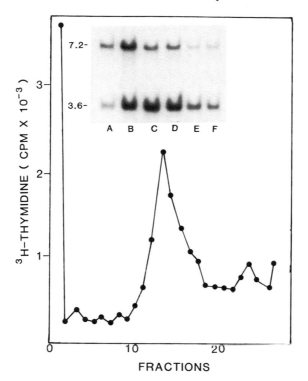

FIGURE 6.   Evidence for concatemers.   Cells were labeled
with [^{3}H]thymidine from 2.5 to 3 h after infection and
cytoplasmic DNA aggregates were purified by sedimentation
into a sucrose cushion (42).   DNA, deproteinized by
digestion with proteinase K and multiple extractions with
phenol and phenol:chloroform, was applied to a 10 to 30%
neutral sucrose gradient formed on top of a 60% sucrose
cushion.   After sedimentation at 193,000 X g for 165 min,
fractions were collected from the bottom of the tube.   The
distribution by scintillation counting and pooled fractions
(A, 1; B 2-9; C, 10-13; D, 14-17; E, 18-21; F, 22-25) were
precipitated with ethanol and digested with SalI.   After
agarose gel electrophoresis, DNA was transferred to
nitrocellulose and hybridized to ^{32}P-labeled SalI end
fragment of mature virion DNA.

To obtain larger amounts of DNA, replication was
synchronized by infecting cells in the presence of
5-fluorodeoxyuridine and then adding thymidine several

hours later. Analysis of cytoplasmic DNA revealed that
maximal amounts of the palindromic sequence were present
at 1 h after addition of thymidine. DNA isolated at this
time was cleaved with SalI and the 7.2 kb fragment was
purified by agarose gel electrophoresis. Several restriction endonucleases were used to compare the structure of
this DNA segment with the 3.6 kb SalI fragment of mature
virion DNA (Fig. 4). In each case, the data were consistent with a palindrome. One apparent anomaly, however, was
noted with RsaI. The palindrome appeared to have an additional site of cleavage not noted in the mature terminus.
An explanation for this was found upon examining the nucleotide sequence (Fig. 2). Although the RsaI recognition
sequence GTAC is present within the hairpin, it is not
base-paired. Fig. 5 shows the presence of RsaI sites in
the predicted nucleotide sequence of a palindrome formed
from the hairpins. Thus, the experimental finding of the
RsaI site supports the proposed structures. Attempts are
being made to clone the palindrome for sequencing and
other purposes.

Although the presence of palindromic sequences suggests the presence of concatemers, attempts to demonstrate
vaccinia virus DNA molecules greater than unit length have
generally been unsuccessful (24). The sedimentation of
cytoplasmic DNA labeled with [^{3}H]thymidine from 2 to 3 h
after infection is shown in Fig. 6. The major peak was
found to co-sediment with [^{14}C]thymidine labeled mature
vaccinia virus DNA (not shown). However, additional DNA
sedimented faster and some was collected in a 60% sucrose
cushion. Pooled DNA fractions were digested with the
restriction endonuclease SalI and anlyzed by agarose gel
electrophoresis. After transfer to nitrocellulose, the DNA
was probed with ^{32}P-labeled SalI end fragment of mature
vaccinia virus DNA. The ratio of 7.2 kb palindrome to
3.6 kb mature end is clearly greater in DNA molecules
sedimenting faster than the unit length (Fig. 6). Thus,
these data support the concept that the palindromes are
parts of concatemers.

Mapping of the DNA Polymerase Gene.

The resistance to phosphonoacetate (PAA) of a DNA
polymerase isolated from cells infected with a PAA^R mutant
of vaccinia virus provided genetic evidence that the enzyme
is virus coded (20). Furthermore, PAA resistance could be

TABLE 1. RESCUE OF PAA RESISTANCE MARKER
USING CLONED HindIII PAA^R FRAGMENTS

PAA^S Virus		TITER	
		Total Virus	PAA^R Virus
	DNA (2 µg)	(PFU/ml X 10^{-8})	(PFU/ml X 10^{-4})
+	—	4.6	2.1
+	intact PAA^R	6.1	160
+	pUC 9	5.2	3.2
+	pHindIII A^R	2.2	3.2
+	pHindIII D^R	5.0	3.5
+	pHindIII E^R	6.4	170
+	pHindIII F^R	4.9	1.7
+	pHindIII G^R	5.0	2.4
+	pHindIII H^R	6.9	4.1
+	pHindIII I^R	4.7	2.8
+	pHindIII J^R	4.5	2.7
+	pHindIII K^R	4.5	4.4
+	pHindIII L^R	3.9	5.1
+	pHindIII M^R	5.3	2.7
+	pHindIII O^R	4.2	4.0

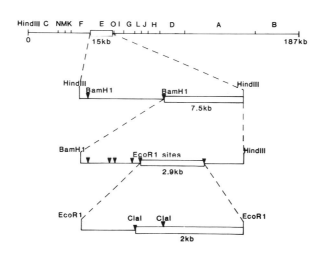

FIGURE 7. Summary of marker rescue experiments used to map
the DNA polymerase gene of vaccinia virus. Cloned DNA
fragments that are boxed contained the PAA^R marker.

used as a dominant marker for mapping the DNA polymerase
gene in a manner similar to that previously used to locate
the vaccinia virus thymidine kinase gene (18). Initial
experiments indicated that drug resistant recombinants
could be isolated from cells infected with wild-type PAA^S
virus following transfection with intact or HindIII
digested DNA from PAA^R mutants. To locate the PAA^R marker,
HindIII fragments of PAA^R DNA were cloned in plasmid and
cosmid vectors. Of these DNA segments, only HindIII E
exhibited marker rescue (Table 1). By successive subclon-
ing, the PAA^R marker was localized to a 2,000 bp DNA fragment within the left one-third of the genome (Fig. 7).
Marker rescue could not be obtained after complete ClaI
digestion of this fragment indicating that the locus of
PAA^R is very close to the ClaI site within the 2,000 bp
segment (Fig. 7).

DISCUSSION

Hairpin loop termini appear to be present in replicating forms of single-stranded parvovirus DNA (32-34), paramecium mitochrondrial DNA (35), tetrahymena rDNA (36) and
yeast chromosomal DNA (37) as well as poxvirus DNA. This
structural feature undoubtedly represents a mechanism for
replication of the ends of linear DNA molecules (38, 39).
A replication scheme similar to those previously
proposed (3,28) accounts for the structural features of
mature and replicating vaccinia virus DNA (Fig. 8). In
this model, we suggest that a nick occurs within the
inverted terminal repetition proximal to the terminal loop.
The 3' end can then serve as a primer for DNA synthesis.
The high A+T composition and the incomplete base-pairing in
this region may facilitate strand displacement. Hairpin
transfer, which occurs at this stage, accounts for
flip-flop inversion of the terminal sequence. After
reannealing of the palindrome, replication can proceed by a
strand displacement mechanism. If initiation occurs at
only one end of a molecule, concatemers would form. These
concatemers could be resolved by nicking and rearrangement.
Alternatively, if initiation occurred at both ends of a
molecule, only monomers would be formed. The findings that
nicks are introduced near the ends of the parental genome
soon after infection (40, 41), replication starts near the
ends of the genome (27), and ligation of nicks to form
uninterrupted covalently linked polynucleotide chains

SELF PRIMING MODEL FOR VACCINIA
VIRUS DNA REPLICATION

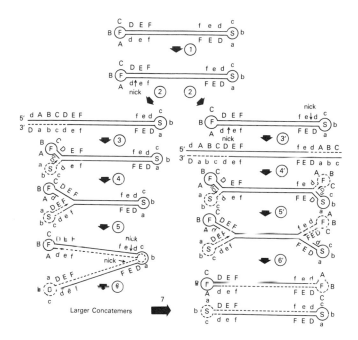

FIGURE 8.   Replication model.

occurs relatively late in infection (23, 32) are consistent
with the above scheme. Moyer and Graves (28) pointed out
that a model similar to this could also account for deletions and rearrangements commonly found in poxviruses. It
is important to stress, however, the over simplification of
the model, since some investigators have provided evidence
for RNA primers and Okazaki-type fragments (23-27).

   DNA polymerase is the second functional gene of
vaccinia virus to be mapped, the first being thymidine
kinase. To confirm that the PAA^R locus corresponds to the
structural DNA polymerase gene, mRNA that hybridizes to
this DNA segment was translated <u>in vitro</u>. The <u>in vitro</u>
translation product had the same electrophoretic mobility
and peptide map as the authentic vaccinia virus DNA
polymerase (our unpublished observations). It should now
be possible to determine the amino acid sequence of the

vaccinia virus DNA polymerase, study the regulation of its synthesis, locate the active sites of the enzyme and produce defined mutations.

REFERENCES

1. Moss B (1974). Reproduction of poxviruses. In Fraenkel-Conrat H, Wagner RR (eds): "Comprehensive Virology, Vol. 3," New York: Plenum Press, p 405.
2. Geshelin P, Berns KI (1974). J Mol Biol 88:785.
3. Baroudy BM, Venkatesan S, Moss B (1982) Cell 28:315.
4. Wittek R, Menna A, Muller, HK, Schumperli D, Bosely PG, Wyler R (1978). J Virol 28:171.
5. Garon CF, Barbosa E, Moss B (1978). Proc Natl Acad Sci USA 75:4863.
6. Wittek R, Moss B (1980) Cell 21:277.
7. Baroudy BM, Moss B (1982) Nucleic Acids Res 10:5673.
8. Pickup DJ, Bastia D, Stone HO, Joklik WK (1982) Proc Natl Acad Sci USA 79:7112.
9. Cairns J (1960). Virology 11:603.
10. Prescott DM, Kates J, Kirckpatrick JB (1971) J Mol Biol 59:505.
11. Pennington TH, Follett EAC (1974) J Virol 13:488.
12. Citarella RV, Muller R, Schlabach H, Weissbach A (1972). J Virol 10:721.
13. Challberg MD, Englund PT (1979). J Biol Chem 254:7812.
14. Sambrook J, Shatkin AJ (1969). J Virol 4:719.
15. Bauer WR, Ressner EC, Kates J, and Patzke JV (1977). Proc Natl Acad Sci USA 74:1841.
16. Pogo BGT, Dales S (1969). Proc Natl Acad Sci USA 63:820.
17. Dubbs DR, Kit S (1964). Virology 22:214.
18. Weir JP, Bajszar G, Moss B (1982). Proc Natl Acad Sci USA 79:1210.
19. Hruby DE, Ball LA (1982). J Virol 43:403.
20. Moss B, Cooper N (1982) J Virol 43:673.
21. Bajszar G, Wittek R, Weir JP, Moss B (1983) J Virol 45:62.
22. Weir JP, Moss B (1983). J Virol 46:530.
23. Esteban M, Holowczak JA (1977) Virology 78:57.
24. Esteban M, Holowczak JA (1977) Virology 82:308.
25. Esteban M, Flores L, Holowczak JA (1977) Virology 78:57.
26. Pogo BGT, O'Shea M (1978). Virology 84:1.

27. Pogo BGT, O'Shea M, Freimuth P (1981).
     Virology 108:241.
28. Moyer RW, Graves RL (1981). Cell 27:391.
29. Baroudy BM, Venkatesan S, Moss B (1983). Cold Spring
     Harbor Symp Quant Biol, in press.
30. Salzman NP (1960) Virology 10:150.
31. Joklik WK, Becker Y (1964). J Mol Biol 10:452.
32. Tattersal P, Ward DC (1976). Nature 263:106.
33. Straus SE, Sebring ED, Rose, JA (1976). Proc Natl Acad
     Sci USA 73:742.
34. Hauswirth WW, Berns KI (1977). Virology 78:488.
35. Pritchard AE, Cummings DJ (1981). Proc Natl Acad Sci
     78:7341.
36. Blackburn EH, Gall JG (1978). J Mol Biol 120:33.
37. Forte MA, Fangman WL (1976). Cell 8:425.
38. Cavalier-Smith T (1974). Nature 250:467.
39. Bateman AJ (1975) Nature 253:379.
40. Pogo BGT (1977). Proc Natl Acad Sci USA 74:1739.
41. Pogo BGT (1980). Virology 101:520.
42. Dahl R, Kates JR (1970). Virology 42:453.

Mechanisms of DNA Replication and Recombination, pages 463–494
© 1983 Alan R. Liss, Inc., 150 Fifth Avenue, New York, NY 10011

FINAL STAGES OF DNA REPLICATION: MULTIPLY
INTERTWINED CATENATED DIMERS AS SV40
SEGREGATION INTERMEDIATES[1]

Alexander Varshavsky, Olof Sundin,[2]
Engin Özkaynak, Richard Pan, Mark Solomon
and Robert Snapka

Department of Biology,
Massachusetts Institute of Technology,
Cambridge, Massachusetts, 02139

ABSTRACT   Terminal stages of SV40 DNA replication, from the latest Cairns structure to the
monomeric supercoiled SV40 DNA I are shown to
proceed via three discrete families of multiply
intertwined catenated DNA dimers. When SV40-
infected cells are placed into hypertonic medium,
newly synthesized SV40 DNA accumulates as form C
catenated dimers. These molecules consist of two
supercoiled monomer circles of SV40 DNA interlocked by one or more topological intertwinings
and are seen as transiently labeled intermediates
during normal replication.   Form C catenated
dimers represent pure segregation intermediates,
replicative DNA structures in which DNA synthesis
is complete but which still require topological
separation of the two daughter circles.   Hypertonic shock seems to block selectively a type II
topoisomerase activity involved in disentangling
the two circles. This is reflected in the fact
that form C catenated dimers that accumulate

[1]This work was supported by a grant to A.V. from
the National Cancer Institute (CA30367).
[2]Present address: Cold Spring Harbor Laboratory,
Cold Spring Harbor, N.Y. 11724.

during the block are highly intertwined, with catenation linkage numbers up to $C_L \cong 20$. While initiation of replication is also inhibited by hypertonic treatment, ongoing SV40 DNA synthesis is not affected, and replication is free to proceed from the earliest Cairns structure through to form C catenated dimers. The block to segregation is rapidly and completely released by shifting the cells back to normal medium. A much slower recovery of DNA segregation takes place on prolonged incubation in hypertonic medium, apparently due to a cellular homeostatic response. These and related findings lead to a detailed view of the final stages of SV40 DNA replication. The results of biochemical and electron microscopic analyses of nucleoprotein organization of the form C catenated SV40 chromosomes suggest that nucleoprotein fibers of the two monomeric circles wrap around each other at several widely separated, apparently random points (non-confined intertwining). We also show that a specific exposure of a ~400 bp long, origin-proximal region in the SV40 chromosomes is "erased" by a passage of a replicaton fork through the exposed region. The exposure of the origin-proximal region is restored later, suggesting that an exposed (nuclease-hypersensitive) region is not inherited "passively" by daughter chromosomes upon their replication and that a specific post-replicative "activation" mechanism is involved.

INTRODUCTION

During lytic infection, simian virus 40 (SV40) DNA exists as a minichromosome within the host nucleus and is replicated in a manner similar to that of the cellular genome. The 5.2-kb circular, double-stranded SV40 DNA directs site-specific synthesis of a single replication "bubble" which is then enlarged through movement of both replication forks (1-3). 10-15 minutes later, the forks converge halfway around the genome and their forward movement stops, leaving ~200 bp of unreplicated parental duplex (ref. 3; Sundin and Varshavsky, unpublished data).

Separate, covalently closed, circular SV40 DNA monomers appear approximately 5 minutes after the forks meet (4). We have recently discovered a new class of SV40 replicative intermediates participating in these latest stages of SV40 DNA replication (5,6). All members of this class are catenated dimers, two circular molecules of SV40 DNA linked topologically by one or more intertwining events. The catenated SV40 DNA dimers comprise three specific families and occur as minichromosomes. They are all rapidly processed in vivo to mature supercoiled SV40 DNA I. We have also found conditions for selective arrest of SV40 replication in vivo at the stage of catenated chromosomes. The experimental approach developed in these studies (5,6) provides a general method for the analysis of complex mixtures of multiply intertwined catenated dimers and leads to a detailed view of the final stages of SV40 DNA replication. Some of the early studies on singly intertwined catenated dimers in mitochondrial, SV40 and other DNAs are those by Rush, Eason and Vinograd (7), Jaenisch and Levine (8), Novick et al. (9) and Kupersztoch and Helinski (10).

What follows is a condensed account of the "DNA-level" work on catenated SV40 dimers (5,6), together with our more recent data on the nucleoprotein organization of SV40 chromosomes during the replication and segregation of daughter DNA nucleoprotein molecules.

CAIRNS STRUCTURES AND CATENATED DIMERS: REPLICATIVE INTERMEDIATES OBSERVED IN PULSE-LABELED SV40 DNA

When SV40-infected CV1 cells are exposed briefly to ^{3}H-thymidine, the label appears in replicating and newly completed DNA molecules. These SV40 DNAs can be extracted from nuclei and resolved by electrophoresis into a variety of distinct species (5,6). The final products of replication, supercoiled (form I) and nicked circular (form II) SV40 monomers serve as prominent landmarks (Figure 1a). Cairns structures in varying stages of completion are resolved by electrophoresis as a continuous smear extending from form I to the LC band, a hesitation point in replication fork movement termed the latest cairns structure (LC) (Figure 1a). Superimposed upon the smear are discrete bands, catenated DNA dimers belonging to different families or "forms". The form A molecule consists of two intertwined, open (relaxed) circles; form B has one open and one supercoiled circle, and form C is composed of two supercoiled SV40 circles (5,6). Discrete heterogeneity within each of these classes is due to differences in the number of

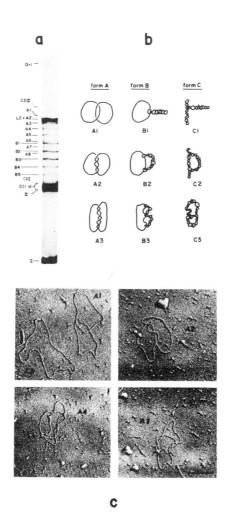

Figure 1.   One-dimensional   electrophoretic

fractionation of pulse-labeled SV40 DNA and electron microscopy of multiply intertwined catenated SV40 dimers. (a) SV40 chromosomes were selectively extracted from CV-1 nuclei, then deproteinized, and electrophoresed in 0.8% agarose-SDS gel (5,6). (A1-A8) series of SV40 form A catenated dimers; (B1-B5) series of form B catenated dimers; (C[1...n]) series of form C catenated dimers (heterogeneity of this band is obscured by overexposure). LC is the latest SV40 Cairns structure. CDI and CDII indicate positions of the supercoiled circular (head-to-tail) SV40 dimer and its nicked form, respectively; I and II are supercoiled and relaxed (nicked or gapped) circular monomeric SV40 DNAs, respectively. (b) Terminology of multiply intertwined catenated DNA dimers. In this diagram single lines represent unit-length circles of double-stranded DNA. One parameter of a catenated dimer is the catenation linkage number, C_L (C_L values from 1 to 3 are shown here: C_L of catenated SV40 DNA dimers was observed to vary from 0 [monomeric DNA circles] to 25-30 in highly intertwined catenated dimers; see Figure 2B). The other independent parameter of a catenated dimer is determined by the state of each of its two double-stranded circular DNA domains. In form A catenated dimers, both circles are nicked or gapped; in form B, one of the circles is covalently closed and therefore supercoiled due to the presence of nucleosomes on SV40 DNA in vivo; and in form C, both circles are covalently closed and supercoiled. Although the degree of supercoiling of monomeric SV40 DNA is not by itself a fixed parameter, it does not vary significantly under normal conditions and is presumed to be constant in the terminology used here. (c) Electron microscopy of catenated SV40 dimers. (CD) Circular head-to-tail dimer; (A1) singly intertwined catenated dimer; (A2) doubly intertwined; (A3) triply intertwined; (A4) quadruply intertwined. All micrographs are printed to the same scale; bar indicates 0.5 μm (for details, see ref. 6).

topological intertwining events that hold the two circles together. This is the catenation linkage number, or C_L (see Figure 1b). Biochemical identification of different catenated SV40 dimers shown in Figures 1 and 2 was carried out using both electrophoretic and isopycnic

sedimentation methods; it has been described in detail
(5,6). Electron microscopic appearance of individual
catenated DNA dimers is shown in Figure 1c. Topological
states corresponding to C_L = 1,2,3 and 4 are readily
observed in form A catenated dimers (A1-A4) (Figure 1c).

HYPERTONIC TREATMENT SELECTIVELY
ARRESTS SEGREGATION

The relative abundance of catenated dimers in
replicating SV40 DNA is insensitive to significant
variations in pH, temperature, thymidine concentration,
method of DNA extraction or time after infection.
However, varying the osmolality of the DME growth medium
used for pulse-labeling strikingly alters the fraction of
pulse-labeled DNA found as catenated dimers. Increasing
the concentraton of DME used for 30 minutes of labeling
from normal (Figure 2A, lane 1.0) to 1.45X (1.45-fold-
concentrated, Figure 2A, lane 1.45) resulted in
accumulation of form C relative to all other forms of
radiolabeled SV40 DNA. On further increase to 1.75X
medium, label was incorporated into form C alone, with
little detectable synthesis of monomeric form I or II
(Figure 2A, lane 1.75). When isoosmotic amounts of
either NaCl, sucrose or glucose were added to 1X DME
medium to make its final osmolality equal to that of 1.7X
DME, results showed that it is the hypertonicity of the
medium and not its specific composition which is
responsible for the segregation arrest (6).

Both initiation and segregation steps of SV40 DNA
replication are severely inhibited by the hypertonic
treatment. In striking contrast, DNA synthesis in the
replicating SV40 chromosomes continues quite efficiently.
This can be demonstrated by pulse-labeling replicating
DNA under normal conditions and then shifting cells to
1.75X medium. The results of such an experiment (6) show
that Cairns structures at various stages of replication
are all converted to covalently closed (C form) catenated
dimers in the 1.75X medium. Thus DNA synthesis, once

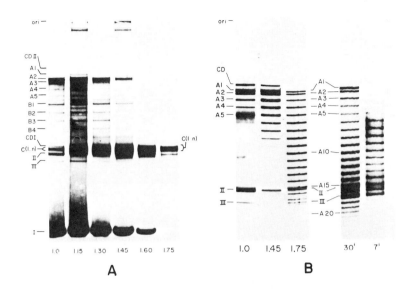

Figure 2. Arrest of SV40 DNA segregation in hypertonic medium and determination of catenation linkage numbers in form C catenated dimers. (A) Effect of medium concentration on the replication of SV40 DNA in vivo. The lanes contain SV40 DNA from cells washed and labeled in 1.0X (DME medium of normal concentration) (control), 1.15X, 1.30X, 1.45X, 1.60X, and 1.75X DME. (B) Analysis of form C catenated DNA dimers. SV40 DNA pulse-labeled for 30 min in 1.0X, 1.45X, or 1.75X DME was obtained from the same samples used in A, and electrophoresis was carried out in the same gel system. The form-C plus form-II region of the gel was excised, and the DNA was electroeluted and nicked with DNase I at 0°C. Equal counts of this DNA were loaded in each lane of an 0.8% agarose gel. The lanes contain nicked "form-C region" DNA labeled in the presence of 1.0X, 1.45X, or 1.75X DME, respectively. The series of bands shown in the third lane (1.75) represents (with the exception of monomeric forms II and III) an ordered array of form A catenated

dimers with catenation linkage states running from C_L = 1(A1) to C_L = 20 (A20) and beyond. In the fourth lane (30') the labeling was in 1.75X DME for 30 min. and in the fifth lane (7') the labeling was in 1.75X DME for 7 min. (for details see ref. 6).

begun, continues to completion under conditions where segregation is fully inhibited. The result is that the complete, covalently closed circular daughter DNA molecules accumulate as form C catenated dimers without segregating into monomers.

ACCUMULATION AND PROCESSING OF MULTIPLY INTERTWINED CATENATED DIMERS

As mentioned above, catenated dimers may be linked by different numbers of intertwining events. What happens to the number of intertwinings in form C molecules during segregation arrest? Electrophoresis as used in Figure 2A cannot clearly resolve form C species bearing different catenation linkage numbers (C_L). To determine C_L numbers, the form C plus form II region was excised after electrophoresis, the DNA was eluted from the gel and nicked extensively by DNase I treatment or γ-irradiation. This converts form C to form A catenated dimers. The C_L number cannot be altered by nicking and is therefore easily visualized as an ordered array of well resolved form A bands (6).

In Figure 2B the form C catenated dimer family derived from pulse-labeling in 1.0X DME (control) contains mostly C1 and C2 by this analysis (Figure 2B, lane 1.0). Under conditions of moderate arrest, represented by 1.45X medium, the form C family contains more heavily intertwined molecules (Figure 2B, lane 1.45). In 1.75X medium where arrest is complete (Figure 2B, lane 1.75), form C catenated dimers have an extremely broad distribution of catenation linkage states, the average being $C_L \cong 10$, with visible range from C_L = 1 to $C_L \cong 25$. Thus treatment with 1.75X DME has not only caused the quantitative accumulation of form C molecules, but also arrested these in a highly intertwined state not characteristic of form C during normal replication in 1.0X medium.

Is the diversity of linkage states observed in form C catenated dimers after 30 min of labeling determined at the time of their synthesis, or is the diversity generated by topoisomerase action on an initially more limited set? To approach this question, labeling of SV40-infected cells was carried out in 1.75X DME medium for either 30 minutes (Figure 2B, lane 30') or 7 minutes (Figure 2B, lane 7') and form C DNA was isolated and nicked to give form A. After 30 minutes of pulse, the intertwining states range from C1 to C22. After 7 minutes of pulse, a narrower distribution, ranging from C5 to C18, is observed (Figure 2B, lane 7'). The less intertwined species C1 to C4 are absent, which suggests that label appears first in the more highly intertwined dimers which are then processed to dimers with lower C_L numbers (see Figure 3).

ARREST OF CHROMOSOME SEGREGATION IS REVERSIBLE

To determine whether the hypertonic arrest of segregation is reversible, replicative SV40 intermediates were labeled for 25 min in 1.70X medium (the extent of segregation arrest is similar in 1.75X and 1.70X media) followed by transfer of cells to 1X medium for a short period of time (1 to 5 min) before rapid lysis of cells and extraction of SV40 DNA. It has been found that upon shift back to normal medium every arrested form C chromosome was segregated into monomeric supercoiled circles at a rate equal to or greater than that observed under normal steady-state conditions (6).

Remarkably, a complete recovery of SV40 chromosome segregation occurs spontaneously after approximately one hour in 1.70X medium (6), presumably through action of a homeostatic mechanism responsible for maintenance of the optimal intracellular ionic environment. Unlike the segregation pathway, initiation of new rounds of SV40 DNA replication does not recover by itself after prolonged incubation in hypertonic medium (6).

PROPOSED MECHANISMS FOR THE TERMINATION OF
DNA SYNTHESIS AND THE SEGREGATION OF
DAUGHTER SV40 CHROMOSOMES

Taken together, our results suggest a set of mechanisms and pathways which the circular SV40 genome follows during its final stages of replication. The pathways are depicted in Figures 3 and 4 and discussed in some detail in the corresponding legends. Briefly, SV40 replication is initiated at the origin site and the replication bubble of the nascent Cairns structure enlarges by movement of both forks until they meet halfway around the genome. The latest Cairns structure, LC, in which the forks are approximately 200 bp apart, is then converted directly to a form A catenated dimer in which the two gapped double-stranded daughter DNA molecules are wrapped around each other (intertwined) many times. The number of intertwinings (catenation linkage number, C_L) of a newly formed catenated dimer is directly related to the number of DNA base pairs in the unreplicated portion of the parental DNA duplex of the LC (see Figure 4). Newly formed catenated dimers (Figures 3 and 4) are then processed along two independent pathways (Figure 3). DNA synthetic machinery fills in and seals each of the two single-stranded gaps at the replication terminus. Simultaneously a DNA topoisomerase activity gradually disentangles the two intertwined daughter DNA molecules, most likely by the type II mechanism of passing one duplex through a transient double-stranded break in the other (11-14).

An essential feature of the proposed scheme (Figures 3 and 4) which is supported by our data is that DNA synthesis and DNA decatenation behave as independent and separable processes. The major paths taken by replicating DNA molecules are therefore heavily dependent on the relative rates of gap-filling DNA synthesis and DNA segregation (decatenation).

Figure 4 (A-C) describes explicitly how the latest Cairns structure (LC) can be converted into a multiply

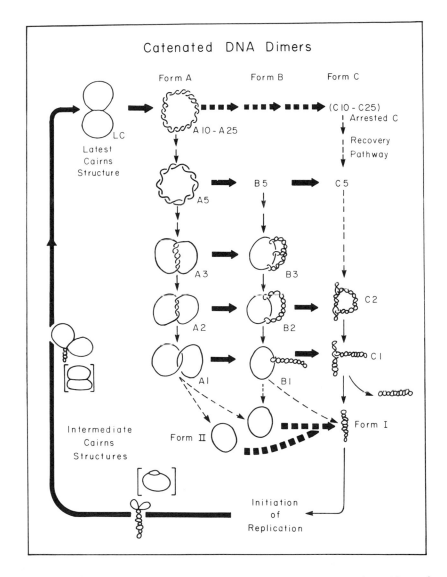

Figure 3. Catenated DNA dimers: proposed paths of
SV40 DNA synthesis and segregation (see text). Thick
horizontal arrows designate pathways of DNA synthesis,

and thin vertical arrows designate pathways of decatenation. During hypertonic shock, little (if any) decatenation activity seems to occur, and A10-A25 are converted directly to highly intertwined form C catenated dimers (thick dashed arrows at the top). On release of block, these are rapidly topoisomerized stepwise to supercoiled monomers (vertical dashed arrows; recovery pathway). The actual paths taken during segregation should be highly dependent on the relative rates of gap-filling DNA synthesis (short heavy arrows) and topoisomerase II action (thin vertical arrows). For example, if the decatenation pathways were much more efficient, one should observe relatively little label in catenated dimers and an increased concentration of gapped circular monomers similar to those described previously (27). This would make the segregation pathway less visible, but would not alter the basic role of multiply intertwined catenated dimers and a topoisomerase II activity in the segregation process. The corresponding set of possible transitions to form II DNA is depicted by dashed thin arrows and dashed thick arrows at the lower right.

intertwined form A catenated dimer. The essential observation is that the two are topologically equivalent, and that the action of DNA polymerase alone is formally sufficient to carry out the conversion.

CHROMATIN STRUCTURE OF INTERMEDIATES IN SV40 REPLICATION AND SEGREGATION

Catenated Chromosome Dimers. Two possible models for catenated SV40 chromosomes are depicted in Figures 4D and 4E. In the first model, nucleoprotein (presumably nucleosomal) fibers may wrap around each other at several widely separated points, whereas in the other model, intertwining of DNA is localized to a small portion of the genome. Physical characterization of isolated catenated chromosomes was carried out as a step towards confirming or excluding such models.

Isolated monomeric SV40 chromosomes (Figure 5A) sediment in a sucrose gradient more slowly than chromosomes pulse-labeled with ^{3}H-thymidine (mostly

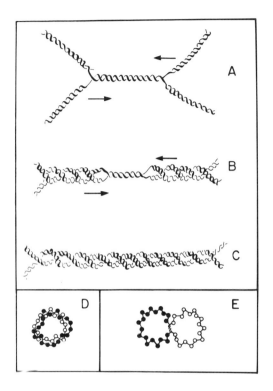

Figure 4. Proposed pathway from latest Cairns structure (LC) to multiply intertwined catenated DNA dimers. The proposed mechanism is based on the topological equivalence of LC and catenated DNA dimers. Consider an arbitrary LC whose blown-up terminus with a few remaining turns of the parental DNA duplex is depicted in Figure 3A. The actual number of double-helical DNA turns in the parental duplex of LC (~200 bp long; unpubl.) appears to be ~20, assuming ~10 bp/turn. We assume that beyond this point in replication the remaining short parental DNA duplex (~200 bp) is no longer susceptible to relaxation by topoisomerase I enzyme because of sterical interference from closely

juxtaposed replication complexes of the two forks. DNA synthesis, however, still proceeds, perhaps solely in the $5' \rightarrow 3'$ leading-strand mode (28), until the two growing DNA strands meet or pass each other ~100 bases later. This produces two daughter DNA circles with a staggered arrangement of noncomplementary gaps at their termini; are therefore no longer base-paired to each other. The above-mentioned inability to relax the last 20 or so turns of parental DNA double helix means, however, that these turns will remain, appearing now not as a Watson-Crick double helix, but as multiple intertwinings of the two catenated double-stranded DNA circles. (A-C) proposed pathway, with a simplification that the intertwined circles are depicted completely double-stranded, i.e., already matured from A-form to C-form catenated dimers (see Figure 3). (D-E) Two extreme ways in which the chromatin structure might accomodate the multiple intertwining of nucleosomal fibers. In model A the fibers cross and wrap around each other at several widely separated points, whereas in model B, all intertwining is confined to one region. In the absence of additional specific constraints, the structure in D (nonconfined intertwining) would be a direct consequence of the process of catenated dimer formation as shown in A-C.

Cairns structures; the leading peak in Figure 5B) or arrested catenated dimer chromosomes (solid curve in Figure 5C). This result is expected since the late replicative forms are almost twice the size of the monomeric chromosomes.

Electrophoresis of SV40 chromosomes in low-ionic-strength agarose gels (15) fractionates replicating and non-replicating chromosomes with a higher degree of resolution (Figures 6 and 8). Figure 6 (e-g) is an autoradiogram of ^{3}H-thymidine labeled SV40 chromosomes electrophoresed from left to right in an 0.8% agarose gel under low ionic strength conditions. Monomeric chromosomes migrate as a single band (Figure 6g), while replicating chromosomes pulse labeled under isotonic conditions (Figure 6f) consist mainly of lower mobility forms corresponding to cairns structure-type nucleo-proteins. SV40 chromosomes labeled during segregation arrest (Figure 6e) migrate in a low ionic strength gel in two broad bands (Figure 6a), one corresponding to the LC and the other to C form catenated chromosomes (Figures 6a and 6b); an overexposed pattern from a similar experiment is shown in Figure 6e). The relative amounts of LC and form C catenated chromosomes labeled during the segregation arrest vary considerably from one experiment to another, from virtually no labeled LC present (Figure 2A, lanes 1.60 and 1.75) to a considerable proportion of labeled LC present (Figures 6a and 6b); the reason for this variability is not understood.

The results of a second-dimension (DNP→DNA) analysis of SV40 chromosomes labeled under conditions of segregation arrest is shown in Figure 6b. The residual latest Cairns structures are observed as a broad streak at the top of the second-dimension gel, while the form C catenated dimers form a well-defined DNA spot (Figure 6b). Note that the catenated dimer chromosomes migrate more rapidly through the agarose gel than the latest Cairns structures, even though they contain approximately the same amount of DNA. One explanation is that catenated dimer chromosomes have a more compact configuration due to the constraints of intertwining events.

In order to determine the location of discrete catenated DNA isomers after electrophoresis of catenated chromosomes in the first (DNP) dimension, the DNA was nicked in situ by γ-irradiation prior to second-dimension analysis of DNA (Figure 6d). One now observes a ladder of form A catenated DNA dimers in the position otherwise occupied by form C dimers, and it is clear from the tilt of the ladder that increased intertwining results in a small but significant increase in electrophoretic mobility of the corresponding DNP particle. This is consistent with the notion that, as with naked form A DNA dimers, an increasing nonconfined intertwining of nucleoprotein fibers (Figure 4D) places additional spatial constraints on the catenated chromosome dimer and makes it more compact. Thus the data of Figures 6b and 6d suggest that either LC differs significantly in protein composition from form C catenated dimer chromosomes, or that a model of non-confined intertwining, such as the one in Figure 4D, is correct since it predicts a greater dependence of the degree of dimer compactness on the catenation linkage number.

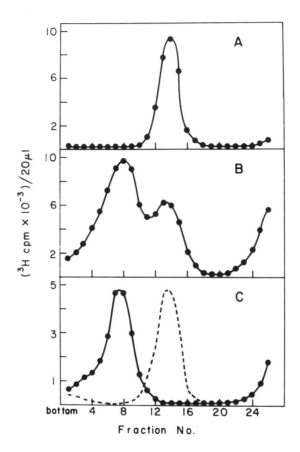

Figure 5. Sedimentation patterns of SV40 chromosomes. (A) SV40-infected CV-1 cells labeled with [³H]thymidine 30–40 hr postinfection in 1X DME, followed by isolation of SV40 chromosomes and sedimentation through a 15–30% sucrose gradient in SW41 rotor (5,6); (B) same as in A but labeled for 20 min in 1X DME; (C) same as in B, but labeled in 1.70X DME. (────) Sedimentation pattern of SV40 chromosomes isolated 5 min after transfer of infected cells from the segregation-arrest conditions (1.70X DME; see solid line in C) to the 1X DME medium.

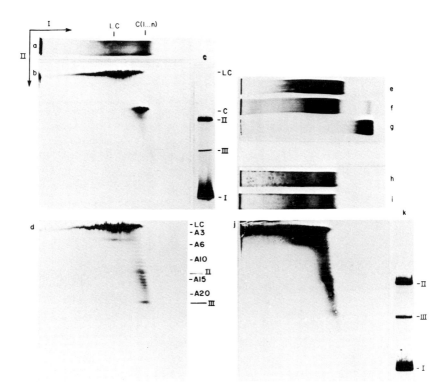

Figure 6. Two-dimensional fractionation of multiply intertwined catenated SV40 chromosomes. Sucrose gradient fractions from the peak of arrested catenated chromosomes (Figure 5C) were subjected to low-ionic-strength electrophoresis in an 0.8% agarose gel. Fractionation in the second (DNA) dimension was then carried out in an 1.3% agarose-SDS gel. (a) Pooled fractions 5-9 from the sucrose gradient (see Figure 5C) electrophoresed in the first (DNP) dimension as described above; (b) second dimension (DNA) electrophoresis of the SV40 chromosomes fractionated in the first (DNP) dimension in a; (c) supercoiled (I), nicked circular (II), and linear (III)

forms of the monomeric SV40 DNA (a marker); (d) same as b, but the first-dimension (DNP) strip was γ-irradiated to nick the DNA before the second-dimension (DNA) electrophoresis; (e) same as a, but from a different experiment (overexposed pattern); (f) pooled fractions 5-10 (see Figure 5B) of replicating SV40 chromosomes electrophoresed as in e; (g) pooled fractions 11-16 (see Figure 5A) of monomeric SV40 chromosomes electrophoresed as in e; (h) same as a, but from a different experiment; (i) same as h, but the SV40 chromosome sample was γ-irradiated to nick the DNA before first-dimension (DNP) electrophoresis; (j) second-dimension (DNA) electrophoresis of the SV40 chromosomes fractionated in the first (DNP) dimension in i; (k) same as c (a marker).

The results of electron microscopic visualization of extensively purified, arrested C form catenated chromosomes (Figures 7d-7i) are also consistent with the model of non-confined intertwining (Figure 4D). However, the following two complications preclude the definitive assignment of the non-confined chromosome intertwining model by the available data (Figures 6 and 7). First, the above results were obtained with isolated chromosomes subjected to in vitro treatments with a low ionic strength buffer. The second, less direct complication stems from the appearance of so-called double minute chromosomes (DMs) at metaphase in stained cytological preparations. These acentric structures represent amplified extrachromosomal subsets of major cellular chromosomes in mammalian cells (30). A single origin of replication is likely to be responsible for the duplication of at least some of the DMs; at the same time, at metaphase they appear to be composed of two compact chromatin domains connected by what may be an unresolved intertwining between the two daughter chromosomal fibers. (Other models of the linkage between the two halves of a DM are also possible). Assuming that the "catenation" model for the organization of DMs at metaphase is correct, their appearance is strongly in favor of confined intertwining (Figure 4E), in apparent

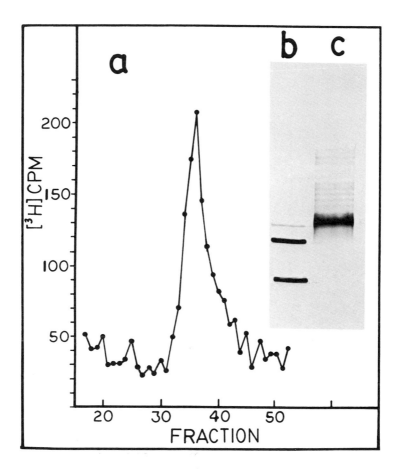

Figure 7. Purification and electron microscopic visualization of arrested, multiply intertwined (form C) SV40 chromosomes. Arrest of segregation and in vivo labeling of form C catenated dimers were carried out as described in the legend to Figure 2. Form C catenated chromosomes were partially purified by sucrose gradient centrifugation (Figure 5C). Peak fractions (6-8) (see Figure 5C) were dialysed against 1 mM Na-EDTA, 0.5 mM Na-EGTA, 10 mM Na-HEPES (pH 7.5), concentrated by ultrafiltration in an Amicon apparatus and loaded onto a

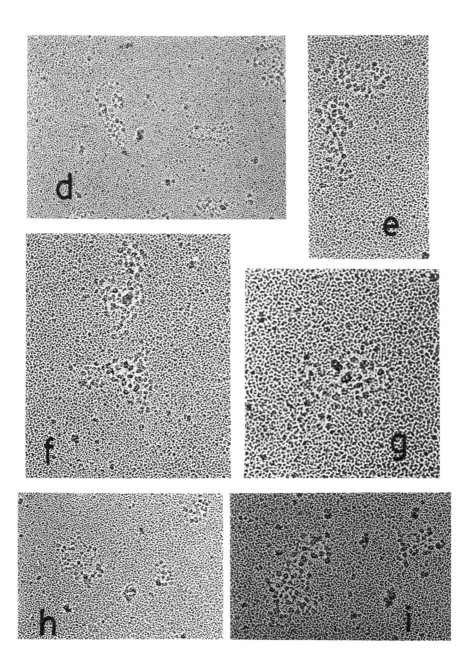

preparative 0.8% agarose tube gel (1 cm in diameter, 8.5 cm in length) with a built-in elution chamber (Bethesda Research Laboratories). Electrophoresis was carried out at 50 V in 1 mM Na-EDTA, 0.5 mM Na-EGTA, 10 mM Na-HEPES (pH 7.5). Fractions (1.2 ml) were eluted in the same buffer at a flow rate of 1 ml/hour. (a) Elution profile of ^{3}H-labeled, multiply intertwined (form C) chromosome dimers. (b) Electrophoretic pattern of forms I, II and III monomeric SV40 DNA (a marker) in an analytical 0.8% agarose gel (fluorography). (c) Electrophoretic patterns of the initial sample of ^{3}H-labeled, catenated (form C) chromosomes partially purified by sucrose gradient centrifugation (fractions 6-8 in Fig. 5C); the major labeled DNA band is that of C form catenated DNA dimers, with a minor DNA "ladder" of B and A form catenated dimers. Although the preparation shown was at least 90% catenated chromosomes <u>radiochemically</u>, it contained a significant proportion of <u>unlabeled</u> monomeric chromosomes (data not shown); these had to be separated by agarose electrophoresis from labeled catenated chromosomes before electron microscopy. The most highly enriched, peak electrophoretic fractions in <u>a</u> had approximately equal molar concentrations of <u>labeled</u> (C form) catenated chromosome dimers and of <u>unlabeled</u> monomeric SV40 chromosomes (electron microscopic data not shown). <u>d</u> through <u>i</u> are representative examples of catenated C form chromosomes (carbon film; Pt/Pd rotary shadowing). (d) magnification x106,000 (two catenated dimers and two monomeric chromosomes); (e) and (f) x159,000 (two catenated dimers in <u>e</u> and in <u>f</u>); (g) x238,000 (a single catenated dimer); (h) x106,000 (one catenated dimer and two monomeric chromosomes); (i) x159,000 (one catenated dimer and one monomeric chromosome).

contradiction with the data on isolated SV40 catenated chromosomes. More direct approaches (in particular, <u>in vivo</u> cross-linking studies on catenated SV40 and $\overline{DM}$ chromosomes) are required to further address this problem.

Whatever the structure of the isolated catenated chromosome dimers, it is not strongly dependent on the covalently closed, supercoiled state of its circular DNA domains; both form C catenated chromosomes and mature monomeric chromosomes may be extensively nicked by

Y-irradiation prior to the first-dimension (DNP)
electrophoresis with no apparent effect on their
electrophoretic mobilities (Figures 6h-6j).

    Cairns Structures. SV40 chromosomes pulse-labeled
in vivo for 20 minutes with ^{3}H-thymidine under isotonic
conditions sediment in a sucrose gradient as a broad peak
consisting of growing Cairns structures and of a smaller
proportion of catenated dimer chromosomes. A peak of
newly formed monomeric chromosomes which have segregated
during labeling, is also observed (Figure 5B). Low-
ionic-strength first-dimension gel electrophoresis of
sucrose gradient-purified replicating chromosomes reveals
a striking "gap" between electrophoretic mobilities of
the monomeric SV40 chromosomes and an earliest detectable
Cairns structure (Figure 8A and 8E; see also Figure 6f).
However, the second-dimension DNA electrophoresis of the
chromosomes fractionated in the first (DNP) dimension
shows that the impression of a gap (Figure 8A) is not due
to a "quantal" decrease in mobility of a SV40 chromosome
upon initiation of DNA replication. While relatively low
fluorographic exposures of the two-dimensional DNP$\rightarrow$DNA
pattern still suggest the existence of a gap (Figures 8C
and 8F), a much longer exposure clearly shows an "arc" of
growing Cairns structures extending into the gap (Figures
8B); upon even longer fluorographic exposures the pattern
of Cairns structures can be seen to start directly from
the first-dimension (DNP) position of the monomeric SV40
chromosomes (data not shown). A steep decrease of the
first-dimension (DNP) electrophoretic mobility of a
Cairns structure once it grows to about 1/3 of its full
(LC) size is one reason for occurence of an apparent gap
in Figures 8B, 8C, and 8F. An additional factor
contributing to the impression of a gap is a decrease in
the rate of replication fork movement at later stages of
SV40 DNA replication (3); the result would be an
enrichment in later Cairns structures in the population
of pulse-labeled chromosomes. Catenated dimer
chromosomes with lower C_L numbers migrate closer to the
LC in the first-dimension (DNP) gel (Figure 8C) and
thereby further enhance the apparent gap (Figures 8A and
8E).

    Blotting hybridization with the ^{32}P-labeled SV40 DNA
probe shows that replicating SV40 chromosomes, while
readily detectable by pulse labeling with ^{3}H-thymidine,
constitute a very small proportion of the total SV40 DNA

in the patterns shown (Figure 8G).

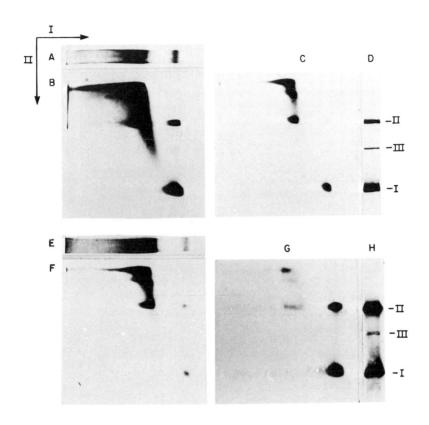

Figure 8.   Two-dimensional fractionation of replicating SV40 chromosomes.

(A) Replicating SV40 chromosomes were labeled in vivo with ^{3}H-thymidine for 20 min in 1.0X DME at 36 hours postinfection, extracted from isolated nuclei (see the legend to Figure 1) and fractionated in a sucrose gradient (se Figure 5B).   Pooled fractions 5-10 (see Figure 5B) were electrophoresed in the first (DNP) dimension as in Figure 6f (same sample, a different electrophoretic run).   (B) Second-dimension (DNA) electrophoresis of the SV40 chromosomes fractionated in the first (DNP) dimension in A.   (C) Same as in B, but a

~3-fold shorter fluorographic exposure. (D) Forms I, II and III of the monomeric SV40 DNA (a marker). (E) Same as in A, but from a different experiment. (F) Second-dimension (DNA) electrophoresis of the SV40 chromosomes fracionated in the first (DNP) dimension in E (same as in B but from a different experiment). (G) DNA from a second-dimension gel run in parallel to the one shown in F was denatured in situ, transferred to a nitrocellulose filter and hybridized to ^{32}P-labeled SV40 DNA. (H) Same as in D (a marker) except that forms I, II, and III of the monomeric SV40 DNA were visualized by hybridization as in G. Only hybridized ^{32}P-DNA but not the transferred ^{3}H-DNA is seen in G and H under autoradiographic conditions used (see ref. 29).

PROBING OF ORIGIN-CONTAINING EXPOSED REGION IN SV40 CHROMOSOMES WITH DNase I AND Hae III: DECREASED EXPOSURE IN REPLICATIVE INTERMEDIATES

Short (<500 bp) "exposed" stretches of DNA apparently devoid of nucleosomes and especially sensitive ("hypersensitive") to either DNase I or appropriately chosen restriction endonucleases are consistently found near the 5' ends of transcribed genes (reviewed in refs. 16, 17). A ~400 bp region which includes the origin of SV40 DNA replication and the promoter regions for both early and late SV40 mRNAs was the first exposed region discovered (18-23).

In the case of SV40, one can isolate ^{3}H-thymidine-labeled chromosomes at different stages during and after replication. It is therefore possible to examine the structure of SV40 chromosomes at these stages for the presence of an origin-containing exposed region. One stringent criterion for the presence of a nucleosome-free, exposed region is the exicison of a specific DNA fragment with an appropriately chosen restriction endonuclease from a chromosome that had been extensively cross-linked by formaldehyde (19).

In Figure 9, isolated SV40 chromosomes were treated with formaldehyde, then digested under limit-digest conditions with Hae III. Thereafter the entire digest was made 1% in SDS and fractionated in a polyacrylamide gel to determine whether any specific Hae III-produced DNA fragments were released from the formaldehyde-fixed

SV40 chromosomes (see the legend to Figure 9 and ref. 19 for details of this approach).

In chromosomes pulse-labeled with ^{3}H-thymidine and then "chased" for 4 hours in vivo, the origin-containing, 300-bp DNA fragment (F-Hae) is selectively excised by Hae III with an efficiency of approximately 20% (Figure 9e). A different result is obtained if SV40 chromosomes are pulse labeled for 20 minutes, isolated immediately thereafter and fractionated by zonal sedimentation into replicative and nascent monomer peaks (as in Figure 5B). Replicative form chromosomes (which are largely Cairns structures plus a much smaller amount of catenated dimer chromosomes) show virtually no specific excision of the origin-containing F-Hae DNA fragment upon digestion with Hae III (Figure 9d). Newly formed (newly segregated) monomeric chromosomes show a much reduced exposure (Figure 9c; cf Figures 9d and 9e).

These results suggest that replicating SV40 chromosomes lack a structural feature which is responsible for the specific exposure of the F-Hae region, while in newly segregated monomeric chromosomes the exposure is clearly present, although considerally below the level seen after additional 4 hours in vivo (Figure 9C-9F). The conclusion therefore is that most SV40 chromosomes acquire the exposure of their origin-proximal region after, rather than during replication. It cannot be ruled out, however, that replicating chromosomes that do express the exposure, are selectively retained in the nucleus during extraction. Catenated dimer chromosomes, the latest replicative intermediates, show a lack of exposure similar to that observed with the total replicative form chromosomes (Figure 9b; cf Figures 9d and 9e).

In Figures 9h-9j, the exposure phenomenon in formaldehyde-fixed SV40 chromosomes was probed by using a mixture of four single-cut restriction endonucleases, Bgl I, MspI, Bcl I, and Bam HI instead of a multiple-cut enzyme, Hae III. Treatment of naked SV40 DNA with Bgl I + Msp I produces a 348-bp fragment spanning the origin-proximal exposed region and a 4895-bp fragment comprising the rest of the SV40 genome. The BclI + Bam HI pair produces a 237-bp DNA fragment which spans the

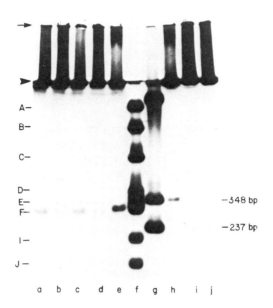

Figure 9. Selective excision of the origin-containing Hae III-F fragment from formaldehyde-fixed SV40 chromosomes by HaeIII: reduced efficiency and selectivity of excision from catenated dimers, Cairns structures, and newly segregated monomer chromosomes.

SV40-infected CV-1 cells were labeled at 36 hr postinfection with [methyl-^{3}H]-thymidine for 25 min in 1.0X DME, followed by a change for 1.70X DME containing 10 µM cold thymidine and an additional incubation for 25 min. This regimen results in labeling of both newly segregated monomeric SV40 chromosomes and catenated form C chromosome dimers arrested at the segregation stage. Cells were lysed, and SV40 chromosomes were extracted and centrifuged in a sucrose gradient as described in Figure 5. Peak fractions from both the monomer (newly segregated chromosomes) and dimer (arrested catenated chromosomes) area were removed for further processing. Samples were fixed with formaldehyde and then digested

with Hae III as described previously (19). Equal amounts
of cpm in each sample were loaded onto a gel consisting
of 7% polyacrylamide (acrylamide:bisacrylamide ratio of
30:1) in 0.1% SDS, 5 mM Na-acetate, and 40 mM Tris-HCl
(pH 8.0), with an overlayer of 1.5% agarose in the same
buffer to prevent accumulation of larger SV40 chromosomal
fragments at the top of a polyacrylamide gel. (a)
HCHO-fixed, HaeIII-digested newly segregated monomer SV40
chromosomes; (b) same as in a, but with arrested
catenated dimer chromosomes; (c) peak monomer fractions
(see Figure 5B) of newly segregated SV40 chromosomes
(labeled in 1X DME) processed for fixation and then for
digestion with Hae III; (d) replicating SV40 chromosomes
(labeled in 1X DME); (e) same as in a, but cells were
returned back to 1.0 X DME in the presence of 10 μM cold
thymidine and incubated for another 4 hours before SV40
chromosome isolation and sucrose gradient centrifugation
(see Figure 5); (f) ^{3}H-labeled SV40 DNA purified from
unfixed monomer chromosomes produced in e and digested to
completion with Hae III (a marker); (g) same as in f, but
digested with a mixture of Bgl I, MspI, BclI, and BamHI
instead of Hae III (a marker); (h) same as in e, but
digested with a mixture of Bgl I, MspI, BclI, and BamHI
instead of Hae III; (i) same as in b, but digested as in
h; (j) same as in a, but digested as in h. Arrow
indicates the origin of electrophoresis; arrowhead
indicates the boundary between the agarose and
polyacrylamide portions of the gel.

replication terminus (24, and references therein). The
results obtained with a mixture of Bgl I, Msp I Bcl I and
Bam HI (Figure 9h-9j) are in complete agreement with
those obtained using HaeIII alone (Figures 9a-9e): while
15 to 20% of "mature" monomeric SV40 chromosomes display
a specific exposure of the origin-containing region
(Figure 9h), little exposure is seen in either Cairns
structures or catenated dimer chromosomes (Figures 9i and
9j; cf Figure 9h).
    Another striking feature of the exposure patterns
seen in Figure 9 is that not only there is a strong
difference in the extent of excision of the
origin-containing F-Hae (Figures 9a-9e) or Bgl I-Msp I
DNA fragments (Figure 9h-9j) from replicating versus
"mature" monomeric chromosomes, but there is also a

strong decrease in <u>selectivity</u> of excision. Specifically, while only F-Hae or Bgl I-Msp I DNA fragments are
excised by the corresponding endonucleases from the
"mature" ("chased") monomeric chromosomes (Figures 9e and
9h), additional DNA fragments located away from the
origin region, are also excised at low efficiency from
both Cairns structures and catenated dimer chromosomes
(Figures 9a-9d and 9j; cf Figure 9e and 9h,
respectively); these DNA fragments are clearly seen at
longer fluorographic exposures (not shown).

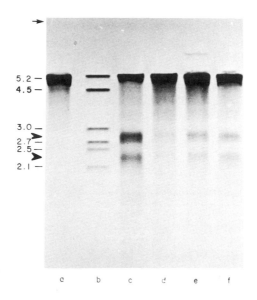

Figure 10. Probing of the origin-containing exposed
region in SV40 chromosomes with DNase I: decreased
exposure in replicative intermediates. Unfixed SV40
chromosomes were digested with DNase I in 0.1 M NaCl 3 mM
$MgCl_2$, 0.5 mM Na-EDTA, and 10 mM Na-HEPES (pH 7.5) at
$30°C$ for 3 min. Several concentrations of DNase I were
tested, and the one that produced ~50% of the nicked
monomer circles was chosen. Relative susceptibilities of
form C catenated dimer chromosomes and of monomer
chromosomes to nicking by DNase I were approximately
equal. DNase-I digested samples were deproteinized and
digested to completion with BamHI, the DNA was denatured,
and equal amounts of cpm in each sample were

electrophoresed in a 1.3% alkaline agarose gel containing
1 mM Na-EDTA and 40 mM NaOH. (a) SV40 chromosomes
obtained as in Figure 9e were deproteinized, digested
with BamHI, and then electrophoresed in alkaline agarose
in the absence of DNase I treatment (a control). (b)
^{3}H-labeled SV40 DNA was digested to completion with
either BamHI + EcoRI, BamHI + MspI, or BamHI + Bgl I.
Equal amounts of DNA from each of these three digests
were mixed to provide a marker. (c) Same as in a, but
before the deproteinization SV40 chromosomes were nicked
with DNase I. (d) Form C catenated (arrested) chromosomes
isolated by sedimentation (Figure 9b) were nicked with
DNase I and analyzed as in c. (e) Same as in d, but
newly segregated monomer chromosomes (Figure 9a) were
nicked with DNase I and analyzed as in c. (f) Cells in
1.70X + DME, containing [^{3}H]thymidine-labeled catenated
SV40 chromosomes, were shifted back to 1.0X DME for 5 min
followed by purification of newly segregated monomer
chromosomes (Figure 5C, dashed line) and DNase I-BamHI
analysis as in c above. Arrowheads indicate diffuse
sublinear bands of single-stranded DNA produced by a
combination of nicks at the DNase-I-hypersensitive sites
in isolated SV40 chromosomes and subsequent BamHI cuts in
deproteinized SV40 DNA. Arrow indicates the electrophoretic origin.

Figure 10 presents the results of using a different
exposure probe, DNase I (20,21,23,25,26). Regions of
SV40 chromosomes hypersensitive to DNase I were mapped by
nicking chromosomes with DNase I, purifying the DNA,
digesting it to completion with a single-cut resticition
endonuclease Bam HI, and analyzing the products on an
alkaline agarose gel (Figure 10). In this approach one
is looking at the position of the first nick rather than
of the first double-stranded DNA cleavage by DNase I.
The major DNase I hypersensitive region is that located
between the Bgl I and Msp I restriction sites near the
origin of replication (Figure 10). Cleavage by Bam HI of
the SV40 DNA previously nicked in the hypersensitive
region produces two single-stranded DNA fragments upon
alkaline gel analysis as seen in Figures 10b-10f. In
agreement with the results obtained using restriction
endonucleases as primary probes (Figures 9), the extent
of exposure as probed by DNase I is strongly decreased in

replicating SV40 chromosomes (Figures 10d, 10e; cf Figure 10c). It is not yet clear whether the residual exposure seen in Figures 10d and 10e is due to a small admixture of monomeric chromosomes in the analyzed samples or to a genuine heterogeneity of replicative SV40 intermediates with respect to the exposure of the origin-proximal region.

One important implication of the above results (Figures 9 and 10) is that a passage of replication fork through an exposed chromosomal region apparently results in "erasure" of the exposure. The exposed region re-appears in SV40 chromosomes some time after their segregation into monomeric circles (Figures 9 and 10), suggesting that an exposed region is not inherited "passively" by daughter chromosomes upon their replication and that a specific post-replicative "activation" mechanism is involved.

One important feature of DNA replication that emerges from the present work is that DNA synthesis and segregation appear to behave as independent processes that can function in isolation from each other. If this property holds for the replication of cellular chromosomes, segregation could be regulated separately from DNA synthesis. Segregation of all replicons might be delayed until a later time, such as the G2 phase of the cell cycle. Alternatively, segregation might be delayed only for certain replicons, such as those at centromeres, to maintain specific associations between sister chromatids.

SV40 is the first replicative system in which multiply intertwined catenated dimers have been directly implicated as segregation intermediates. It is possible that analogous mechanisms operate during the final stages of replication in a variety of other eucaryotic and procaryotic replicons.

REFERENCES

1. Danna KJ, Nathans D (1972). Bidirectional replication of SV40 DNA. Proc Nat Acad Sci USA 69: 3097-3100.
2. Fareed GC, Garon CF, Salzman NP (1972). Origin and direction of SV40 DNA replicaton. J Virol 10: 484-491.

3. DePamphilis ML, Wassarman PM (1980). Replication of eukaryotic chromosome: a close-up of the replication fork. Ann Rev Biochem 49: 627-666.

4. Levine AF, Kang H, Billheimer FE (1970). DNA replication in SV40-infected cells. I. Analysis of replicating SV40 DNA. J Mol Biol 50: 549-568.

5. Sundin O, Varshavsky A (1980). Terminal stages of SV40 DNA replication proceed via multiply intertwined catenated dimers. Cell 21: 103-114.

6. Sundin O, Varshavsky A (1981). Arrest of segregation leads to accumulation of highly intertwined catenated dimers: dissection of the final stages of SV40 DNA replication. Cell 25: 659-669.

7. Rush, MG, Eason R, Vinograd J (1970). Studies on the replication of mitochondrial DNA. Biochim Biophys Acta 228: 585-594.

8. Jaenisch P, Levine A (1972). DNA replication in SV40-infected cells. V Circular and catenated oligomers of SV40 DNA. Virology 44: 480-493.

9. Novick RP, Smith K, Sheely RJ, Murphy E (1973). A catenated intermediate in plasmid replication. Biochem Biophys Res Commun 54: 1460-1465.

10. Kupersztoch YM, Helinski D (1973). A catenated DNA molecule as an intermediate in the replication of the resistance transfer factor R6K in E.coli. Biochem Biophys Res Commun 54: 1451-1459.

11. Liu LF, Liu CC, Alberts BM (1980). Type II DNA topoisomerases: enzymes that can unknot a topologically knotted DNA molecule via a reversible double-strand break. Cell 19: 697-707.

12. Gellert M (1981) DNA topoisomerases. Ann Rev Biochem 50: 879-910.

13. Krasnov MA, Cozzarelli NR (1982). Catenation of DNA rings by topoisomerases. J Biol Chem 257: 2687-2693.

14. Goto T, Wang J. Yeast DNA topoisomerase II. J Biol Chem 257: 5866-5872.

15. Varshavsky A, Bakayev VV, Chumackov PM, Georgiev GP (1976). SV40 minichromosomes: presence of histone H1. Nucl Acids Res 3: 2101-2112.

16. Elgin SCR (1982). DNase I-Hypersensitive sites of chromatin. Cell 27: 413-415.

17. Weisbrod ST (1982b). Active chromatin. Nature 297: 289-295.

18. Varshavsky A, Sundin O, Bohn M (1978). SV40 viral minichromosome: preferential exposure of the origin of replication as probed by restriction endonucleases. Nucl Acids Res 5: 3469-3478.

19. Varshavsky A, Sundin O, Bohn M (1979). A stretch of late SV40 viral DNA about 400 bp long which contains the origin of replication is specifically exposed in SV40 minichrosomes. Cell 16: 453-466.

20. Scott WA, Wigmore DJ (1978). Sites in SV40 chromatin which are preferentially cleaved by endonucleases. Cell 15: 1511-1518.

21. Cremisi S (1981). The appearance of DNase I hypersensitive sites at the 5' end of the late SV40 genes is correlated with transcriptional switch. Nucl Acids Res 9: 5949-5964.

22. Jacobovitz EB, Bratosin S, Aloni Y (1982). Formation of a nucleosome-free region in SV40 minichromosomes is dependent upon a restricted fragment of DNA. Virology 120: 340-348.

23. Saragosti S, Moyne G, Yaniv M (1980). Absence of nucleosomes in a fraction of SV40 chromatin between the origin of replication and the region coding for the late leader RNA. Cell 20: 65-73.

24. Tooze J (ed) (1981). DNA tumor viruses. Cold Spring Harbor, NY.

25. Wu C, Yong YC, Elgin SCR (1979b). The chromatin structure of specific genes. II. Disruption of chromatin structure during gene activity. Cell 16: 807-814.

26. Stalder J, Larsen A, Engel JD, Dolan M, Groudine M, Weintraub H (1980). Tissue-specific DNA cleavages in the globin chromatin domain introduced by DNase I. Cell 20: 451-460.

27. Chen ME, Birkenmeyer E, Salzman NP (1975). SV40 DNA replication: characterization of the gaps in the termination region. J Virol 17: 614-621.

28. Kornberg A (1980). DNA replication. Freeman and Co San Francisco.

29. Levinger L, Barsoum J, Varshavsky A (1981). Two-dimensional hybridization mapping of nucleosomes: comparison of DNA and protein patterns. J Mol Biol 146: 287-304.

30. Cowell J (1982). Ann Rev Genet 16: 21-59.

Mechanisms of DNA Replication and Recombination, pages 495–510
© 1983 Alan R. Liss, Inc., 150 Fifth Avenue, New York, NY 10011

THE DNA POLYMERASE-PRIMASE COMPLEX
OF DROSOPHILA MELANOGASTER EMBRYOS[1]

Laurie S. Kaguni, Jean-Michel Rossignol,
Ronald C. Conaway and I. R. Lehman

Department of Biochemistry, Stanford University
School of Medicine, Stanford, California 94305

ABSTRACT An intact DNA polymerase-primase complex has
been purified from embryos of Drosophila melanogaster.
The near homogeneous enzyme consists of at least three
polypeptides with M_rs of 182,000, 60,000 and 50,000.
These are related antigenically to the α (M_r 148,000),
β (M_r 58,000) and γ (M_r 46,000) subunits, respectively, of the DNA polymerase described previously
(1). The α subunit (M_r 182,000) has a molecular
weight indistinguishable from that observed in extracts of freshly harvested embryos and presumably
present in vivo. The ratio of primase to polymerase
remains constant throughout the purification. Thus,
the primase is an integral component of the Drosophila
DNA polymerase α. A factor which stimulates the
polymerase on single-stranded DNA templates has been
identified and partially purified, and may represent a
polymerase accessory protein removed during purification.

INTRODUCTION

The Drosophila melanogaster embryo is attractive for
studies of DNA replication for several reasons. In early
embryonic development Drosophila chromosomal DNA is replicated in about 4 min and the nuclei divide every ten min
until several thousand nuclei are formed. Maintenance of
this rapid replication rate is dependent on activation of

[1]This work was supported by research grants from the
National Institutes of Health (GM 06196) and the National
Science Foundation (PCM-7904638).

multiple replication origins. Because regulation of replication in this system resides in the initiation phase, the
Drosophila embryo provides a model system for studying
initiation events. Further, because of the high density of
replication forks, the early embryo is an abundant source
of replication enzymes. For example, the relative abundance of DNA polymerase is approximately 30 fold greater in
Drosophila embryos than in cultured human cells (unpublished).

We can envision two approaches to the study of
Drosophila DNA replication in vitro. The first is to
obtain and fractionate embryo extracts which support bidirectional replication initiating at Drosophila origins.
Although we have no defined origin sequences from Drosophila,
and therefore have no means to assess the authenticity of
such reactions, recent development of a transformation
system for Drosophila makes possible the identification of
replication origins in the near future (2). The second
approach, which is the one that we have taken, is to purify
and characterize replication enzymes with the intent of
reconstituting a replication complex or "replisome" which
might act at replication forks in vivo. Here, we present
our work on the isolation and characterization of an intact
DNA polymerase-primase complex from Drosophila embryos, and
on the identification and partial purification of a factor
which stimulates DNA polymerase activity on primer-templates
containing long stretches of single-stranded DNA.

A striking feature of eukaryotic DNA polymerase α as
purified from many sources is its physical heterogeneity.
α polymerase, believed to be the replicative polymerase
(3), has been isolated from several sources as a multisubunit
protein consisting of a catalytic core of M_r ~150,000 in
association with 3 or 4 smaller subunits whose M_r's range
from 40-60,000 (1,4-9).

Several years ago, purification to near-homogeneity of
the DNA polymerase α from early Drosophila embryos was
described (1). This enzyme consisted of subunits with M_r's
of 148,000 (α), 58,000 (β), 46,000 (γ), and 43,000 (δ), and
was associated with a DNA primase activity (10,11). However,
because of a history of proteolysis during purification of
the Drosophila polymerase, and because the enzyme was the
product of a substantial number of purification steps,
subunit specific antisera were used to examine the fate of
the various subunits during the course of purification
(12). These and more recent data indicate that while the

60,000, 46,000 and 43,000 dalton polypeptides have probably
not undergone degradation, the 148,000 dalton α subunit is
derived from larger polypeptides of M_r 182,000 and 160,000
by in vitro proteolysis. We describe a purification procedure, starting with freshly harvested embryos, that yields
a DNA polymerase α with an α subunit of M_r 182,000. DNA
primase activity remains quantitatively associated with the
polymerase throughout the purification.

RESULTS

We have devised a rapid purification procedure for the
isolation of an intact form of DNA polymerase α from freshly
harvested Drosophila melanogaster embryos. The procedure,
which is shown in Table 1, involves 3 chromatographic steps
and glycerol gradient sedimentation and requires 4 days.
The polymerase is purified approximately 1000-fold with a
yield of 6%, representing a 6-fold increase in yield relative to our previous procedure. At the same time, the time
required has been reduced by a factor of two.

The DNA primase activity previously shown to be associated with the Drosophila polymerase (11) remains quantitatively associated with the polymerase through each of the
purification steps, indicating that it is an integral
component of the multisubunit Drosophila polymerase α.
This finding represents a significant departure from that
observed with the well characterized prokaryotic DNA polymerases which are not capable of priming DNA synthesis.
Whether the Drosophila polymerase primes at replication
origins and/or in Okazaki fragment synthesis in vivo remains to be determined.

Electrophoresis of the purified enzyme (Fraction VI)
on a 5-10% SDS-polyacrylamide gel (Fig. 1, lane 2) yielded
predominantly five polypeptides, with M_rs of 215,000,
182,000, 73,000, 60,000 and 50,000. Chromatography on blue
A agarose (Fraction VII) prior to electrophoretic analysis
(Fig. 1, lane 1) resulted in removal of most of the minor
contaminants and in a 40% decrease in the relative amount
of the M_r 215,000 polypeptide; the relative abundance of
the remaining four major polypeptides was unchanged. Remarkably, although the M_r 60,000 and M_r 50,000 polypeptides
had the molecular weights of the β and γ subunits of the
enzyme described previously (1), the M_r 148,000 α subunit
was absent. Instead, a M_r 182,000 polypeptide was found

TABLE 1

PURIFICATION OF DNA POLYMERASE-PRIMASE FROM FRESHLY HARVESTED EMBRYOS OF D. MELANOGASTER

Fraction	Volume ml	Protein mg	Polymerase[a]		Primase[a]		Polymerase/Primase
			Activity units($\times 10^2$)	Specific Activity units/mg	Activity units($\times 10^2$)	Specific Activity units/mg	
I. S-100	360	2581	1792	69.4	792	30.7	2.3
II. Phosphocellulose	493	261	704	269	226	86.5	3.1
III. Ammonium sulfate	50	65	772	1,188	189	291	4.1
IV. Hydroxylapatite	41	8.3	284	3,425	66.4	800	4.3
V. DNA-cellulose	5	0.78	142	18,256	34.6	4,431	4.1
VI. Glycerol gradient	1.5	0.19	98.3	51,737	24.6	12,947	4.0

[a]Assays of polymerase and primase activity were as described (13).

FIGURE 1. SDS-polyacrylamide gel electrophoresis of
D. melanogaster DNA polymerase-primase. Fraction VI, prior
to (5.0 µg, lane 2) and after (2.0 µg, lane 1) chromatography on blue A agarose, was denatured and electrophoresed
in a 5-10% linear gradient SDS-polyacrylamide slab gel.
Marker proteins electrophoresed in adjacent lanes and
indicated by their molecular weights (x 10^{-3}) were: myosin,
E. coli RNA polymerase β' and β subunits, E. coli β-galactosidase, rabbit muscle glycogen phosphorylase, bovine serum
albumin, rabbit muscle pyruvate kinase, E. coli alkaline
phosphatase, and bovine carbonic anhydrase.

whose molecular weight is consistent with that predicted
for an undegraded α subunit based on an immunological
analysis of embryo extracts (12). Densitometric scanning
of the stained gel revealed that the relative abundance of
the M_rs 182,000, 73,000, 60,000 and 50,000 polypeptides in
the two preparations shown was 1.0/1.0/1.6/1.2, respectively; these four polypeptides together account for about
85% of the protein applied (Fig. 1, lane 1). The amount of
M_r 215,000 polypeptide was variable in these and other
preparations; it is very likely a contaminant.

   To determine which of the polypeptides revealed by
SDS-polyacrylamide gel electrophoresis are related antigenically to the previously described DNA polymerase, the
polypeptides were transferred electrophoretically from a

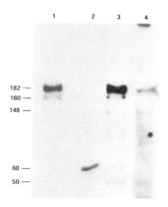

FIGURE 2. Antigenicity of polypeptides in the purified DNA polymerase-primase and in an embryo extract.
Fraction VI (2.5 µg, lanes 1-3) and extract derived from
homogenization of freshly harvested embryos in SDS (150 µg,
lane 4) were electrophoresed in a 5-10% linear gradient
SDS-polyacrylamide gel and the polypeptides were transferred electrophoretically to diazotized paper and probed
with α subunit (lanes 1 and 4), β subunit (lane 2), and DNA
polymerase-primase-specific (lane 3) IgGs as described by
Reiser, et al. (14). The paper was then incubated with
[125]I-labeled protein A and autoradiographed. The molecular
weights of the radioactive polypeptides are as indicated (x
10^{-3}).

gel to diazotized paper. The paper was probed with antisera directed against the α and β subunits as well as the
intact DNA polymerase and then was incubated with [125]I-
labeled Staphylococcus aureus protein A and autoradiographed. When probed with α-subunit-specific IgG (Fig. 2,
lane 1) one predominant and one minor species were observed, with M_rs of 182,000 and 160,000, respectively.
Incubation with β-subunit-specific IgG (lane 2) revealed a
single band of M_r 60,000. Anti-DNA polymerase-specific IgG
reacted with polypeptides of M_rs 182,000, 60,000, and
50,000 and, to a lesser extent, species of M_rs 160,000 and
148,000 (lane 3). The latter two were also found upon

longer exposure with the α-subunit-specific IgG and are
proteolytic products of the M_r 182,000 α subunit.
    In summary, three polypeptides with M_rs of 182,000,
60,000, and 50,000 related antigenically to the α, β, and γ
subunits, respectively, of the earlier DNA polymerase α (1)
were identified in the new enzyme preparation. Of these,
the β and γ subunits had unaltered molecular weights,
whereas that of the α subunit increased from M_r 148,000 (1)
to M_r 182,000. Neither the M_r 215,000 nor the M_r 73,000
polypeptide observed in the SDS-polyacrylamide gel (Fig. 1)
was related antigenically to the previous enzyme nor were
the other minor polypeptides. A parallel experiment was
performed with freshly harvested embryos homogenized with
SDS and using the α subunit-specific IgG as a probe (Fig.
2, lane 4). In this case, a single polypeptide of M_r
182,000 was observed, indicating that this species, present
in embryo extracts, represents the intact α subunit.
    An examination of the purified polymerase-primase for
additional enzymatic activities revealed that the enzyme
contained no detectable exonuclease activity using either
single-stranded or double-stranded DNA as substrate, no
detectable endonuclease activity, no RNase H-like activity
and no DNA-dependent ATPase activity.
    The $(NH_4)_2SO_4$, NaCl, $MgCl_2$ and pH optima for the DNA
polymerase activity of the intact enzyme were nearly identical to those found for the previous enzyme (Table 2); the
same was found for the reaction requirements. Further, the
resistance to dideoxy TTP, and sensitivities to aphidicolin
and N-ethylmaleimide were unchanged. In fact, among the
biochemical properties examined, only the Km for dNTPs
differed between the two forms of the enzyme. The Km for
dTTP was 17.5 μM for the degraded enzyme, and 3.7 μM for
the undegraded form (Fig. 3). This change may signify more
profound differences in the enzymatic capacity of the
polymerase and its interactions with other DNA replication
proteins.
    The previous enzyme preparation, though highly active
on DNase I-treated (activated) DNA utilized primer-templates
containing long stretches of single-stranded DNA very
inefficiently (15,16). The intact enzyme was similarly
inefficient in replicating RNA-primed M13 DNA, thereby
identifying a requirement for polymerase accessory proteins
which may have been removed during purification. A comparison of the relative activity of various fractions of the
intact polymerase on RNA-primed M13 versus activated calf

TABLE 2
COMPARISON OF THE BIOCHEMICAL PROPERTIES OF
D. MELANOGASTER POLYMERASE α PURIFIED BY THE
REVISED AND THE FORMER PROCEDURES

	Revised	Former
Optimum conditions		
$(NH_4)_2SO_4$	35 mM	38 mM[a]
NaCl	50 mM	60 mM[a]
$MgCl_2$	12 mM	12 mM[a]
pH	8.5-9	8.5[a]
Reaction requirements		
Standard reaction mix	100%	100%[a]
+ $(NH_4)_2SO_4$ at optimal concentration	170%	300%[a]
+ NaCl " " "	135%	180%[a]
- $MgCl_2$ + $(NH_4)_2SO_4$ 20 mM	0%	0%[a]
Km for dTTP	3.7 μM	17.5 μM
Effect of inhibitors		
No inhibitor	100%	100%
+ N-ethylmaleiimide 0.2 mM	37%	24%
+ N-ethylmaleiimide 1 mM	0%	0%
+ Aphidicolin 0.5 μg/ml	58%	42%
+ Aphidicolin 5 μg/ml	18%	10%
+ dideoxy TTP 50 μM	88%	N.D.

Polymerase activity was assayed under standard conditions (1) except that $(NH_4)_2SO_4$ was omitted for determination of salt optima and reaction requirements. Dideoxythmyidine triphosphate was used at a final molarity of 50 μM - at a ratio of 1/1 with dTTP

[a]Values taken from Banks, et al. (1).

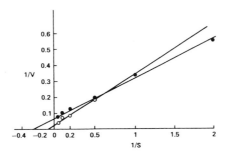

<u>FIGURE 3</u>.   Decreased Km for dTTP in intact pol α.   The undegraded polymerase-primase (fr VI) and the degraded form (fr VI, ref 1) were assayed under standard conditions for polymerase activity (1) except that the dTTP concentration was varied, and incubation was for 15 min.   The resulting data are presented in the form of a Lineweaver-Burk plot. ●-●, undegraded pol α; o-o, degraded pol α.

thymus DNA revealed that the relative ability to replicate long stretches of single-stranded DNA decreased as purification proceeded (Fig. 4).   The largest drop occurred after $(NH_4)_2SO_4$ precipitation following phosphocellulose chromatography.   This finding suggested that the required factor might be present in the supernatant fluid from the $(NH_4)_2SO_4$ precipitation. This fraction did, in fact, contain a heat-labile factor that stimulated the replication of RNA-primed M13 DNA by the nearly pure α polymerase.   The stimulatory factor was partially purified by $(NH_4)_2SO_4$ fractionation and blue A agarose chromatography and its activity was characterized using several primer-templates.

Stimulation by the partially purified factor was 10-20 fold on multi-RNA primed M13 and 3- to 8-fold on denatured activated DNA; a titration of the factor on these primer-templates is shown in Fig. 5.   The multi-RNA primed M13 contains approximately 3-5 primers per 10,000 nucleotide genome while the denatured DNA, though uncharacterized, very likely contains a higher primer concentration.   A time course of stimulation on denatured DNA indicated that the

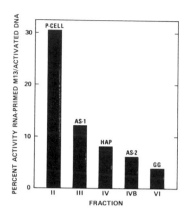

FIGURE 4. Loss of pol α activity on RNA-primed M13 during purification. The indicated polymerase α fractions (approximately 0.7 unit) were assayed under standard conditions with activated calf thymus DNA (1) and with multi-RNA primed M13 DNA (230 pmol) in reaction mixtures (0.05 ml) containing 40 mM Hepes (pH 8.0), 8 mM $MgCl_2$, 5 mM dithiothreitol, 40 μM each dATP, dGTP, dCTP and $[^{32}P]$dTTP (4000 cpm/pmol).

extent of stimulation was the same at early and late times of incubation (Fig. 6).

A differential stimulation was observed with singly-and multi-RNA primed M13 (Fig. 7). On multi-RNA primed M13, the polymerase was stimulated ten-fold, allowing complete replication of the input DNA. On the other hand, only 3-fold stimulation of synthesis on the singly-RNA primed M13 was found. The differential synthesis on the two defined templates by the purified polymerase may reflect different sites of priming relative to DNA template-specific barriers to replication. Whereas the single RNA primer is located at the M13 complementary DNA strand origin, the multiple RNA primers, synthesized by E. coli RNA polymerase in the absence of single-stranded DNA binding protein (SSB) are likely situated at hairpin sequences, thereby removing these hairpins as impediments to DNA

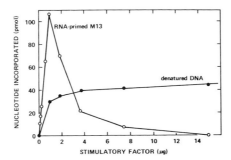

FIGURE 5. Stimulation of pol α activity on RNA-primed M13 and denatured calf thymus DNA by partially purified factor. The stimulatory factor was recovered from the supernatant fluid resulting from $(NH_4)_2SO_4$ precipitation of the fraction II polymerase-primase (Table 1) by increasing the $(NH_4)_2SO_4$ concentration to 75% of saturation. The resulting precipitate was chromatographed on blue A agarose and stimulatory activity was eluted at 1 M NaCl; active fractions were concentrated by $(NH_4)_2SO_4$ precipitation at 75% of saturation. The stimulatory factor was assayed under the conditions described in the legend to Figure 4 with multi-RNA primed M13 (230 pmol) or denatured activated calf thymus DNA (8 nmol, heated at 100°C for 4 min) in the presence of pol α (fraction VI, 0.4 units). Incubation was for 30 min at 37°C.

synthesis. Because the stimulatory factor does not support equal utilization of the two primer-templates, its role is probably not helix destabilization. The nascent strands resulting from synthesis on the two primer-templates, with increasing stimulatory factor added, were examined by electrophoresis in a 4% polyacrylamide-7 M urea gel after incubation of the reaction products in alkali (Fig. 8). In both cases, though more apparent in the case of the singly-primed template, there was no increase in the length of product strands in the presence of the added factor. With the singly-primed template, the polymerase paused at discrete sites on the template with or without added factor as was observed previously with the degraded polymerase (16).

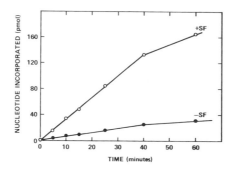

FIGURE 6. Time course of stimulation of pol α with denatured DNA template. Reactions conditions were as described in the legend to Figure 5; stimulatory factor, when added, was present at a saturating level.

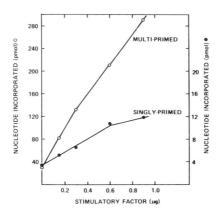

FIGURE 7. Differential stimulation of pol α on multi- and singly-primed M13 DNA. The stimulatory factor was assayed under the conditions described in the legend to Fig. 5 with multi-RNA primed M13 (230 pmol) or singly-RNA primed M13 (230 pmol).

MULTI-PRIMED M13   SINGLY-PRIMED M13
1 2 3 4 5   6 7 8 9 10

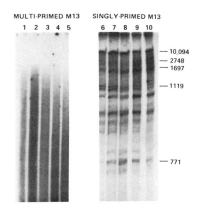

— 10,094
— 2748
— 1697

— 1119

— 771

FIGURE 8. Electrophoresis of products of DNA synthesis with multi- and singly-primed M13 DNA. Nascent DNA
strands were prepared for electrophoresis by incubation in
0.15 N NaOH for 4 hr at 37°C to release them from the
template and to degrade the RNA primers. After neutralization, the samples were precipitated with ethanol and aliquots containing approximately equal amounts of radioactivity were electrophoresed in a 4% polyacrylamide/7 M
urea gel. Lanes 1-5: products of synthesis with multi-RNA
primed M13 DNA with 0, 0.15, 0.3, 0.6 and 0.9 µg stimulatory factor added, respectively. Lanes 6-10: as for 1-5
except with singly-RNA primed M13 DNA. Molecular weight
markers whose sizes are indicated in nucleotides were
electrophoresed in lanes not shown; the M13 DNA used has a
genome size of 10,094 nucleotides (16).

These data support the conclusion that the stimulatory
factor is neither an SSB-like protein nor a protein which
increases the processivity of the enzyme, both of which
would be expected to allow the polymerase to synthesize
longer DNA chains. In fact, E. coli SSB stimulated synthesis beyond that observed with the factor alone, even though
E. coli SSB did not stimulate polymerase α appreciably in
the absence of the factor.

We suggest that the stimulatory factor may be involved
in either enzyme recycling or in primer recognition, and
that a Drosophila counterpart to the E. coli SSB will still

be required for the intact polymerase to efficiently replicate primed single-stranded DNAs.

DISCUSSION

Purification of a DNA polymerase α from $\underline{D}$. $\underline{melano}$-
$\underline{gaster}$ embryos consisting of a M_r 182,000 α subunit and two
or three smaller subunits has relied both on the development
of a rapid purification procedure and on the use of freshly
harvested embryos. Immunological analysis of enzyme fractions obtained at various stages in the previous purification procedure showed that proteolysis had occurred in all
but the final steps (12). Further, such studies with
extracts of embryos, frozen both before and after dechorio-
nation, have shown that the M_r 182,000 α subunit is only a
minor species in frozen embryos (unpublished data; ref
12), indicating that it is exceedingly sensitive to proteolysis.

The subunit structure of the DNA polymerase-primase
purified by the procedure shown here is similar to that of
the earlier enzyme. However, the immunological studies
show that the α subunit has a M_r of 182,000 rather than a
M_r of 148,000, while the molecular weights of the β and γ
subunits have remained at M_r 60,000 and M_r 50,000, respectively. An additional polypeptide of M_r 73,000, which
is not related antigenically to the previous enzyme, now
routinely copurifies with the polymerase and may have
dissociated, possibly as a result of diminished interaction
with the partially degraded enzyme. On the other hand, the
M_r 43,000 δ subunit (1) appears not to be associated with
our current polymerase preparation. This polypeptide is
highly abundant in embryo extracts and in early enzyme
fractions (12) and may have been a contaminant in purified
enzyme fractions prepared by the previous procedure. These
issues will be resolved only when functions can be ascribed
to the various polypeptides.

DNA primase copurifies quantitatively with the DNA
polymerase. Therefore, it is very likely associated with
one or more of the subunits of the enzyme. The undegraded
polymerase possesses no endo- or exo-deoxyribonuclease
activity, no RNaseH-like activity and no DNA-dependent
ATPase activity.

Finally, we have identified and partially purified a
factor which stimulates the activity of the intact polymer-

ase on primer-templates containing long stretches of single-stranded DNA. While its mechanism of action is not known, it may be involved in enzyme recycling or in primer utilization. A helix destabilizing protein is also very likely required in addition to the stimulatory factor, and efforts are underway to isolate such a protein from Drosophila embryos.

ACKNOWLEDGMENTS

We are grateful to Ms. Yee Lan Wang for technical assistance. L.S.K. was supported by American Cancer Society Grant PF-1964. J.-M.R. was supported by a European Molecular Biology Organization fellowship. R.C.C. was supported by National Institutes of Health Training Grant 5T32 GM 07599.

REFERENCES

1. Banks GR, Boezi JA, Lehman IR (1979). A high molecular weight DNA polymerase from Drosophila melanogaster embryos. Purification, structure and partial characterization. J Biol Chem 254:9886.
2. Rubin GM, Spradling AC (1982). Genetic transformation of Drosophila with transposable element vectors. Science 218:348.
3. Weissbach A (1979). The functional roles of mammalian DNA polymerases. Arch Biochem Biophys 198:386.
4. McKune K, Holmes AM (1979). Further studies on partially purified calf thymus DNA polymerase-α. Nucleic Acids Res 6:3341.
5. Mechali M, Abadiedebat J, de Recondo A-M (1980). Eukaryotic DNA polymerase-α. Structural analysis of the enzyme from regenerating rat liver. J Biol Chem 255:2114.
6. Grosse FG, Krauss G (1981). Purification of a 9S DNA polymerase α species from calf thymus. Biochemistry 20:5470.
7. Yamaguchi M, Tanabe K, Takahashi T, Matsukage A (1982). Chick embryo DNA polymerase α. Polypeptide components and their microheterogeneity. J Biol Chem 257:4484.
8. Masahi S, Koiwai O, Yoshida S (1982). 10S DNA polymerase α of calf thymus shows a microheterogeneity in its large polypeptide component. J Biol Chem 257:7172.

9. Albert W, Grummt F, Hübscher U, Wilson SH (1982).
    Structural homology among calf thymus α-polymerase
    polypeptides. Nucleic Acids Res 10:935.
10. Villani G, Sauer B, Lehman IR (1980). DNA polymerase
    α from Drosophila melanogaster embryos. Subunit
    structure. J Biol Chem 255:9479.
11. Conaway RC, Lehman IR (1982). A DNA primase activity
    associated with DNA polymerase α from Drosophila
    melanogaster embryos. Proc Natl Acad Sci USA 79:2523.
12. Sauer B, Lehman IR (1982). Immunological comparison
    of purified DNA polymerase α from embryos of Drosophila
    melanogaster with forms of the enzyme present in vivo.
    J Biol Chem 257:12394.
13. Kaguni LS, Rossignol J-M, Conaway RC, Lehman IR (1983).
    Isolation of an intact DNA polymerase-primase from
    embryos of Drosophila melanogaster. Proc Natl Acad
    Sci USA 80: in press.
14. Reiser J, Stark GR (1983). Immunologic detection of
    specific proteins in cell extracts by fractionation in
    gels and transfer to paper. Methods Enzymol, in
    press.
15. Villani G, Fay PJ, Bambara RA, Lehman IR (1981).
    Elongation of RNA-primed DNA templates by DNA polymer-
    ase α from Drosophila melanogaster embryos. J Biol
    Chem 256:8202.
16. Kaguni LS, Clayton DA (1982). Template-directed
    pausing in in vitro DNA synthesis by DNA polymerase α
    from Drosophila melanogaster embryos. Proc Natl Acad
    Sci USA 79:983.

Mechanisms of DNA Replication and Recombination, pages 511–516
© 1983 Alan R. Liss, Inc., 150 Fifth Avenue, New York, NY 10011

DNA PRIMASE FROM MOUSE CELLS[1]

Ben Y. Tseng[2] and C. N. Ahlem

Department of Medicine
University of California San Diego
La Jolla, California 92093

ABSTRACT. A DNA primase has been purified from
mouse hybridoma cells to near homogencity. The
activity correlates with two proteins of 56 kDa
and 46 kDa, determined by SDS-gel electrophoresis,
that co-purify in a 1 to 1 stoichiometry. The size
of the native enzyme determined by glycerol gradient
sedimentation (5.5 S) and gel filtration is consistent with a size of one 56 kDa and one 46 kDa
subunit per molecule. The average size of RNA
products synthesized on synthetic single stranded
DNA templates is 10 and 20 nucleotides and on single
stranded natural DNA templates approximately 50
nucleotides. The products are, interestingly,
synthesized with a periodicity of approximately
10 nucleotides. DNA polymerase α reduces the
amount of RNA synthesized by primase, possibly by
complexing with the enzyme. RNA that primes
DNA synthesis is found to be approximately 10
nucleotides long.

[1]This work was supported by research grants
  GM 29091 and CA 11705 from the National
  Institutes of Health.
[2]Leukemia Society of America Scholar

INTRODUCTION

Initiator RNA for the priming of Okazaki fragments in mammalian cells has been characterized from studies of cellular and papovavirus DNA synthesis. The size of the RNA primer from studies both in vivo and in broken cell systems indicated a relatively specific RNA of 7-10 nucleotides (see 1, for review). This size remains relatively constant even with perturbations such as omission of 1-2 ribonucleotides from in vitro incubations. Two mechanisms have been suggested that mitigate a requirement for a full complement of ribonucleotides, incorporation of cognate deoxynucleotides (2,3) and substitution by other ribonucleotides (3). These features suggest that the size of the primer may be an important aspect of its synthesis and is a characteristic that distinguishes it from other RNA polymerases. We have examined cell extracts for the presence of an activity with properties expected of a mammalian DNA primase, synthesis of a distinct size oligoribonucleotide that primes DNA synthesis.

RESULTS AND DISCUSSION

We have previously reported the detection and partial purification of an activity from a human lymphoblastoid cell line that has properties of a DNA primase (4), synthesis of relatively specific sized oligoribonucleotides approximately 10 long that prime DNA synthesis. Primase initiated DNA synthesis also exhibited unusual properties, synthesis of DNA on poly(dIT) template occurred in multiples of approximately 10 nucleotides and was aphidicolin resistant.

We have used the procedures developed from our previous study (4) to examine mouse hybridoma cells for a similar activity. The assay we have developed is the synthesis of oligoribonucleotides approximately 10 long on the synthetic template poly(dIT), a random polymer with equivalent amounts of dIMP and dTMP. A DNA primase has been purified over 10,000-fold from mouse cells to near homogeneity by conventional purification steps (5). Activity corresponds with two proteins of 56 kDa and 46 kDa, determined by SDS polyacrylamide gel electrophoresis, that appear in a 1 to 1 stoichiometry.

Sedimentation analysis indicates a value of 5.5S and
gel filtration (Figure 1) indicates a molecular weight
of approximately 90 kDa. Both measurements are
consistent with a native enzyme structure of one 56 kDa
and one 46 kDa subunit.

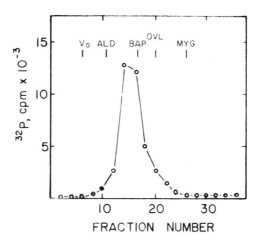

FIGURE 1. DNA primase analyzed by gel filtration.
Primase from mouse hybridoma cells was eluted from a
Sephacryl S-200 column and assayed as described (4)
Elution positions of protein standards were determined
from a separate run (ALD, aldolase; BAP, bacterial alkaline phosphatase; OVL, ovalbumin; MYG, myoglobin). V_o
indicates void volume and the included volume was at
fraction 35. Column buffer was 20% Me_2SO, 10% ethylene
glycol, 50 mM Tris, pH 7.9, 0.5 mM DTT, 0.1 mM EDTA,
0.2 M KCl, 0.01% NP-40.

Synthesis of oligoRNA. Primase requires a single
stranded DNA template for activity. Synthesis of RNA
was examined with a variety of DNA templates. The
average size of product differed with different
templates and, interestingly, synthesis was observed in
multiples of approximately 10 nucleotides. The products
observed with random polymers poly(dIT) and poly(dCT)
and the homopolymer poly(dT) are shown in Figure 2. On
poly(dIT) the predominant products are approximately 10
nucleotides long whereas on poly(dCT) and poly(dT)

products of 10 and 20 nucleotides long are synthesized. The chain lengths of oligo (AG) may appear to be longer due to the high G content in the oligonucleotide. The RNA products on single stranded DNA templates, fd DNA or denatured DNAs of salmon sperm, calf thymus, and E.coli, were longer with an average size of 50 nucleotides although similarly, the product showed a periodicity of 10 nucleotides. One mechanism we suggest for the periodicity of synthesis is that there is one site on the enzyme for the initial binding of template (anchor site) and a second site for polymerization (catalytic site). The anchor site would release the template after approximately 10 nucleotides are synthesized, similar to the movement of an inch worm. More generally, this may be related to the processivity of polymerization enzymes; that is, enzymes may differ in their inter- action at the anchor site that manifests as differences in the processivity of an enzyme.

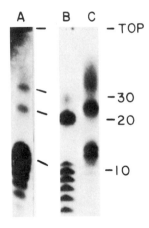

FIGURE 2. OligoRNA products synthesized by DNA primase on single stranded DNA templates. Primase was incubated with A) poly(dIT), B) poly(dT) or C) poly (dCT) templates and α-^{32}PrATP and complementary rNTP. The products were precipitated with ethanol and analyzed by electrophoresis on 20% polyacrylamide gels (4).

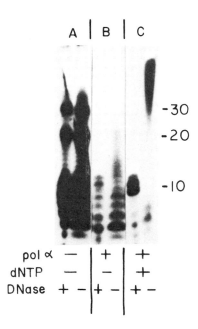

FIGURE 3. ^{32}P-oligoRNA priming of DNA synthesis.
A.   Primase (10 units) was incubated under standard
conditions with rATP, ^{32}P-rCTP and poly(dIT); B. as in
(A) but with addition of DNA polymerase α (0.5 units); C
as in (A) but with addition of DNA polymerase α (0.5
units) and complementary dNTPs.   The products were
collected by ethanol precipitation, divided, and 2/3 of
the sample was analyzed directly and 1/3 was digested
with pancreatic DNase (4) before analysis on 20% polyacrylamide gels.

<u>Interactions</u> <u>With</u> <u>DNA</u> <u>Polymerase</u>. No DNA polymerase activity is detected in the most purified fraction of mouse primase in contrast to the co-purification of Drosophila melanogaster DNA polymerase α and primase (6). Some preliminary studies on the initiation of DNA synthesis by primase are presented.

We have followed DNA synthesis by examining the extension of the labeled RNA primer by DNA polymerase α. In the absence of pol α, primase incorporated ^{32}P-rCTP into RNA of approximately 10 nucleotides (Figure 3A). When pol alone was added, RNA synthesis was inhibited by 80% (Figure 3B). When pol α and dNTPs were added, labeled RNA was observed at longer chain lengths that, after pancreatic DNase digestion, were reduced in size to approximately 11 and 12 nucleotides long (Figure 3C). The size of these molecules includes one to two deoxynucleotides on the RNA not removed by pancreatic DNase digestion and indicates a RNA primer length of approximately 10 nucleotides. (DNase digestion had no effect on RNA alone, Figure 3A,B)

The results suggest that polymerase α inhibits excess primer RNA synthesis, possibly by forming a complex with primase, and that only oligoRNA approximately 10 long prime DNA synthesis.

The isolation of a mammalian DNA primase will facilitate further studies on reconstruction of eukaryotic DNA replication complex.

REFERENCES

1. DePamphilis ML, Wasserman PM (1980). Annual Rev Biochem 49:627-666.
2. Eliasson R, Reichard P (1978). Nature 272:184-185.
3. Tseng BY, Goulian M (1980). J Biol Chem 255:2062-2066.
4. Tseng BY, Ahlem CN (1982). J Biol Chem 257:7280-7283.
5. Tseng BY, Ahlem CN (1983). J Biol Chem (In Press).
6. Conaway RC, Lehman IR (1982). Proc Natl Acad Sci USA 79:2523-2527.

Mechanisms of DNA Replication and Recombination, pages 517–526
© 1983 Alan R. Liss, Inc., 150 Fifth Avenue, New York, NY 10011

MAMMALIAN DNA POLYMERASE α HOLOENZYME FUNCTIONING ON

DEFINED IN VIVO-LIKE TEMPLATES

Ulrich Hübscher and Hans-Peter Ottiger

Institute for Pharmacology and Biochemistry
School of Veterinary Medicine, University of Zürich
CH-8057 Zürich, Switzerland

ABSTRACT   In analogy to the E.coli replicative DNA polymerase III we define two forms of DNA polymerase α: the
core enzyme and the holoenzyme. A DNA polymerase α holoenzyme from freshly harvested calf thymus has been identified and purified by using defined in vivo-like templates. This enzyme complex has been compared to the
corresponding homogeneous DNA polymerase α core from the
same tissue. The holoenzyme is able to use long single-
stranded DNA templates such as (i) single-stranded parvoviral DNA which has short hairpin structures at each
end and therefore only one 3'hydroxylgroup per genome to
serve as a starting point and (ii) single-stranded M13
DNA with a single RNA primer. The core enzyme, on the
other hand, although active on DNA treated with deoxyribonuclease to create random gaps, is unable to act on
these long single-stranded DNAs. The most purified holoenzyme contains eleven polypeptides as judged from SDS-
polyacrylamide gel electrophoresis. The high molecular
weight catalytic subunit of either the holoenzyme or the
core enzyme contains the polymerase and the primase activity. Finally a protein factor has been separated from
the DNA polymerase α holoenzyme by chromatography on hydroxyapatite. This factor is able to restore holoenzyme
activity on in vivo-like templates and it is designated
as a DNA polymerase α accessory protein.

INTRODUCTION

Using the DNA of small bacteriophages of E.coli as model replicons to understand the host DNA replication events,
it was discovered that multienzyme systems are involved (1, 2).
The DNA elongation step needs DNA polymerase III, the major
replicase of prokaryotes. DNA polymerase III functions in the
form of a multipolypeptide complex called DNA polymerase III
holoenzyme (3). This holoenzyme is required for the elongation
of primed single-stranded DNA, while the catalytic DNA polymerase III enzyme, called the core enzyme (3), is only active
on DNAs treated with deoxyribonuclease to create random gaps
(1, 2). To date at least four additional DNA polymerase factors, called the DNA polymerase III holoenzyme subunits β (4),
γ (5), δ (6) and τ (7) have been isolated and characterised.
   The idea that replicative DNA polymerases have been conserved through evolution (8) prompted us to search for a DNA
polymerase α holoenzyme in a mammalian tissue. For this purpose we have used in vivo-like templates : (i) single-stranded
parvoviral DNA, which has short hairpin structures at each end
and therefore only one starting point for DNA polymerases and
(ii) M13 DNA : after incubation with single-stranded binding
protein (SSB) of E.coli and the RNA polymerase holoenzyme from
the same cell, this DNA has one RNA primer per circle at the
origin of replication (9). We give evidence that (i) a DNA polymerase α holoenzyme can be identified and purified, (ii) the
calf thymus primase appears to be part of the high molecular
weight DNA polymerase α polypeptide and (iii) a DNA polymerase
accessory protein, that is partially able to restore holoenzyme activity on single-stranded templates, can be separated
from the intact holoenzyme.

METHODS

Isolation of calf thymus DNA polymerase α holoenzyme
was preformed as described, using single-stranded parvoviral
DNA as the template (10). Two additional chromatographic steps
(Bio-Rex-70 and a second phosphocellulose) were added in order
to improve purification (see Result and Discussion). Primase
activity was determined as described (11). The activities of
DNA polymerase and primase following SDS-polyacrylamide gel
electrophoresis and in situ renaturation of the enzymes were

carried out as described (12) by using a miniature gel method (11). The DNA polymerase accessory protein was separated from the holoenzyme by chromatography on hydroxyapatite (see Results and Discussion). It was determined in a reaction mixture with a final volume of 25 μl containing the following components : 20 mM Tris-HCl (pH 7.5), 4% (w/v) sucrose, 8 mM DTT, 80 μg/ml BSA, 10 mM $MgCl_2$, 5 mM ATP, 10 mM KCl (omitted in the M13 assay), dATP, dGTP and dCTP each at 48 μM, 18 μM [^{3}H]dTTP (100-200 cpm/pmol), 200 pmol of single-stranded parvoviral DNA or singly RNA-primed M13 DNA (20 nmol activated DNA), 0.2 units (determined as DNA polymerase α) of hydroxyapatite eluted fraction and hydroxyapatite flow-through fraction to be assayed. Incubation was at 37°C for 60 min and the TCA precipitable radioactivity was determined as described (6). Details of this assay and the isolation of the DNA polymerase accessory protein will be published elsewhere.

RESULTS AND DISCUSSION

Isolation of a DNA polymerase α holoenzyme functioning on defined in vivo-like templates. We have shown that a calf thymus DNA polymerase α form can be purified by using long single-stranded templates (10). The purification procedure has now been extended to an additional Bio-Rex-70 and a second phosphocellulose step (Figure 1). The final DNA polymerase α holoenzyme complex contains at least eleven major polypeptides as judged after SDS-polyacrylamide gel electrophoresis (Figure 1). A different protocol was elaborated for the isolation of the core enzyme (Figure 1). The DEAE-cellulose chromatography separated the DNA polymerase α holoenzyme into four different polymerase forms[1] and these distinct α polymerases all have partially lost the holoenzyme activity on long single-stranded DNA templates[1]. The most prominent peak was then further purified by a single-stranded DNA cellulose column and by chromatofocussing. This DNA polymerase is designated as the core enzyme. It contains two high molecular weight polypeptides of 150,000 and 125,000 dalton and three polypeptides in the Mr range of 50,000-60,000. The two high Mr bands contain both DNA polymerase activities (see Figure 2, lane 2), while the lower Mr bands derive from the higher Mr catalytic polypeptides, possibly by proteolysis (for the rational see ref. 13).

[1] Ottiger, H.P. and Hübscher, U., unpublished results

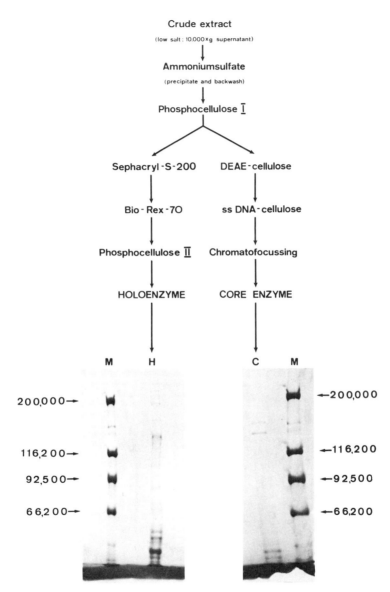

FIGURE 1. Protocol for purification of DNA polymerase α holo- and core enzyme followed by the analysis of the two isolated enzymes by SDS-polyacrylamide electrophoresis in a 8% gel. Markers (M) were: Myosin (Mr 200,000), β-galactosidase (Mr 116,200), phosphorylase B (Mr 92,500), and BSA (Mr 66,200). H: Holoenzyme; C: Core enzyme.

Table 1

Properties of DNA polymerase α holo- and core enzyme

Property	DNA polymerase α	
	Holoenzyme	Core enzyme
Template specificity		
activated DNA	100[1]	134
denatured DNA	100	21
poly(dA)·(dT)$_{12-18}$ (10:1)	100	82
parvoviral DNA	100	11
RNA-primed M13 DNA	100	6
Apparent Mr (kilodalton)	>500	200-250
Sedimentation coefficient(s)	> 11.3	7.4
Mr of catalytic subunit		
(kilodalton)	125	125/150
Inhibition by		
aphidicolin	+	+
d$_2$TTP	−	−
N'ethylmaleimide	+	+
monovalent kations	+	+

[1]The activity measured with the holoenzyme is taken as 100%.

Some properties of the DNA polymerase α holo- and core enzyme are summarised in Table 1. The holoenzyme, but not its core, is active on templates with long single-stranded regions (denatured DNA, parvoviral DNA and RNA-primed M13 DNA). The holoenzyme and the core enzyme have different sedimentation coefficients (> 11.3 versus 7.4 S) and apparent native Mr determined on a polyacrylamide gel under non-denaturing conditions (> 500,000 versus 200,000 to 250,000). Both have typical criteria for a DNA polymerase α and contain a catalytically active polypeptide of 125,000.

This functional distinction of a mammalian DNA polymerase α holo- and core enzyme on defined single-stranded DNA templates provides a basis to begin the characterisation, dissection and reconstitution of a functionally operating replicase. Two first aspects will be covered in the subsequent sections.

   The mammalian primase appears to be part of a high molecular weight DNA polymerase α polypeptide. In the course of
studies with DNA polymerase α holoenzyme a primase activity
was detected and this enzyme could never be separated from the
polymerase by the usual chromatographic and sedimentation techniques (11). Even our apparent homogeneous DNA polymerase α,
the core enzyme, always contained the primase activity (11).
This raised the possibility that the primase and the polymerase activities might reside on the same polypeptide. To test
this we electrophoresed a homogeneous DNA polymerase α core

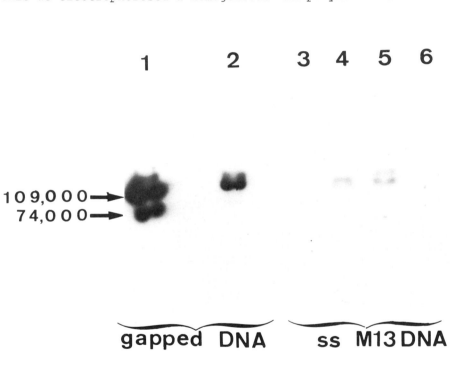

FIGURE 2. Autoradiogram of the activities of DNA polymerases and primase following SDS-polyacrylamide gel electrophoresis in miniature gels (47 x 30 x 0.8 mm) and in situ renaturation of the enzymes. The lanes contained the following
DNA (lanes 1 and 2 gapped DNA; lanes 3 to 6 single-stranded
M13 DNA) and enzymes (lanes 1 and 3 a mixture of E.coli DNA
polymerase I and the Klenow fragment; lanes 2, 4, 5 and 6 apparently homogeneous calf thymus DNA polymerase α core enzyme.
Further explanation see text.

enzyme on SDS-polyacrylamide gels, and renatured enzyme activities in the gels (10, 11). Both enzymes had identical mobilities and coincide with the high Mr catalytic subunit (125,000) of DNA polymerase α (Figure 2). The DNA polymerase α was electrophoresed in SDS-polyacrylamide miniature slab gels containing either gapped DNA or single-stranded M13 DNA. E.coli DNA polymerase I and its Klenow fragment were used as size markers and controls. Lanes 1 and 2 of Figure 2 are positive controls. They show that in the presence of gapped DNA, representative of a multiply primed template, deoxyribonucleotide polymerisation occurred with E.coli DNA polymerase I (109,000 dalton) and its Klenow fragment (74,000 dalton) on the one hand and with calf thymus DNA polymerase α on the other. With the latter a high Mr ($\geq$ 125,000) activity band was observed. A faint activity band above the 125,000 dalton activity could possibly represent the 150,000 dalton band seen in our purified DNA polymerase α core enzyme (Figure 1). When these enzymes were tested on unprimed, single-stranded M13 DNA (lanes 3-6) in the presence of all four ribo- and [α^{32}P]-deoxyribonucleoside triphosphates, the E.coli enzymes were no longer capable of deoxynucleotide polymerisation (lane 3). In contrast, the polymerising activity of calf thymus DNA polymerase was not affected by the absence of a preformed primer as long as the four ribonucleoside triphosphates were available (lane 5) but was suppressed when the latter were omitted from the reaction mixture (lane 6). This indicated that the high Mr activity band produced its own primer for DNA synthesis. That this was the case was shown by directly assaying DNA polymerase α for primase activity by incubating the polymerase with [α^{32}P]-ribonucleoside triphosphates in the presence of cold deoxyribonucleoside triphosphates. This produced a single band of RNA polymerase activity (lane 4), which exactly corresponded to the band of DNA polymerase activity visible in lane 5.

Separation of a DNA polymerase accessory protein from the DNA polymerase α holoenzyme. A partially purified DNA polymerase α holoenzyme (ref. 10, Sephacryl S-200 fraction) was bound to a hydroxyapatite column containing 50 mM Tris-HCl (pH 7.5), 5 mM potassium phosphate (pH 6.8), 5 mM DTT, 25% (v/v) glycerol, 10 mM sodium-bisulfite, 1 µM pepstatin and 1 M KCl. The DNA polymerase was eluted by raising the potassium phosphate concentration to 100 mM. The eluted DNA polymerase was virtually unable to replicate single-stranded parvoviral DNA and

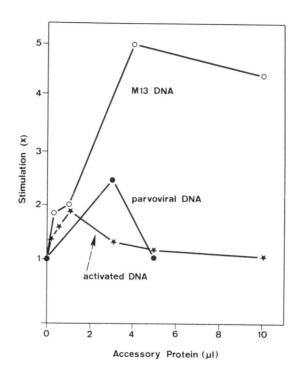

FIGURE 3. Stimulation of DNA polymerase α by a hydroxy-apatite flow-through fraction. Increasing amounts of the desal-ted hydroxyapatite flow-through fraction were added to the DNA polymerase α eluted from the same column, and holoenzyme acti-vity, detected as stimulation, was measured as described in Methods.

RNA-primed M13 DNA. If, however, the desalted flow-through fraction of the same hydroxyapatite column was added to the eluted DNA polymerase a 2 to 5 fold stimulation of replication was evident on the two long single-stranded DNA templates (Fi-gure 3). This assay provides a basis to isolate this DNA poly-merase accessory protein. Preliminary experiments suggest that it is an extremely unstable protein with an acidic isoelectric point and a tendency to aggregate.

CONCLUSIONS

In summary we presented evidence that (i) a DNA polymerase α holoenzyme with a very complex polypeptide structure can be isolated from freshly harvested calf thymus and this enzyme can functionally be distinguished from a homogeneous DNA polymerase α, called the core enzyme, (ii) the primase activity appears to be part of a high Mr DNA polymerase α polypeptide and (iii) a DNA polymerase accessory protein can be separated from the DNA polymerase α holoenzyme.

ACKNOWLEDGEMENTS

This work was supported by the Swiss National Science Foundation, Grant 3.006-0.81, and by the Sandoz Stiftung zur Förderung der medizinisch-biologischen Wissenschaften.

REFERENCES

1. Kornberg, A. (1980) DNA Replication, published by W.H. Freeman and Co., San Francisco, Cal.
2. Kornberg, A. (1982) DNA Replication, supplement, published by W.H. Freeman and Co., San Francisco, Cal.
3. McHenry, C.S. and Kornberg, A. (1982) in: The Enzymes (Boyer, P., ed.) Academic Press, N.Y. 14, 39-50.
4. Johanson, K.O. and McHenry, C.S. (1980) J. Biol. Chem. 255, 10984-10990.
5. Hübscher, U. and Kornberg, A. (1980) J. Biol. Chem. 255, 11698-11703.
6. Hübscher, U. and Kornberg, A. (1980) Proc. Natl. Acad. Sci. USA 76, 6284-6288.
7. McHenry, C.S. (1982) J. Biol. Chem. 257, 2657-2663.
8. Hübscher, U., Spanos, A., Albert, W., Grummt, F. and Banks, G.R. (1981) Proc. Natl. Acad. Sci. USA 78, 6771-6775.
9. Villani, G., Fay, P.I., Bambara, R.A. and Lehman, I.R. (1981) J. Biol. Chem. 256, 8202-8207.
10. Hübscher, U., Gerschwiler, P. and McMaster, G.K. (1982) EMBO J. 1, 1513-1519.
11. Hübscher, U. (1983) EMBO J. 2, 133-136.

12. Spanos, A. and Hübscher, U. (1983) in: Methods in Enzymology (Hirs, C.H.W. and Timasheff, S.N., eds.) Academic Press, N.Y. 91, 265-277.
13. Albert, W., Grummt, F., Hübscher, U. and Wilson, S.H. (1982) Nucl. Acids Res. 10, 935-946.

Mechanisms of DNA Replication and Recombination, pages 527–552
© **1983 Alan R. Liss, Inc., 150 Fifth Avenue, New York, NY 10011**

IN VITRO RECONSTITUTION OF YEAST 2-μm PLASMID
DNA REPLICATION

Akio Sugino[1], Akira Sakai[1], Francis Wilson-Coleman,
Josef Arendes[2], and Kwang C. Kim

Department of Molecular and Population Genetics
University of Georgia, Athens, GA 30602
and
Laboratory of Genetics
National Institute of Environmental Health Sciences
National Institutes of Health
P.O. Box 12233 Research Triangle Park, NC   27709

ABSTRACT

A soluble extract from exponentially growing
yeast Saccharomyces cerevisiae initiates semiconservative replication of exogenously added yeast
2-μm and ARS (autonomously replicating sequences)
containing plasmid DNAs. This system utilizes
more than 50% of input DNA template and its DNA
replication initiates at the in vivo origin and
proceeds bidirectionally as does in vivo.
Extract prepared from cdc8-1 (a temperature-
sensitive cell division cycle mutant in yeast) is
defective for both 2-μm plasmid and ARS-1 plasmid
DNA replication in vitro at restrictive temperatures, but replication activity can be restored by
addition of appropriate purified fractions derived
from wild-type cells. Purified cdc8-1 complementing activity consists of the 38Kd yeast single-
stranded DNA binding protein (1).

[1]Present address: Department of Molecular and Population Genetics, University of Georgia, Athens, GA 30602.
[2]Present address: Johannes Gutenberg - Universität,
Physiologisch - Chemisches Institut, 6500 Mainz, West
Germany.

The crude extract has been fractionated into several components and in vitro reconstitution of the activity has been successfully achieved. The components which have been identified and purified in the course of this study are a single-stranded DNA binding protein (ySSB - CDC8 gene product), yeast DNA polymerase I, DNA primase, and a type II DNA topoisomerase. The reconstituted activity replicates both 2-μm and ARS plasmid DNAs.

"Preinitiation complex" from crude extracts has also been isolated and characterized. This complex formation is salt-sensitive, requires ATP and Mg^{++}, but does not require a specific DNA template. However, it is accelerated by specific DNA templates including 2-μm DNA and ARS-DNA. The size of the complex(es) could be greater than 5,000,000 daltons and includes ySSB, DNA polymerase I, DNA primase, type II topoisomerase, and possibly type I topoisomerase and RNaseH.

Native 2-μm and ARS plasmid DNAs are methylated at the specific sites. Most of the sites are at or near the origin of DNA replication, implicating that the methylation is regulating initiation of DNA replication in yeast.

INTRODUCTION

The complex events that occur during the replication of eucaryotic genomes are not yet understood. By analogy with knowledge of prokaryotic systems, it is expected that much of the basic information about chromosomal replication might be obtained by studying the replication of the genomes of extrachromosomal elements and viruses. The same analogy predicts that great progress in this area will be achieved by use of in vitro replication systems that carry out faithful replication of such genomes as in vivo replication.

The low eucaryote, yeast Saccharomyces cerevisiae, has many advantages over other eucaryotes. The system is much less complex. It is very easy to prepare large amounts of cells and to do enzymology. It has one of the best genetics and a better DNA transformation system (2). Moreover, some of yeast strains have plasmid-like extrachromosomal DNAs and the best known DNA is the 2-μm plasmid, whose replication is regulated by the same mechanism as is chromosomal replication (3). ARSes have been isolated from yeast chromosome, are maintained extrachromosomally in yeast (4), and their DNA replication is regulated by the same mechanism as is chromosomal DNA (5). These DNAs are believed to be in vivo origins of chromosomal DNA replication. Therefore, these systems are thought of as model systems for understanding of eucaryotic chromosomal DNA replication.

Various in vitro replication systems for yeast 2-μm and ARS plasmid DNAs have been described (6-8) which appear to mimic in vivo DNA replication. Recently using such an in vitro system, some of the DNA replication proteins have been identified and purified (1,8). Therefore, it is now possible to understand more precisely the mechanism of eucaryotic DNA replication. However, the efficiency of utilizing DNA templates in these systems is less than 10% (6-8). In some cases, the efficiency was less than a few percent (6,8). These low efficiencies of DNA replication encumber subsequent study of the fractionation of various replication components and reconstitution of the activity. Furthermore, such low efficiency of DNA synthesis may be accounted for by DNA strand elongation activity utilizing pre-existing contaminated primer in DNA templates. Much improvement of such in vitro systems was needed.

In this paper, we summarize our progress of the improvement of our in vitro DNA replication system. The improved system now synthesizes 2-μm and ARS-1 plasmid DNAs very efficiently and more than 50% of the input DNA can be faithfully replicated. Using the system, we have fractionated and reconstituted in vitro 2-μm and ARS-1 DNA replication. We will also describe some attempts of isolation and characterization of a "preinitiation complex" of the replication and, finally, we will discuss a possible role of DNA methylation on regulation of initiation of DNA replication in yeast.

MATERIALS AND METHODS

Bacterial and Yeast Strains
     All strains used in this work are listed in Table 1 and growth conditions have been described (7).

DNA and Enzymes

Bacterial plasmid DNA was isolated by repeated CsCl/ ethidium bromide centrifugation (9). Sometimes, alkaline-extraction method (10) was employed to eliminate possible contamination of small RNA or DNA associated with covalently closed supercoiled DNA. The purity of supercoiled plasmid DNAs was more than 95%. Various restriction endonucleases were purchased from Bethesda Research Laboratories and used as recommended by the supplier. HpaII methylase was from New England Biolabs. Zymolyase 60,000 was from Miles.

Preparation of Crude Extracts for In vitro DNA Replication

The protocol for making crude extracts from S. cerevisiae was essentially the same as described before (7) except that 0.5M KCl was used during extraction of proteins. After centrifugation of cell lysate, supernatants were applied on DEAE-cellulose column (10-15 ml) equilibrated with 0.5M KCl-buffer A (50 mM Tris-HCl, pH 7.5 - 1mM EDTA - 10 mM 2-mercaptoethanol - 10% glycerol - 1 mM PhMeSO$_2$F) and the proteins were eluted with the same buffer. The pass-through protein fractions were collected and solid ammonium sulfate was added to 45-50% saturation over 20 min at 0-4°C. The precipitates were collected by centrifugation at 10,000 rpm for 10 min in a Sorvall SS34 rotor, suspended into 1-2 ml buffer A, and dialyzed against 1 l of buffer A, twice for 2 hrs at 4°C. The dialyzed crude extract was divided into 200 μl aliquots, transferred to 1.5 ml plastic Eppendorf tubes, frozen and stored in liquid N$_2$.

Assay for 2-μm and ARS-1 Plasmid DNA Replication

The reaction mixture (0.1 ml) contained 35 mM Hepes buffer, pH 7.8, 10 mM MgCl$_2$, 1 mM dithiothreitol, 1 mM spermidine, 2 mM ATP, each of the other three rNTPs at 0.2 mM, all four dNTPs at 50 μM (dTTP was [α-^{32}P] labeled, 400 cpm/pmole), 0.1 mg/ml bovine serum albumin, 1 mM DPN, 1 mM phosphoenolpyruvate, 21.3 mM phosphocreatine, 10 mg/ml creatine kinase from rabbit muscle, 25% glycerol, 1 μg (about 2.85 mμmoles total nucleotides) of the chimeric plasmid pJDB36 (pMB9 + 2 μm) DNA (13) or pLC544 DNA (pBR313 + TRP-1) (14) and crude extract (40 μl) (about 2 mg protein). Incubation was at 30°C for the indicated period, unless otherwise indicated in the Figures. The reaction was terminated by the addition of 2 ml cold trichloroacetic acid (TCA) - 1% sodium pyrophosphate and acid insoluble materials were collected on a Whatman GF/C glass filter.

TABLE 1
YEAST AND E. coli STRAINS

Strains	Genotype	References
a) Yeast		
Saccharomyces cerevisiae A364A	a ade1,2 ura1 his7 lys2 tyr1 cir$^+$	Yeast Stock Center
ts198	a ade1,2 ura1 his7 lys2 tyr1 cir$^+$ cdc8	Dr. V.A. Zakian
ts124	a ade1,2 ura1 his7 lys2 tyr1 cdc7-1 cir$^+$	Yeast Stock Center
AS15	a ade1,2 ura1 his7 lys2 tyr1 aph^{s}15 cir$^+$	(11)
20B-12	α pep4-3 trp1 cir$^+$	(12)
b) E. coli		
HB101	F$^-$ hsdS20 recA13 ara-14 proA2 lacY1 galK2 rpsL20 xy1-5 mt1-1 supE44 λ	E. coli Genetic Stock Center
GM215	dam-3 endA1 rna-1 thi-1 supE44 λ	"
GM31	dcm-6 thr-1 leuB6 hisG4 thi-1 ara-14 lacY1 galK2 galT22 xy1-5 mt1-1 strA136 tonA31 tsx-78 supE44 λ	"
ED8767(pJDB36)	recA56 metB hsdS supE supF (pJDB36)	Dr. J. Beggs (13)
HB101(pLC544)	HB101(pLC544)	American Type Culture Collection

RESULTS AND DISCUSSION

Further Improvement of the In vitro 2-μm and ARS Plasmid
DNA Replication System

We have shown that a crude extract from exponentially
growing yeast S. cerevisiae cells catalyzes de novo DNA
synthesis on exogenously added yeast 2-μm plasmid DNA, E.
coli chimeric plasmids containing 2-μm DNA, and ARS (autonomously replicating sequence) from yeast chromosomal
DNA (7). This DNA synthesis mimics in vivo replication.
However, efficiency of utilization of added template DNA
was less than 10% (7). Such a low level of DNA synthesis
may be explained by the activity which might elongate the
contaminating oligonucleotides associated with supercoiled
DNA templates (J. Scott, personal communication). Therefore it was necessary to improve the system in order to
prove that the system in fact initiates DNA synthesis de
novo.

As described in the MATERIALS AND METHODS, introduction of DEAE-cellulose column chromatography and subsequent
dialysis steps in making crude extracts from yeast increased the efficiency of DNA synthesis more than 2-fold.
More importantly, we are able to use larger volumes of
crude extract than before. The further drastic improvement
has come from changing DNA templates. E. coli chromosomal
and plasmid DNAs are fully methylated at -GATC- sites by
deoxyadenosine methylase in wild-type cells (15). On the
other hand, yeast chromosomal and 2-μm plasmid DNA are not
methylated at these sites. Therefore, we have tested
whether or not the methylated DNA templates are as good as
unmethylated DNA templates for our in vitro DNA replication
system.

As shown in Fig. 1, the chimeric plasmid pJDB36 (pMB9
+ 2-μm form A DNA) (13) and pLC544 (pBR313 + yeast ARS-1 +
TRP-1) (14) from E. coli dam-3 (deoxyadenosine methylase⁻)
supported in vitro DNA synthesis at least 3-4 times as much
as the DNAs from dam⁺ E. coli, although a few percentage of
-GATC- sites in DNAs were still methylated in dam⁻ strain
(15 and our unpublished results). Under the condition
shown in Fig. 1, the in vitro system synthesized more than
1 nmoles of DNA on 2.8 nmoles of added template DNA in 60
min at 30°C, representing more than 35% of input DNA.

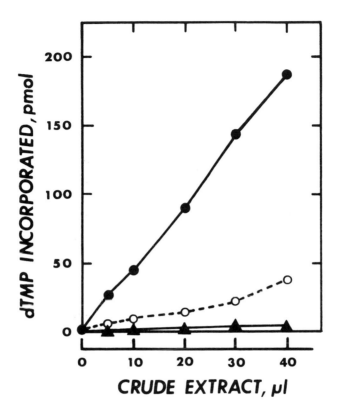

Fig. 1.   DNA synthesis activity in crude extract from
S. cerevisiae.
    DNA   synthesis   in   crude   extract   from   S. cerevisiae
A364A cells was measured as MATERIALS AND METHODS using the
hybrid plasmid pJDB36 (pMB9 plus 2-µm yeast plasmid form A)
from dam-3 E. coli (-●-), from dam+ (wild-type) E. coli
(-o-),   or pJDB36 DNA from wild-type E. coli methylated in
vitro by Haemophilus parainfluenzae HpaII methylase (-▲-).
After incubation for 40 min at 30°C, acid insoluble radioactivity was measured as before (7).   40 µl crude extract
roughly corresponded to 2 mg proteins.

    As   shown   in   Fig.   2,   DNA   synthesis   proceeds   linearly
for at least 60 min at 30°C, then levels off.   After 2 hrs
incubation, the system synthesized nearly 70% of input DNA.
The K_m^{app} under   the   standard   conditions   is   about   14 µM (or
5 µg/ml) (Fig. 3).

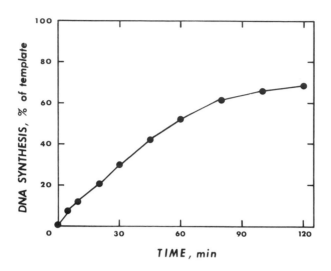

Fig. 2. Time course of DNA synthesis in the crude extract.

DNA synthesis activity on the hybrid plasmid pJDB36 DNA from dam-3 E. coli was measured in the reaction mixture (1.5 ml) containing 15 μg of the plasmid DNA and 0.6 ml crude extract from S. cerevisiae A364A cells. After incubation at 30°C for the indicated periods, 100 μl aliquotes were taken and acid insoluble radioactivity was measured.

Furthermore, more than 50% of input DNA molecules after brief incubation with crude extract had typical Θ form structure under the electron microscope (data not shown). These results clearly diminish the possibility that our in vitro DNA replication system is only able to elongate DNA strands using pre-existing primers. The characteristics of our improved in vitro DNA replication system are summarized in Table 2.

DNA replication origin(s) have been re-examined as before (7) using pJDB36 DNA as a template. In vitro origin and directionality of the DNA replication were the same as before (7). However, the specificity of 2-μm DNA replication was increased (≥90% of the molecules begin DNA replication at 2-μm in vivo origin, whereas ≤10% of the molecules begin at pMB9 DNA) (data not shown).

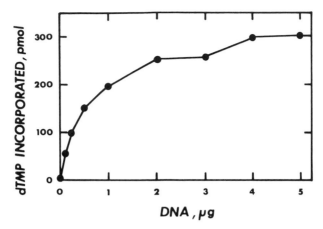

Fig. 3.   DNA concentration dependency on in vitro DNA replication.

The reaction conditions were the same as Fig. 2, except that the concentration of the template plasmid DNA was varied as indicated. Since DNA synthesis under the condition in the figure proceeds linearly at least for 60 min at 30°C (Fig. 2), it is possible to estimate the rate of DNA synthesis and the K_m^{app} for the template DNA concentration from this experiment.

In eucaryotes, it is known that some of -CG- sequences are methylated, and in some cases, such methylation has been correlated to inactive and actively transcribing genes and it has been speculated that DNA methylation regulates gene expression (16,17). Yeast is one of the organisms which contains very few 5-methyl deoxycytosine residues in DNA (18).

Total yeast DNA was extracted from stational phase cells, restricted with various restriction endonucleases, and electrophoresed through agarose gels. DNAs were then transferred on a nitrocellulose using the Southern Method (19), and hybridized to ^{32}P-nick-translated cloned 2-μm from A DNA. As shown in Fig. 4, the expected 2-μm DNA fragments generated by restriction endonucleases showed up after autoradiography. However, the DNA digested with restriction endonuclease HpaII, which recognizes -CCGG- sequence and cut between two Cs, but does not cut -CC^{m}GG- sequence, showed the unexpected bands (Fig. 4, lane h and i). On the other hand, such unexpected bands could not be observed in the case of MspI restriction endonuclease which is able to cut both -CCGG- and -CC^mGG- sites. This strongly indicates that 2-μm DNA is methylated at the specific sites. As the total 2-μm DNA sequence is known, the sites resistant to the digestion by HpaII are able to be assigned on its physical map. As shown in Fig. 5, one of the two sites is mapped in the replication origin (20, 21) and the other is in the REP1 gene of 2-μm plasmid (20,22).

TABLE 2
REQUIREMENT FOR in vitro DNA REPLICATION

Omissions and Additions	Activity[a]	% of template
Complate	200	28.6
-ATP	10	1.4
-rNTPs	2	0.3
-CP	75	10.7
-CPKase	85	12.1
-Glycerol, 25%	63	9.0
-Glycerol + 6% PEG 6,000	110	15.7
-Glycerol + 6% PVA	151	21.6
-Spermidine	230	32.9
-DNA(pJDB36)	1	0.1
-pJDB36, + pLC544 (1μg)	205	29.3
+Type II topoisomerase antibody	131	18.7
+Type I topoisomerase antibody	210	30.0
+E. coli DNA gyrase subunit A antibody	139	19.9

[a] dTMP incorporated (pmoles/1μg DNA template).

DNA synthesis activity of the crude extract from S. cerevisiae A364A was measured in the standard assay described in METHODS. 2.85 mμ moles (total nucleotides concentration) of pJDB36 which consists of pMB9 and 2-μm form A DNA and 30 mg/ml of protein (40 μl) from crude extract were used and the incubation was 40 min at 30°C. 200 pmole of dTMP incorporation corresponds to about 30% of synthesis added DNA template.

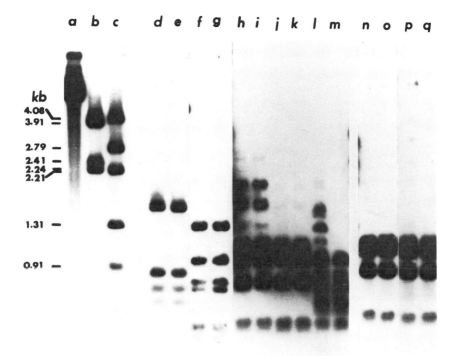

Fig. 4. 2µm plasmid DNA is methylated at the specific sites in S-phase yeast[1].

Total DNA was extracted from yeast <u>Saccharomyces</u> <u>cerevisiae</u> ts124 (<u>a</u> <u>adel,2</u> <u>ural</u> <u>his7</u> <u>lys2</u> <u>tryl</u> <u>cdc7-1</u> <u>cir</u>$^+$) synchronized cells, digested with various endonucleases for 2 hrs at 37°C and analyzed by electrophoresis on 1.2% agarose gel. Yeast total DNA (~2 µg) was applied. After electrophoresis, the DNA was blotted to nitrocellulose (19) and hybridization was performed in the solution containing 5x Denhardt solution, 6xSSC, 1% SDS, 100 µg/ml heat-denatured calf thymus DNA for 5 hr with pJDB36 DNA probe labeled with ^{32}P by nick-translation (1 x 10^8 cpm/µg DNA, 5 x 10^6 cpm/filter). The filter was washed with 250 ml of 2xSSC - 0.5% SDS at 68°C for 1 hr and with 250 ml of 2xSSC - 0.5% SDS at room temperature for 30 min and auto-radiographed for 8 hr at -80°C with an intensity screen. <u>Cdc7-1</u> cells growing exponentially (2 x 10^7 cell/ml) were

[1]Sugino, A., Sakai, A., and McClelland, <u>in</u> <u>preparation</u>.

synchronized by the procedure described before (5). The culture was then shifted down to 25°C with ice and incubated at 25°C (0 min G_1 arrested cells).

200 ml of the culture was taken at 40 min (S-phase) and 100 min (G_2 phase) after temperature shift-down and the DNA was extracted as before. Lane a, yeast total DNA from late S-phase cells; b, a plus EcoRI (10 units); c, a plus HindIII (10 units); d and e, a plus HhaI (5 and 10 units); f and g, a plus MboI (5 and 10 units); h and i, a plus HpaII (5 and 10 units); j and k, a plus MspI (5 and 10 units); l, a plus EcoRI (10 units) and HpaII (10 units); m, a plus EcoRI (10 units) and MspI (10 units); n and o, yeast DNA from G_1 arrested cells digested with HpaII (5 and 10 units); p and q, yeast DNA from G_2 phase cells digested with HpaII (5 and 10 units).

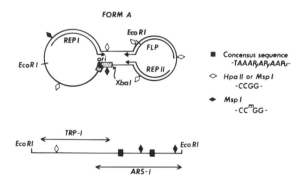

Fig. 5. Methylated sites in yeast 2-μm plasmid DNA and the TRP-1-ARS-1 gene of yeast chromosome.

The methylated sites in 2-μm plasmid DNA (Fig. 4) and TRP-1-ARS-1 gene were assigned on their physical maps. ◇ shows the sites susceptible to HpaII digestion and ♦ shows the sites protected against HpaII digestion. ■ shows the concensus sequence of various ARSes and 2-μm plasmid DNA replication origins (23,24, and Broach, J.R., personal communication).

Total DNA was isolated from the cells at the end of G_1 phase (just before the beginning of S phase) or the cells which just enter S-phase and analyzed by the same manner as described above. As shown in Fig. 4, lane n and o, no HpaII-resistant sites in 2-μm DNA could be detected.

This indicates the methylation at the specific sites in 2-μm DNA is cell-cycle-dependent. A similar phenomenon could be observed in the case of TRP-1-ARS-1 DNA (Fig. 5). An interesting point is that methylated sites could be detected in or near DNA replication origin, assuming ARS-1 is an in vivo DNA replication origin. A further remarkable coincidence is that the regions which are defined as DNA replication origins in various ARSes always have -CCGG- sequence(s) (23-25). Furthermore, the frequency of appearance of -CpG- dinucleotide at or near origins is very high (unpublished results). Although it is too early to draw any conclusions from these results, they strongly suggest that methylation at -CCGG- sites might regulate initiation of DNA replication.

Interestingly enough, if all -CCGG- sites of the chimeric plasmid pJDB36 and pLC544 DNAs were fully methylated by Haemophilus parainfluenza HpaII methylase (24) in vitro, the DNA is totally inactive on our in vitro DNA replication system under the standard conditions (Fig. 1). The mechanism of inhibition, however, is not yet understood. One possibility is that methylation causes a change in the number of negative superhelical twists in a plasmid (Fig. 6), probably due to conformational change in the DNA (from ordinal B form DNA to left-handed or Z form DNA). As a result, the DNA replication complex fails to recognize DNA replication origin. Another possibility is that methylation controls transcription and such transcription activates origin of DNA replication.

Purification of CDC-8 Gene Product by Complementation Assay (1)

Crude extracts from S. cerevisiae cell-division-cycle mutant, cdc8-1, and some of the newly isolated temperature-sensitive DNA synthesis mutants in our laboratory could not synthesize 2-μm DNA at restrictive temperatures (1,7,8, and our unpublished data). However, supplementation of appropriate purified fractions from wild-type crude extracts resumed in vitro DNA synthesis activity as much as wild-type crude extracts (1). This result enabled us to purify the complementing activity from both wild-type and cdc8-1 mutant cell crude extracts (1). The most purified fraction was almost 90% pure and contained mainly a Mr 38,000 polypeptide under the denatured condition. This protein fraction exclusively contained a single-stranded DNA binding activity and each protein molecule binds every

dam⁻ dam⁺
a b c d e f

Fig. 6. Supercoiling difference in the plasmid DNA from dam⁻ and dam⁺ E. coli.

Native supercoiled pLC544 DNA (pBR313 + TRP-1-ARS-1) was prepared from both dam-3 and wild-type (dam⁺) E. coli cells as before (9) and the difference of supertwists in both DNAs was analyzed on 0.85% agarose gel electrophoresis in the presence of 2.5 µg/ml chloroquine (in both gel and gel running buffer). After electrophoresis at 80 volts/gel and room temperature for 24 hours, the gel was stained with 0.5 µg/ml ethidium bromide, destained with water, and photographed under UV lamp. a, 2 µg of the pLC544 DNA from dam-3 cells; b, 1 µg of the DNA from dam-3 cells; c, 0.5 µg of the DNA from dam-3 cells; d, 2 µg of the DNA from wild-type cells; e, 1 µg of the DNA from wild-type cells; f, 0.5 µg of the DNA from wild-type cells. The faster moving broad bands are monomer of the pLC544 DNA and the slower moving bands are the dimer of the DNA.

10-20 bases of single-stranded DNA. The binding is not cooperative, unlike T4 gene 32 protein. The same protein has been purified from yeast by using a simple DNA binding assay and has the cdc8-1 complementing activity. Therefore, we designated the CDC-8[1] gene product as a yeast

[1]Wild-type gene represents capital letter (CDC), while mutant gene is used as small letter (cdc).

single-stranded DNA binding protein (ySSB). However, final proof that this ySSB is the CDC-8 gene product should be determined upon completion of the cloning and the nucleotide sequence determination of the CDC-8 gene and the determination of amino acid sequence of ySSB. Nevertheless, the ySSB specifically stimulates the reaction catalyzed by yeast DNA polymerase I, a true DNA replicase inside the cell (11), but not the reaction catalyzed by yeast DNA polymerase II. It also increases the processivity of DNA polymerase I reaction at least 10-fold (1).

Fractionation and Reconstitution of 2-μm Plasmid In vitro DNA Replication Activity

Although various temperature-sensitive mutants have been isolated (27), few mutants are DNA-replication mutants in yeast S. cerevisiae. So far, only a cdc8-1 mutant crude extract shows temperature-sensitive in vitro DNA replication on 2-μm and ARS-plasmid DNAs (1,7,8). By analogy to prokaryotic systems, it is expected that many DNA replication proteins should be required for yeast chromosomal and 2-μm plasmid DNA replication, and many more DNA replication mutants should be isolated in yeast. We, as well as others (28,29), have been isolating a new set of DNA replication mutants using S. cerevisiae AS15. Crude extracts from some of the mutants which we have isolated have been shown to be temperature-sensitive on in vitro DNA synthesis of 2-μm and ARS-1 plasmids (unpublished results). This is very encouraging for purification of the DNA replication proteins by complementation assay. In the mean time, we have tried to fractionate our in vitro replication activity of 2-μm plasmid DNA into various components and to reconstitute the activity using the separated fractions.

First, a single-stranded DNA-cellulose column was used to separate the activity into two fractions (pass-through and 1 M NaCl eluate fractions). After making sure that the pass-through and the retained fractions fully reconstituted the in vitro 2-μm plasmid DNA replication activity, the retained fractions were further fractionated on phosphocellulose columns. The five different pools shown in Fig. 7 were obtained and reconstitution experiments have been performed (Table 3). The P1, P3, P4, P5, and single-stranded DNA cellulose pass-through fractions reconstituted fully the activity and any combinations missing one of these six fractions did not show full activity of DNA synthesis. The reconstituted DNA synthesis activity could not be distinguished from unfractionated crude extract by the product analyses (e.g., gel electrophoresis, electron microscopic analysis, and alkaline sucrose gradient sedimentation) and is thought to mimic in vivo replication. As shown in Fig. 7, P3 contained most of yeast DNA polymerase I activity and P4 contained cdc8-1 complementing activity. Therefore, P3 and P4 were substituted by the highly purified DNA polymerase I which was composed of at least 4 major polypeptides (1) and the CDC8 gene product (ySSB) (1) and reconstitution experiments have been carried out. However, virtually no in vitro replication activity could be recovered. This indicated that additional protein factors had been lost during purification of DNA polymerase I and ySSB.

From a P3 fraction, the activity which is able to reconstitute in vitro DNA replication activity on 2-μm DNA from the inactive combination between the ssDNA-cellulose column pass-through fraction, P1, P5, DNA polymerase I and ySSB could be assayed and purified extensively[1]. The most purified fractions contained a major polypeptide of about 65,000 and some minor polypeptides on a polyacrylamide gel in the presence of sodium dodecyl sulfate. Under native conditions, the molecular weight of this activity has been estimated to be about 65,000 (Fig. 8). Therefore, assuming Mr 65,000 major protein is the complementing activity, the protein is a monomer polypeptide of the Mr 65,000.

This protein preparation synthesizes quite a discrete size of oligoribonucleotides in the presence of ribonucleoside 5'-triphosphates on a single-stranded DNA. Major products of this reaction are either 8 or 9 mer and each oligonucleotide has a triphosphate end. In the presence of large amounts of deoxyribonucleoside 5'-triphosphates, this activity also synthesizes deoxyoligomers, mainly 4-5 mer (Fig. 9), but at less efficiency. Yeast DNA polymerase I is capable of utilizing oligonucleotides synthesized by this protein as primers and synthesizes DNA on a single-stranded DNA template, but yeast DNA polymerase II cannot use oligoribonucleotides as a primer efficiently. Although it is already known that typical RNA polymerases I, II, and III reaction products are utilized as primers by DNA polymerases (30), the activity purified in this study is well distinguished from these RNA polymerases by different column chromatographic behaviors, α-amanitin inhibition, apparent molecular weight, and reaction products. Therefore, we named this activity a yeast DNA primase.

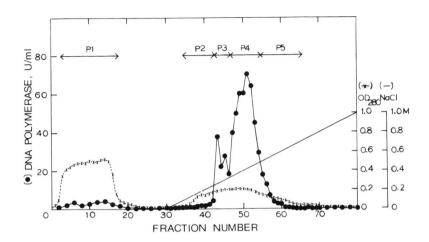

Fig. 7. Phosphocellulose column chromatogram of 1 M eluate from ssDNA-cellulose column.

Crude extract (200 ml) from S. cerevisiae 20B-12 cells (200 g) was applied on 100 ml of ssDNA cellulose column equilibrated with Buffer A-25 mM NaCl, the column was washed with 500 ml of the same buffer (the pass-through fractions were collected, the proteins were precipitated with 75% saturation of $(NH_4)_2SO_4$, dissolved into 10 ml of the buffer A-25 mM NaCl-50% glycerol, and stored at -80°C; the pass-through fraction), and the proteins retained were eluted with 200 ml of 1 M NaCl-buffer A. The proteins were dialyzed against Buffer A (2 1 x twice) for 5 hr at 0°C, and applied on 25 ml of phosphocellulose column equilibrated with Buffer A, and the column was washed with 50 ml of the same buffer. The proteins were eluted with 500 ml of 0-1 M NaCl linear gradient in Buffer A. A_{280} and DNA polymerase activity in the presence of activated DNA as a template were measured. The five different fractions indicated in the Fig. were pooled, dialyzed, concentrated, and stored in 50% glycerol Buffer A at -80°C.

TABLE 3

RECONSTITUTION OF IN VITRO REPLICATION OF YEAST 2-μm PLASMID DNA

	experiments												
Crude extract	+	−	−	−	−	−	−	−	−	−	−	−	−
ssDNA cellulose pass-through	−	+	+	+	+	+	+	+	+	+	+	+	+
ssDNA cellulose 1 M eluate	−	−	+	−	−	−	−	−	−	+	+	+	−
P-1	−	−	−	+	−	−	−	−	+	+	+	+	+
P-2	−	−	−	−	+	−	−	−	−	+	+	−	+
P-3	−	−	−	−	−	+	−	−	+	+	+	+	+
P-4	−	−	−	−	−	−	+	−	−	−	+	+	+
P-5	−	−	−	−	−	−	−	+	−	−	−	+	+
DNA synthesis, pmoles of dTMP incorporated	180	2.3	14.8	1.8	1.0	5.8	4.5	1.5	2.0	15.8	25.3	210	190
(%)	(100)	(1.3)	(8.2)	(1.0)	(0.6)	(3.2)	(2.5)	(0.8)	(1.1)	(8.8)	(14.1)	(117)	(105)

(+) indicates addition of the fraction, while (−) indicates omission of the fraction. The DNA synthesis was measured as Fig. 1 using pJDB36 DNA from dam−3 E. coli.

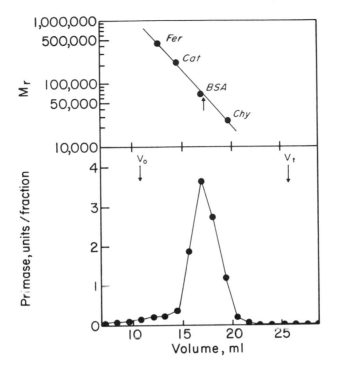

Fig. 8.   Sephacryl S200 Gel filtration of DNA primase[1]
    The DNA primase preparation (0.5 ml, 25 units) (Heparin-Sepharose Fraction) was applied on 25 ml of Sephacryl S200 gel column equilibrated with Buffer A-0.15 M NaCl and eluted with the same buffer.   In vitro DNA synthesis reconstitution activity was measured using the pass-through fraction of ssDNA cellulose column, P1, P5, the highly purified DNA polymerase I, and ySSB.   One unit represents the activity which reconstitutes 1 nmole of 2-μm plasmid DNA synthesis at 30°C for 30 min; V_o, the excluded volume and V_t, the included volume.   In the upper panel, the column was calibrated with Ferritin (Fer), Catalase (Cat), Bovine serum albumin (BSA), chymotrypsinogen A (Chy), and ^{3}H-thymidine.

[1]F.   Wilson-Coleman   and   A.   Sugino,   in   preparation.

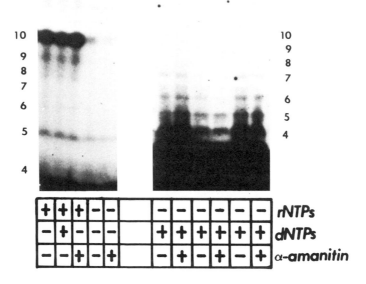

Fig. 9. Autoradiograms of 20% polyacrylamide gel electrophoresis of the products made by the reaction of DNA primase[1].

The reaction products by the DNA primase reaction at 30°C for 30 min on the ØX174 single-stranded DNA in the presence of rNTPs (one of rNTPs was [d-^{32}P]ATP) (A) or dNTPs ([α-^{32}P]dATP was included) (B) were analyzed on 7 M urea-20% polyacrylamide gel (36). Autoradiograms were taken for 8 hours (A) or 48 hours (B) at -80°C. (+) and (-) represent addition and omission of rNTPs, dNTPs, or α-amanitin, respectively. Aa-e and Ba-d contained DNA primase; B-e and f contained DNA polymerase I of yeast (Fig. 10 c-b) and 100 µg/ml aphidicolin. B, c, and d were subjected to alkaline digestion (in 0.3 N NaOH for 15 hr at 37°C.

[1]F. Wilson-Coleman and A. Sugino, in preparation.

In other eucaryotic systems (31-34), it is well documented that DNA primase activity is tightly associated with DNA polymerase α or α-like DNA polymerase. In yeast, DNA primase activity can be well separated from DNA polymerases I and II as shown in Fig. 10. This might be due to less tight association with DNA polymerase I, as a small amount of DNA primase activity could be detected in the most purified typical DNA polymerase I preparation (Fig. 10).

In the mean time, the P5 fraction has been further fractionated into 2 components by hydroxyapatite and Sephacryl S300 column chromatographies. One component was associated with typical type II topoisomerase activity (35), which relaxes negative supercoiled DNA, catenates DNA rings and decatenates DNA in the presence of ATP. Furthermore, this topoisomerase activity was inhibited by the antibody against yeast type II topoisomerase purified by Goto and Wang (35). By addition of the first component, only less than 50% stimulation of DNA synthesis has been observed. Similar stimulation could be obtained by addition of highly purified yeast type II topoisomerase (A. Sugino, unpublished) which is inhibited by antibody against yeast type II topoisomerase purified by Goto and Wang. This small stimulation of DNA synthesis by type II topoisomerase is consistent with the fact that only 20-30% of 2-μm DNA replication catalyzed by unfractionated crude extracts is inhibited by antibody against type II topoisomerase (Table 2). By analogy with prokaryotes, it is reasonable to think that various topoisomerases, DNA dependent and/or independent ATPase, and DNA helicase(s) are required for advance of replication forks. If one of these components is missing, the DNA synthesis still occurs although the rate of DNA synthesis might decrease. This might explain why the requirement of type II topoisomerase for 2-μm DNA synthesis in vitro is not so strict. In any case, further proof of the need for a type II topoisomerase for in vitro 2-μm and ARS-1 plasmid DNA replication is required.

Isolation of Preinitiation Complex of Yeast Plasmid DNA Replication

By analogy with prokaryotic systems (37), it is reasonable to assume that yeast DNA replication proteins form a complex which is able to initiate DNA replication (de novo initiation) and/or to elongate DNA strands. One way of understanding complex DNA replication reactions is to isolate and characterize a DNA replication initiation complex. This kind of study has been already initiated and one report has appeared recently (38). To isolate a complex capable of de novo initiation of 2-μm DNA replication, we have used a Bio-Gel A5m column. If the crude extract was incubated with rATP and $MgCl_2$, at 25°C for 2 min and applied on Bio-Gel A5m equilibrated with buffer $A-rATP_0 -MgCl_2$, 2-μm plasmid DNA replication activity was totally excluded as shown in Fig 11. At the same time, other enzymatic activities were co-isolated by the column. The isolated complex is greater than Mr 5,000,000 and very salt-sensitive (50 mM NaCl or $(NH_4)_2SO_4$ destroyed this complex formation (data not shown)). The DNA accelerated the formation of the complex (data not shown). The activity found in the complex closely resembled the observed reaction in crude extract (namely, specific initiation of DNA replication at in vivo origin of 2-μm plasmid and TRP-1 plasmid DNAs, bidirectional and discontinuous DNA replication).

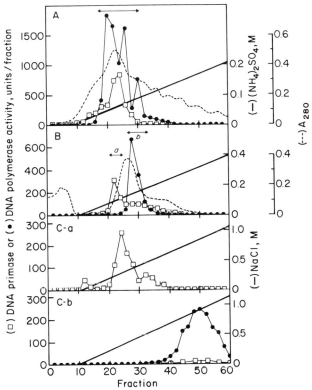

Fig. 10.   Purification of yeast DNA primase.
Crude extract from 500 g of <u>S. cerevisiae</u> 20B-12
(<u>pep4</u>) A364A was applied on ssDNA cellulose column as
before (Fig. 7) and the retained proteins were eluted with
0.5 M NaCl-Buffer A.   The proteins were then dialyzed
against buffer A, applied on a DEAE-Sephadex A-25 column as
Fig. 7, and eluted with 0-0.5 M $(NH_4)_2SO_4$ linear gradient
in Buffer A.   The active fraction indicated in the figure
was pooled, dialyzed against Buffer A, reapplied on phosphocellulose column equilibrated with Buffer A, and eluted
with 0-0.4 M NaCl linear gradient in Buffer A.   DNA polymerase (b) and DNA primase (a) activities were separately
pooled and applied on Heparin-Sepharose column equilibrated
with Buffer A, and eluted with 0-1.5 M NaCl linear gradient
in Buffer A.   A, DEAE-Sephadex A25 column chromatogram; B,
phosphocellulose column chromatogram of DNA primase fraction of A; C-a, Heparin-Sepharose column chromatogram of
pool (a) of phosphocellulose column (A); C-b, heparin-
Sepharose column chromatogram of pool (b) of phosphocellulose column.

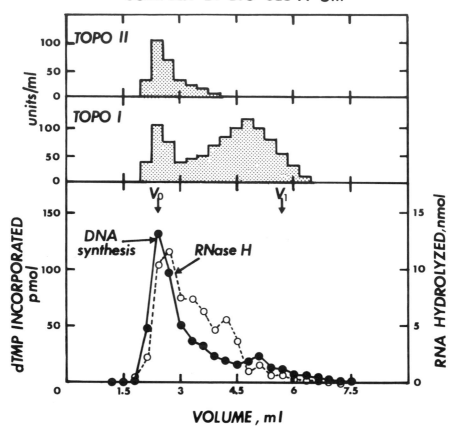

Fig. 11. Isolation of DNA replication complex by Bio-Gel A5m.

Crude extract (0.3 ml) from S. cerevisiae A364A cells was incubated with 1 mM ATP and 10 mM $MgCl_2$ for 2 min at 25°C, applied on the Bio-Gel A5m column (6 ml), equilibrated with buffer A-1 mM ATP-10 mM $MgCl_2$, and the complex was eluted with the same buffer. DNA replication activity was measured using pJDB36 DNA from dam-3 E. coli as Fig. 1, RNase H activity was measured using $poly(dT)_n$. ^{3}H-poly $(rA)_m$ (39), type I DNA topoisomerase activity (35) was measured using supercoiled pMB9 plasmid DNA as a substrate, and type II DNA topoisomerase activity (35) was measured using knotted P4 phage DNA (40). The column was calibrated using T4 phage ($\bar{V}_0$) and ^{3}H-thymidine (V_1).

REFERENCES

1. Arendes J, Kim KC, Sugino, A (1983). Proc Natl Acad Sci USA 80:673-677.
2. Hinnen A, Hicks JB, Fink GR (1978). Proc Natl Acad Sci USA 75:1929-1933.
3. Broach JR (1982). Cell 28:203-204.
4. Stinchcomb DT, Struhl K, Davis RW (1979). Nature 282:39-43.
5. Fangman WL, Hice RH, Chlebowicz-Sledziewska E (1983). Cell 32:831-838.
6. Jazwinski SM, Edelman G (1979). Proc Natl Acad Sci USA 76:1223-1227.
7. Kojo H, Greenberg BD, Sugino A (1981). Proc Natl Acad Sci USA 78:7261-7265.
8. Celniker SE, Campbell JL (1982). Cell 31:201-213.
9. Standenbauer W (1976). Mol Gen Genet 145:273-280.
10. Birnboim HC (1983). Methods Enzymol in press.
11. Sugino A, Kojo H, Greenberg BD, Brown PO, Kim KC (1981). "The Initiation of DNA Replication." New York, Ray and Fox, p 529-553.
12. Peebles CL, Gegenheimer P, Abelson J (1983). Cell 32:525-536.
13. Beggs JD (1978). Nature 275:104-109.
14. Hsiao CL, Carbon J (1979). Proc Natl Acad Sci USA 76:3829-2833.
15. Geier GE, Modrich P (1979). J Biol Chem 254:1408-1413.
16. Razin A, Riggs AD (1980). Science 210:604-610.
17. Wigler MH (1981). Cell 24:285-286.
18. Fehér Z, Kiss A, Venetianer P (1983). Nature 302:266-268.
19. Southern EM (1975). J Mol Biol 98:503-517.
20. Broach JR, Hicks JB (1980). Cell 21:501-508.
21. Newlon CS, Devenish RJ, Suci PA, Rohfis CJ (1981). "The Initiation of DNA Replication." New York, Ray and Fox, p 501-516.
22. Hartley JL, Donelson JE (1980). Nature 286:860-864.
23. Tschumper G, Carbon J (1981). "The Initiation of DNA Replication." New York, Ray and Fox, p 489-500.
24. Tschumper G, Carbon J (1980). Gene 10:157-166.
25. Feldmann H, Olah J, Friedenreich H (1981). Nucleic Acids Res 9:2949-2959.
26. Mann MB, Smith HO (1977). Nucleic Acids Res 4:4211-4221.

27. Mortimer RK, Schild D (1980). Microbiol Rev 44:519-571.

28. Johnston LH, Thomas AP (1982). Mol Gen Genet 186:439-444.

29. Dumas LB, Lussky JP, McFarland EJ, Shampay J (1982). Mol Gen Genet 187:42-46.

30. Plevani P, Chang LMS (1977). Proc Natl Acad Sci USA 74:1937-1941.

31. Conaway RC, Lehman IR (1982). Proc Natl Acad Sci USA 79:2523-2527.

32. Kozu T, Yagura T, Seno T (1982). Nature 298:180-182.

33. Tseng BY, Ahlem CN (1982). J Biol Chem 257:7280-7283.

34. Shioda M, Nelson EM, Bayne M, Benbow RM (1982). Proc Natl Acad Sci USA 79:7209-7213.

35. Goto T, Wang JC (1982). J Biol Chem 257:5866-5872.

36. Maxam AM, Gilbert W (1980. "Methods Enzymol." New York, 65:499-560.

37. Arai K, Low R, Kobori J, Shlomai J, Kornberg A (1981). J Biol Chem 256:5273-5280.

38. Jazwinski SM, Edelman GM (1982). Proc Natl Acad Sci USA 79:3428-3432.

39. Arendes J, Carl PL, Sugino A (1982). J Biol Chem 257:4719-4722.

40. Liu LF, Davis JL, Calendar R (1981). Nucleic Acids Res 9:3979-3989.

Mechanisms of DNA Replication and Recombination, pages 553–562
© 1983 Alan R. Liss, Inc., 150 Fifth Avenue, New York, NY 10011

TERMINI FROM MACRONUCLEAR DNA OF CILIATED PROTOZOANS CAN
PROVIDE TELOMERE FUNCTION FOR YEAST PLASMIDS IN
MITOSIS AND MEIOSIS[1]

Ginger M. Dani, Ann F. Pluta and Virginia A. Zakian

Division of Genetics, Hutchinson Cancer Research Center,
Seattle, Washington 98104

ABSTRACT Circular plasmids containing an ARS
(Autonomously Replicating Sequence) can be maintained
as self-replicating molecules in transformed yeast
cells. In contrast, transformation by linearized
plasmids normally occurs by integration of the molecule
into chromosomal DNA. Termini of naturally occurring
DNA molecules from two different ciliated protozoans,
extrachromosomal ribosomal DNA (rDNA) from Tetrahymena
thermophila and Bam HI restriction fragments from
macronuclear DNA of Oxytricha fallax, enable
recombinant DNA-ARS plasmids to persist as linear,
extrachromosomal molecules in mitotic yeast cells, even
though the sequence and structure of these termini
differ from each other. Moreover, the Tetrahymena rDNA
termini can also provide telomere function for
recombinant DNA plasmids during yeast meiosis.

INTRODUCTION

Eukaryotic chromosomes are linear molecules whose
physical ends (telomeres) are specialized structures which
are essential for the maintenance of chromosome form and
function. The natural ends of all examined linear DNAs are
specialized in some way, presumably, at least in part, to

[1]This work was supported by a grant from the National
Institutes of Health (to V.A.Z.). G.M.D. was supported by
postdoctoral fellowships from the American Cancer Society
and the NIH. A.F.P. is an NIH predoctoral fellow.

provide a solution to the special problems associated with
replication of DNA termini. Certain viruses such as λ
and T7 have sticky or terminally redundant ends and can
therefore form circles or concatamers during replication
(1). Human adeno viruses and Bacillus phage ϕ29 contain
covalently linked proteins at each end of the DNA molecule,
and these proteins are required for viral DNA replication
(2). A hairpin structure is found at the termini of
vaccinia viral DNA, replicating Paramecium mitochondrial
DNA and possibly yeast chromosomes (3-5). The ends of both
the rDNA and bulk macronuclear DNA of Tetrahymena contain
multiple copies of the simple sequence $5'-C_4A_2-3'$ (6,7;
Figure 1). The structure at the extreme end of the rDNA
molecule is believed to be a small hairpin (9). Specific
single base gaps are found every few repeats (5). The
termini of all DNA molecules in the macronucleus of
Oxytricha fallax (12) share a common sequence ($5'-C_4A_4-3'$)
which is related to that found at Tetrahymena rDNA termini
(Fig. 1). However the structure deduced for Oxytricha
termini differs markedly from that seen in Tetrahymena
(Fig. 1).

METHODS

The following strains were used: fH8 (a ade2-1 ade8-18
trp1 ura3-52 leu2-1 his3; from S. Henikoff), ED109-16D (α
ade2-1 ade8-10 ura3; from S. Henikoff), 3482-16-1 (a met2
his3 γ -1 leu2-3 leu2-112 trp1-289 ura3-52; from L.
Hartwell). A derivative of 3482-16-1 called 34 cir^o ρ^o
which lacks both 2um DNA, an endogenous yeast plasmid, and
mitochondrial DNA was constructed by B. Veit and K. Keegan.
Recombinant DNA plasmids used or constructed in this work
are diagrammed in Figure 1. Methods for yeast cell growth,
DNA extractions, and hybridization conditions are described
in refs. 15 and 16. Two dimensional gel electrophoresis
was carried out using the method of Bell and Byers as
described in Zakian and Kupfer (17). For copy number
experiments, DNA extracted from transformed cells was
digested with Eco RI, subjected to electrophoresis in 0.7%
agarose gels, transferred to nitrocellulose, and hybridized
to a nick-translated 834 bp Eco RI-Hind III fragment
containing the chromosomal locus ARS1, a sequence also
found on each of the linear plasmids (Figure 1). To
determine plasmid copy number, the amount of hybridization
to the fragment derived from plasmid DNA was normalized to

A.

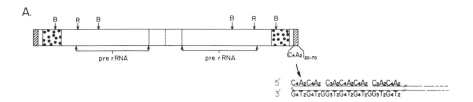

5′ C₄A₂C₄A₂ C₃A₂C₄A₂C₄A₂ C₃A₂C₄A₂
3′ G₄T₂G₄T₂GG₃T₂G₄T₂G₄T₂GG₃T₂G₄T₂

B.

5′ C₄A₄C₄A₄C₄ ▨▨▨ G₄T₄G₄T₄G₄T₄G₄T₄G₄ 3′
3′ G₄T₄G₄T₄G₄T₄G₄T₄G₄ ▨▨▨ C₄A₄C₄A₄C₄ 5′
    ‿‿‿‿‿‿   ‿‿‿‿‿‿   ‿‿‿‿
    16 bases   20 bps   0.4-20 kb

C.

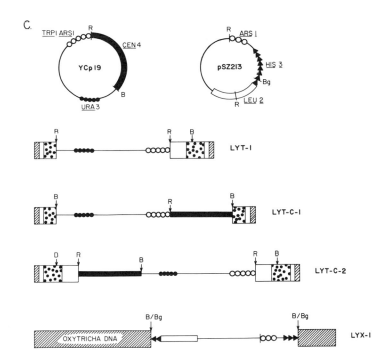

FIGURE 1. Structure of DNAs.

(A) Macronuclear rDNA from Tetrahymena thermophila strain C3V. Tetrahymena macronuclear rDNA exists in extrachromosomal, palindromic molecules of about 20 kb (8,9). The terminal 120 to 420 bps at each end of the molecule are comprised of $5'-C_4A_2-3'$ repeats (hatched region) with single nucleotide gaps every few repeats (6). The structure at the extreme end of the molecule is believed to be a small hairpin (10).

(B) Termini of Macronuclear DNA from Oxytricha fallax. Macronuclear DNA molecules of O. fallax range in size from 0.4 to 22 kb (11). The termini of most or all macronuclear DNA molecules are identical (12): they contain a 16 base single-stranded tail $(G_4T_4)_2$ at both 3' ends followed by a duplex region of 20 bps.

(C) Recombinant DNA Plasmids. YCp19 was obtained from D. Stinchcomb and pSZ213 from J. Szostak (described in refs. 13 and 14, respectively). LYT-1, LYT-C-1, and LYT-C-2 were constructed by ligating terminal restriction fragments from Tetrahymena macronuclear rDNA to linearized YCp19 (or to YCp19 from which CEN4 was removed). LYX-1 was constructed by ligating Bam HI digested total O. fallax macronuclear DNA to Bgl II digested pSZ213. Symbols used in linear molecules are the same as those identified for YCp19 and pSZ213. Recognition sites for restriction enzymes are B, Bam HI; R, Eco RI; and Bg, Bgl II.

that from the single copy 1.4 kb Eco RI fragment containing ARS1 from chromosomal DNA. Meiotic studies were carried out as described in ref. 16.

RESULTS

Natural Termini from the DNA of Ciliated Protozoans can Provide Telomere Function in Mitotic Yeast Cells.

Szostak and Blackburn (14) have shown that terminal restriction fragments from macronuclear rDNA of Tetrahymena allow the maintenance of ARS plasmids as linear molecules in mitotic yeast cells. We have determined that Tetrahymena rDNA termini can also be used to construct small linear centric chromosomes (16). The properties of

these linear mini-chromosomes have been measured to see if
their stability is similar to that displayed by a true
yeast chromosome (16). Two plasmids containing yeast
centromere (CEN) DNA (LYT-C-1 and LYT-C-2, Figure 1) and
Tetrahymena rDNA termini were constructed in vitro. Both
plasmids were maintained as linear molecules in mitotic
yeast cells at about one copy per cell (16). Although both
plasmids were more stable than LYT-1, a linear plasmid
lacking a centromere (Fig. 1), both were three to four
times less stable in mitotic cells than circular CEN
plasmids (16). These results may reflect constraints on
overall chromosome size or on telomere-centromere spacing,
constraints which presumably must be satisfied to achieve
maximal stability of chromosomes in mitosis.

The entire macronuclear genome of the ciliated
protozoan Oxytricha fallax consists of gene-sized pieces
which range in size from about 0.4 to 22 kb (11). All of
the termini of O. fallax macronuclear DNA molecules appear
to share a common sequence and structure (Figure 1). In
order to determine if Oxytricha termini can function as
telomeres in yeast, we ligated Bgl II-digested pSZ213 (Fig.
1) to total Bam HI digested O. fallax macronuclear DNA and
used the ligation mixture to transform yeast strain fH8.
DNA was isolated from individual Leu+ transformants and
examined by Southern blot hybridization using
nick-translated pBR322 as a probe. Two transformants (LYX-1
and LYX-2) displayed hybridization to a single discrete
band, a pattern expected for cells carrying a linear
plasmid.

The linearity of LYX-1 and LYX-2 was confirmed by two
dimensional agarose gel electrophoresis (Figure 2;
conditions described in Figure 2 legend). As seen in the
ethidium bromide stained profile of the second dimension
(Fig. 2B), these conditions permit the separation of linear
molecules from circular DNAs. Most yeast cells contain 50
to 100 copies of a 6.32 kb circular plasmid called 2µm DNA.
Thus, the positions expected for circular DNAs can be
deduced from the arc of DNA produced by the multiple forms
of 2µm DNA. A second arc with mobility greater than that
of circular molecules is defined both by bulk yeast DNA and
by the Bam HI restriction fragments from Adeno2 DNA which
were added as molecular weight markers (Fig. 2). When DNA
in the second dimension gel was transferred to
nitrocellulose and hybridized to nick-translated pBR322
DNA, hybridization was detected to a single locus on the
arc of linear DNA molecules (Fig. 2). A similar

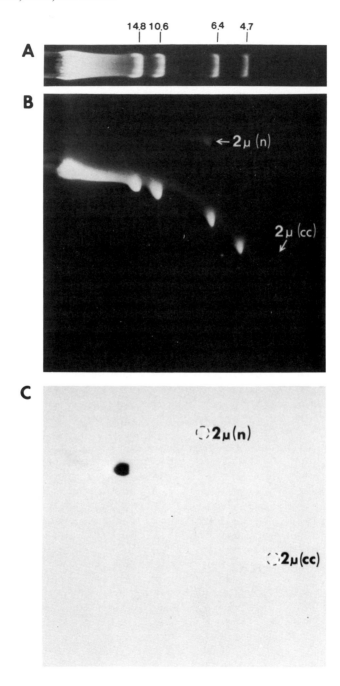

FIGURE 2. Two dimensional agarose gel electrophoresis of total DNA from fH8 cells carrying LYX-1.

(A) Total undigested DNA from fH8/LYX-1 was mixed with Bam HI digested Adeno2 DNA and subjected to electrophoresis in a 0.35% agarose gel containing 0.5 µg/ml ethidium bromide at 1 volt/cm. Sizes are in kilobase pairs.

(B) The lane in panel A was removed from the gel, rotated 90°, embedded in 1% agarose containing 0.5 µg/ml ethidium bromide, and subjected to electrophoresis at 2 volts/cm. The nicked (n) and covalently closed (cc) forms of the endogenous plasmid 2µm DNA are visible in this ethidium bromide stained profile. The bulk of the yeast DNA and the Adeno2 restriction fragments delimit an arc of linear DNA molecules.

(C) The DNA in panel B was transferred to nitrocellulose and hybridized to nick-translated pBR322 DNA. Hybridization is detected only on the linear arc indicating that LYX-1 migrates as a linear DNA molecule.

hybridization pattern was seen for LYX-2 (unpublished data). In contrast, hybridization of pBR322 to DNA fractionated by two dimensional gel electrophoresis from cells carrying the circular plasmid pSZ213 occurs on the arc defined by 2µm DNA (unpublished results). We conclude that LYX-1 and LYX-2 are linear molecules. In experiments to be published elsewhere, we show that LYX-1 and LYX-2 both contain Oxytricha DNA and that a probe homologous to Oxytricha termini hybridizes to both plasmids. Thus, we conclude that Oxytricha termini can provide telomere function in mitotic yeast cells.

The efficiency with which Tetrahymena and Oxytricha termini function in yeast will presumably be reflected in the doubling times of transformed cells growing under selection for maintenance of plasmid DNA. The doubling time for a culture of transformed cells is a complex function of the copy number, replication efficiency, and segregation rate of a given plasmid. We have determined the copy number of three linear ARS plasmids (LYT-1; Tetrahymena ends; LYX-1 and 2; Oxytricha ends) (Fig. 3; Table I). Like circular ARS plasmids, the linear molecules are found in multiple copies per cell. Culture doubling times, plasmid copy number, and percent of cells with plasmid can provide indirect measures of the efficiency of plasmid propagation. By these criteria, we conclude that all three linear plasmids function at least as well as circular <u>ARS</u> plasmids such as pSZ213 (Table I) or YRp12 (16). Indeed our data on copy number and percent of cells with plasmid suggest that unlike many circular plasmids (Table I; refs. 16 & 17), LYX-1 and LYX-2 may often segregate to both progeny cells at mitosis.

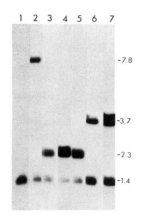

Figure 3: DNA isolated from various strains was digested with Eco RI, subjected to electrophoresis in 0.7% agarose gels, transferred to nitrocellulose, and hybridized with a nick-translated fragment containing <u>ARS1</u>. DNAs are from untransformed 3482-16-1 (lane 1), 3482-16-1/LYT-1 (lane 2), 3482-16-1/LYX-1 (lane 3), 34cir^{o} ρ^{o}/LYX-1 (lane 4), 3482- 16-1/LYX-2 (lane 5), 3482-16-1/pSZ213 (lane 6), 34cir^{o} ρ^{o}/pSZ213 (lane 7). Sizes are in kilobase pairs.

Tetrahymena rDNA Termini can Provide Telomere Function During Yeast Meiosis.

During meiosis circular <u>CEN</u> plasmids like YCp19 segregate predominately 2+:2- and usually do so at the first meiotic division (13,18). We have shown that LYT-C-2 functions at least as well as YCp19 in meiosis (16). Briefly 3482-16-1 cells carrying LYT-C-2 were mated to ED-109-16D cells. The diploids were sporulated and tetrads analyzed for presence of plasmid DNA (Ura+ phenotype): in 62% of the tetrads which contained plasmid (i.e., excluding 0+:4- tetrads), LYT-C-2 segregated 2+:2- (compared to 45%

TABLE 1
MITOTIC PROPERTIES OF TRANSFORMED CELLS

Strain	dt	% cells with plasmid	Copies/cell
3482-16-1/LYT-1	2.7	40 $\pm$ 3	4 $\pm$ 0.3
3482-16-1/LYX-1	2.2	59 $\pm$ 9	5 $\pm$ 1
34cir$^{\circ}$ ρ°/LYX-1	2.1	72 $\pm$ 2	10 $\pm$ 2
3482-16-1/LYX-2	2.6	96 $\pm$ 3	6 $\pm$ 2
3482-16-1/pSZ213	2.3	33 $\pm$ 3	8 $\pm$ 0.5
34cir$^{\circ}$ ρ°/pSZ213	4.3	11 $\pm$ 5	33 $\pm$ 6

The doubling time (dt) in hours of each strain was determined at 30°C in complete medium (17) lacking either uracil (LYT-1) or leucine (rest of plasmids). For comparison, 3482-16-1 and 34cir$^{\circ}$ ρ° have doubling times of 1.7 and 1.9 hrs, respectively in complete medium. 3482-16-1/LYT-1 cells were maintained in log phase growth, sonicated briefly and then spread on both complete plates or complete minus uracil to determine the percent of cells with plasmid. DNA was extracted and plasmid copy number determined (Fig. 3). For the five other strains, cells were grown to early stationary phase (O.D. 660 = 7) in complete medium lacking leucine. Cells were then plated to determine the per cent of cells with plasmid and DNA extracted and used for copy number measurements. Copy numbers are presented for those cells in the culture which actually contain plasmid. Standard deviations are given for both percent of cells with plasmid and plasmid copy number.

2+:2- tetrads for YCp19). Moreover, plasmid DNA extracted from colonies derived from spores in which the plasmid had undergone 2+:2- segregation displayed the same size and pattern of digestion with restriction enzymes that was present prior to mating. Similar results were found with LYT-C-1. Therefore we conclude that LYT-C-1 and LYT-C-2 segregate properly and are transmitted apparently unaltered during yeast meiosis.

DISCUSSION

A variety of solutions to the problems of replication and maintenance of chromosome termini appear to be utilized in eukaryotic cells. We have shown that termini from two different ciliated protozoans, termini which differ from each other in both sequence and structure, can provide telomere function for recombinant DNA plasmids in yeast. Further studies will be necessary to determine which structural features are essential for telomere function in yeast.

REFERENCES

1. Watson JD (1972). Nature New Biol 239:197.
2. Wimmer E (1982). Cell 28:199.
3. Geshelin P, Berns KI (1974). J Mol Biol 88:785.
4. Pritchard AE, Cummings DJ (1981). Proc Natl Acad Sci USA 78:7341.
5. Forte MA, Fangman WL (1979). Chromosoma 72:131.
6. Blackburn EH, Gall JG (1978). J Mol Biol 120:33.
7. Yao M-C, Yao C-H (1981). Proc Natl Acad Sci USA 78:7436.
8. Engberg J, Anderson P, Leick V, Collins J (1976). J Mol Biol 104:455.
9. Karrer KM, Gall JG (1976). J Mol Biol 104:421.
10. Blackburn EH, Chiou S-S (1981). Proc Natl Acad Sci USA 78:2263.
11. Lawn RM, Heumann JM, Herrick G, Prescott DM (1978). Cold Spring Harbor Symp Quant Biol 42:483.
12. Pluta AF, Kaine BP, Spear BB (1982). Nucl Acids Res 16:8145.
13. Stinchcomb DT, Mann C, Davis RW (1982). J Mol Biol 158:157.
14. Szostak JW, Blackburn EH (1982). Cell 29:245.
15. Zakian VA, Scott JF (1982). Mol Cell Biol 2:221.
16. Dani GM, Zakian VA (1983). Proc Natl Acad Sci USA 80: in press.
17. Zakian VA, Kupfer DM (1982). Plasmid 8:15.
18. Clarke L, Carbon J (1980). Nature 287:504.

Mechanisms of DNA Replication and Recombination, pages 563–580
© **1983 Alan R. Liss, Inc., 150 Fifth Avenue, New York, NY 10011**

XENOPUS EGGS AS A MODEL-SYSTEM FOR
DNA REPLICATION IN EUKARYOTES

Marcel Méchali, Richard Harland, and Ronald Laskey

Laboratory of Molecular Biology, MRC Centre,

Hills Road, Cambridge CB2 2QH, England.

ABSTRACT   In vitro and in vivo studies on DNA
replication using Xenopus eggs as a model system are
described.   In vitro, a cell-free system derived from
the eggs of the frog Xenopus laevis, reproduces the
enzymatic events presumed to act at the replicative
fork during chromosomal DNA replication.   Complete
complementary   strand   synthesis   occurs   with
exceptional efficiency on single stranded circular
DNA templates.   Priming of DNA synthesis is made by
an RNA primer of 10±1 nucleotides, synthesised by a
DNA primase-like enzyme.   DNA polymerase-α is
responsible   for   the   elongation   process.   ATP
participates both in initiation and elongation of
nascent DNA chains.   RNase, ligase, and chromatin
assembly activities allow the final synthesis of a
supercoiled covalently closed molecule.   Initiation of
DNA synthesis or double stranded circular DNA is
undetectable however.
    In vivo, we studied the regulated replication of
microinjected DNA in Xenopus eggs, in the presence of
the tumour promoter TPA.   TPA induced an increase in
the frequency of initiation of injected DNA, through
a membrane-mediated process.   No effect was observed
when TPA was injected directly into the cell.   TPA
appears to act through a structure precommitted to
DNA replication and is not itself a trigger for
replication
    In view of these overall results, it is possible
that initiation of double stranded DNA replication in
eukaryotes   is   limited   by   a   specific   cellular
structure.   Such   a   structure   could   be   lost   or

strongly disturbed by membrane alteration or disruption. This structural organisation would be less important for the process of elongation of DNA chains.

INTRODUCTION

Analysis of the mechanism and control of chromosomal DNA replication in eukaryotic cells requires the examination of some essential parameters in the choice of an experimental system with the following properties. First, cell division and DNA synthesis must be sufficiently amplified to detect replicating structures and the different components involved in DNA replication. Second, due to the complexity of the cellular genome, we have to use small DNA probes which are able to replicate in the cell system chosen, and whose DNA synthesis is preferably coupled to the host DNA synthesis. Third, good synchrony of cell division will be necessary to analyse the sequence of events.

The use of Xenopus eggs as a model system appears to respect all these major requirements. We describe here results from in-vitro as well as in-vivo studies using this system. In-vitro we use extracts from Xenopus eggs to derive a system for studying DNA replication in eukaryotes. This system appears to reproduce the events occurring at the replicative fork during chromosomal replication. In-vivo, we show that Xenopus eggs can be used to study the control of initiation of replication. We describe the effect of a potent tumor promoter, the phorbol diester TPA, on the control of replication of microinjected DNA into the egg.

RESULTS

An in-vitro system for DNA replication in eukaryotes.

Eggs of <u>Xenopus</u> <u>laevis</u> cleave every 35 min during early development, a rate comparable with the rate of cleavage of bacteria. We used this unusual potential for cell division to derive a cell free system for DNA replication. An egg extract free of double stranded or single stranded nuclease activity was prepared from homogenised eggs by differential centrifugation (1).

When double stranded DNA was incubated in the egg extract a very low background DNA synthesis is observed The same result was obtained with extracts from either unfertilised eggs or fertilised cleaving eggs (unpublished results). Agarose gel electrophoresis shows products migrating at the position of form I, and II DNA. However, density labelling experiments show that these products were due to a repair reaction. In contrast single stranded M13 DNA was a very efficient template for DNA synthesis (Table 1).

TABLE 1
TEMPLATE UTILISATION IN VITRO

Template	DNA synthesis pmol dTMP incorporated/hr
M13 DNA SS	68
M13 DNA RF	0.8
pBR322 DNA	1
_ DNA	0.9
SV40 DNA	1.4
Denatured calf thymus DNA	15
Activated calf thymus DNA	9

Various DNA templates (10 µg/ml) were incubated with 15µl extract in a total volume of 20 µl. 10 µl aliquots were taken at 30 and 60 min for determination of acid insoluble material (1). Incorporation with calf thymus DNA is resumed within the first 30 min incubation.

The product from a single stranded circular template was full-length double stranded DNA form I, the supercoiled molecule, and form II and III, the circular and linear molecules (1). The fate of parental DNA converted to the double-stranded form was determined with a prelabelled M13 template. As shown in Figure 1 this in-vitro system allows the complete conversion of parental ^{32}P-DNA to the double-stranded forms. Alkaline gels done in parallel show that the reaction did not arise by priming on DNA hairpin from broken single stranded circular molecules. The complete dependance on the presence of ribonucleoside triphosphates (see below) confirms this result.

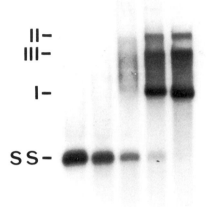

FIGURE 1. Fate of parental DNA during M13 DNA synthesis. M13 (^{32}P) DNA, 10 μg/ml, was incubated in the extract and aliquots analysed by agarose gel electrophoresis (1).

The rate of DNA synthesis in-vitro is very high. 100 μl of the undiluted egg extract can convert up to 1.5 μg M13 DNA into a double stranded form in 1 hr at 22°C. This corresponds to 1.5 ng DNA made by the extract equivalent to 1 egg in 1 hr at 22°C. This data is a minimum estimate obtained by choosing incubation conditions which are closest to the in-vivo conditions.

The kinetics of the in-vitro reaction show that a significant lag is observed before DNA replication takes place (Figure 2). This lag could correspond to the assembly of the components necessary for DNA replication of the template. The sigmoid character of the curve is consistent with the saturation of a multi factor system necessary for the priming event.

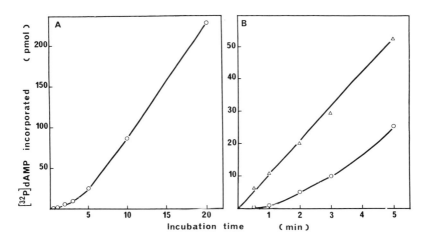

FIGURE 2. Lag in incorporation of dNTPs at the beginning of in vitro replication. The extract was preincubated and reaction started by addition of 12 µg/ml M13 DNA and (^{32}P)dATP (o--o). In B, the same experiment was done, except that (^{32}P)dATP was added 10 min after the addition of DNA (Δ--Δ).

The products synthesised were also characterized by a density labelling procedure in the presence of BrdUTP (Figure 3). The DNA synthesised migrates at the position of heavy light hybrid molecules containing one parental "light" strand and one "daughter" "heavy" strand after one round of replication. No heavy-heavy molecules were detected and this indicates that replication stops when the double stranded molecule RFI is synthesised. The lack of replication with a double stranded molecule is confirmed by the analysis of the reaction products made in presence of SV40 DNA as template. Very low incorporation occurs (Table 1) and the products migrate at the position of light-light DNA (Figure 3B). This result is totally consistent with a repair reaction, introducing the labelled precursor into nicks or small gaps present in the original parental SV40 DNA preparation.
We tried to determine if a specific start was used for DNA replication on the M13 DNA by pulse labelling of the DNA I synthesised (2), or by alkaline gel and poly-

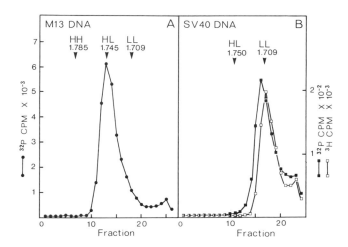

FIGURE 3. Equilibrium density gradients of DNA products made in presence of single stranded or double stranded DNA. A) single stranded M13 DNA. B) double stranded (^{3}H) SV40 DNA. Incubation was in presence of (^{32}P) dCTP and BrdUTP.

acrylamide urea gel electrophoresis of nascent DNA. The results showed that the replication start points on the template were not unique. However populations of nascent DNA chains of discrete size length were detected. This suggests that multiple discrete initiation sites and probably stop sites are preferentially used on the template.

DNA polymerase-α is involved in complementary strand synthesis.

DNA polymerase-α was the major enzyme present in the extract, together with a small amount of DNA polymerase-β (1). The nature of the functional enzyme involved in complementary strand synthesis in the in-vitro reaction was investigated by testing the effect of a wide variety of known inhibitors. Drugs that inhibit DNA polymerase-α, like aphidicolin and N-Ethylmaleimide, inhibit in vitro DNA synthesis (Figure 4A and B). A similar result was observed with arabinofuranosyl cytosine 5'-triphosphate (ara CTP). In contrast when a known inhibitor of DNA polymerase-β or -γ, the dideoxynucleotide ddTTP, was used, no effect was observed (Figure 4C). Taken altogether these results indicate that DNA polymerase-α is the enzyme responsible for complementary-strand synthesis.

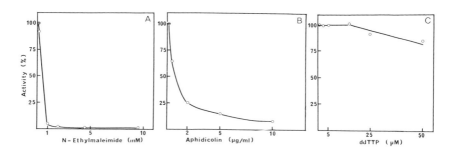

FIGURE 4. Effect of DNA polymerase inhibitors. M13 DNA was incubated with (^{32}P)dATP in the extract with the inhibitors shown.

Priming of DNA synthesis by the synthesis of a decaribonucleotide.

The ribonucleotide triphosphate dependance of the system was assayed with a dialysed extract (1). DNA synthesis is completely dependant on the addition of both dNTPs and rNTPs. In presence of one or three ribonucleoside triphosphates, very limited DNA synthesis occurs after a lag (1). This could indicate partial deoxyribonucleotide substitution for the formation of the primer RNA (3). The system requires a relatively high concentration of ATP, maintained by the addition of an ATP regenerating system in the medium (1). In the absence of ATP, only a few DNA chains are initiated and accumulate as short nascent DNA molecules. (Figure 5). Therefore in addition to participation in synthesis of the primer RNA, ATP is probably also needed for the elongation reaction. This conclusion is confirmed by the inhibition of the reaction by the ATP analogs β-α imido adenosine 5' triphosphate (AMP-PNP) and α-β methyladenosine 5' triphosphate (AMP-CPP).

The RNA primer was further characterised by analysis of the nascent DNA chains. After a short pulse labelling, nascent DNA chains were purified and fractionated by polyacrylamide-urea gel electrophoresis (1). Short DNA

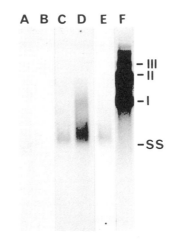

FIGURE 5. rNTP dependance of in vitro DNA synthesis. M13 DNA, 10 µg/ml, was incubated with the dialysed extract, dNTPs, and (^{32}P)dATP under various conditions. A) With dNTPs. B) with rNTPs. C) with dNTPs + GTP. D) with dNTPs + ATP. E) with dNTPs + GTP + UTP + CTP. (F) with dNTPs + rNTPs. Analysis was done by agarose gel electrophoresis.

chains of defined length were isolated and treated with sodium hydroxide, RNAase T2 or RNAase A. As shown in Figure 6, 40 to 50% of the population of nascent DNA chains were shortened by 9 nucleotides after these treatments. The new faster moving band after RNA degradation has the same length range as the original population of nascent DNA. This result demonstrates the unique length of the RNA primer used for DNA replication.
    RNA priming was also analysed by direct labelling of RNA primers, fractionation of early replicative intermediates, and size analysis of the primer RNA after DNase digestion. The size of the RNA primer was found here to be 10 to 11 nucleotides. The presence of RNA primers at the 5' end of the DNA was further assayed by the spleen exonuclease assay (4). Nascent DNA chains of 160-350 nucleotide length were purified and isolated. They were phosphorylated with polynucleotide kinase and ATP and further treated with alkali. Hydrolysis of the RNA primer would leave an OH group at the 5' end of the DNA. The presence of 5' OH DNA was then revealed by

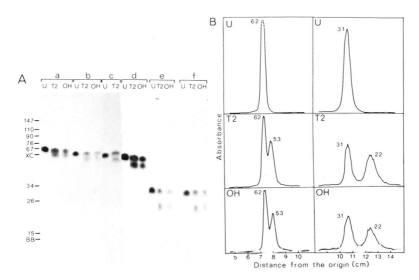

FIGURE 6. Analysis of RNA primers from nascent DNA chains. Single stranded nascent DNA of defined length was digested with either RNAse T2 or NaOH, and further subjected to polyacrylamide-urea gel analysis (1). A) autoradiograph of the gel. Lanes U: untreated sample. Lane T2: after RNase T2 treatment. Lanes OH: after alkali treatment. Position of markers in the range of 15 to 197 nucleotides are indicated. B) microdensitomer scans of the gel containing the 62 nucleotide or 32 nucleotide DNA chains before (U) and after treatment with RNase T2 (T2) or alkali (OH).

susceptibility to spleen exonuclease. We found that 50 to 70% of nascent DNA chains contained RNA at their 5' ends, in agreement with previous results.

To obtain more information about the priming mechanism, drugs affecting transcription were assayed. Actinomycin D inhibits replication in vitro at relatively low concentrations (Figure 7A). More interesting is the lack of inhibition by α-amanitin, a potent inhibitor of RNA polymerase II and III (5). These results suggest the participation of a eukaryotic primase in the in-vitro reaction. Since the DNA polymerase-α contains an associated primase activity (6), it could be involved in the process in-vitro. However other possibilities still remain.

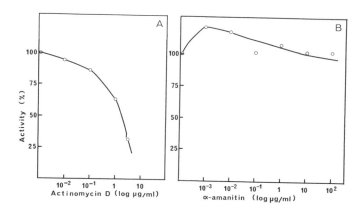

FIGURE 7. Effect of RNA polymerase inhibitors. Extracts were incubated in the presence of M13 DNA and various concentration of actinomycin D or α-amanitin. Activity of α-amanitin was checked both in vitro and in vivo in known transcription systems (1).

The use of this cell-free system to study eukaryotic DNA replication and its fidelity.

All the properties of the cell-free system discussed here are shared with the mechanism of replication in chromosomes (7 for review). This system may be easily manipulated to study events of initiation and elongation, and therefore may provide new insights into eukaryotic replication. This system can be also used to study the factors necessary for fidelity of the replication. Using a bio-assay to examine the progeny of the newly replicated strand, the error rate during complementary strand synthesis was determined (8). Single stranded viral DNA from an amber mutant of ϕX174 was replicated in the cell-free system, and the error rate was determined from the frequency at which wild-type phage are formed on transfection of E. coli spheroplasts. The frequency of revertants was found to be 2.3 to 3.3 x 10^{-4}, a value close to that expected in the absence of error correction. Conditions necessary for an increase in the fidelity observed may allow a reconstitution of the different steps necessary for accuracy of DNA replication at the replicative fork.

Microinjection of DNA into Xenopus eggs: a system for studying regulation of DNA replication in-vivo.

Although double stranded DNA replication cannot still be studied in the in-vitro system derived from Xenopus eggs, these cells allow the replication in vivo of microinjected DNA. Indeed microinjected double stranded DNA replicates in-vivo with a relatively high efficiency (9,10). The ability to initiate replication on double stranded molecules arises after hormonal maturation of the oocyte into an egg (10). The replication of injected DNA follows the timing of replication of the endogenous nucleus and is subject to cellular control, so that DNA is replicated only once within the same cell cycle (10). A wide variety of DNA molecules can replicate efficiently in the eggs and efficiency of replication appears to a large extent proportional to the length of the molecules (unpublished observations). Since the frequency at which injected DNA is replicated is tightly controlled by the host cell, factors which might alter this control can be assayed. We used this capacity to test a potent tumour promoter, TPA, for stimulation of initiation of DNA replication.

The tumour promoter TPA increases frequency of initiation of replication of microinjected DNA.

Tumour promoters such as the phorbol diester TPA are not themselves mutagenic or carcinogenic, but they markedly increase the incidence of neoplastic transformation and tumours when applied after an initiating carcinogen (11). Recently it was proposed that TPA was acting by facilitating gene amplification (12), in accordance with its induced cell proliferation and tumour promotion. This phenomenon could alter gene dosage and would create conditions for expression of an initial transformed gene. Such gene amplification could be the result of an increase in the frequency of replication of a DNA molecule during the cell cycle. We tested the effect of TPA on the level of initiation of replication by analysis of the replication of microinjected DNA in eggs incubated in presence of TPA. We injected a large DNA, the phage λ DNA in order to detect more easily any

increase in DNA replication induced by TPA. Figure 8
shows that the replication of microinjected DNA increases
in eggs in the presence of TPA, in a dose dependant
manner. The TPA concentrations used are among the lowest
shown to be biologically active in-vivo (11). Agarose gel
electrophoresis analysis shows that this stimulation is
due to effective λ DNA synthesis (13).

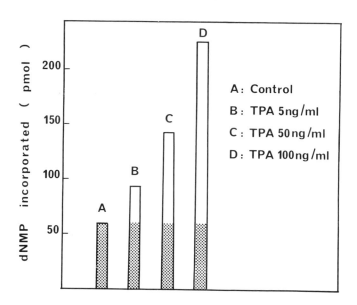

FIGURE 8. Dose-dependent stimulation of in vivo DNA
replication by TPA. Groups of 50 eggs were injected with
λ DNA and (^{32}P)dATP, and incubated in the presence of
various concentrations of TPA. Total dNMP incorporated
was derived from acid insoluble material.

The reaction products were also analysed by density
labelling in the presence of BrdUTP, followed by CsCl
equilibrium density gradients (Figure 9 A and B). Panel A
shows the result obtained in the absence of TPA. The DNA
migrated at the density of "heavy-light" (HL) molecules
which had replicated once, and "heavy-heavy" (HH)
molecules, which had replicated twice or more. In the
presence of 100 ng/ml TPA (panel B) an increased
incorporation occurs in both HL and HH molecules. This
clearly indicates that TPA induced DNA synthesis was true
semi conservative replication and not a repair reaction.

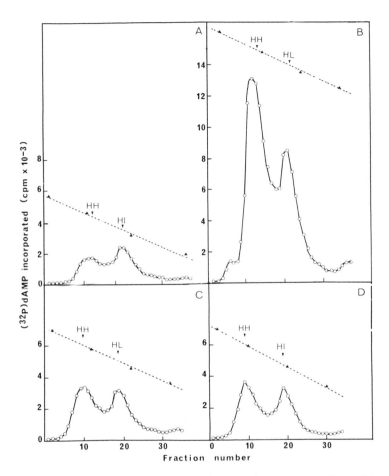

FIGURE 9. Effect of TPA in the external medium or injected into the cell. In A and B, eggs were injected with λ DNA, ^{32}P(dATP) and BrdUTP. They were further incubated in the absence (A), or in the presence of 100 ng/ml TPA (B). In C and D, eggs were injected with λ DNA, (^{32}P)dATP and BrdUTP, without (C), or with addition of 1μg /ml TPA (D). They were further incubated in a medium without TPA. Analysis of the purified DNA products was by CsCl equilibrium density gradients. (Δ – – Δ) density gradient. The density of HH and HL peaks is 1.788 and 1.761 respectively.

This result also indicates that TPA induced replication was the result of two phenomena. First, TPA increases the percentage of parental molecules entering the replication pool (HL DNA). Second, TPA increases the probability of reinitiation of a molecule which has already replicated (HH DNA). This second phenomenon seems more effective than the first one, since the ratio of HH DNA to HL DNA increases as a result of TPA induction.

Application of TPA outside the cell membrane is necessary for stimulation of DNA replication.

Time course experiments indicated that no effect of TPA was detected before the first cycle of DNA replication (13). Since TPA is a lipophilic molecule, it could be delayed in entering the cell. This possibility was tested by microinjection of TPA together with DNA into the egg. Surprisingly no effect on DNA replication was then observed, even if TPA was injected at concentrations 20 to 50 times more than those previously used in the external medium (Figure 9 C and D). Thus TPA increases the frequency of initiation of replication of microinjected DNA through a membrane mediated process. This stimulation is not a general property of the phorbol family since TPA analogs which are not tumour promotors, like phorbol or 4-0 methyl TPA, do not stimulate DNA replication (13).

TPA acts on a structure already precommitted to DNA replication.

Cycloheximide blocks protein synthesis, the cell cycle clock, and the reinitiation of replication of a DNA molecule which has already replicated once (10). In the presence of cycloheximide the frequency of initiation of the injected DNA is still increased by TPA (13). However no reinitiation is observed. Therefore TPA, although able to stimulate initiation of replication, cannot bypass the need for protein synthesis in reinitiation of replication.
    The ability to induce DNA replication arises only after hormonal maturation of the oocyte into an egg (10). TPA cannot induce oocyte maturation, nor the replication of DNA injected into an oocyte (13). Taken together these results indicate that TPA alone cannot commit a quiescent cell to DNA replication. However TPA is able to alter the regulation of DNA replication of a cell which is already engaged or precommitted to the process of cell division.

This interpretation is in accordance with the synergistic interaction between TPA and growth factors in stimulating DNA synthesis (14).

DISCUSSION

We have shown that a cell free system derived from Xenopus eggs allows replication of a single stranded DNA molecule with an unusual efficiency. Events acting in-vitro are those presumed to act in-vivo at the replicative fork during chromosomal DNA replication. DNA polymerase-α is the enzyme required for the polymerisation reaction. RNA primers of unique length 10 ± 1 nucleotides are used to initiate the replication process, in a reaction which is α-amanitin insensitive. ATP is required for both the initiation and elongation reactions. RNase, ligase and chromatin assembly activities allow the formation of a supercoiled covalently closed molecule.

In contrast with the efficiency of the system for replicating single stranded DNA, no replication was observed with double stranded DNA, even with extracts from cleaving eggs. This striking discrepancy adds to the repeated failures to initiate DNA replication in numerous eukaryotic in-vitro systems. In some cases initiation of replication in-vitro was inferred by electron microscopy experiments (15-17). However, true semi-conservative DNA replication was not shown by density labelling experiments, and this field remains controversial. Some of the most detailed in vitro analyses of bidirectional eukaryotic replication are from studies on SV40 or polyoma virus DNA replication in-vitro (7). Even in these systems, where essential components of the replicon (18) like the origin of replication and the initiating protein are extensively characterised, the initiation of replication in-vitro is not detected in spite of numerous efforts in many laboratories. In egg extracts the exceptional efficiency of complementary strand DNA synthesis and the lag observed before replication, suggest that a multicomponent structure is acting in-vitro, and that its formation is a limiting factor in this system. The absence of specific component(s) necessary for double stranded replication or their inactivation during the preparation of the in-vitro system can explain the results observed. Another possibility could be that all components necessary for initiation of DNA replication are present but a specific organisation is lost during homogenisation of the cell. In accordance with this hypothesis it is observed that initiation of double stranded replication appears to involve the same events used to initiate the synthesis of Okazaki fragments during the elongation reaction (19). These events act in-vitro in most characterised eukaryotic systems and are particularly efficient in the egg extract. However this is not sufficient to observe initiation of double stranded DNA replication in-vitro.

The hypothesis of a specific organisation for initiation of DNA replication is also suggested by the effect of the tumour promoter TPA on DNA replication in-vitro. We showed that TPA induced stimulation of replication of micro-injected double stranded DNA. The mechanism of induction appears to be an increase in the frequency of initiation of replication through a membrane mediated process. No effect was observed by direct injection of the tumour promoter inside the cell. Our other observations suggest that TPA acts through a structure which is already committed to DNA replication and is not itself a trigger for DNA replication. Injection of DNA into the egg is followed by rapid assembly into chromatin. Organisation at a higher level of structure could be possible in view of in-vitro experiments analysing the sedimentation complex of DNA incubated with an egg extract (U.K. Laemmli, personal communication and unpublished results). The presence of a specific organisation for DNA replication is also suggested by other observations. Only a fraction of injected DNA is used for initiation of replication and is then preferentially used during the following cycles (10,13). The total amount of DNA synthesised is then similar to the amount of DNA normally synthesised during the same time in fertilised eggs.

In view of our results, together with observations made in other laboratories, we suggest that initiation of replication depends on a specific structural organisation of the whole cell, rather than on isolated cell components. Interaction of TPA with the cytoplasmic membrane could alter this structure so that the control of the frequency of initiation is altered. A close relationship between the cell membrane and this specific organisation could be important for restriction of initiation of replication to only one round of DNA synthesis per replicon per cell cycle. Factors disturbing this relationship like tumour promoters, or possibly gene products from transforming viruses, could then lead to a partial uncoupling between cell division and DNA replication. Such a critical structure could explain the lack of initiation of double stranded DNA outside in-vivo conditions, even if all the individual components necessary are present in an active form. This critical structure would not be important for the elongation reaction and the synthesis of Okazaki fragments, which are observed at high efficiency in-vitro. On the other hand, the destruction of this structure would completely abolish initiation on double stranded DNA even if specific components, like for example T antigen for papovaviruses, are still present in the reaction medium. Finally a model involving a specific cell structure for initiation of replication is also tempting to explain how a cell can discriminate between unreplicated and replicated DNA within one cell cycle. If the replicated DNA escapes from this structure as a direct consequence of its replication, this would give a very efficient way to control initiation of replication. Obviously this hypothesis remains to be tested, as well as the nature of the transmembrane signal generated by the interaction of TPA with the membrane.

REFERENCES

1. Méchali M, Harland R (1980). DNA synthesis in a cell-free system from Xenopus eggs: priming and elongation on single-stranded DNA in vitro. Cell 30:93-101.
2. Danna KJ, Nathans D (1972). Bidirectional replication of simian virus 40 DNA. Proc Natl Acad Sci USA 69:3097-3100.
3. Eliasson R, Reichard P (1978). Primase initiates Okazaki pieces during polyoma DNA synthesis. Nature 272:184-185.
4. Kurosawa Y, Ogawa T, Hirose S, Okazaki T, Okazaki R (1975). Mechanism of DNA chain growth. RNA-linked nascent DNA pieces in Escherichia Coli strains assayed with spleen exonuclease. J Mol Biol 96:653-664.
5. Chambon P (1975). Eukaryotic nuclear RNA polymerases. Ann Rev Biochem 44:613-638.
6. Conaway RC, Lehman IR (1982). A DNA primase activity associated with DNA polymerase-α from Drosophila melanogaster embryos. Proc Natl Acad Sci USA 79:2523-2527.

7. De Pamphilis ML, Wassarman P (1980). Replication of eukaryotic chromosomes: a close up of the replication fork. Ann Rev Biochem 49:627–666.
8. Coles AM, Méchali M, Fersht AR (1983). DNA polymerase-α does not proofread in a eukaryotic cell extract. Submitted for publication.
9. Gurdon JB, Birnsteil ML, Speight VA (1969). The replication of purified DNA introduced into living egg cytoplasm. Biochim Biophys Acta 174:614–628.
10. Harland R, Laskey R (1980). Regulated replication of DNA microinjected into eggs of Xenopus laevis. Cell 21:761–771.
11. Diamond L, O'Brien TG, Baird WM (1980). Tumor promoters and the mechanism of tumor promotion. In Advances in Cancer Research 32:1–73.
12. Varshavsky A (1981). Phorbol ester dramatically increases incidence of methotrexate-resistant mouse cells: possible mechanisms and relevance to tumor promotion. Cell 25:561–572.
13. Méchali M, Méchali F, Laskey R (1983). Phorbol ester TPA stimulates replication of DNA microinjected into Xenopus eggs through a membrane mediated process. Submitted to publication.
14. Dicker P, Rozengurt E (1978). Stimulation of DNA synthesis by tumour promoter and pure mitogenic factors. Nature 276:723–726.
15. Benbow RM, Krauss MR, Reedes RH (1978). DNA synthesis in a multi-enzyme system from Xenopus leavis eggs. Cell 12:191–204.
16. Chambers JC, Watanabe S, Taylor JK (1982). Dissection of a replication origin of Xenopus DNA. Proc Natl Acad Sci USA 79: 5572–5576.
17. Celniker SE, Campbell JL (1982). Yeast DNA replication in-vitro: initiation and elongation events mimic in vivo processes. Cell 31:201–213.
18. Jacob F, Brenner S, Cuzin F (1963). On the regulation of DNA replication in bacteria. Cold Spring Harb Symp Quant Biol 28: 329–348.
19. Hay RT, De Pamphilis ML (1982). Initiation of SV40 DNA replication in vivo: location and structure of 5' ends of DNA synthesized in the ori region. Cell 28:767–779.

Mechanisms of DNA Replication and Recombination, pages 581–593
© 1983 Alan R. Liss, Inc., 150 Fifth Avenue, New York, NY 10011

THE INITIATION OF MAMMALIAN MITOCHONDRIAL DNA REPLICATION AT TWO DISTINCT ORIGINS[1]

Douglas P. Tapper and David A. Clayton

Department of Pathology
Stanford University School of Medicine
Stanford, California 94305

ABSTRACT Mammalian mitochondrial DNA (mtDNA) contains two distinct origins of DNA replication. The heavy (H) strand origin of replication is maintained as a stable displacement loop (D-loop) structure in mtDNA. The 5' ends of D- loop strands have been precisely located on the mouse mtDNA genome. The same distribution of 5' ends can be demonstrated on nascent H strands isolated from replicative intermediates. The 5' ends of nascent light (L) strands from mouse mtDNA replicative intermediates are located in a 32 nucleotide region separated from the H strand origin by 67% of the genome in the direction of H strand synthesis. The 5' ends of both nascent H and L strands are characterized by a single abundant species with a clustering of minor 5' ends around the dominant 5' end. The presumptive initially synthesized residues at both origins have been identified by the specific labeling of 5' triphosphate residues on nascent strands. For nascent H strands, these positions are found from approximately 10 to 165 nucleotides upstream from the major 5' end previously identified and, for nascent L strands, from approximately 43 to 80 nucleotides upstream from the major 5' end of processed L strands.

[1]This work was supported by NIH grant CA-12312 and American Cancer Society, Inc., grant NP-9.

INTRODUCTION

The replication of mammalian DNA (mtDNA) proceeds by an asynchronous mode of replication whereby the origin of H (heavy) strand replication is separated both temporally and spatially from the L (light) strand origin of replication (see review by Clayton (1)) (Fig. 1). These origins, being distinct both structurally and mechanistically, and presumably representing divergent evolution, act in concert to replicate the circular mtDNA molecule. This report describes our current state of knowledge of these two origins of DNA replication.

RESULTS AND DISCUSSION

The Origin of Light Strand Replication.

The origin of light strand replication (O_L) is only activated when the nascent heavy strand traverses this region, which is located approximately 10 kilobases (kb) away (in the direction of H–strand synthesis) from the H–strand origin of replication (2). The precise origins of L strands have been determined in mouse (3,4) and human (5) mtDNA. The 5' ends of nascent L strands from both mouse and human mtDNA map to positions within a unique stem and loop structure that lies among a cluster of five tRNA genes. This region of about 36 nucleotides has no known coding function and is more divergent than the surrounding sequences that code for tRNA genes. In spite of this divergence, all changes have preserved a striking stem and loop structure that remains conserved in mouse, human and bovine mtDNA (6,7,8).

The precise location of 5' ends from human mtDNA have been mapped by 5' end labeling nascent L strands with subsequent cleavage by restriction endonucleases. These experiments demonstrated two unique 5' ends: one at the base of the stem defining O_L and the other within the anticodon stem and loop of tRNA^{Cys}. $_L$ Both ends have 5' dAMP residues. There is no evidence of any ribonucleotides either at the 5' end or located internally. In contrast to these results, the 5' ends of nascent L strands from mouse mtDNA, determined in an identical way, were distributed heterogeneously around a major 5' end that mapped in the stem of the O_L noncoding region (Fig. 1,2). This heterogeneity spanned a region of 37 nucleotides. The majority of these ends contain two 5' end ribonucleotide residues; the major 5' end ribonucleotide coincides with the recognized site of alkali sensitivity observed in mature mouse mtDNA molecules (9).

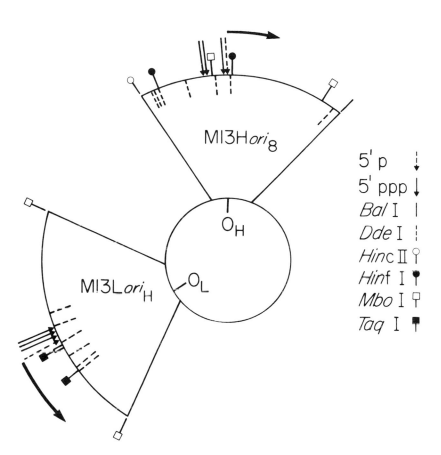

Figure 1. Location and partial restriction maps of the two origins of mouse mtDNA replication. 5'p represents the positions of the majority of nascent H and L strands identified by labeling with polynucleotide kinase. 5'ppp represents the positions of nascent H and L strands identified by labeling with vaccinia virus guanylyltransferase.

These apparent differences in the processing of RNA primers between these two species is also reflected at the origin of heavy strand replication (O_H) (see below). There is currently no information regarding the 5' ends of nascent L strands from bovine mtDNA.

The data obtained for the map positions of nascent L strands from mouse L cells did not distinguish whether the heterogeneity reflected a single point of initiation with numerous points of switch-over from RNA to DNA synthesis with subsequent processing, or initiation occurring at many sites within a 37 nucleotide region with a two nucleotide primer. A mechanism between these two extremes is also possible. An approach to this is to identify nascent L strands with 5' triphosphate termini. By definition, these ends must reflect primary initiation events with the triphosphate containing nucleotide being the first nucleotide of the RNA primer. For the purpose of identifying all nascent 5' ends at O_L, mtDNA containing replicative intermediates was 5' end labeled with $[\gamma - ^{32}P]$ATP and polynucleotide kinase following phosphatase treatment. The $[^{32}P]$ DNA was annealled to M13Lori$_H$ (a 1600 nucleotide Mbo I fragment spanning O_L (Fig. 1)) and digested with Hinc II. This treatment generates a prominent band of 270 nucleotides. This fragment and the region around it (plus and minus 50 nucleotides) was extracted electrophoretically from a 4% polyacrylamide-50% urea gel. This DNA was again annealled to an excess of M13Lori$_H$, digested with Taq I and the resulting digest was analyzed on a 20% polyacrylamide -7 M urea gel. This procedure generates a prominent band 34 nucleotides in length (Fig. 2). Placement of this species in the noncoding region of O_L is consistent with the position of the Taq I site relative to the Hinc II site (Fig. 1). On close inspection, microheterogeneity

---

Figure 2. (right) Electrophoresis of 5' end labeled mouse mtDNA before and after cleavage with restriction endonucleases. mtDNA containing replicative intermediates were cleaved with Mbo I and single stranded DNA from 500-800 nucleotides was isolated by preparative gel electrophoresis. The DNA was labeled with either $[\gamma - ^{32}P]$ATP and polynucleotide kinase (A,B) or with $[\alpha - ^{32}P]$GTP and vaccinia virus guanylyltransferase (D through H). Labeled DNA was annealled to either M13Lori$_H$ (A,B,F,G,H) for analysis of nascent L strands or to M13Hori$_8$ (D,E) for analysis of nascent H strands. The autoradiogram of the 20% polyacrylamide-7M urea gel contains 5' end labeled DNA treated as follows: A, Taq I, B, untreated, D, untreated, E, Hinf I, F, Taq I, G, Dde I, H, untreated. Marker DNA fragments are C, $[^{32}P]$ human mtDNA Hpa II fragments and I, $[^{32}P]$M13mp7 Hpa II fragments.

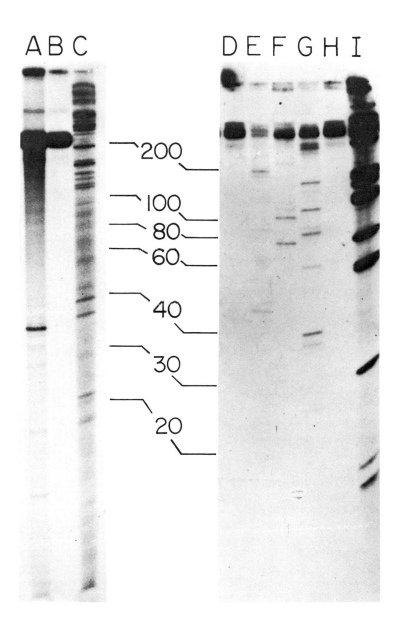

around this band can be detected. Each band is separated by a single nucleotide and each has two 5' end ribonucleotides (4).

Triphosphate terminated nascent strands can be specifically labeled with the enzyme guanylyltransferase isolated from vaccinia virus virions. In the presence of $[\alpha\text{-}^{32}\text{P}]\text{GTP}$ this enzyme will specifically label ribonucleoside triphosphate terminated nucleic acids but not those terminated with monophosphate containing residues (10). The substrate used for this reaction was a fraction of single stranded DNA 500-800 nucleotides in length from an Mbo I digest of LA9 mtDNA. This fraction contains both nascent D-loop strands (see below) and nascent L strands whose 5' ends could be up to 140 nucleotides upstream of the Taq I site defining the 34 nucleotide fragment described above. This DNA was labeled with guanylyltransferase and $[\alpha\text{-}^{32}\text{P}]\text{GTP}$ (Fig. 3), annealled to M13Lori_H (Fig. 1), digested with Taq I and analyzed on a 20% polyacrylamide-7 M urea gel (Fig. 2). Three prominent bands can be observed: one at 77, one at 104 and a faint band at 115 nucleotides. These ends all lie within the tRNA^{Asn} gene. To confirm these results, the same sample was digested with Dde I (Fig. 2). The analogue of the Taq I bands can be observed at 60, 84 and 89 nucleotides. The difference in location of the Dde I site in the stem at O_L and the Taq I site is 20 nucleotides. The other bands in the Dde I lane cannot arise from initiations upstream since the fractionation procedure excludes strands that would have 5' ends 120 nucleotides from the Dde I site. Thus, these data cannot exclude other initiation events upstream from the sites defined in these experiments. On the other hand, sites downstream of the Taq I site are unlikely to be involved in the initiation of L strands since no nascent 5' ends were observed between a position 10 nucleotides upstream of the Taq I site and the downstream Hinc II site 232 nucleotides away (4).

These data suggest that the region defined as O_L is required for the processing of RNA primers and the switch-over from RNA to DNA synthesis. The size of the RNA primer at mouse O_L would then be, on the average, between 43 and 85 nucleotides in length. The processing step then efficiently removes all but two ribonucleotides of the primer. This last process is either slow or incomplete in the mouse mitochondrial system, accounting for the significant incidence of retained ribonucleotide residues at O_L (9).

The Origin of Heavy Strand Replication.

At the origin of heavy strand replication there is maintenance of a set of stable nascent DNA strands, termed D-loop strands. These strands have been well characterized in

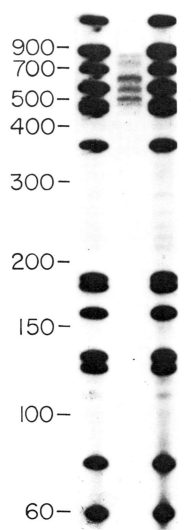

Figure 3. Electrophoresis of mtDNA 5' end labeled with $[\alpha-^{32}P]$GTP and vaccinia virus guanylyltransferase. mtDNA containing replicative intermediates was digested with Mbo I and the DNA was fractionated electrophoretically. The fraction containing single stranded DNA from 500–800 nucleotides was labeled with $[\alpha-^{32}P]$GTP and guanylyltransferase. A portion of the sample was analyzed on a 4% polyacrylamide–7M urea gel (middle lane). The outside lanes are $[^{32}P]$M13mp7 Hpa II fragments.

human (12,13) and mouse (13,14) mtDNA. In human, there are at least four distinct D-loop species, all differing at the 5' end. The 3' ends are unique. It has been shown that nascent H strands from human mtDNA replicative intermediates have identical 5' ends as do D-loop strands (5). Thus, the mechanism of initiation is likely the same for D-loop strands and for H strands involved in DNA replication. It is not known whether pre-existing D-loop strands can serve as primers for DNA replication.

In contrast to human D-loop strands, D-loop strands from mouse mtDNA contain a family of 5' ends clustered around a single major 5' end (14). These 5' ends contain a mixture of ribo and deoxyribonucleotides. Thus as at O_L, the mouse system seems to be less precise in removal of nascent ribonucleotides when compared to human mitochondria. This accounts for the increase in alkali sensitivity at O_H in mouse mtDNA, although a sequence specificity of ribosubstitution is not apparent as it is at O_L (9). In addition to the 5' end differences between mouse and human D-loop strands, there is a difference when comparing the 3' ends. Mouse D-loop strands contain four unique 3' ends, human D-loops terminate at a unique site. The number of 3' end stop sites correlates with the number of termination associated sequences in each D-loop region (four in mouse, one in human) (13,15).

Here we address two questions regarding the mouse H strand origin of replication. The first is whether mouse H strands from replicating molecules contain the same 5' termini as D-loop strands, as is the case in human mitochondria. The other fundamental question we have begun to approach is the identification of the true point of initiation of nascent H strands.

To determine the 5' ends of nascent H strands, LA9 mtDNA containing replicative intermediates was dephosphorylated and labeled with $[\gamma-^{32}P]$ATP and polynucleotide kinase. The labeled DNA was then annealed to a ten-fold excess of L strands and digested with Bal I restriction endonuclease (Fig. 1). This enzyme has three recognition sites in mouse mtDNA, one site occurring 830 nucleotides downstream from the major 5' end of the D-loop strands. Thus by fractionating the Bal I digested $[^{32}P]$DNA on a 4% polyacrylamide–50% urea gel, two fractions can be obtained: one between 500 and 700 nucleotides representing D-loop strands and another fraction from 800-900 nucleotides representing nascent H strands from replicating mtDNA molecules. The two fractions were electro-eluted, annealed to an excess of L strands and digested with Hinf I (Fig. 4). The pattern obtained from purified D-loop strands is identical to that previously described (16). A major band is observed at 31 nucleotides with a series of minor bands of varying intensity of both larger and smaller sizes each separated by a single nucleotide. An identical pattern is

A B C D E

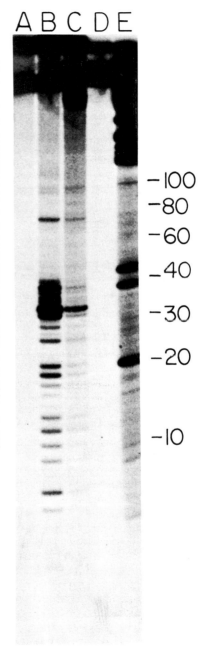

Figure 4. Electrophoresis of 5' end labeled D-loop strands and nascent H strands from mouse mtDNA. Mouse mtDNA containing replicative intermediates was 5' end labeled with $[\gamma\text{-}^{32}P]ATP$ and polynucleotide kinase. The labeled DNA after annealling to L strand template was digested with Bal I restriction endonuclease and fractionated on a 4% polyacrylamide 7M urea gel into two fractions: one containing single stranded DNA 500-700 nucleotides (D-loop fraction) and one containing single stranded DNA from 800-900 nucleotides (nascent H strand fraction). Each fraction was then annealled to excess L strands and then analyzed electrophoretically on a 20% polyacrylamide-7M urea gel before (A,D) and after (B,C) digestion with restriction endonuclease, Hinf I. A,B, D-loop fraction, C,D, nascent H strand fraction, E, $[^{32}P]$ human mtDNA Hpa II fragments.

−100
−80
−60
−40
−30
−20
−10

obtained when the 800-900 nucleotide fraction is digested with Hinf I (Fig. 4). This is not due to contaminating D-loop strands, since prolonged exposure of an analytical 4% polyacrylamide-50% urea gel of the 800-900 nucleotide fraction reveals no contamination with D-loop strands (data not shown). These data cannot exclude the existence of novel 5' ends of nascent H strands outside of the region examined, but they do show conclusively that the 5' ends of nascent H strands that lie within the D-loop region contain the same distribution of 5' ends as D-loop strands. As in the case of human nascent H strand synthesis, it remains to be shown whether pre-existing D-loop strands serve as primers for DNA replication.

As for mouse O_L, it is presumed that the distribution of 5' ends of D-loop strands represents the end point of processing and does not reflect the true point of initiation. To demonstrate this, total mtDNA was digested with Mbo I and the fraction from 500-800 nucleotides was isolated and triphosphate termini were specifically labeled with guanylyltransferase as described above (Fig. 3). Mbo I does not have a recognition site within the D-loop although there is an Mbo I site 181 nucleotides upstream from the Hinf I site (Fig. 1). Even if there are 5' ends beyond this site these will not be digested with Mbo I since a single cleavage in a mtDNA molecule containing a D-loop sized strand will release the nascent strand (17). One additional factor should be mentioned, that being the absence of any knowledge about the steady-state relationship of the initiation event and subsequent elongation. For example, if initiation occurs upstream of the major 5' end identified, it is possible that processing occurs either before or after the 3' end reaches its termination point, as defined above. Thus the fractionation procedure may be excluding molecules containing triphosphate-terminated 5' ends.

The labeled fraction was run on an analytical 4% polyacrylamide-50% urea gel (Fig. 3). The gel shows numerous species superimposed on a background between 500-800 nucleotides. Because the triphosphate terminated species from O_L would be at 741, 770 and 779 nucleotides, the dark bands must arise from triphosphate terminated D-loop strands.

This DNA fraction was annealed to L strand sequence from M13Hori[8] (containing sequences 756 nucleotides downstream from the Hinf I site defining O_H and 715 nucleotides upstream from this site (Fig. 1)). This was then digested with Hinf I and analyzed on a 20% polyacrylamide-7 M urea gel (Fig. 2). Three species are observed at 200, 185 and 45 nucleotides. These assignments are considered as only tentative because confirmatory evidence was obtained by digestion of the same fraction, as single strands, with Dde I. Dde I does digest single stranded DNA (D. Tapper, unpublished observation) and this accounts for the unexpected extra bands when the "capped" sample annealed to O_L template sequence was digested with Dde I (Fig. 2). Extra bands were observed at 25, 36, 39, 112, 156 and 163 nucleotides. In addition, a doublet at approximately 350 nucleotides was observed. There are two Dde I recognition sites near the 5' end region of the D-loop (Fig. 1). One is 27 nucleotides upstream of the Hinf I site, the other is 130 nucleotides upstream. The Dde I bands at 25, 156 and 163 nucleotides match with the three Hinf I bands. If the other Dde I site was recognized, then bands at 55 and 70 nucleotides would be predicted. These are not observed and may be due to a lack of secondary structure at this site which would facilitate cleavage of the single stranded nucleic acid. Still unexplained are the Hinf I doublets at 500 nucleotides and the Dde I doublet at 350 nucleotides and the last set of Dde I bands at 36, 39 and 112 nucleotides. These could all correspond to initiations occurring up to 500 nucleotides upstream of the Hinf I site. Exact assignment of these species remains to be determined.

Although preliminary, the data presented represent a striking analogy to the situation at O_L. Initiation of DNA synthesis occurs upstream of the major 5' end of the stable D-loop species. This site is at least 10 nucleotides and as much as 165 nucleotides upstream of the major D-loop 5' end. In analogy to O_L, the processing of the initial primer results in a major 5' end exhibiting microheterogeneity containing a mixture of ribo and deoxyribonucleotide terminated 5' ends.

In conclusion, this report began as a contrast of the two distinct origins of replication contained within a single mtDNA molecule. Although strikingly different in DNA sequence and in the temporal ordering of initiation, the data presented here suggest that there are possibly more similarities than initially expected. Both origins appear to initiate upstream from the position defined by the stable 5' ends of nascent H and L strands. The mechanism of processing of the 5' ends of nascent strands appears rather similar at the two origins. The stable 5' ends of both nascent H and L strands exhibit microheterogeneity clustered around a single major site. In addition, both contain 5' ribonucleotides, possibly reflecting the same enzymatic machinery used for processing RNA primers at both mouse O_H and O_L.

REFERENCES

1. Clayton DA (1982). Replication of animal mitochondrial DNA. Cell 28:693.
2. Berk AJ, Clayton DA (1974). Mechanism of mitochondrial DNA replication in mouse L cells: asynchronous replication of strands, segregation of circular daughter molecules, aspects of topology and turnover of an initiation sequence. J Mol Biol 86:801.
3. Martens PA, Clayton DA (1979). Mechanism of mitochondrial DNA replication in mouse L-cells: localization and sequence of the light-strand origin of replication. J Mol Biol 135:327.
4. Tapper DP, Clayton DA (1982). Precise nucleotide location of the 5' ends of RNA-primed nascent light strands of mouse mitochondrial DNA. J Mol Biol 162:1.
5. Tapper DP, Clayton DA (1981). Mechanism of replication of human mitochondrial DNA. Localization of the 5' ends of nascent daughter strands. J Biol Chem 256:5109.
6. Bibb MJ, Van Etten RA, Wright CT, Walberg MW, Clayton DA (1981). Sequence and gene organization of mouse mitochondrial DNA. Cell 26:167.
7. Anderson S, Bankier AT, Barrell BG, deBruijn MHL, Coulson AR, Drouin J, Eperon IC, Nierlich DP, Roe BA, Sanger F, Schreier PH, Smith AJH, Staden R, Young IG (1981). Sequence and organization of the human mitochondrial genome. Nature 290:457.
8. Anderson S, deBruijn MHL, Coulson AR, Eperon IC, Sanger F, Young IG (1982). Complete sequence of bovine mitochondrial DNA. Conserved features of the mammalian mitochondrial genome. J Mol Biol 156:683.
9. Brennicke A, Clayton DA (1981). Nucleotide assignment of alkali-sensitive sites in mouse mitochondrial DNA. J Biol Chem 256:10613.
10. Monroy G, Spencer E., Hurwitz J (1978). Characteristics of reactions catalyzed by purified guanylyltransferase from vaccinia virus. J Biol Chem 253:4490.
11. Margolin K, Doda JW, Clayton DA (1981). Mechanism of mitochondrial DNA replication in mouse L-cells: localization of alkali-sensitive sites at the two origins of replication. Plasmid 6:332.
12. Brown WM, Shine J, Goodman HM (1978). Human mitochondrial DNA: analysis of 7S DNA from the origin of replication. Proc Nat Acad Sci USA 75:735.

13. Walberg MW, Clayton DA (1981). Sequence and properties of the human KB cell and mouse L cell D-loop regions of mitochondrial DNA. Nucl Acids Res 9:5411.
14 Gillum AM, Clayton DA (1978). Displacement-loop replication initiation sequence in animal mitochondrial DNA exists as a family of discrete lengths. Proc Nat Acad Sci USA 75:677.
15. Doda JN, Wright CT, Clayton DA (1981). Elongation of displacement-loop strands in human and mouse mitochondrial DNA is arrested near specific template sequences. Proc Nat Acad Sci USA 78:6166.
16. Gillum AM, Clayton DA (1979). Mechanism of mitochondrial DNA replication in mouse L cells: RNA priming during the initiation of heavy-strand synthesis. J Mol Biol 135:353.
17. Robberson DL, Clayton DA (1973). Pulse-labeled components in the replication of mitochondrial DNA. J Biol Chem 248:4512.

Mechanisms of DNA Replication and Recombination, pages 595–604
© 1983 Alan R. Liss, Inc., 150 Fifth Avenue, New York, NY 10011

REPLICATION IN PARAMECIUM MITOCHONDRIAL DNA: FUNCTIONAL
SEQUENCE REQUIREMENTS IN THE ORIGIN REGION[1]

Arthur E. Pritchard and Donald J. Cummings

Univ. of Colorado Health Sciences Center,
Denver, CO 80262

ABSTRACT   Replication of the linear mitochondrial DNA
from Paramecium is initiated at a unique cross-linked
terminus. The monomer cross-link is several hundred
bases of single-stranded DNA arranged in an array of
A+T rich direct tandem repeats. Because of the cross-
link, dimer length molecules are generated during replication. Restriction fragments containing the dimer
form of the initiation region have been cloned and sequenced for five different species and several stocks
within each species. The inter-species comparison reveals surprising sequence diversity. The function of
this sequence as a cross-link or as a replication origin apparently does not require a specific sequence.
Adjacent to the cross-link sequence is a transcribed
region which is highly conserved among several species. The 5' end of an RNA closest to the replication
origin has been located in this sequence. The processing of dimer to monomer molecules was studied by
determining the terminal sequence of a cross-linked
monomer molecule. Two monomer isomers, inverted complements of each other, were found. The results are
discussed in regard to mechanisms for dimer-monomer
conversion.

INTRODUCTION

The 41 kilobase pair (Kbp) linear mitochondrial (mt)
DNA from Paramecium has a unique origin of replication at
one terminus. The duplex strands at the initiation terminus are covalently joined or cross-linked (1). Because of

[1]This work was supported by a grant from the U.S.
Public Health Service (GM 21948)

the cross-link, a head-to-head dimer length molecule is a
replicative intermediate. Restriction enzyme fragments
containing the dimer form of the initiation region have
been cloned and sequenced (2,3). This dimer initiation region is palindromic (the sequence reads the same from both
5' ends) except for a central, nonpalindromic, A+T rich sequence several hundred base pairs long. The nonpalindrome
region is composed, in part, of direct tandem repeats. The
single stranded form of this nonpalindrome region was identified as the monomer cross-link (3). This region of the
genome functions both as a replication origin and as a
structure for completing replication at the terminus of a
linear DNA. Similar models have been proposed for replication of the ends of eukaryotic chromosomes (4).

     Because of the importance of the initiation end of the
mt DNA genome, we have undertaken an extensive inter- and
intra-species comparison with the hope of identifying essential features by their conservation. We found striking
diversity between species in the size and sequence of the
nonpalindrome region. In contrast, there are highly conserved sequences located in the flanking palindrome region
which is transcribed. Finally, we have investigated the
mechanism for converting the dimer intermediate to the
monomer form. As proposed in Fig. 1, the dimer molecule is
processed by staggered single strand nicks, on opposite
strands, near the ends of the nonpalindrome region. The
two strands of this region are then separated and each
folds back to form the two isomers of the monomer. The
cross-link sequences of the two are inverted complements
(see Fig. 1). Restriction fragments containing these isomers were isolated and sequenced.

                          RESULTS

Inter- and Intra-species comparison

     Restriction fragments containing the dimer form of the
initiation region for five species were cloned and sequenced. For four of these species, at least two races (or
stocks) of each were analyzed. Techniques for the cloning
and sequencing protocols are given elsewhere (2,3,5). The
sequence for two races of sp. 1 are shown in Fig. 2. Sequences of the equivalent region for stocks and races of
Sp. 4,5,7 and 8 are published elsewhere (5). A comparison
of these sequences shows surprising inter-species diversity

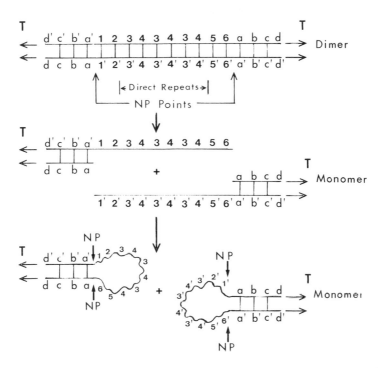

Figure 1. The relation between the sequences in the dimer and monomer molecules is depicted. The dimer molecule is produced by unidirectional replication of the monomer (not shown) from the initiation end loop. The non-palindrome region, with numbered complementary bases, is bounded by the non-palindrome points (NP points). The palindrome region, with lettered complementary bases, is distal to these points. Arrows indicate the direction of the termination end of the molecule (T). A possible mechanism for the formation of two monomer isomers is shown in the two steps.

although an intra-species comparison exhibits relative homogeneity.

As shown in Fig. 2, the sequences can be divided into sub-regions. First, following the nonpalindrome (NP) point, there is a 46 bp unique A+T rich sequence (sp. 1, position -485 to -440).Other species either do not have a corresponding unique sequence (Sp. 7), a shorter one (13 bp, Sp.5) or a longer one (171 bp, Sp. 4,51). In this region, there is no significantly long sequence which is common to all species.

```
                              .              .                 .          -450.              .
SP.1,513        ▼ATTTTTATTTTATAAATATAAATTTTTATTATATAATTTTATAATT
SP.1,168    AAT---------------------------------------------
                ▲

    .             .             .          -400.                .*             .                    .
ATATAATTAATTATAAATTTAATTTTTTAATAAATTATAT  TATTTATATACATATA     TTATAATTATTTTT
-C-----------------------------------------  -------------G--   ---------------
    *        .*     *     .-350        .          .               .                        -300.
ATACAATTAACTATAAGTTTAATTTTTTAATAAATTATAT  TATTTATATATATATA     TTATAATTATTTTT
-------------------------------------G-  ----------C-----   ---------------
    *           .             .             .                    -250.          .
ACATAATTAATTATAAATTTAATTTTTTAATAAATTATAT  TATTTATATATATATATATTATAATTATTTTT
-T-C-----------G---------------------------A---A----------------------------
    *         .             .        -200.            .          .        .
ACATAATTAATTATAAATTTAATTTTTTAATAAATTATATAT   *
-T-----------------------------------------G-  -ATAT  TATATATAT
.  ** *    *.           **  .-150   *  .             .    *
ATAATAATAAATATAAATAAAATTTTTTATTAAAT      ATTTATTTATTTAT
-----------------------------------      --------------
    .* *                       .          .   *  -100. ▼
ATACAT                              TTATATTTATATATATAC
------                              -----------------TAC
                                                        ▲
```

```
        .            .             .             .         -50.          .            .
                    ATTTACTATAATTTAAACACGCGGCGGCATAGCGGCGCTGCGTACAAGC
ATTTTTAATTTAATATATAAT--------------------------------------------------
        .             .-10       ↓ Msp
ACCCGCACCCACCAGGGCGGGCTTCCG
    --------------------------
```

FIGURE 2. Sequences of the dimer non-palindrome region for two races of
sp. 1 are shown in the 5' to 3' direction. The first line shows a unique
sequence (to position -440) followed by a repeat region (bracketed)
arranged in block form to show the direct tandem repeats. The last two
lines are unique sequences. An asterisk above a base in the repeat
indicates a base change at that position in the repeat pattern. Vertical
filled arrowheads identify the non-palindrome points. Sequences of
of sp. 1,168 are shown as a dash if they are identical to sp.1,513, base
changes are written, and a blank indicates the absence of a base. Gaps in
the sequences are for alignment in the repeat region. Position Ø at the
end of the sequence is contiguous with the first base shown in Fig. 3.
There are a few corrections to this previously published sequence (5).

Following this unique sequence is a region of direct tandem
repeats (sp. 1, positions -439 to -98 in Fig. 2). For Sp. 1,
the repeats are not all identical but the first 3 repeats
show little variation. All species have direct tandem repeats but the size, number and sequence of the repeat unit
vary considerably. The repeat units for sp. 1,4 and 8 are
72, 34 and 28 bp, respectively. The total number of base
pairs in the repeat region also varies from 71 bp in sp. 8
to 449 bp in some races of sp. 4. Different sp. 4 mutants
and races contain from 6 to 14 repeat units and even a single harvest of paramecia contains dimer molecules differing by integral numbers of repeat units. An inter- and
intra-species comparison of repeat unit sequence reveals
that the longest common perfect homology is 5'TATTTATA 3'
(positions -399 to -392 in Fig. 2), whose complement, 5'
TATAAATA 3', is identical to the Goldberg-Hogness box which
is postulated to be involved in eukaryotic transcriptional
initiation. A third subdivision of the cross-link sequence

is between the end of the repeating units and the second NP point. For sp. 1 (Fig. 2), this division is not precise since the last three repeat units are only partial and numerous "mismatches" are present. These sequences could be considered unique or imperfect repeats. Sequences of other species exhibit the same characteristic to varying degrees. In an inter-species comparison, there is considerable diversity in the size and sequence of this subregion.

Gene Coding Region

The central nonpalindrome region, bounded by the NP points, is flanked by a palindromic area whose sequence exhibits inter-species homology. A comparison between sp. 1 and 4 is shown in Fig. 3 (comparisons of sp. 5, 7 and 8 have been published (5)). The homologous region does not start at the NP point. There are 76 bp between the NP point and the homologous region in sp. 1, 513 (Fig. 2) but this distance varies from species to species. This homologous region is transcribed. Restriction fragments containing the dimer initiation region hybridize to Northern blots of mt RNA (5). The 5' end of an RNA transcribed from this region has been identified by S1 mapping (6) and its position is shown in Fig. 3.

```
        .10        .          .          .         .50
CGCTGCACGGGCTCGTTTTTTGTATGGGGGGGCATAAAATTAGGGGGTAC  SP.4
G-G -    -----T--AAG-------—--C---------------CAAA---  SP.1

↓ S1         .          .          .         .100
CCAGGTGCTGTTTAATTCTCTAGCTTTAAGGATAAGCCTCAATCAAATCC
-------T--------------------AAG------C-C-A-GGCC--

        .          .          .          .        .150
AACGCGCCTGCTTCTGCAGCGGGGGGGTTAAAAAAAAAAGAAGCCCAAG
-G---G---TT-TT--T----A-----C---G--G--G-----------

        .          .          .          .       . 200
TACCGCTTTGGCTTCCCCATGGCCCTTTATTTTTTTTTTGAGAAGCTAAA
--------C----------------G--------------------
```

FIGURE 3. A comparison of sp. 1,513 and 4,51 sequences in the palindrome region. The arrow labeled S1 identifies the 5' end of an RNA located by S1 mapping. This RNA continues to the right and its 3' end lies beyond the sequences shown.

Processing of Dimer to Monomer

A proposed mechanism for the conversion of the dimer replicative intermediate to the cross-linked monomer is shown in Fig. 1. The dimer experiences staggered single strand nicks on opposite strands at or distal to the NP points. The strands composing the nonpalindrome region are somehow separated and each folds back to be ligated to the duplex strands located distal to the NP points. This results in two isomers of the monomer with terminal cross-linking loops.

To test the proposed mechanism for processing, two sp. 1 monomer isomers were isolated and sequenced. First, it was necessary to identify a small restriction fragment which contains the initiation end of the monomer molecule. Sequences of the dimer initiation region show an MspI site located 73 bp distal to the NP points (Fig. 2). The next closest MspI site is located at a Sma I site, approximately 2 Kbp away (unpublished). A Southern blot of MspI digested mt DNA was hybridized to a ^{32}P labeled dimer initiation region probe, a 2 Kbp fragment (corresponding to ∿ 1 Kbp, monomer size) previously cloned and sequenced. This probe detected 1) a small fragment, which migrates at ∿ 290 bp on a 1.8% agarose gel, corresponding to the monomer form of the terminal Msp fragment, (2) the dimer form of the terminal fragment which migrates at 540 bp, and (3) a large fragment (adjacent to the terminal fragment) which migrates at 2.0 Kbp. The small monomer fragment was isolated from a preparative agarose gel and end labeled with ^{32}P. When this small fragment was heated, quick cooled and examined on a 3.5% acrylamide gel, two bands were observed which migrated at 1.1 and 1.3 Kbp. Without heating, this fragment migrates as a single band at 1.5 Kbp. Sequence analysis of these fragments showed that they were the two isomers of the monomer initiation end depicted in Fig. 1. Their sequences are identical in the palindrome region from the Msp I site to the NP point where the sequences become inverted complements of each other. One fragment is the "right end" of the dimer non-palindrome region NP↓GATATATATAAA...(complementary to the sequence leftward from position -99 in Fig. 2); the other is the "left" end of this region, NP↓ATTTTTA...(rightward from position -485, Fig. 2). The exact sequence could not be read for the entire length of the fragment but it was evident that the entire non-palindrome region was present on each isomer. This result agrees with the model presented in Fig.1. Details of these experiments will be published elsewhere.

DISCUSSION

Sequence diversity in the non-palindrome region.

Inter-species comparison of the non-palindrome initiation region shows striking sequence diversity. All species
contain A+T rich, direct, tandem repeats but the sizes and
sequences are very different. This result is surprising
since there is at least partial conservation of sequence,
size, or structure among replication origins of different
bacteria (7), related E. coli plasmids (8), papovaviruses
BKV and SV40 (10), related lambdoid bacteriophages (10), and
different parvoviruses (11). On the other hand, there are
examples of replication systems which do not require a single specialized sequence. In many of the examples cited
above, there are origin regions which show complete nonhomo-
logy as well as regions of extensive homology. Unlike the
pattern of moderate to high homology between various mammalian mt DNA over most of the genome, the DNA sequence in the
D loop region, which contains the origin of heavy strand
synthesis, shows only slight inter-species homology (12). A
similar situation may exist in the origin region of Drosophila mt DNA (13). Plasmids containing two specific regions
of the extrachromosomal rDNA of Tetrahymena are capable of
autonomous replication in yeast, yet neither of these regions share obvious sequence homology with the origin of
replication of the yeast 2 - μm circle plasmid or with ars
1, a yeast chromosomal replicator (14). A variety of eukaryotic DNA fragments injected into unfertilized eggs of
Xenopus are capable of replication, suggesting that a specialized DNA sequence is not essential for replication initiation (15).
We emphasize the size and sequence diversity in the
Paramecium mt origins but there may also be some essential
features which are conserved. The conserved "Goldberg-
Hogness" box may be one, but since this region contains only
A and T residues, the probability of this sequence occurring
by chance in all species studied is relatively high. Conserved secondary structure may also be an essential requirement but it is very difficult to predict secondary structure
in such a large region containing only A and T. The sequences
immediately distal to the NP points are conserved among all
species (Fig. 3), and it is possible that part of this region plays a role in replication initiation.
In contrast to the inter-species diversity, there is
substantial intra-species homogeneity (Fig. 2 and ref. 5).

We suggest that this homogeneity is produced by unequal re-combination (16). A variety of eukaryotic repeated sequences display the same characteristic of a greater intra-species homogeneity compared with inter-species homogeneity. The process producing this phenomena is known as concerted evolution (17), and unequal cross-over is one proposed mechanism. There are several similarities between predictions of this model and our observed sequences (see ref. 5).

Palindrome Region

An inter-species comparison of the palindrome region, which flanks the nonpalindrome region in the dimer molecules, reveals highly conserved sequences (Fig. 3 and ref. 5). The sequence homology, at least for sp. 1 and 4, continues several hundred base pairs beyond those shown in Fig. 3 (unpublished). Most of this region is transcribed; Fig. 3 shows the 5' end of the first RNA is located 52 bp from the beginning of the homologous region. Whether this region has a role in replication initiation is not known.

Dimer Processing

The proposed mechanism for the dimer to monomer conversion is shown in Fig. 1. Evidence for it was obtained by isolating and sequencing the two predicted cross-linked monomer isomers. Similar models involving site-specific nicking and re-ligation to form terminal loops have been proposed for eukaryotic telomeres (18), and experimental evidence for terminal hairpins or cross-links have been found for vaccinia virus (19), and the parvovirus genome (11). A significant difference between the above examples and Paramecium mt DNA is the lack of terminal palindromic sequences capable of forming base-paired hairpins. Examination of the various Paramecium mt DNA nonpalindrome region sequences (the monomer terminal loop), does not reveal an obvious base-paired structure for the terminal loop. Our model predicts site-specific nicking of dimer and ligation to form the two terminal loops. This process would be facilitated if the site-specific nicking occurred not at the NP points, as depicted in Fig. 1, but distal to these points in the palindrome region shown in Fig. 3. This would allow for complementary base pairing in the formation of the terminal loops. This mechanism may explain completion of replication of this terminus of the mt DNA but some questions of replication initiation remain unanswered. It is possible that RNA primers are formed in the terminal loop to initiate replication but there is, at present, no supporting experimental evidence.

REFERENCES

1. Goddard JM, Cummings DJ (1975). Structure and Replication of Mitochondrial DNA from Paramecium aurelia. J Mol Biol 97:593.
2. Pritchard AE, Herron LM, Cummings DJ (1980). Cloning and Characterization of Paramecium Mitochondrial DNA Replication Initiation Regions. Gene 11:43.
3. Pritchard AE, Cummings DJ (1981). Replication of Linear Mitochondrial DNA from Paramecium: Sequence and Structure of the Initiation-End Cross-Link. Proc Natl Acad Sci USA 78:7341.
4. Cavalier-Smith T (1974). Palindromic Base Sequences and Replication of Eukaryotic Chromosome Ends. Nature 250:467.
5. Pritchard AE, et. al. (1983). Inter-species Sequence Diversity in the Replication Initiation Region of Paramecium Mitochondrial DNA. J Mol Biol 163:1.
6. Berk AJ, Sharp PA (1977). Sizing and Mapping of Early Adenovirus in DNAs by Gel Electrophoresis of S1 Endonuclease Digested Hybrids. Cell 12:721.
7. Zyskind JW, et al. (1983). Chromosomal replication origin from the Marine Bacterium Vilrio harveyi functions in E. coli:ori C Consensus Sequence. Proc Natl Acad Sci USA 80:1164.
8. Selzer G, Som T, Itoh T, Tomizawa J (1983). The Origin of Replication of Plasmid p15A and Comparative Studies on the Nucleotide Sequences around the Origin of Related Plasmids. Cell 32:119.
9. Dhar R, Lai C-J, Khoury G (1978). Nucleotide Sequence of the DNA Replication Origin for Human Papovavirus BKV: Sequence and Structural Homology with SV40. Cell 13:345.
10. Grosschedl R, Hobom G (1979). DNA Sequences and Structural Homologies of the Replication Origins of Lambdoid bacteriophages. Nature 277:621.
11. Astell CR, et al. (1979). Structure of the 3' Hairpin Termini of Four Rodent Parvovirus Genomes: Nucleotide Sequence Homology at Origins of DNA Replication. Cell 17:691.

12. Walberg MW, Clayton DA (1981). Sequence and Properties of the Human KB Cell and Mouse L Cell D-Loop Regions of Mitochondrial DNA. Nucl Acids Res 9:5411.
13. Fauron CM-R, Wolstenholme DR (1980). Intraspecific Diversity of Nucleotide Sequences within the Adenine + Thymine Rich Region of Mitochondrial DNA Molecules of Drosophila mauritiana, Drosophila melanogaster and Drosophila simulans. Nucl Acids Res 8:5391.
14. Kiss GB, Amin AA, Pearlman RE (1981). Two Separate Regions of the Extrachromosomal Ribosomal DNA of Tetrahymena thermophila Enable Autonomous Replication of Plasmids in Saccharomyces cerevisiae. Mol Cellular Biol 1:535.
15. Harland RM, Laskey RA (1980). Regulated Replication of DNA Microinjected into Eggs of Xenopus laevis. Cell 21:761.
16. Smith GP (1976). Evolution of Repeated DNA Sequences by Unequal Crossover. Science 191:528.
17. Dover G (1982). Molecular Drive: A Cohesive Mode of Species Evolution. Nature 229:111.
18. Bateman AJ (1975). Simplification of Palindromic Telomere Theory. Nature 253:379.
19. Baroudy BM, Venkatesan S, Moss B (1982). Incompletely Base-Paired Flip-Flop Terminal Loops Link the Two DNA Strands of the Vaccinia Virus Genome into One Uninterrupted Polynucleotide Chain. Cell 28:315.

Mechanisms of DNA Replication and Recombination, pages 605–614
© 1983 Alan R. Liss, Inc., 150 Fifth Avenue, New York, NY 10011

ISOLATION OF THE INITIATION LOCUS OF THE AMPLIFIED
DIHYDROFOLATE REDUCTASE DOMAIN IN CHO CELLS

Joyce L. Hamlin,[1] Nicholas Heintz, and Jeffrey Milbrandt

Department of Biochemistry, University of Virginia
Charlottesville, Virginia 22908

ABSTRACT   We have been attempting to isolate a
mammalian chromosomal origin of DNA synthesis that can
be demonstrated to function in vivo. We have developed
a Chinese hamster ovary line (CHOC 400) which contains
1000 copies of a 135 kb region that includes the gene
for dihydrofolate reductase (DHFR) (Milbrandt et al.,
PNAS 78, 6043, 1981). Pulse-labeling studies in
synchronized CHOC 400 cells have shown that only 5 of
the 30-35 Eco Rl restriction fragments that comprise
the amplified DHFR domain become labeled during the
onset of the S period (Heintz and Hamlin, PNAS 79,
4083, 1982). In order to isolate the initiation locus
(loci), we excised early-labeled fragments (ELFs) from
preparative agarose gels, nick translated them with
^{32}P-dCTP, and used them as hybridization probes on a
genomic cosmid library prepared from CHOC 400 DNA. We
found that all the ELFs map together in a single
recombinant cosmid (S21) derived from a region 30 kb
from the 3' end of the DHFR gene, indicating that these
fragments represent a single origin of DNA replication
and its flanking sequences. Subclones of the DHFR
initiation locus are being tested for: 1) autonomous
replication in CHO cells, 2) the ability to increase
the rate of transformation of TK⁻ CHO cells to the TK⁺
phenotype, and 3) ars activity in yeast.

[1]This work was supported by NIH grants GM26108 and AM22125,
and a grant from the March of Dimes to J.L.H. N.H.H. and
J.D.M. were supported by NIH postdoctoral fellowships, and
J.L.H. is the recipient of an American Cancer Society
Faculty Research Award.

INTRODUCTION

In eukaryotic cells, each chromosomal DNA fiber is synthesized through the agency of thousands of tandemly-arranged replicons (1). Each replicon probably functions at a characteristic time within the DNA synthetic (S) period (2). DNA synthesis proceeds bidirectionally from the origin in each replicon, and the termini are operationally defined by the position at which forks from adjacent replicons meet (1). It is not known whether eukaryotic chromosomal origins are precisely defined genetic sequences, as they are in micro-organisms and in viruses that infect eukaryotic cells. However, genomic sequences from yeast have been shown to support the autonomous replication of recombinant plasmids when introduced back into yeast, and may represent bone fide chromosomal origins of DNA replication (3,4). Cloned fragments which appear to contain the origin of the rDNA locus in Xenopus have also been shown to replicate autonomously in frog eggs (5), although there is some question about the specificity of initiation in this system (6).

We are attempting to determine whether fixed origins of replication exist in mammalian cells, and, if so, what molecular properties of the origin and the initiation event determine time-ordered replication of mammalian chromosomes. Our approach has been: 1) to first identify a chromosomal sequence that behaves as an origin in situ; 2) to characterize the origin and the rest of the replicon with regard to sequence arrangement, chromatin conformation, and organization in the nucleus, and 3) to isolate an origin and any associated sequences required for autonomous replication in order that initiation of DNA synthesis may be studied in vitro. The system we have developed for these studies is a methotrexate-resistant Chinese hamster ovary cell line (CHOC 400) that has amplified the dihydrofolate reductase gene and large amounts of flanking sequence approximately 1,000 times (7). The multiple copies of the amplified sequence (amplicon) are arranged in tandem, primarily on a single chromosome, and are manifested as a homogenously-staining region (HSR) in G-banded preparations of metaphase chromosomes. We have previously shown that DNA synthesis in this region begins in very early S, and initiation occurs simultaneously at multiple loci along the length of the HSR (7,8). This result suggests that each amplicon contains an origin of DNA synthesis in addition to the dihydrofolate reductase gene.

RESULTS

Sequence Specific Initiation in the DHFR Amplicon

Because there are approximately 1,000 copies of the amplicon per cell, we are able to visualize the restriction pattern of the repeated sequence in digests separated on ethidium bromide-stained agarose gels. By summing the lengths of individual amplified restriction fragments, we estimate the amplicon in CHOC 400 to be 130-150 kb in length (7), which is in the range of estimates for mammalian replicons (1). In order to determine where DNA synthesis initiates within the amplicon, we pulse-labeled synchronized cells for short intervals in the beginning of S, transferred digests of the genomic DNA to DBM paper, and subjected these transfers to autoradiography. Our results showed clearly that labeling in the first 20-40 min after release from aphidicolin is restricted primarily to 3-5 of the 30 Eco Rl fragments derived from the amplicon (Fig. 1 and Ref. 9).

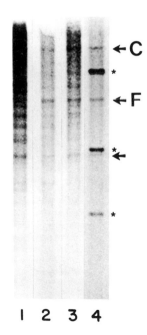

FIGURE 1. Autoradiographic identification of the early-labeled fragments in the DHFR amplicon. CHOC 400 cells were synchronized with either aphidicolin (lane 2) or hydroxyurea (lane 3) as described (9), and were labeled with ^{14}C-thymidine (^{14}C-TdR) for 40 min at the onset of S. DNA in lane 3 was labeled by allowing cells to enter S in the presence of ^{14}C-TdR and 2.5 ug cytosine arabinoside/ml. DNA from experimental cultures was isolated, was digested with Eco Rl, and was electrophoresed along with a log-labeled control digest (lane 1) prior to transfer to DBM paper and autoradiography. Arrows indicate prominent Eco Rl ELFs. Fragments C and F used for screening the CHOC 400 cosmid library are indicated. The asterisks denote fragments of mitochondrial DNA whose synthesis is insensitive to cytosine arabinoside. Reproduced from _Biochemistry_ (Ref. 10) with permission.

Furthermore, regardless of the drug used to collect the cells at the G1/S boundary (aphidicolin, hydroxyurea, or cytosine arabinoside), the same early-labeled fragments (ELFs) were detected (Fig. 1 and Ref. 10). We conclude from these studies that DNA synthesis does not initiate randomly within the amplicon, but rather begins at the same defined locus or loci within each repeated unit.

Isolation of the Initiation Locus from the DHFR Domain

We then addressed the question whether the ELFs represented one or more than one origin of DNA synthesis. The former possibility would predict that the three ELFs would represent an origin-containing fragment and flanking fragments that would map colinearly with one another in the genome. To isolate the locus or loci containing these fragments, we excised the two most prominent ELFs (C and F in Fig. 1) from preparative gels, nick-translated them with ^{32}P-dCTP, and used them separately as hybridization probes on a cosmid library prepared from CHOC 400 DNA. We isolated a total of 22 colonies that reacted with one or both probes. One of these cosmids (S21) contains a 28 kb insert that hybridizes to all of the fragments that are labeled in the first 40 min of the S period in CHOC 400. and therefore must contain the single early-replicating origin of DNA synthesis in this domain (Fig. 2 and Ref. 11). This cosmid was mapped with restriction enzymes, and the repetitive sequences within this region were localized by utilizing ^{32}P-labeled CHO genomic DNA as a hybridization probe (Fig. 4).

Location of the Initiation Locus Relative to the DHFR Gene

We have isolated several recombinant cosmids from the CHOC 400 genomic library that react with a murine DHFR cDNA. Three of these cosmids (H2, H1, and 7-6) overlap one another. and together span approximately 70 kb of DNA (Fig. 3). One of the cosmids (H1) was shown to contain the entire 25 kb DHFR gene by its ability to rescue DHFR-deficient CHO cells (12). Utilizing these cosmids, we have mapped the gene with restriction enzymes, and have determined the positions of the exons, and the orientation of the gene within the restriction map (12). In order to locate the initiation locus with respect to the gene, we asked whether S21 overlapped any of the cosmids spanning the DHFR locus. In hybridization studies utilizing radioactive S21 as a probe, we found that S21 hybridized to 7-6, which contains the 3' end of the DHFR gene. Based on the restriction maps of S21 and 7-6, the distance between the 3' terminus of the gene and fragment F (the most prominent Eco Rl ELF) is estimated to be approximately 30 kb; this result was confirmed with a second cosmid (S14-XK1) that overlaps both S21 and 7-6.

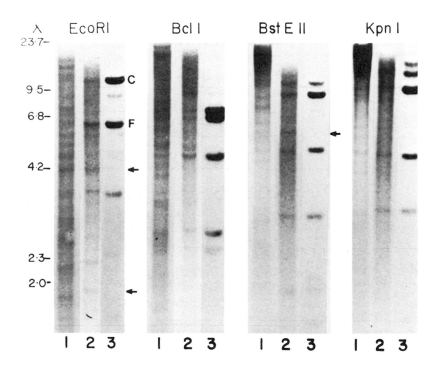

FIGURE 2. Cosmid S21 spans the initiation region of the DHFR amplicon. CHOC 400 DNA labeled with ^{14}C-TdR during log phase growth (lanes 1) or during the first 40 minutes of S (lanes 2) was digested with the indicated enzymes, electrophoresed on a 0.8% agarose gel, and was transferred to nitrocellulose paper. The *in vivo* ^{14}C-TdR incorporation patterns were obtained by exposure to X-ray film. The blot was then hybridized with ^{32}P-labeled cosmid S21 and re-exposed to film (lanes 3). ELFs that map in cosmids adjoining S21 are denoted by arrows. Eco Rl ELFs C and F are indicated. Reproduced from *Nature* (Ref. 11) with permission.

By connecting the DHFR gene and the initiation locus, we have now constructed a contiguous map of more than 90 kb of DNA in the DHFR amplicon, with the initiation locus positioned at the rightward (3') end of this map (Fig. 3). This has allowed us to follow the pattern of replication of the restriction fragments in this domain. By pulse-labeling synchronized CHOC 400 cultures during sequential intervals in S, we have observed that ELFs C and F are labeled before detectable labeling occurs in fragments situated at the 3' end of the gene. It is also clear that the replication fork moves from 3' to 5' through the DHFR gene, since no labeling of a 5' DHFR fragment is detected at a time when 3' fragments are clearly labeled (approximately 90 min into S). Our preliminary data therefore suggest that a replication fork moves in a 3' direction from an origin close to Eco R1 ELF-F, proceeds through the region represented by cosmid S14-XK1 into the DHFR gene, and then through the gene itself (a total of 60 kb of DNA). Since we have not as yet isolated cosmids mapping to the right of S21, we cannot

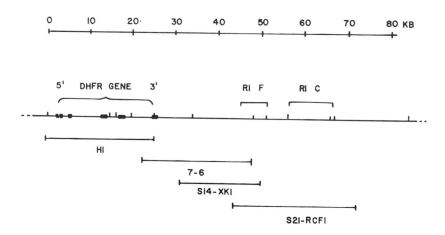

FIGURE 3. The initiation locus lies 30 kb to the 3' end of the DHFR gene. The DHFR gene represented in cosmid H1 was isolated by phenotypic rescue and its organization was determined by restriction enzyme mapping and cDNA hybridization studies (12). Recombinant cosmids overlapping the DHFR gene and the initiation locus were isolated as described in the text. Broad lines indicate exons in the DHFR gene; the locations of ELFs C and F are indicated.

presently monitor the movement of the right hand fork emanating from this origin. However, if one assumes that replication forks travel at equal rates bidirectionally, then this particular origin would control replication of approximately 120 kb of DNA, and would therefore likely be the only origin in the DHFR amplicon.

Characterization of the Initiation Locus

Our *in vivo* labeling studies indicate that initiation of DNA synthesis within the DHFR amplicon occurs close to or within Eco Rl ELF-F (Fig. 1). In order to locate the origin more precisely, we have begun to characterize subcloned fragments of S21 for their ability to replicate autonomously either in CHO cells or in yeast.

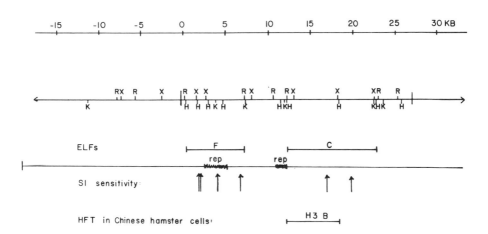

FIGURE 4. Functional organization of the DHFR initiation locus. The organization of cosmid S21 and its corresponding region in the CHOC 400 genome was determined by restriction enzyme mapping and hybridization studies. Repetitive DNA elements (rep) were located by probing digests of S21 with nick-translated whole genomic CHO DNA. S1 sensitive sites in the Hind III subclones of S21 were determined *in vitro* as described (18). A fragment giving high frequency transformation of Chinese hamster ovary cells (H3 B) is indicated. Abbreviations for restriction enzyme sites are: Eco Rl, R; Hind III, H; Xba I, X; Kpn I, K.

We first cloned the nine Hind III fragments of S21 into a poison-minus derivative of pBR322 that was constructed in our laboratory (J. Hamlin, unpublished). This vector (pBR⁻) also contains a 3.4 kb Bam Hl fragment carrying the selectable Herpes thymidine kinase gene (13). In initial experiments, each subclone was introduced into CHO TK⁻ cells by the spheroplast method of gene transfer (14), and the rates of transformation to the TK⁺ phenotype for each subclone were compared to the rate for the TK/pBR⁻ control. This procedure is analogous to an assay used to identify autonomously-replicating yeast chromosomal sequences (15). It is based upon the observation that the frequency of transformation of an auxotrophic yeast strain by a cloned selectable marker is greatly enhanced by the presence of a chromosomal replicator in the same plasmid (16).

When we compared the rates of transformation of CHO TK⁻ cells to wild type by individual Hind III subclones, there were marked differences among the clones (Fig. 4). At the earliest time that TK⁺ colonies could be observed on the culture dishes (ca. 7 days), we measured a 30-fold enhancement of transformation rate by Hind III fragment B when compared to TK/pBR⁻. Other fragments gave slight increases or values similar to the vector itself, suggesting that increased transformation is not simply a result of the presence of a genomic insert. After a month in culture, many of the colonies transformed with Hind III B died off, and the survivors were shown to have integrated the plasmid into high molecular weight DNA. We have not been able to detect autonomously-replicating entities in the low molecular weight DNA fraction of these CHO transformants, whether measured in the first few hours or days after transformation, or after a month in culture. It is possible that the percentage of cells transformed initially is so low (1-5X10⁻⁴) that a plasmid maintained autonomously (even at moderate copy number) in the early stages after transformation would be below the limits of detection by Southern blotting and hybridization procedures.

We have also begun to examine S21 for sequences that have ars activity in yeast. We first replaced the Eco Rl-Bam Hl fragment of pBR322 with the Eco Rl-Bam Hl fragment of pUB110 that contains a single site for Xba I. We then introduced the ura3 gene of YIp31 into the Bam Hl site of this hybrid vector. We have subcloned the Xba I fragments of S21 into this vector, and have begun transformation studies with uracil auxotrophs to determine if sequences from the DHFR initiation locus confer autonomous replication on plasmids in yeast.

To date, the highest rates of transformation both of CHO cells have been obtained from fragments located in the region of Eco Rl ELF-C (Fig. 4), whereas high resolution synchrony experiments suggest that the fragment labeled earliest in the S period is Eco Rl ELF-F. Our results could be interpreted to mean that ELF-C may contain an enhancer sequence analogous to those found close to origins in eukaryotic viruses (17). Our inability to observe sustained autonomous replication of any subclone of S21 would then be due to the fact that we had separated the origin from a required enhancer element. We are currently testing this possibility by preparing larger, random subclones of S21 by Sau 3a partial digestion; some of these clones may contain all the elements required for autonomous replication.

We have also begun to characterize the initiation locus by several other approaches that have been used to examine the structure of regulatory regions (e.g., promoters and origins) in DNA. Larsen and Weintraub have shown that particular sequences at the 5' end of the globin gene are especially sensitive to S1 nuclease digestion both _in vitro_ and _in vivo_, presumably due to the presence of single-stranded regions in hairpin structures or A/T rich DNA (18). We have used this enzyme to look for analogous regions in subclones of S21, and have mapped the location of specific S1 sites as indicated in Fig. 4. The presence of corresponding sites in chromatin has not yet been determined. These studies are part of a larger effort to use selected nucleases to examine the chromatin conformation of the DHFR domain at various times during the cell cycle.

In summary, we hope to prepare a complete physical map of the initiation locus, and to identify sequences within this locus that are able to support autonomous replication of plasmids when introduced into CHO cells. We will then be able to correlate the structural features of this region with the functional characteristics of a mammalian origin of DNA synthesis.

ACKNOWLEDGEMENTS

We wish to thank Carlton White and Kay Greisen for expert technical assistance, and Mitch Smith and Amy Bouton for collaborative efforts in the yeast transformation studies.

REFERENCES

1. Edenberg HJ, Huberman JA (1975). Ann Rev Genet 9:245.
2. Hand R, (1978). Cell 15:317.
3. Struhl K, Stinchcomb D, Scherer S, Davis RW (1979). PNAS 76:1035.
4. Chan CSM, Tye B-K (1980). PNAS 77:6329.
5. Hines PJ, Benbow RM (1982). Cell 30:459.
6. Harland RM, Laskey RA (1980). Cell 21:761.
7. Milbrandt JD, Heintz NH, White WC, Rothman SM, Hamlin JL (1981). PNAS 78:6043.
8. Hamlin JL, Biedler JL (1981). J Cell Physiol 107:101.
9. Heintz NH, Hamlin JL (1982). PNAS 79:4083.
10. Heintz NH, Hamlin JL (1983). Biochemistry: in press.
11. Heintz NH, Milbrandt JD, Greisen KS, Hamlin JL (1983). Nature:in press.
12. Milbrandt JD, Azizkhan JC, Greisen KS, Hamlin JL (1983). Mol Cell Biol:in press.
13. Wigler M, Silverstein S, Lee L-S, Pellicer A, Cheng Y-C, Axel R (1977). Cell 11:223.
14. Schaffner W (1980). PNAS 77:2163.
15. Stinchcomb D, Struhl K, Davis RW (1979). Nature 282:39.
16. Hsiao CL, Carbon J (1979). PNAS 76:3829.
17. Gruss P, Dhar R, Khoury G (1981). PNAS 78:943.
18. Larsen A, Weintraub H (1982). Cell 29:609.

VI. SITE-SPECIFIC RECOMBINATION

Mechanisms of DNA Replication and Recombination, pages 617–636
© 1983 Alan R. Liss, Inc., 150 Fifth Avenue, New York, NY 10011

THE MECHANISM OF PHAGE LAMBDA SITE-SPECIFIC RECOMBINATION: COLLISION VERSUS SLIDING IN att SITE JUXTAPOSITION

Nancy L. Craig and Howard A. Nash

Laboratory of Neurochemistry
National Institute of Mental Health, Bldg 36/Rm 3D-30
Bethesda, Maryland 20205

ABSTRACT   We have examined in vitro some of the properties of bacteriophage lambda excisive recombination. We have found that excisive resolution takes place in both the presence and absence of Xis but that excisive inversion and intermolecular recombination require Xis. We have also found that the products of Xis-promoted excisive resolution are topologically linked but that the products of Xis-independent excisive resolution are topologically unlinked. These results indicate that there are mechanistic differences between Xis-promoted and Xis-independent excisive recombination and we conclude that Xis does not enhance excisive recombination by simple stimulation of the Xis-independent pathway. We suggest that during Xis-promoted excisive recombination, att sites come together by three-dimensional diffusion and random collision but that during Xis-independent excisive recombination they come together through one-dimensional diffusion. We also propose that the mechanism by which att sites come together is determined by the higher order structure of complexes of att sites and recombination proteins and that the higher order structure of such complexes can be influenced by both the nature of the att sites and the recombination proteins which comprise them.

INTRODUCTION

The integration/excision cycle of bacteriophage lambda is mediated by site-specific recombination. Both the proteins and DNA sequences that participate in these reactions have been identified and characterized (for recent reviews, see 1,2). Because these recombination reactions proceed efficiently in vitro with purified components, they are ideally suited to biochemical dissection.

Integrative recombination occurs by reciprocal recombination between a specific site in phage DNA, attP (POP'), and a specific site in bacterial DNA, attB (BOB'). att sites are tripartite structures. All att sites contain a 15 bp sequence called the core (O) in which the breakage and reunion events which result in strand exchange occur. Each att site also contains unique sequences called arms (P,P',B and B') which flank the core. The products of integrative recombination are attL (BOP') and attR (POB'). Integrative recombination is promoted by a phage-encoded protein, Int, and an E. coli-encoded protein, Integration Host Factor (IHF). Excisive recombination occurs by reciprocal recombination between attL and attR and regenerates attP and attB. Under physiological conditions, excisive recombination is promoted by Int, IHF and Xis, a second phage-encoded protein. However, excisive recombination can be promoted by Int and IHF alone; this Xis-independent reaction is inefficient in vivo but can proceed efficiently in vitro at low ionic strength (3).

How do att sites locate each other and come together as a prelude to recombination? It has been demonstrated (4) that during integrative recombination attP and attB come together by three-dimensional diffusion and random collision. In this paper, we examine how attL and attR come together in excisive recombination. We present data which shows that during Xis-promoted excisive recombination, attL and attR also come together by three-dimensional diffusion and random collision. However, our results indicate that during Xis-independent excisive recombination, attL and attR do not come together by three-dimensional diffusion and random collision. We suggest that under these conditions attL and attR come together by one-dimensional diffusion. Our finding that attL and attR can come together by different mechanisms in Xis-promoted and Xis-independent excision demonstrates that these are distinct recombination pathways.

RESULTS

When att sites that will undergo recombination are located on separate DNA molecules, recombination is intermolecular. Figure 1 shows the substrates and products of intermolecular excisive recombination between plasmids that separately contain attL and attR. The substrates and products of recombination can be distinguished by restriction enzyme digestion and gel electrophoresis. The experiment of Figure 2 demonstrates that Xis-promoted excision can proceed efficiently as an intermolecular reaction (lanes 4,5) but that intermolecular Xis-independent excisive recombination (lanes 6,7) is very inefficient. We have examined excisive recombination at a variety of substrate DNA

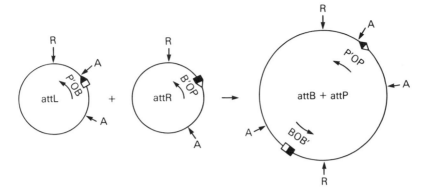

Figure 1. Genetic and physical map of substrates and products of intermolecular excisive recombination. The plasmid containing attL is pMM82 and plasmid containing attR is pMM83 (26). Intermolecular recombination produces a single circular DNA molecule. The approximate locations of restriction sites for Eco RI (R) and Ava I (A) are shown.

concentrations, protein concentrations and ionic strengths; we find that although these variations alter the absolute amount of intermolecular Xis-promoted excisive recombination that occurs, intermolecular Xis-independent excisive recombination is always undetectable (data not shown). It has previously been demonstrated that intermolecular integrative recombination between attP and attB is an efficient reaction (5). We have directly compared intermolecular integrative and Xis-promoted excisive recombination under conditions similar to those of the experiment in Figure 2 and find that these reactions are comparably efficient (data not shown).

If att sites that will undergo recombination are located on a single DNA molecule, recombination can be intramolecular. There are two types of intramolecular recombination, distinguished by the relative orientations of the att sites on the substrate DNA molecule. If the att sites are oriented so that their cores form a direct repeat, intramolecular recombination produces two product circles; this reaction is called resolution. If the att sites are oriented so that their cores form an inverted repeat, intramolecular recombination inverts one segment of the DNA that lies between the att sites with respect to the other; this reaction is called inversion. Figure 3 shows the substrates and products of intramolecular excisive resolution and inversion. The substrates and products of recombination can be distinguished by restriction enzyme digestion and electrophoresis in agarose gels.

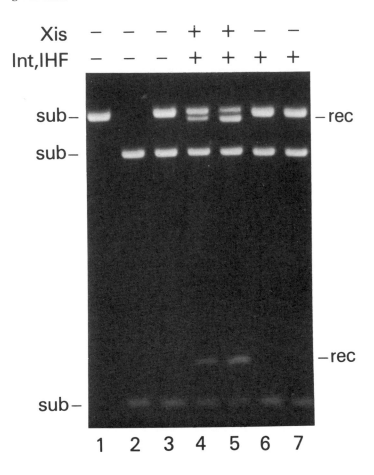

Figure 2. Agarose gel analysis of intermolecular excisive recombination reactions. Supercoiled plasmid DNA, Int and IHF were prepared as previously described (20). Xis (25) was a generous gift of Dr. Sue Wickner (National Institutes of Health, Bethesda, MD). Recombination mixtures (20 ul) included: 48.5 mM Tris HCl (pH 7.6), 1.5% (v/v) glycerol, 30 mM KCl, 3.5 mM NaCl, 2.3 mM KPO_4, 1 mM sodium EDTA, 0.1 mM DTT, 5 mM spermidine, 14.5 ug bovine serum albumin (Pentex) and, as indicated, 0.2 ug pMM82, 0.2 ug pMM83, 56 ng Int, 74 ng IHF and 50 ng Xis. Reaction mixtures were incubated at 25°C for 45 minutes or 180 minutes as indicated, and then stopped by incubation at 65°C for 5 minutes. $MgCl_2$ was then added to 10 mM and the mixtures treated with 20 units Ava I restriction enzyme (Bethesda Research Laboratories) at 37°C for 60 minutes. SDS was added to 0.5%, Ficoll-400

(Figure 2. Continued) to 3.8% and the mixtures electrophoresed through 1% agarose gels in 89 mM Tris, 89 mM boric acid, 2.5 mM sodium EDTA.   DNA was visualized by staining with ethidium bromide and photographed under UV light. DNA fragments derived from substrate (sub) and recombinant (rec) are indicated. Lane 1 = pMM83, 180 mins; lane 2 = pMM82, 180 mins; lane 3 = pMM82 + pMM83, 180 mins; lane 4 = pMM82 + pMM83 + Int + IHF + Xis, 45 mins; lane 5 = pMM82 + pMM83 + Int + IHF + Xis, 180 mins; lane 6 = pMM82 + pMM83 + Int + IHF, 45 mins; lane 7 = pMM82 + pMM83 + Int + IHF, 180 mins.

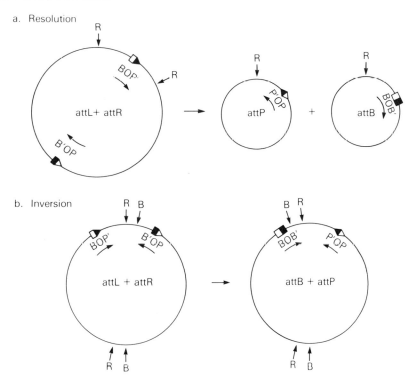

a. Resolution

b. Inversion

Figure 3. Genetic and physical maps of substrates and products of intramolecular excisive recombination. a. Resolution The plasmid containing attL and attR in direct orientation is pLR110 (27). Recombination produces two circular DNA molecules (arbitrarily drawn as not catenated), one of which contains attP and one of which contains attB.   The approximate locations of restriction site for Eco RI (R) are shown. b. Inversion The plasmid containing attL and attR in inverted orientation is pLR90 (6). Recombination inverts one segment that lies between the att sites with respect to the other. The approximate locations of restriction sites for Eco RI (R) and Bam HI (B) are shown.

The experiment shown in Figure 4 demonstrates that excisive resolution is executed very efficiently by both the Xis-promoted and Xis-independent pathways (lanes 2,3). However, this experiment also demonstrates that excisive inversion is executed only by the Xis-promoted pathway (lane 6); Xis-independent excisive inversion is a very inefficient reaction (lane 5). We have examined excisive resolution at a variety of substrate DNA concentrations, protein concentrations and ionic strengths; we find that although the amount of Xis-promoted excisive inversion is altered by these variations, Xis-independent excisive inversion is always undetectable (data not shown,6). It has previously been demonstrated that the integrative pathway efficiently promotes both resolution (7) and inversion (8). We have directly compared integrative and Xis-promoted excisive resolution and inversion under conditions similar to those of the experiment in Figure 4 and find that all four reactions are comparably efficient (data not shown).

In Figure 3, the products of resolution are drawn as separate circles which are not topologically linked. However, when supercoiled DNA is used as the substrate for integrative resolution, electron microscopic and electrophoretic analysis has demonstrated that the product circles are topologically linked as multiply intertwined catenanes of considerable complexity (4). Thus, in integrative resolution, no free circular products are detected although restriction enzyme digestion reveals that recombination has occurred. In the experiment of Figure 5, the topological relationship of the products of excisive resolution was examined by electrophoretic analysis in the absence of restriction enzyme digestion. No free circular DNA molecules were detected as products of Xis-promoted excisive resolution (lane 2) although restriction enzyme digestion (Figure 4, lane 3) revealed that extensive recombination had occurred. This suggests that the products of Xis-promoted excisive resolution are topologically linked. We have directly examined topological linkage of the product circles by electrophoretic analysis using the methods of Sundin and Varshavsky (9) and find that they are catenanes, ranging from simple catenanes to multiply intertwined catenanes of considerable complexity (data not shown). By contrast, electrophoretic analysis in the absence of restriction enzyme digestion of Xis-independent excisive resolution reveals that the products are free circular molecules (lane 3). To quantitate this observation, we used radioactively labeled substrate DNA and measured both the amount of recombination as revealed by restriction enzyme digestion and the amount of free circular product in the absence of restriction enzyme digestion by cutting out appropriate gel slices and counting them; essentially all the

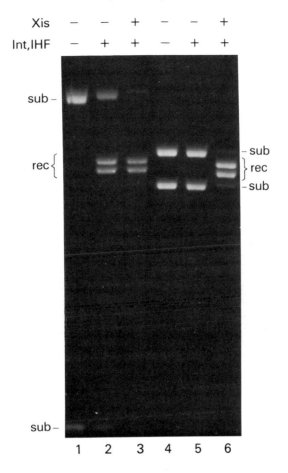

Figure 4.   Agarose gel analysis of intramolecular excisive recombination reactions. Recombination reactions (20 ul) included the following: 48.5 mM Tris HCl (pH 7.6), 1.7% (v/v) glycerol, 30 mM KCl, 4.5 mM NaCl, 4.5 mM KPO$_4$, 1 mM sodium EDTA, 0.1 mM DTT, 5 mM spermidine, 14 ug bovine serum albumin and, as indicated, 0.2 ug pLR110, 0.2 ug pLR90, 56 ng Int, 74 ng IHF and 100 ng Xis.   Reaction mixtures were incubated at 25°C for 45 minutes and then stopped by incubation at 65°C for 5 minutes. For the reactions in lanes 1-3, Tris HCl (pH 7.4) was added to a final concentration of 100 mM, MgCl$_2$ to 5 mM and, NaCl to 50 mM and the mixtures were then treated with 100 units of Eco RI restriction enzyme (Boehringer Mannheim) at 37°C for 60 minutes.   For the reactions in lanes 4-6, MgCl$_2$ was added to a final concentration of 8.9 mM and NaCl to 49.9 mM and the mixtures were then treated with 25 units of Bam HI restriction enzyme (Bethesda Research

(Figure 4. Continued) Laboratories) at 37°C for 60 minutes. The mixtures were analyzed by electrophoresis as described in Figure 2. DNA fragments derived from substrate (sub) and recombinant (rec) are indicated. Lane 1 = pLR110; lane 2 = pLR110 + Int + IHF; lane 3 = pLR110 + Int + IHF + Xis; lane 4 = pLR90; lane 5 = pLR90 + Int + IHF; lane 6 = pLR90 + Int + IHF + Xis.

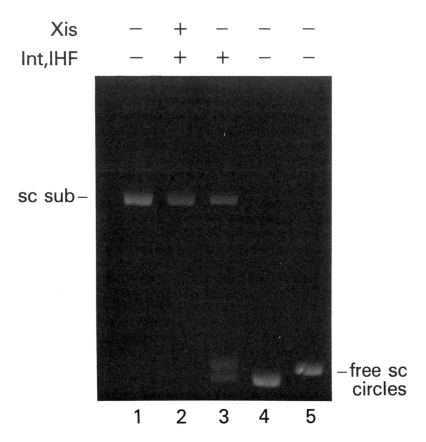

Figure 5. Agarose gel analysis of the topological relationship of the products of excisive resolution. pB110 and pP110 are the products of excisive resolution of pLR110 and were obtained by transformation with recombination reactions. Recombination was carried out as described in Figure 4. After incubation at 65°C, SDS was added to the mixtures to 0.5%, Ficoll-400 was added to 3.8% and the mixtures electrophoresed through a 1% agarose gel as described in Figure 2. Lane 1 = pLR110; lane 2 = pLR110 + Int + IHF + Xis; lane 3 = pLR110 + Int + IHF; lane 4 = pB110; lane 5 = pP110.

products of Xis-independent excisive resolution are free circles (data not shown). By similar means, we have also examined the kinetics of recombination and of the appearance of free circular products; these events occur simultaneously (data not shown). Furthermore, we have found that all the products of Xis-independent excisive resolution are free circles at a variety of substrate DNA concentrations, protein concentrations and ionic strengths (data not shown). Thus, the production of free circular products appears to be an intrinsic property of Xis-independent excisive resolution.

When the substrate DNA is supercoiled as in the experiment of Figure 5, the products of xis-independent excisive resolution comigrate with supercoiled molecules (lanes 4,5). We have also analyzed the products of Xis-independent excisive resolution of supercoiled substrates by CsCl-ethidium bromide density gradient analysis and find that the superhelical density of the recombinant products does not differ detectably from the superhelical density of unrecombined substrate DNA (data not shown). If the Xis-independent resolution substrate is nicked circular DNA, the products are free nicked circles (data not shown).

DISCUSSION

Several properties of the three Int-dependent site-specific recombination reactions that are studied in this work are summarized in Table 1. The properties of the integrative pathway are consistent with the hypothesis that attP and attB come together during recombination by three-dimensional diffusion and random collision (4). 1) Intermolecular recombination which requires collision of separate DNA molecules is efficient. 2) The integrative pathway promotes intramolecular resolution and inversion equally well and is thus insensitive to the relative orientation of att sites on the substrate DNA molecule. 3) The products of integrative resolution are topologically linked as multiply interwined catenanes, reflecting the random interwrapping of segments of the substrate DNA. Although Xis-promoted excisive recombination uses different DNA substrates and recombination proteins than does integrative recombination, the properties of the Xis-promoted pathway suggest that attL and attR come together by three-dimensional diffusion and random collision in this pathway as well. Xis-promoted recombination is an efficient intermolecular reaction, can promote both intramolecular resolution and inversion, and the products of resolution are topologically linked.

TABLE 1

Properties of Bacteriophage Lambda Site-Specific Recombination Reactions

	Efficient Recombination			Free Circles After Resolution of Supercoiled DNA
	Intermolecular	Inversion	Resolution	
Integration	Yes	Yes	Yes	No
Xis-promoted Excision	Yes	Yes	Yes	No
Xis-independent Excision	No	No	Yes	Yes

By contrast, although Xis-independent excision uses the same proteins as integration and the same DNA substrates as Xis-promoted excision, the properties of the Xis-independent excision pathway are very different from those of the integrative and Xis-promoted pathways. Xis-independent excision is inefficient as an intermolecular reaction, promotes resolution but not inversion, and the products of resolution are topologically unlinked. Interestingly, the Xis-independent pathway closely resembles resolvase-dependent site-specific recombination which efficiently promotes resolution but does not promote inversion or intermolecular recombination (10,11).

To explain the unique properties of Xis-independent excision, we suggest that during this reaction, attL and attR do not come together by three-dimensional diffusion and random collision but rather come together by one-dimensional diffusion. One-dimensional diffusion has been invoked to account for the rapidity with which specific DNA binding proteins such as lac repressor and Eco RI endonuclease locate their specific target sites. It is proposed that these proteins execute a one-dimensional random walk mediated by non-specific protein-DNA interactions along a DNA molecule and thereby "slide" to their specific target sites (12,13). We suggest that during Xis-independent excision, a complex of recombination proteins and a prophage att site executes a one-dimensional random walk along the DNA substrate so that when two prophage att sites are located on the same DNA molecule they are brought into juxtaposition by sliding. Furthermore, we suggest that whether att sites come together by three-dimensional diffusion and random collision or by one-dimensional diffusion is determined by the higher order structure of the complexes of recombination proteins and att sites. Sliding models also have been invoked to explain the topological properties of other site-specific recombination reactions (14-16).

Higher Order Structures of att Sites and Recombination Proteins.

Several pieces of evidence suggest that complexes of att site and recombination proteins can be organized into higher order structures. Hsu et al (17) have found that partial digestion of complexes of Int protein and attP DNA with DNase results in enhanced digestion of DNA at 10-11 basepair intervals, a pattern often observed when DNA lies on a surface. By electron microscopic analysis of complexes of Int protein and att sites, Better et al (18) have provided evidence that Int can condense attP and have suggested that Int wraps attP into a nucleosome-like structure. From their study of the topology of the products of integrative recombination, Pollock and Nash (6) have independently proposed that an essential component of this reaction is the wrapping of an att site into a nucleosome-like structure. attP is the most attractive candidate for the wrapped site because it contains a multitude of specific binding sites for both Int (17) and IHF (1,19); it seems reasonable to suggest that these proteins associate to form a complex around which attP would be wrapped. The precise topological path of DNA in such a wrapped structure would likely be determined by multiple protein-DNA and protein-protein interactions. Figure 6, panel a shows a diagram of a nucleosome-like structure formed by the wrapping of attP around a complex of Int and IHF.

Int and IHF promote both integration and Xis-independent excision. What is the higher order structure of a complex of Int, IHF and attL or attR? Figure 6, panel b shows a diagram of a complex of Int, IHF and attR. We suggest that this complex does not assume a complete nucleosome-like structure because only one arm of this att site, the P arm, contains specific recognition sites for Int and IHF so that only one arm has the potential for strong, specific interactions with a complex of Int and IHF. Similarly, we

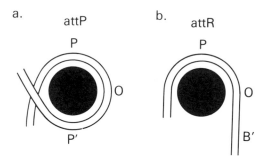

Figure 6. Nucleosome-like structures formed by recombination proteins and att sites. The filled circle represents a complex of Int and IHF.

suggest that attL may only form a partial, asymmetric nucleosome-like structure with Int and IHF because attL also has only one arm, the P' arm, which contains specific Int and IHF recognition sequences. However, both Int and IHF can interact nonspecifically with DNA that does not contain specific recognition sequences for Int and IHF (20,21). Therefore, we presume that the B and/or B' prophage arms can also interact with a complex of Int and IHF although this interaction will be less strong because it is non-specific in nature. Thus, we suggest that although prophage att sites do not form nucleosome-like structures with Int and IHF of the kind formed with attP, they can form less stable or asymmetric nucleosome-like structures which are characterized by strong interactions with the P and P' prophage arms and less strong, non-specific interactions with the B and/or B' prophage arms.

The Higher Order Structure of Prophage att Site Complexes Allows Sliding.

We propose that complexes of Int and IHF and prophage att sites derive their potential for sliding from their asymmetric nucleosome-like stucture as diagrammed in Figure 7. Panel a shows the strong interaction between a P or P' prophage arm and a complex of Int and IHF. Panel b shows an aymmetric nucleosome-like structure formed by strong, specific interactions with a P or P' prophage arm and less strong, nonspecific interactions with a B or B' prophage arm. The position of the P or P' arm with respect to the complex is likely fixed because of strong, specific interactions. However, we propose that because the nonspecific interactions between the B or B' arm and the protein complex are less strong, this DNA can translocate or slide with respect to the protein-att site complex as shown in panel c. This sliding would effectively be one-dimensional diffusion of the protein-att site complex along the adjacent DNA. As shown in panel d, this sliding pathway could bring into juxtaposition the protein-att site complex and other DNA sequences that are distant from the att site. Thus, sliding allows a complex of proteins at one att site to scan adjacent DNA for a second att site. It has been estimated that the lac repressor can "scan" about 10^3 bp in one second (12) and therefore it seems reasonable to suggest that sliding can effectively juxtapose att sites in the substrates we have examined.

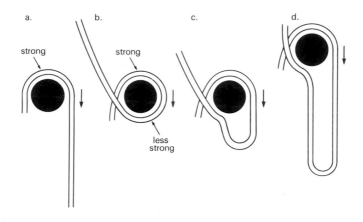

Figure 7. A complex of Int, IHF and a prophage att site facilitates sliding. The downward pointing, vertical arrow indicates the core of an att site.

Sliding Can Account for the Properties of Xis-independent Excision.

The hypothesis that attL and attR are brought into juxtaposition during Xis-independent excisive recombination by sliding accounts well for the properties of this pathway. Xis-independent excisive recombination is not an efficient intermolecular reaction. Complexes of Int, IHF and prophage att sites may be too unstable to allow collision to effectively bring them together to promote recombination and, if attL and attR are located on separate DNA molecules, sliding does not juxtapose these sites. Alternatively, it may be that intermolecular recombination is inefficient because, as the result of sliding, a site essential to recombination on a complex of Int, IHF and a prophage att site is blocked by occupation with non-att DNA.

In intramolecular recombination, both att sites are located on the same DNA molecule and it would be here that the effects of sliding would likely be most profound. Xis-independent excisive resolution is an efficient reaction. Figure 8, panel a diagrams how a sliding pathway could bring the att sites in a resolution substrate into juxaposition. Panel a-1 shows a resolution substrate with its directly repeated att sites. As shown in panel a-2, an asymmetric nucleosome-like structure forms at one att site. (Of course, both att sites may be capable of forming such complexes and each may carry out sliding but we have no information about this point.) The two att sites in the resolution substrate are brought together and closely juxtaposed by sliding (panel a-3). Note that the att sites, once juxtaposed, are in parallel alignment. Evidence suggests that this is the configuration of att sites in which recombination occurs (22,23).

The other intramolecular recombination reaction is inversion. We have demonstrated that Xis-independent excisive inversion is a very inefficient reaction. The hypothesis that Xis-independent excision utilizes a sliding pathway to bring att sites together provides an explaination for the failure of this reaction. Figure 8, panel b shows a sliding pathway on an inversion substrate. Panel b-1 shows the substrate with the att sites as inverted repeats. Panel b-2 shows the formation of an asymmetric nucleosome-like structure at one att site that promotes sliding which results in the juxtaposition of att sites shown in panel b-3. When juxtaposed, the att sites are in anti-parallel alignment, a configuration thought to be unfavorable for recombination.

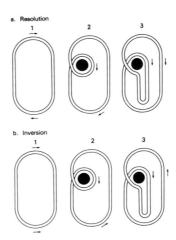

Figure 8. Sliding on resolution and inversion substrates. In each panel, two arrows mark the positions and relative polarities of a pair of prophage att sites.

A striking feature of Xis-independent excision is that the products of resolution reactions that use supercoiled DNA as substrate are free supercoiled circles and thus are not topologically linked. This contrasts sharply with the fact that when a supercoiled substrate is used, the products of integrative resolution, a reaction also promoted by Int and IHF, are multiply intertwined catenanes. It has been proposed that this catenation reflects the structure of the substrate DNA as an interwound helix (4); catenation results from collision between att sites which traps the random interwraps of segments of the substrate DNA. The proposal that attL and attR come together by sliding explains how topologically independent circles can be produced (Figure 9). Panels a-c show how the att sites on a resolution substrate are brought into juxtaposition by sliding. In this figure, we have not shown the interwraps that would be expected from the flexibility of DNA or from writhing caused by supercoiling. However, as suggested by others (15,16), we imagine that sliding would segregate such interwraps and produce a molecule that is topologically equivalent to that in Figure 9, panel c. Once the att sites are juxtaposed, the next step is the breakage of the parental DNAs and their rejoining in novel combinations to produce

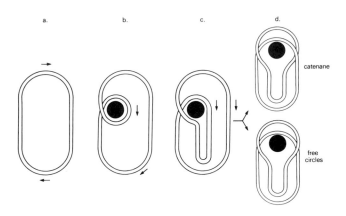

Figure 9. Topological consequences of recombination.

recombinant molecules. As diagramed in Figure 9, panel d, rejoining can occur so that products have either of two configurations: the products can be either free circles or singly catenated circles. If breakage and reunion were random, we would expect that 50% of the product would be free circles and 50% of the product would be singly catenated circles. We have determined that essentially all of the products of Xis-independent excisive recombination are free circles. We note that essentially all of the products of resolvase-dependent resolution are singly catenated circles (16) and that it has been proposed that res sites are brought together by sliding during this reaction (14,16). Thus, although the detailed mechanisms of recombination site juxtaposition and strand breakage and reunion may be different in Int-dependent and resolvase-dependent site-specific recombination, both these reactions must proceed through highly ordered and defined steps.

The Failure of Integrative Recombination to Use Sliding.

Integration and Xis-independent excision are both promoted by Int and IHF. Why is sliding apparently operative only in Xis-independent excision? We have postulated that the capacity of a complex of recombination proteins and an att site to carry out sliding lies in the asymmetry of the interactions between att site arms and the recombination proteins. Because both arms of attP have the capacity for strong, specific interactions with a complex of Int and IHF, we postulate that the nucleosome-like structure is a stable one wherein the positions of both arms of attP are fixed with respect to the complex of Int and IHF. Furthermore, we believe it is unlikely that a nucleosome-like complex forms between Int, IHF and attB. IHF does not bind specifically to attB (19) and although Int does bind specifically to the core of attB (24), there are no Int recognition sequences in the arms that might provide the strong and specific interactions which we suggest mediate the formation of a nucleosome-like structure. Indeed, no stable nucleosome-like structures between attB and recombination proteins have been observed by electron microscopy (25). Thus, we imagine that during integration, a fixed nucleosome-like complex of Int, IHF and attP comes together with attB (which could also have Int specifically bound to it) through three-dimensional diffusion and random collision. The differences in the integrative and Xis-independent pathways illustrate how the nature of the DNA substrate can influence the higher order structure of att sites and their associated proteins and thereby influence the recombination mechanism.

The Role of Xis.

The properties of Xis-promoted and Xis-independent excision are different although these reactions utilize the same att sites. This demonstrates that these reactions proceed through different mechanisms and that Xis does not enhance excisive recombination by simply stimulating the Xis-independent reaction. To account for the differences we have observed, we propose that during Xis-dependent excision, attL and attR come together by three-dimensional diffusion and random collision whereas attL and attR come together during Xis-independent excision by sliding. We suggest that Xis activates complexes of Int, IHF and prophage att sites so that collision between att sites provides effective precursors to recombination and, furthermore, that Xis alters the higher order structure of complexes of Int, IHF and prophage att sites so that sliding does not occur. Xis may stabilize the asymmetric nucleosome-like structure formed at one (or both) prophage att site by strengthening the non-specific interactions so that sliding is blocked. Alternatively, Xis may sufficiently weaken the non-specific interactions between the B or B' arms of a prophage att site and a complex of Int and IHF such that sliding no longer occurs. Although Better et al (25) have observed that Xis stimulates the binding of Int to attR, this finding could support either proposal. The differences in the Xis-promoted and Xis-independent pathways illustrates how the proteins involved can influence the higher order structure of complexes of att sites and recombination proteins and thereby determine the pathway of recombination.

ACKNOWLEDGEMENTS

We thank Carol Robertson for expert technical assistance, Dr. Sue Wickner for her generous gift of Xis and Drs. Cozzarelli, Sherratt, Sternberg and their coworkers for communication of their results prior to publication.

REFERENCES

1. Nash H (1981). Integration and excision of bacteriophage λ: The mechanism of conservative site specific recombination. Ann Rev Genet 15:143.
2. Weisberg R and Landy A (1983). In The Bacteriophage Lambda II (eds. Hendrix R, Roberts J, Stahl F and Weisberg R) Cold Spring Harbor Laboratory, Cold Spring Harbor, N. Y. (in press).

3. Abremski K and Gottesman S (1981). Site-specific recombination: Xis-independent excisive recombination of bacteriophage lambda. J Mol Biol 153:67.

4. Mizuuchi K, Gellert M, Weisberg R and Nash H (1980). Catenation and supercoiling in the products of bacteriophage λ integrative recombination in vitro. J Mol Biol 141:485.

5. Mizuuchi K and Mizuuchi M (1979). Integrative recombination of bacteriophage λ : in vitro study of the intermolecular reaction. Cold Spring Harbor Symp Quant Biol 43:1111.

6. Pollock T and Nash H (1983). Knotting of DNA caused by a genetic rearrangement: Evidence for a nucleosome-like structure in site-specific recombination of bacteriophage lambda. J Mol Biol (manuscript submitted)

7. Nash H (1975). Integrative recombination of bacteriophage lambda DNA in vitro. Proc Natl Acad Sci USA 72:1072.

8. Mizuuchi K, Fisher L, O'Dea M and Gellert M (1980). DNA gyrase action involves the introduction of transient double-strand breaks into DNA. Proc Natl Acad Sci USA 77:1847.

9. Sundin O and Varsharsky A (1980). Terminal stages of SV40 DNA replication proceed via multiply intertwined catenated dimers. Cell 21:103.

10. Reed R and Grindley N (1981). Transposon mediated site-specific recombination in vitro: DNA cleavage and protein-DNA linkage at the recombination site. Cell 25:721.

11. Sherratt D, Arthur A and Burke M (1981). Transposon-specified, site-specific recombination systems. Cold Spring Harber Symp Quant Biol 45:275.

12. Berg O, Winter R, and von Hippel P (1982). How do genome-regulatory proteins locate their DNA target sites? Trends Biochem Sci 7:52.

13. Jack W, Terry B and Modrich P (1982). Involvement of outside DNA sequences in the major kinetic path by which Eco RI endonuclease locates and leaves its recognition sequence. Proc Natl Acad Sci USA 9:4010.

14. Kitts P, Symington L, Dyson P, and Sherratt D (1983). Transposon-encoded site specific recombination: nature of the Tn3 DNA sequences which constitute the recombination site res. Embo J (in press).

15. Abremski K, Hoess R and Sternberg N (1983). Studies on the properties of P1 site-specific recombination: Evidence for topologically unlinked products following recombination. Cell 32:1301.

16. Krasnow M and Cozzarelli N (1983). Site-Specific Relaxation and Recombination by the Tn3 Resolvase: Recognition of the DNA Path between Oriented res Sites. Cell 32:1313.

17. Hsu P-L, Ross W, and Landy A (1980). The lambda att site: functional limits and interaction with Int protein. Nature 285:85.

18. Better M, Lu C, Williams R and Echols H (1982). Site-specific DNA condensation and pairing mediated by the Int protein of bacteriophage λ . Proc Natl Acad Sci USA 79:5837.

19. Craig N and Nash H (1983). (manuscript in preparation).

20. Kikuchi Y and Nash H (1978). The bacteriophage λ Int gene product. A filter assay for genetic recombination, purification of Int, and specific binding to DNA. J Biol Chem 253:749.

21. Nash H and Robertson C (1981). Purification and properties of the E. coli protein factor required for λ integrative recombination. J Biol Chem 256:9246.

22. Mizuuchi K, Weisberg R, Enquist L, Mizuuchi M, Buraczynska M, Foeller C, Hsu P L, Ross W and Landy A (1981). Structure and function of the phage λ att site: size, Int-binding sites and location of the crossover point. Cold Spring Harbor Symp Quant Biol 45:429.

23. Weisberg R, Enquist L, Foeller C and Landy A (1983). A role for DNA homology in site-specific recombination: the isolation and characterization of a site affinity mutant of coliphage λ . J Mol Biol (in press).

24. Ross W, Landy A, Kikuchi Y and Nash H (1979). Interaction of Int protein with specific sites on λ att DNA. Cell 18:297.

25. Better M, Wickner S, Auerbach, J and Echols H (1983). Role of the Xis protein of bacteriophage λ in a specific reactive complex at the att R prophage attachment site. Cell. 32:161.

26. Mizuuchi M and Mizuuchi K (1980). Integrative recombination of bacteriophage λ : extent of the DNA sequence involved in attachment site function. Proc Natl Acad Sci USA 77:3220.

27. Abremski K and Gottesman S (1982). Purification of the bacteriophage λ xis gene product required for λ excisive recombination. J Biol Chem 257:9658.

Mechanisms of DNA Replication and Recombination, pages 637–659
© **1983 Alan R. Liss, Inc., 150 Fifth Avenue, New York, NY 10011**

SITE-SPECIFIC RECOMBINATION BY Tn3 RESOLVASE: MODELS FOR
PAIRING OF RECOMBINATION SITES [1]

Mark A. Krasnow[*][†], Martin M. Matzuk[*][2], Jan M. Dungan[*]
Howard W. Benjamin[*], and Nicholas R. Cozzarelli[*]

[*]Department of Molecular Biology, University of California
Berkeley, California 94720

[†]Department of Biochemistry, University of Chicago
Chicago, Illinois 60637

Recombination is a complex process requiring pairing of
the recombination sites (synapsis), four polynucleotide
strand scissions, coordinated rearrangement of the eight
broken ends, and four ligation events. During the
rearrangement, the tertiary and quaternary structure of the
DNA is typically altered as well as the primary structure.
The entire set of reactions has been achieved in vitro for
site-specific recombination (1-3), where the crossover
occurs between two defined DNA sequences that are identical
or nearly so. For the well-studied bacterial systems only
one or two proteins are required and these are usually
encoded adjacent to the recombination site.

The first in vitro recombination system modeled
site-specific integration of phage lambda DNA into the host
chromosome and excision of the prophage (reviewed in 1,4,5).
Integration occurs at the bacterial attachment site (attB)
via a reciprocal recombination with the phage site (attP)
and requires the phage-encoded Int protein and integrative
host factor (IHF) of Escherichia coli. The left and right
prophage sites (attL and attR) formed by the recombination
are the substrate for the reverse reaction of excision that
typically requires, in addition, Xis protein of the phage.

[1]This work was supported by National Institutes of
Health grant GM 31655.
[2]Present address: Washington University School
of Medicine, Washington University, St. Louis, MO 63110

Reaction between intramolecular att sites has very
different consequences depending on whether the homologous
portions of the att sites are in the same or opposite
orientation along the DNA. For directly repeated att sites
in a circular substrate, the products are two rings
corresponding to the sequences between the crossover points,
and they are usually catenated. If the att sites are in
opposite orientation, the sequences between the crossover
points are inverted in the single product ring which is
usually knotted.

The focus of this paper is the second system developed,
the site-specific recombination that completes transposition
of E. coli Tn3 and the related transposon, gamma delta
(2,6). In the first stage of transposition between two
circular replicons, Tn3 is duplicated and the donor and
recipient rings are fused via transposon bridges (7,8). The
transposon-encoded resolvase breaks down this cointegrate by
recombination between a sequence within each of the directly
repeated transposons called res, so that both separated
replicons now contain one copy of Tn3. Recombination in
vitro requires only the 21,000 dalton resolvase; a
supercoiled substrate is converted into two catenated
product rings (2,6).

Site-specific recombination has also recently been
achieved in vitro with a partially purified extract from
phage P1-infected E. coli (3). Recombination mediated by
the phage protein Cre at the phage loxP site breaks down
concatameric forms of P1 generated by generalized
recombination and thereby facilitates the faithful
partioning of this unit copy plasmid at cell division (9).

The requirements of the in vitro recombination
reactions are strikingly simple. Plasmids containing cloned
copies of the recombination sites are effective substrates,
and only the purified protein(s) and a di- or polyvalent
cation cofactor such as Mg^{2+} or spermidine are necessary
(10,11,2; K. Abremski, pers. comm.). The reactions are
conservative in that all nucleotides and the majority of
parental supercoils are retained in the reciprocal
recombination (12,3). The latter result requires that the
enzymes hold fast to the cut ends. The formation of a
covalent bond to one end of the DNA stores the energy of the
phosphodiester bond transiently broken during recombination,
and allows resealing of the break in the absence of an
external energy source. The controlled reversible breakage
of the DNA backbone and formation of covalent protein-DNA
intermediates are fundamental features of DNA

topoisomerases, enzymes that pass one DNA segment through a
transient enzyme-bridged interruption in another
and alter tertiary and quaternary structure without
rearranging the primary sequence (13-15). Both Int (16,1)
and resolvase (6) are also type 1 topoisomerases, and this
has provided a simpler means of analyzing their breakage and
resealing activity.

We have characterized the specificity of relaxation and
recombination by Tn3 resolvase and the topological
interlinking of the recombinant products and conclude that
not only the res sites are recognized, but also at least 4
global features of the substrate DNA:

1. whether res sites are on the same DNA molecule,
2. the relative orientation of sites along the DNA,
3. the relative position of sites along the DNA of a
   substrate containing more than two sites, and
4. the topological segregation of the DNA domains that
   are separated by recombination.

We discuss in detail models of how resolvase might
distinguish these features, give several simple topological
tests, and compare resolvase to the two other in vitro
recombination systems. This report is intentionally
speculative and heuristic; in some cases preliminary data is
used to obtain the broadest picture possible.

Resolvase has two DNA breakage and reunion activities,
recombination and relaxation. Recombination in vitro was
first demonstrated with the resolvase of gamma delta (2) and
then with Tn3 (6). These enzymes share about 80% amino acid
homology and cross-complement in vivo (17-19). The
topoisomerase activity was first found for the Tn3 resolvase
using plasmids such as pRR51 that contain gamma delta res
sites. We have now shown that relaxation is not an artifact
of the heterologous system but a consistent resolvase
activity. First, Tn3 resolvase relaxed pCM102, containing
directly repeated Tn3 sites (19), as well as or better than
pRR51. Second, purified gamma delta resolvase, supplied by
N. Grindley, relaxed pRR51 as well as Tn3 resolvase did.
Relaxation was not observed in the original analysis of
gamma delta resolvase products (2) because ethidium bromide
included in the electrophoresis running buffer positively
supercoiled both substrate and product.

The recombination reaction can be increased by raising
the concentration of NaCl to 150 mM or $MgCl_2$ to 20 mM in the
standard reaction (6). Under these conditions nearly

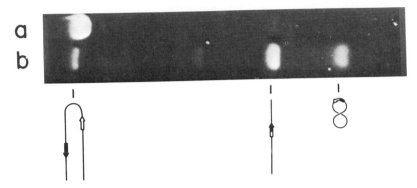

FIGURE 1. Effect of substrate supercoiling on recombination. Recombination between the directly repeated res sites of pRR51 (5800 bp) generates 2 catenated daughter rings, mono51A (3200 bp) and mono51B (2600 bp). 0.2 µg of pRR51 nearly completely relaxed by rat liver topoisomerase I (lane a) or native pRR51 (lane b) were treated with 0.1 µg of resolvase at 37°C in a 20-µl reaction containing 20 mM Tris-HCl (pH 7.45), 150 mM NaCl, 10 mM $MgCl_2$, and 1 mM dithiothreitol. After 3 hrs, the incubation was continued for 15 min in the presence of excess Bam HI restriction endonuclease to linearize the parental plasmid and mono51B. The products were separated by electrophoresis through a 0.8% agarose gel (from left to right as shown) and stained with ethidium bromide. The positions of linear pRR51, linear mono51B, and supercoiled mono51A are indicated. The faint band in lane a between pRR51 and mono51B is open circular mono51A. Recombination of the supercoiled substrate is essentially complete because the residual parental DNA is mostly from the 5% contaminating open circular form. About 10% of the relaxed substrate recombined.

complete recombination was achieved (Fig 1, lane b). Native levels of supercoiling are not absolutely required for recombination since a small amount of recombination was seen with pRR51 containing only a few supercoils (Fig 1, lane a). The nicked substrate, however, was inert (data not shown).

The recombination and relaxation activities are manifestations of the same breakage and reunion function because they show a similar time course, occur under the same ionic conditions, and, most importantly, have the same substrate specificity that is unique in two ways (6). First, both relaxation and recombination are specific for intramolecular res sites. The inability to relax plasmids containing just a single res site is particularly surprising, because topological transformations require breakage and reunion at only a single site. Because the reaction is bimolecular with respect to res sites, we tested whether the concentration of res sites was limiting in the intermolecular reaction. Two single-site plasmids were catenated using DNA gyrase to maintain the res sites at the same local concentration as in the standard substrate. Still neither relaxation nor recombination occurred (6). Thus, there is a specific proscription against intermolecular reactions, and this ensures that resolution is irreversible, as has been found in vivo (20). Second, resolvase distinguishes the relative orientation of intramolecular sites because in vitro, and under some conditions in vivo, both activities are efficient only on directly repeated res sites (2,6). This is surprising because a difference of a single twist in the thousands of base pairs separating the sites could normalize the relative orientation in space of direct and inverted sites. Thus, either the res sites are always directed together in a manner which reflects their orientation in the primary sequence or resolvase faithfully traces the DNA path between the sites.

The defined nature of the pairing event is underscored by an analysis of the catenated recombinant rings. Why are the products of recombination not just free circles? Nash, Mizuuchi, and their colleagues have noted that supercoiling can interwind the parental DNA domains that will be separated by recombination and these windings will become catenane interlinks between the daughter rings (12,21). The interwinding of the segments is maximal when the sites are antipodal (Fig 2A) and is absent when they are adjacent (Fig 2B). This model makes four strong predictions that qualitatively are fulfilled for Int-mediated recombination but are categorically untrue for resolvase. First, the products should be extensively interlinked because there are many supercoils between the sites in the native substrate, but about 99% of the resolvase products are in fact interlinked only once (Fig 3A). Second, the interlinking should be heterogenous given the expected dynamic distribution in the position of the res sites between the extremes depicted in A and B of Fig 2 and the Gaussian distribution in supercoil number (22,23), but the resolvase recombinants are homogenous. Third, since the number of supercoils between sites is proportional to the distance

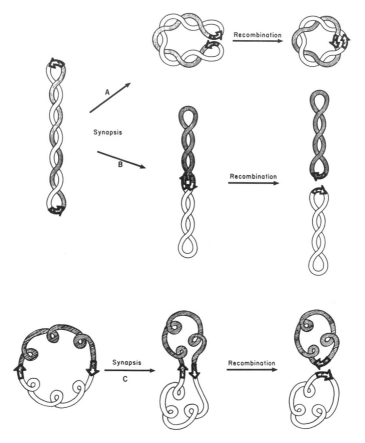

FIGURE 2. Effect of supercoil structure and res site
position on complexity of recombinant catenanes. Directly
repeated recombination sites (open arrows) divide the
parental plasmid into two domains (white and shaded) that
will be separated by the recombination. Any interwinding of
the domains at synapsis will be maintained in the product as
interlinking of the two recombinant rings. Plectonemic
supercoiling of the parental DNA (top) causes extensive
interwrapping of the domains when the sites are antipodal
(A) and results in complex catenanes. A conveyor-belt
motion of the DNA will change the relative position of the
res sites in space until they are directly across from each
other (B). Here the supercoil interwraps are all within and
not between the two domains and the products are free rings.
Forms of writhing such as toroidal supercoiling prevent
extensive interwinding of the domains at synapsis (C) and

separating them in the primary sequence, the complexity of daughter catenanes should reflect this. However, when this distance varied over an order of magnitude, from 500 bp (pCM102) to 5800 bp (pRattL^2), the interlinking of the products was unchanged. Fourth, the extent of product interlinking should be proportional to the superhelical density of the parental plasmid, but was constant for resolvase over the entire range tested from nearly relaxed to almost twice the native density of pRR51. Thus the single interlink is not a passive consequence of supercoiling but must be an intrinsic feature of the resolvase reaction.

There is a subtle but powerful test of any passive scheme for catenation. These models predict a negative topological sign for catenane interwraps, the sign of the parental supercoils. The sign of singly-interlinked rings cannot yet be determined for technical reasons but the topology of a recombinant catenane generated in low amounts by resolvase has been worked out (24). When the products of the resolvase reaction were nicked and then analyzed by high resolution agarose gel electrophoresis, three minor bands migrating faster than the singly-interlinked recombinant were detected (Fig 3A). Electron microscopy of RecA coated molecules showed that the middle band had the novel form of a figure-8 catenane (Fig 3B). It has two features which are incompatible with passive models for catenation. First, the two intermolecular interlinks of a figure-8 catenane are not uniformly (-), but one must be (+) and one (-). Second, the supercoil entrapped in the resolvase product is always (+). Thus, although it is not yet known how this structure is generated, it cannot be the result of the simple conversion of parental supercoils to interlinks between recombinants.

The question for resolvase is now just the opposite: what prevents the extensive interlinking of the daughter rings in the reaction? Either random collision synapsis (Fig 2A) is precluded or the supercoils are toroidal (Fig 2C) or extensively branched and not plectonemically

---

FIGURE 2 (continued)

also give rise to simple product rings. Neglected in this figure is interwinding of the daughter rings introduced during strand exchange. This would be additive to the changes shown here and could give, for example, singly interlinked catenanes in B and C instead of the free rings shown.

A   B

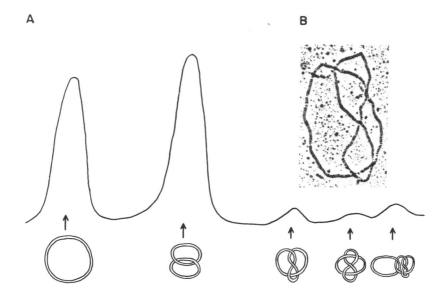

FIGURE 3. High resolution analysis of the resolvase
reaction products.
   A. 0.45 μg of p51A^2, a 6400 bp plasmid containing
directly repeated res (6), was treated with resolvase. The
products were nicked with DNase I and separated by
electrophoresis (from left to right as shown) through a 0.8%
agarose gel at 1 volt/cm for 2 days. A densitometric scan
of the ethidium bromide stained gel is shown, and the
structure of the DNA in each band as determined by
electron microscopy is diagrammed. They are, in order of
increasing electrophoretic mobility: parental plasmid,
singly-interlinked recombinant catenane, figure-8 parental
knot, figure-8 recombinant catenane, and singly-interlinked
recombinant catenane containing one ring with a figure-8
knot.
   B. Electron micrograph of a RecA-coated figure-8
catenane generated by resolvase. The figure-8 catenanes
were coated with RecA, shadowed, and viewed with an
electron microscope (24). The striations on the DNA
correspond to right-handed helical coating by RecA (25) and
provide an internal control for the correct reading of the
photographic negatives.

interwound as shown. Both alternatives would ensure that the supercoil windings are maintained within rather than between the daughter domains at synapsis. Each alternative could arise as a consequence of the structure of DNA in solution or resolvase could actively intervene. Consequently, four classes of models to explain the simple products must be considered. (We assume that the single interlink connecting the daughter rings is introduced during strand exchange).

    I.   Supercoils in solution are not interwound.
   II.   The DNA structure prevents the path shown in Fig 2A.
  III.   Resolvase converts the DNA to a non-interwound form.
   IV.   Resolvase prevents the path shown in Fig 2A.

    Recent experiments make class I models the least likely. Although the evidence on the structure of supercoils in solution is inconclusive, the preponderant view is that they are plectonemic rather than toroidal at high salt concentrations and supercoil density (26,27). Even when the resolvase reaction was carried out at twice the standard ionic strength and a range of supercoiling up to nearly twice the native density, the recombinants were still about 99% singly interlinked. We cannot rule out, however, a dynamic equilibrium between competent and incompetent DNA structures that recombination pulls toward a minor parental form. Another class I model envisions interwound but extensively branched supercoils that would minimize interwrapping of distant DNA segments and therefore tend to keep the catenanes simple. It is difficult, however, to understand how class I models could give rise to homogenous products in face of random foldovers in the substrate and dynamic variations in supercoil structure.
    In class II, the other type of DNA-mediated models, the path in Fig 2A must be unavailable or unproductive. It is unlikely that the substrate is simply too rigid to allow the bending for the direct synapsis required in path A, because this would require a persistence length for pRattL^2 at least ten times that of linear DNA (28). Alternatively, recombination could require a particularly intimate wrapping of the two res sites that is fostered by plectonemic supercoils. Recombination would then occur only after a conveyor-belt motion of the DNA had brought the sites into register as shown in Fig 2B. This would also explain the

specificity of resolvase reactions. Intermolecular pairing, even between res sites on singly-interlinked catenated rings, would not provide tight wrapping and thus would be unproductive. Because the res sites would always face each other in a fixed relationship reflecting their orientation in the primary sequence, direct and inverted res sites would be distinguished.

Although this last alternative is the most reasonable of the DNA-mediated models, there are two difficulties for any scheme based solely on the structure of DNA. First, as discussed below, resolvase preferentially recombines neighboring res sites in a four-site substrate even though recombination between the distal sites is normal when the intervening sites are removed. To explain this in terms of DNA structure alone requires further assumptions, but neighboring site preference is a natural consequence of any model in which the enzyme searches along the DNA for res sites. Second, if the structure of DNA has a pervasive influence, then one expects an overriding similarity between the three in vitro recombination systems, resolvase, Int, and Cre. However, the three systems, at least under most conditions, are radically different.

A comparison of the three systems highlights the individual personality of each (Table 1). The complexity of lambda integration reaction products is explained qualitatively by the Nash-Mizuuchi scheme. The catenanes contain several interlinks, the distribution is broad, and recombination of a relaxed substrate generates about 15% free rings (12,21; H. Nash, pers. comm.). Quantitatively, though, the theory is imperfect. Although some interlinked rings are expected for a relaxed substrate because of random twisting of the DNA, the fraction found is much more than predicted (21). Moreover, the complexity of the catenated lambda products was clearly less than expected if all the supercoils were plectonemic. Lambda integrative recombination is also distinct from resolvase since inverted sites are efficiently recombined. This inverts the DNA segment between the recombination sites and can consequently knot the recombinant ring. Here the data on the topology of the product is more extensive, and the predicted increase in knot complexity with increased supercoiling and distance separating the sites is seen (29,30). For a 9,400 bp plasmid with 1570 bp between the att sites, all the recombinants are in knots that contain up to 9 crossing segments. With a nicked substrate, only half the products are knotted and these are all simple trefoils that contain

Table 1

Comparison of Substrate Site
Preference and Product Topology
For in vitro Recombination Systems

Recombination System		Configuration of DNA Sites		
		Intermolecular	Intramolecular	
Enzyme (refs)	Site		Inverted Repeats	Direct Repeats
resolvase (6,14)	res x res	No	No	Yes (singly-linked catenanes)
Int, IHF (12,21)	attB x attP	Yes	Yes (knots, mostly complex)	Yes (catenanes, mostly complex)
Int, IHF Xis (12,21)	attL x attR	Yes	Yes (knots, mostly complex)	Yes (catenanes, mostly complex)
Int, IHF (34)	attL x attR	No	No	Yes (free rings)
Cre (3)	loxP x loxP	Yes	Yes	Yes (50% catenanes 50% free rings)

just 3 crossovers. Similarly, when the distance between the sites is trimmed, knot complexity diminishes until at 350 bp the product is almost all trefoil.

The P1 system (3) clearly differs from the other two systems, but is particularly different from resolvase. As with Int, intermolecular recombination can occur and intramolecular recombination is insensitive to the relative orientation of loxP sites. The Cre reaction is distinct, however, in that about half of the products are free rings when the sites are 2000 bp apart, and linear DNA is an effective substrate.

The differences between the three recombination systems argue against a strict DNA structural model to explain the resolvase results, but there are three caveats. First, the reactions were performed under very different conditions and with different substrates. For instance, the lambda reactions contain high concentrations of spermidine and stoichiometric amounts of Int and IHF, each of which can dramatically affect DNA structure (31,32). Second, under special conditions, resolvase and Int can appear quite similar. Although attL x attR recombination generally requires Xis, the reaction can proceed in its absence at low ionic strength (33). Craig and Nash (34) have found that attL x attR recombination in the absence of Xis generates only free rings, and thus like resolvase the products are homogenous and simple. Furthermore, inverted sites are inert, as for resolvase. Third, even if the structure of the DNA is identical in the three systems, the enzymes might recognize or interact with different features of the DNA. If this is correct, then both DNA structure and enzyme mechanism are important in determining the path to synapsis.

In class III and IV models, the enzyme alone mediates synapsis. The simplest but perhaps least appealing mechanism for formation of simple catenanes is the class III wrapping model. Wrapping of DNA in a left-handed fashion around resolvase forming a nucleosome-like structure would remove free supercoils and prevent extensive interwinding of the daughter domains. No further binding can occur after each parental supercoil has been converted to this toroidal form or compensatory positive coiling would cause domain intertwining. Thus, a delicate balance is required between the energy of resolvase binding and the energetic cost of transforming a plectonemic supercoil to a toroidal form. Moreover, prodigious quantities of resolvase would be necessary if the coiling is comparable to that in a

nucleosome. Finally, it is hard to explain how resolvase
can relax the parental plasmid because the model requires no
free supercoils.

In class IV models, resolvase itself ensures that
synapsis occurs only with side-by-side sites as in Fig 2B.
Resolvase could trace along the DNA backbone between sites
and in this way ensure not only the segregation of daughter
domains at synapsis but also intramolecular reactions and
the proper orientation of res sites. One example of such a
tracking or looping model is shown in Fig 4A.

Resolvase is shown as a dimer of identical subunits,
each containing a DNA binding site. One subunit of
the enzyme binds nonspecifically to the DNA and translocates
randomly along it just as other repressors are envisioned to
find their operators (35). Once a res sequence is found, it
is bound tightly and translocation by that subunit ceases.
The second subunit binds to the adjacent DNA in a unique
orientation and thereby traps a small loop of DNA between
the res sites. As the second subunit diffuses along the DNA
searching for a second res site, the loop expands and
contracts. When the second subunit encounters a res site in
the proper orientation, it is tightly bound, and the enzyme
changes conformation activating breakage and reunion.

There are two critical features of the model. First,
the entrapped DNA loop must initially be sufficiently small
relative to the stiffness of DNA such that the DNA is
uniquely oriented with respect to the two resolvase binding
sites. Second, once the orientation is established, it must
be maintained until both res sites are bound. This is most
easily achieved if the enzyme never falls off in its
1-dimensional walk between the sites, or if the enzyme does
dissociate, it restarts where the search originated.

Other parts of the model can vary. The non-specific
binding of DNA to the second subunit need not await binding
of the first subunit to a res site. Both binding sites
could first be filled non-specifically. Second, there is
good evidence that recombination requires more than one
resolvase dimer. Footprinting data imply a minimum of 3
resolvase molecules (36) and stoichiometry measurements,
though incomplete, suggest perhaps twice that number of
protomers (2,6). It is likely, therefore, that there will
be more than one resolvase molecule tracking on a substrate
and it is unknown when and where the active ensemble is
formed. There may be two or more expanding loops which
coalesce to form the active recombination complex. The
topological consequences are unchanged as long as no DNA is

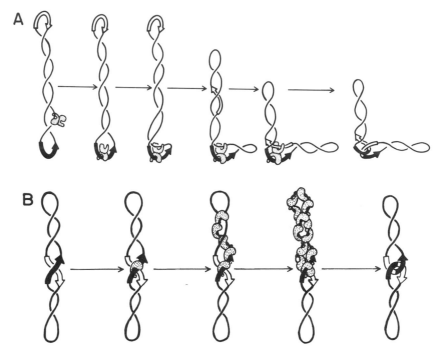

FIGURE 4. Class IV models for synapsis of res sites.
     A. Looping. A functionally dimeric resolvase
(stippled) binds DNA nonspecifically and translocates
randomly along the DNA until reaching a res site filled
arrow) where it binds tightly and specifically. The second
subunit binds the adjacent DNA in a single orientation,
trapping a small loop of DNA between the two resolvase
binding sites. One-dimensional diffusion of the second
subunit along the DNA causes the loop to expand and
contract. When this subunit reaches a second res site (open
arrow) in the proper orientation, the enzyme changes
conformation, activating breakage and reunion. At synapsis,
parental supercoils are maintained within each inter-res
site domain and there is no net intertwining of the two
domains. Strand exchange is directed by resolvase with a
specific geometry, resulting in singly interlinked
recombinants.
     B. Protein scaffold. Cooperative interactions between
asymmetric resolvase molecules bound to DNA results in their
polymerization and formation of a protein scaffold along the
DNA. Scaffold construction could initiate anywhere along
the DNA including a res site as shown. The resolvase

passed over when the ensemble is formed.

There are two tests of the looping model that can distinguish it from the other models described. First, any tracking scheme predicts an orderly search for a second site such that the enzyme encounters sites sequentially according to their position along the DNA. With p51[2], which contains four _res_ sites in the same orientation, adjacent sites should recombine before opposing sites. Recombination between opposing sites will generate catenanes of equal size daughter rings, whereas adjacent site recombination will generate linked small and large daughter rings (Fig 5). It is difficult to distinguish these catenanes by gel electrophoresis. They have the same total size and mobility, and after freeing the products by restriction enzyme digestion the results are obscured by multiple recombination events that yield some of the same progeny rings. Therefore, we used electron microscopy to distinguish product ring size while maintaining topological linkage. We found that more than 90% of the primary recombination events occurred between neighboring _res_ sites. This result, although still preliminary, implies that resolvase recognizes the spatial arrangement of _res_ sites and, if it traces along the DNA, only rarely passes a site.

There is another test of the looping model. As pointed out above, after resolvase has bound one DNA segment, it must bind an adjacent segment of DNA to initiate the search for the second site. Thus, the entrapped loop must initially be small. If two non-specific reporter rings are catenated to the standard substrate, they should be excluded from this small domain (Fig 6). As the loop expands during the search for the second site, the rings should remain excluded since resolvase maintains continuous contact with the DNA. The looping model thus predicts that the two reporter rings would always segregate together. ϕX174 RF I DNA rings were catenated to the standard resolvase substrate

---

FIGURE 4 (continued)

molecules would be aligned in a single orientation along the DNA fixed by the initial resolvase interaction, and in this manner would only recognize sites on the same molecule and in the same orientation. The model does not specify the geometry of the synapsed structure, but here we show the sites brought close by the plectonemic parental supercoils to account for the simple structure of recombinant catenanes.

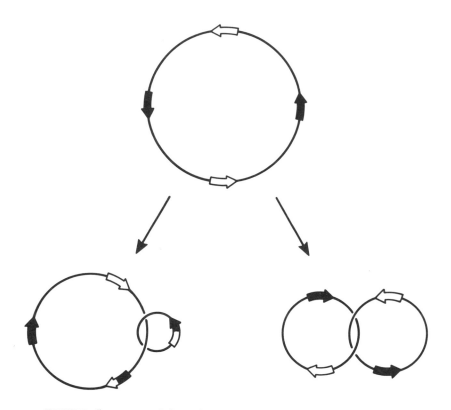

FIGURE 5. Recombination of a four-site plasmid. A plasmid with four directly repeated res sites is shown with the sites alternately open and filled for ease of distinguishing crossover points. Recombination can occur between either neighboring sites (lower left) or opposing sites (lower right). If the sites are spaced regularly as shown, opposing site recombination will give rise to equal size catenated daughter rings, whereas recombination between adjacent sites will generate catenanes with one ring triple the size of its sibling. Tracking mechanisms require that neighboring sites recombine preferentially.

by DNA gyrase, and the product containing one pRR51 and two φX rings was isolated. This species was treated with resolvase and the products analyzed by electron microscopy. The analysis of the first experiment showed about an equal number of recombinants in which the reporter rings segregated together and those in which one had gone to each

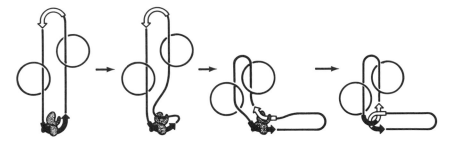

FIGURE 6. Reporter ring test of the looping model. An abbreviated form of the recombination scheme depicted in Fig 4A is shown, except that two rings are catenated to the resolvase substrate. The initial DNA loop between the res sites is small enough to exclude the reporter rings and they remain so as the loop expands, and both res sites are finally bound. After recombination, both reporter rings will therefore be linked to only one of the daughter rings. In the other models considered, the reporters segregate independently.

daughter. This is direct evidence against the looping model.

There are speculative modifications of the looping model that can explain the reporter ring results. If resolvase not only slides along DNA but also can move by direct transfer from one DNA segment to another or by hopping (35), then a reporter ring could enter the expanding loop generated during tracking. Resolvase would still have to maintain a specific orientation relative to the DNA as it moves and transfer only to segments within the ring. Otherwise it would lose the ability to distinguish direct from inverted and intra- from intermolecular sites. This is disturbing though because it requires that resolvase allow a ring to pass through (the reporter ring) but never hop on to a second ring (the catenated single res site plasmid). Furthermore, the jumps must be high enough for a ring to pass but short enough that resolvase does not readily skip a site on the four res site plasmid. Another possibility is that the enzyme is doughnut-shaped with its two DNA binding sites inside. It could begin its search for res sites by slipping onto a hairpin end of a DNA circle, just as one places a napkin ring on a napkin. Each side of the DNA hairpin would slide through one binding site until a res site is encountered. The DNA would become fixed at that

site but continue moving through the other site until a
second res sequence is reached. This model shares all the
properties of the simple looping model except the
predictions of the reporter ring experiment. Resolvase
could first slip onto a reporter ring, search unsuccessfully
for res sites and then pass directly onto the substrate DNA
by a short jump. This would trap a single reporter ring
inside the domain, and the ratio of isolated to segregated
ϕX rings in the product would reflect the number of times
resolvase initiated its search on a ϕX circle versus the
substrate circle.

    Resolvase could also move by cooperative polymerization
along the DNA to form a protein scaffold between the sites
(Fig 4B); this type of cooperative coating of the DNA has
been seen with Int (32). Resolvase would then only utilize
intramolecular res sites. If the interaction between
resolvase molecules were asymmetric, the chain would be
oriented with respect to the DNA and would act only on res
sites in the same orientation. This model also explains the
preference for neighboring res sites and is consistent with
the results of the reporter ring experiment since an isolated domain is not formed. It does not explain the singly-
linked products because it lacks the appealing feature of
tracking which allows resolvase to direct and maintain the
two res sites close to each other at synapsis. Moreover,
footprinting studies with resolvase show binding only at the
res sites and not in between (36). Finally, if each protomer
extends over 20 bp then 300 would be required to coat the
entire distance between res sites in pRattL^2 and this is an
order of magnitude greater than the observed stoichiometry.

    The features of resolvase that each of the models
explains are summarized in Table 2. No model is free of
contradictory observations (X) or several unexplained
properties (-). Since some of the data is preliminary it is
possible that that the discrepancies will disappear with
corroboratory experiments although we doubt it. The simple
toroidal supercoil (class I), wrapping (class III), and
protein scaffold models (class IV) have little in their
favor. The most appealing part of the class IV looping
model is its power in explaining the widest range of
observations and perhaps a variant can avoid the pitfalls of
the example considered. The class II conveyor belt model
explains less but the contrary indications are less serious;
it should be tested directly. Perhaps most reasonable are
hybrid models that incorporate features of both enzyme and
DNA structure to extend their explicatory power.

Table 2

Evaluation of Models for Resolvase[a]

RESOLVASE PROPERTY	Class I Toroidal supercoils	II Conveyor belt	III Wrapping	IV Looping	IV Protein scaffold
No intermolecular reaction	−	+	−	+	+
Directly repeated sites required	−	+	−	+	+
Singly-interlinked catenane products	+	+	+	+	−
Catenane complexity independent of σ	X	+	+	+	−
Neighboring sites preferentially recombined	−	−	−	+	+
Reporter rings segregate independently	+	+	+	X	+
Recombination requires about ten protomers per substrate	0	0	X	0	X
Recombination pattern is very different from Int and Cre	−	−	+	+	+

KEY   + = model explains property
0 = property not specified by model but easily could be
− = model does not explain property
X = property contradicts model

[a]The models are explained more fully in the text. The "toroidal supercoils" model postulates that the daughter DNA domains are not intertwined in the substrate and therefore recombination results in simple catenanes (Fig 2C). The substrate is postulated in the "conveyor belt" model to have the normal plectonemic supercoils, but recombination only occurs when res sites have moved to a side-by-side position where they are interwrapped and the daughter domains are not intertwined (Fig 2B). In the "wrapping" model the substrate supercoils become toroidal as a result of enzyme binding. In the "looping" model a dimeric resolvase traps a small loop of DNA between its DNA binding sites. Translocation along the DNA causes the loop to expand, and when both subunits have bound res sites recombination occurs (Fig 4A). In the "protein scaffold" model the res sites are brought together after resolvase molecules have contiguously bound to the substrate between res sites (Fig 4B).

ACKNOWLEDGMENTS

We thank Andrzej Stasiak and Theo Koller for the electron microscopy of resolvase products, Richard Kostriken for a helpful discussion on alternatives to the looping model, and Elisabeth Lindheim for invaluable assistance in preparing the manuscript.

REFERENCES

1. Nash HA (1981). Integration and excision of bacteriophage λ: the mechanism of conservative site specific recombination. Ann Rev Genet 15:143.
2. Reed RR (1981). Transposon-mediated site-specific recombination: a defined in vitro system. Cell 25:713.
3. Abremski K, Hoess R, Sternberg N (1983). Studies on the properties of P1 site-specific recombination: evidence for topologically unlinked products following recombination. Cell 32:1301.
4. Echols H, Guarneros G (1983). The lysogenic program: control of integration and excision. In Hendrix R, Weisberg R, Stahl F, Roberts J (eds): "Lambda II," Cold Spring Harbor: Cold Spring Harbor Laboratory, in press.
5. Weisberg RA, Landy A (1983). In Hendrix R, Weisberg R, Stahl F, Roberts J (eds): "Lambda II," Cold Spring Harbor: Cold Spring Harbor Laboratory, in press.
6. Krasnow MA, Cozzarelli NR (1983). Site-specific relaxation and recombination by the Tn3 resolvase: recognition of the DNA path between oriented res sites. Cell 32:1313.
7. Shapiro JA (1979). Molecular model for the transposition and replication of bacteriophage Mu and other transposable elements. Proc Nat Acad Sci USA 76:1933.
8. Heffron F (1983). Tn3 and its relatives. In Shapiro JA (ed): "Mobile Genetic Elements," New York: Academic Press, in press.
9. Austin S, Ziese M, Sternberg N (1981). A novel role for site-specific recombination in maintenance of bacterial replicons. Cell 25:729.
10. Nash HA, Robertson CA (1981). Purification and properties of the Escherichia coli protein factor required for λ integrative recombination. J Biol Chem 256:9246.

11. Abremski K, Gottesman S (1982). Purification of the bacteriophage λ xis gene product required for λ excisive recombination. J Biol Chem 257:9658.
12. Mizuuchi K, Gellert M, Weisberg RA, Nash HA (1980). Catenation and supercoiling in the products of bacteriophage λ integrative recombination in vitro. J Mol Biol 141:485.
13. Cozzarelli NR (1980). DNA gyrase and the supercoiling of DNA. Science 207:953.
14. Gellert M (1981). DNA topoisomerases. Ann Rev Biochem 50:879.
15. Wang JC, Liu LF (1979). DNA topoisomerases: enzymes that catalyze the concerted breaking and rejoining of DNA backbone bonds. In Taylor, JH (ed): "Molecular Genetics," New York: Academic Press, part 3, p 65.
16. Kikuchi Y, Nash H (1979). Nicking-closing activity associated with bacteriophage λ int gene product. Proc Nat Acad Sci USA 76:3760.
17. Heffron F, McCarthy BJ, Ohtsubo H, Ohtsubo E (1979). DNA sequence analysis of the transposon Tn3: three genes and three sites involved in transposition of Tn3. Cell 18:1153.
18. Reed R, Shibuya GI, Steitz JA (1982). Nucleotide sequence of γδ resolvase gene and demonstration that its gene product acts as a repressor of transcription. Nature 300:381.
19. Kostriken R, Morita C, Heffron F (1981). Transposon Tn3 encodes a site-specific recombination system: identification of essential sequences, genes and actual site of recombination. Proc Nat Acad Sci USA 78:4041.
20. Muster CJ, MacHattie LA, Shapiro JA (1983). pλCM system: observations on the roles of transposable elements in formation and breakdown of plasmids derived from bacteriophage lambda replicons. J Bacteriol 153:976.
21. Pollock TJ, Nash HA (1980). Catenation of the products of integrative recombination: comparison of supertwisted and nonsupertwisted substrates. In Alberts B (ed): "Mechanistic Studies of DNA Replication and Genetic Recombination," New York: Academic Press, p 953.
22. Depew RE, Wang JC (1975). Conformational fluctuations of DNA helix. Proc Nat Acad Sci USA 72:4275.

23. Pulleybank DE, Shure M, Tang D, Vinograd J, Vosberg H-P (1975). Action of nicking-closing enzyme on supercoiled and non-supercoiled closed circular DNA: Formation of a Boltzmann distribution of topological isomers. Proc Nat Acad Sci USA 72:4280.

24. Krasnow MA, Stasiak A, Spengler SJ, Dean F, Koller Th, Cozzarelli NR. Determination of the absolute handedness of knots and catenanes of DNA., submitted.

25. DiCapua E, Engel A, Stasiak A, Koller Th (1982). Characterizations of complexes between RecA protein and duplex DNA by electron microscopy. J Mol Biol 157:81.

26. Gray, HB (1967). Sedimentation coefficient of polyoma virus DNA. Biopolymers 5:1009.

27. Brody GW, Fein DB, Lambertson H, Grassian V, Foos D, Benham CJ (1983). X-ray scattering from the superhelix in circular DNA. Proc Nat Acad Sci USA 80:741.

28. Hagerman PJ (1981). Investigation of the flexibility of DNA using transient electric birefringence. Biopolymers 20:1503.

29. Nash HA, Pollock TJ (1983). Site-specific recombination of bacteriophage lambda: the change in topological linking number associated with exchange of DNA strands. J Mol Biol, in press.

30. Pollock TJ, Nash HA (1983). Knotting of DNA caused by a genetic rearrangement: evidence for a nucleosome-like structure in site-specific recombination of bacteriophage lambda. J Mol Biol, in press.

31. Krasnow MA, Cozzarelli NR (1982). Catenation of DNA rings by topoisomerases: mechanism of control by spermidine. J Biol Chem 257:2687.

32. Better M, Lu C, Williams RC, Echols H (1982). Site-specific DNA condensation and pairing mediated by the int protein of bacteriophage λ. Proc Nat Acad Sci USA 79:5837.

33. Abremski K, Gottesman S (1981). Site specific recombination: Xis-independent excisive recombination of bacteriophage lambda. J Mol Biol 153:67.

34. Craig NL, Nash HA (1983). The mechanism of phage lambda site-specific recombination: collision versus sliding in att site juxtaposition. ICN-UCLA Symposium on Molecular and Cellular Biology, this volume.

35. Winter RB, Berg OG, von Hippel PH (1981). Diffusion-driven mechanisms of protein translocation on nucleic acids. 3. The Escherichia coli lac repressor-operator interaction: kinetic measurements and conclusions. Biochemistry 20:6961.

36. Grindley NDF, Lauth MR, Wells RG, Wityk RJ, Salvo JJ, Reed RR (1982). Transposon-mediated site-specific recombination: identification of three binding sites for resolvase at the res site of γδ and Tn3. Cell 30:19.

Mechanisms of DNA Replication and Recombination, pages 661 –670
© **1983 Alan R. Liss, Inc., 150 Fifth Avenue, New York, NY 10011**

IDENTIFICATION OF INHIBITORS OF TN3 RESOLVASE

Michael A. Fennewald[+] and John Capobianco[*]

+Department of Microbiology, University of Notre
Dame, Notre Dame, Indiana   46556

*Abbott Laboratories, North Chicago,
Illinois

ABSTRACT   We have devised a genetic screening assay for
the identification of inhibitors of site-specific recombination carried out by Tn3 resolvase. This procedure employs the λXJS845 phage which carries a gene
for streptomycin resistance between two Tn3 resolvase
recombination sites. In the presence of resolvase,
site specific recombination deletes the streptomycin
gene. We have tested over 6,000 compounds for the
ability to maintain the streptomycin resistance inside λXJS845-infected cells carrying Tn3 resolvase,
and have discovered 26 that produced streptomycin-
resistant colonies. All of these 26 were then tested
for their ability to inhibit site-specific recombination in vitro using homogeneous Tn3 resolvase. 5 of
26 inhibited the reaction in vitro. In addition to
site-specific recombination, at least one of the compounds also inhibits the topoisomerase activity of
Tn3 resolvase. This is the first report of the isolation of specific inhibitors of site-specific recombination.

INTRODUCTION

Tn3 is a transposon of 4957 base pairs that can dupli-
cate itself and insert a copy into a different DNA site
(for a review, see ref. 1). In addition to insertion, Tn3
also mediates a number of genome rearrangements such as
deletions, inversions and replicon fusions. Tn3 encodes
three proteins: 1) a beta-lactamase that hydrolyzes peni-
cillins such as ampicillin or carbenicillin and makes the
cell resistant to these drugs, 2) a "transposase" (the tnpA

gene product) that is absolutely required for transposition, and 3) a resolvase (the tnpR gene product) that carries out site-specific recombination and also represses synthesis of itself and transposase. Resolvase was formerly known as the "Tn3 repressor" because the repressor activity was uncovered first.

Transposition of Tn3 seems to proceed in two steps, as formulated by Shapiro (2). First, the replicons containing the original and new insertion sites fuse with an exact duplication of Tn3 at the junctions. These two Tn3s are oriented as direct repeats. The only Tn3 protein known to be required for replicon fusion is the transposase (3,4). The resolvase is not directly involved in this step and only influences it by repressing the level of transposase. The 38 base pair inverted repeats of Tn3 are required in cis for this step to occur (5). The second step is the resolution of the replicon fusion containing two Tn3s into two separate molecules each with one copy of Tn3. This involves site-specific recombination carried out by the resolvase and the point of recombination is located between the tnpA and tnpR genes and has been termed the "res" or the "IRS" site (6,7). This recombination site also is the site where resolvase represses synthesis of the tnpA and tnpR genes. The res site consists, at most, of only several hundred base pairs between tnpA and tnpR. No other Tn3 DNA is needed for site-specific recombination to occur (8,9). The two steps for Tn3 transposition are independent of each other, and the replicon fusions are a stable intermediate, especially in a recA strain.

Tn3 is related by sequence homology to several other transposons that have been useful in the investigation of Tn3 transposition. TnA is very closely related to Tn3 and has 3 very similar genes that complement their Tn3 counterparts. The transposon γδ is also related to Tn3 but not as closely as TnA. The transposon γδ doesn't have a beta-lactamase and the tnpA genes will not complement (R. Reed, personal communication); but the tnpR genes will complement efficiently and either res site can be used. These two tnpR genes share substantial DNA sequence and amino acid sequence homology (10). The γδ resolvase has been purified to homogeniety and performs recombination at two, directly repeated res sites in vitro. Resolvase requires only Mg^{++} as a cofactor and does not require any proteins or high energy cofactors (8).

We report here a search for inhibitors of resolvase. We devised a screening procedure using a bacteriophage λ carrying two directly repeated <u>res</u> sites. We have screened over 6,000 compounds at random. 26 of these inhibited <u>in vivo</u>, and 5 also inhibited pure Tn3 <u>in vitro</u>. This is the first report of inhibitors for site-specific recombination.

MATERIALS AND METHODS

Screening Assay for Isolation of Inhibitors. This assay uses MF211, a K12 <u>E. coli</u> that is a C600 derivative containing the plasmid RSF1050 which is a high copy plasmid containing Tn3 and which produces high levels of resolvase (11). The bacteriophage used was λXJS845, which has been described by Muster et al. (12). An overnight culture of MF211, grown in LB, was diluted 1/10 into LB with 0.2% maltose, and grown for about 4 hours at 32° until the O.D. 600 was 6.5. This culture was then diluted 1/1 with 10mM MgSO$_4$. Then 1.0 ml of a λXJS845 lysate of ~3x10^4 trans- ductants/ml was mixed with 2.5 ml of soft agar and then overlayed onto an LB chloramphenicol (40 mg/ml) strepto- mycin (25 mg/ml) plate. After 10 min at room temperature, 1.0 ml of the diluted cell culture was mixed with 2.5 ml of soft agar and overlaid on the first layer.

The compounds to be tested were dissolved in DMSO at 2-5 mg/ml, and spotted onto a sterile filter disc which was placed on top of the second layer. Approximately 20-50 mg were thus placed on each disc. 12 discs were put on each plate, including one control with DMSO alone. The plates were then incubated at 32° and scored after 24, 48, and 72 hours for the presence of visible colonies.

In Vitro Assay of Tn3 Resolvase. These were performed as described by Reed (8) with the following changes: 1) Tn3 resolvase in place of γδ resolvase, 2) 1% DMSO present (the inhibitors were dissolved in DMSO and diluted to 10% DMSO before addition to the reaction), and 3) 1-2mM EDTA was present to bind any metal ions that might be present in the compounds. The reaction was stopped with 0.4% SDS and the products were separated by electrophoresis through 1% agarose gels.

RESULTS

The screening procedure for the inhibitors of Tn3 resolvase employs the λXJS845 phage. This phage, isolated by
Muster et al. (12), is a defective lambda that replicates
as a plasmid inside the cell because it lacks λN function.
It also has a Sm^R (streptomycin resistance) gene which is
flanked by two directly repeated copies of TnA ΔAp. TnA
ΔAp contains a <u>res</u> site but is deleted for its resolvase
gene. λXJS845 also contains a gene for Cm^R (chloramphenicol resistance) (Figure 1). In the presence of resolvase,
site-specific recombination occurs between the two <u>res</u>
sites and the Sm^R gene is deleted from the rest of the λ.
Because the DNA between the two <u>res</u> sites does not contain
an origin of replication, the resolution product with the
Sm^R gene is very quickly lost <u>in vivo</u>. This process of
resolution and loss of the Sm^R gene is so efficient that
cells with resolvase that are infected with λXJS845 are Cm^R
but Sm^S. Infected cells without resolvase are Cm^R and Sm^R.
Reed (7) has also observed that resolution and loss of
intervening DNA is fast <u>in vivo</u>. The logic of the screen is
that if resolvase is present but inhibited, then the cells
infected with λXJS845 will be $Cm^R Sm^R$ whereas a cell with
uninhibited resolvase will be $Cm^R Sm^S$.

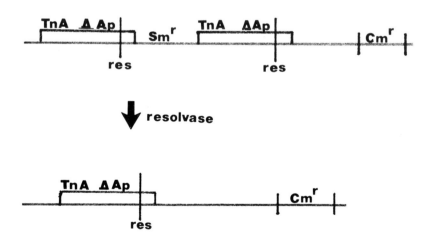

FIGURE 1: Structure of λXJS845 and the effect of the
Tn3 resolvase upon it. Cm^r = chloramphenicol resistance.
Sm^r = streptomycin resistance. Res is the site of resolvase-
mediated recombination.

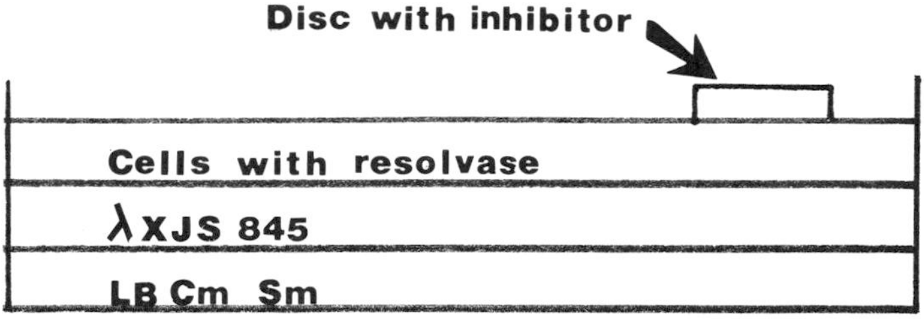

Disc with inhibitor

FIGURE 2. The Resolvase Inhibition Assay. This is the cross section of an agar plate with the inhibition assay. The top two layers are soft agar overlays with the cells or phage. The bottom layer is a regular agar plate with chloramphenicol at 40μg/ml and streptomycin at 25μg/ml. Inhibitors were dissolved in DMSO at 2-5 mg/ml.

The procedure employs two soft agar overlays (Figure 2). The bottom layer contains the λXJS845 phage and the top layer contains cells with resolvase. Compounds to be tested were put onto a filter disc which was placed on top of the overlays. Inhibitors of resolvase produce Sm^R colonies if they are taken up into the cell and inhibit resolvase more than cell growth.

Using this assay, we tested, at random, over 6,000 compounds in a collection maintained by Abbott Laboratories. 26 of these compounds produced 1 or more Sm^R colonies. The best in vivo inhibitor, using the plate assay, was compound A20832 (Figure 3). The 26 compounds that were positive in the in vivo assay were then tested in an in vitro assay using homogeneous Tn3 resolvase. The in vitro assay used is very similar to that described by Reed (8) with homogeneous Tn3 resolvase used instead of γδ resolvase. This assay measures intramolecular site-specific recombination with the pRR51 plasmid which contains two res sites. This reaction requires only Mg^{++}, but the dependence of recombination on enzyme concentration is sigmoidal, with about 15-20 resolvase monomers per res site required. This non-linear result is not due to formation of a resolvase multimer and has been observed by both Reed (8) and Krasnow and Cozzarelli (personal communication). Also, enzyme turnover has not been observed. This makes the measurement of inhibition of resolvase much more difficult than for other, more conventional, enzymes and has precluded accurate K_I measure-

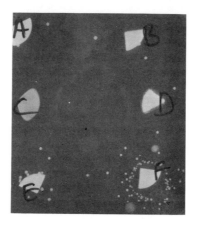

FIGURE 3. In Vivo Inhibition of Site-Specific Recombination. This is the double layer assay as diagrammed in Figure 2 with λXJS845 (1X10⁷ transducing particles) in the bottom layer. The inhibitors were dissolved in DMSO at 100 mg/ml. A = DMSO alone, B = A1062, C = A9387, D = A18947, E = A20812, and F = A20832.

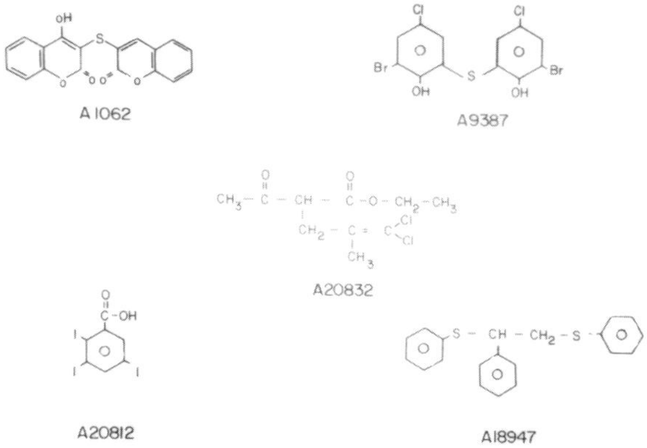

A 1062

A9387

A20832

A20812

AI8947

FIGURE 4. Structures of the Resolvase Inhibitors. These structures were first determined when the compounds were first synthesized at Abbott Labs and have since been confirmed.

ments. The approach we have taken is to measure the amount of inhibitor needed to give a 50% reduction in product with a set amount of DNA and enzyme.

5 of the 26 compounds inhibited Tn3 resolvase in vitro (Figure 4). 3 of these compounds show some structural similarity (A1062, A9387, A18947) but the other two A20812 and A20832 are not very similar to any of the others. We are now investigating analogues of these compounds to try to establish a structure/function relationship. The other 21 showed no inhibition at $\geq$ 200 µg/ml of the drug.

Although most of the compounds were negative for the in vitro assay, we do not conclude that these compounds are not inhibitors. It could be that the cell metabolizes these in vivo to an inhibitor or that we are not assaying resolvase under the proper conditions to disclose the inhibition. Two ways that this screen can possibly produce false positives are: 1) mutagens that makes the cells Sm^R or 2) compounds that cure the plasmid containing the tnpR gene. We have tested 10 colonies each from A1062 and A20832 and none of them were Sm^R and all still contained RSF1050, the plasmid with the tnpR gene. We have not yet detected compounds producing mutations to Sm^R, possibly because resistance to Sm^R is a very rare chromosomal mutation.

The results with inhibitor A1062 show that the 50% inhibition point is ˜10 µg/ml, with >90% inhibition at 50 mg/ml (Figure 5). In addition to a site-specific recombination activity the Tn3 resolvase also is a Type 1 DNA topoisomerase (Krasnow and Cozzarelli, personal communication). The topoisomerase activity is also strongly inhibited at 50 µg/ml. Thus A1062 inhibits both the site-specific recombination activity and the topoisomerase activity. A1062 is not a general protein inhibitor because it does not effect EcoR1 or BamH1 at 50 mg/ml. For A20832, the compound that produces the largest number of colonies in vivo, the 50% inhibition point is ˜100 mg/ml. Compound A20832 does not effect BamH1 or EcoR1 at 200 mg/ml.

We have not yet determined the mechanisms of inhibition of these compounds. The inhibition of A1062 and A20832 seem to be competitive with the amount of enzyme and not the DNA substrate, but the sigmoid enzyme concentration curve prevents us from measuring this precisely. We also do not know if this inhibition is irreversible or not; nor do we know at what stage in the reaction inhibition by these compounds occurs. These approaches are currently under investigation.

This is the first report of the isolation of inhibitors of recombination. These inhibitors specifically affect the site-specific recombination mediated by the Tn3 resolvase. They also inhibit the topoisomerase activity of the resolvase. We have not yet investigated whether these inhibitors will also effect other types of site-specific recombination that are related to Tn3 resolvase, such as phase variation in Salmonella or G-loop flipping in the bacteriophage Mu.

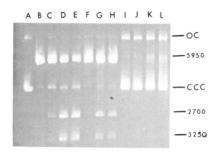

FIGURE 5.   Inhibition of Tn3 Resolvase by A1062.
Tn3 resolvase was incubated as described in Materials and Methods.   The recombined products are a linear 2700 base pair molecule and a 3250 base pair supercoiled molecule.
A = no resolvase, no BamH1; B = no resolvase, BamH1; C = 1.3 u of resolvase, BamH1; D = 2.6 u of resolvase, BamH1;
E = 5.2 u of resolvase, BamH1; F = 2.6 u of resolvase, BamH1, 50μg/ml A1062; G = 2.6 u of resolvase, BamH1, 10μg/ml A1062; H = 2.6 u of resolvase, BamH1, 1μg/ml A1062; I = 2.6 u of resolvase, no BamH1, 50μg/ml A1062; J = 5.2 u of resolvase, no BamH1, 50μg/ml A1062; K = 2.6 u of resolvase, no BamH1, 10μg/ml A1062; L = 5.2 u of resolvase, no BamH1, 10μg/ml of A1062.

ACKNOWLEDGEMENTS

We would like to thank N.R. Cozzarelli and James McAlpine for helpful discussion, Mark Krasnow for purified Tn3 resolvase, and Carol Muster for strains and phages.

REFERENCES

(1)　Grindley NDF　(1983).　Transposition of Tn3 and Related Transposons.　Cell 32:3-5.

(2)　Shapiro JA　(1979).　Molecular model for the transposition and replication of bacteriophage mu and other transposable sequences.　Proc Nat Acad Sci USA 76:1933-1937.

(3)　Chou J, Lemaux PG, Casadaban M and Cohen N　(1979).　Transposition protein of Tn3: identification and characterization of an essential repressor controlled product.　Nature 282:801-806.

(4)　Gill R, Heffron F and Falkow S　(1979).　Identification of the protein encoded by the transposable element Tn3 which is required for its transposition. Nature 282:797-801.

(5)　Gill R, Heffron F, Douglas G and Falkow S　(1978).　Analysis of sequence transposed by complementation of two classes of transposition deficient mutants of Tn3.　J Bacteriol 136:742-756.

(6)　Arthur A and Sherratt D　(1979).　Dissection of the transposition process:　a transposon-encoded site-specific recombination system.　Mol Gen Genet 175: 267-274.

(7)　Reed RR　(1981).　Resolution of cointegrates between transposons γδ and Tn3 defines the recombination site. Proc Nat Acad Sci USA 78:3428-3432.

(8)　Reed RR　(1981).　Transposon-mediated site-specific recombination:　a defined in vitro system.　Cell 25: 713 719.

(9)　Grindley NDF, Lauth MR, Wells RG, Wityk RJ, Salvo JJ and Reed RR　(1982).　Transposon-mediated site-specific recombination:　Identification of three binding sites for resolvase at the res sites of γδ and Tn3.　Cell 30:19-27.

(10)　Muster Carol J, MacHattie Lorne A and Shapiro James A (1983).　pλCM system:　Observations on the roles of transposable elements in formation and breakdown of plasmids derived from bacteriophage lambda replicons. J Bacteriol 153:976-990.

(11)   Heffron F, Bedinger P, Champoux JJ and Falkow S
       (1977). Deletions affecting the transposition of
       an antibiotic resistance gene. Proc Nat Acad Sci
       USA 74:702-706.
(12)   Reed Randall R, Shibuya Grant I and Steitz Joan A
       (1982). Nucleotide sequence of $\gamma\delta$ resolvase gene
       and demonstration that its gene product acts as a
       repressor of transcription. Nature 300:381-383.

Mechanisms of DNA Replication and Recombination, pages 671–684
© 1983 Alan R. Liss, Inc., 150 Fifth Avenue, New York, NY 10011

THE P1 lox-Cre SITE-SPECIFIC RECOMBINATION
SYSTEM: PROPERTIES OF lox SITES
AND BIOCHEMISTRY OF lox-Cre INTERACTIONS[1]

N. Sternberg, R. Hoess and K. Abremski

Basic Research Program-LBI
Frederick Cancer Research Facility
Frederick, Maryland 21701

ABSTRACT   In vivo and in vitro studies on the P1
lox-Cre site-specific recombination system indicate
that: 1. loxP consists of a 34-bp sequence containing a 13-bp inverted repeat separated by an 8-bp
spacer region. DNA sequences flanking this region
of dyad symmetry are not essential for recombination
and, under certain circumstances, a segment of one
of the 13-bp repeats can be dispensed with; 2. The
site of crossover in a recombination event between
loxP-and a site in the bacterial chromosome, loxB,
is located asymmetrically in the 8-bp spacer; 3. A
P1 gene, cre, encodes the only protein (Cre) necessary for recombination between lox sites; 4. Cre
protects the 34-bp region of dyad symmetry and 1-2
flanking bases from nuclease degradation; 5. Segments
of the loxP sequence necessary for binding Cre and
for recombination are separable as shown by the
existence of a lox mutant (loxR) that binds normally
but recombines poorly; and 6. In an in vitro lox-Cre
recombination reaction, Cre produces a high yield
of free supercoiled circular recombinant molecules
when promoting recombination between directly
repeated loxP sites on a supercoiled circle.
Several models are presented that can account for
these results.

[1]Research sponsored by the National Cancer Institute,
DHHS, under contract No. NO1-CO-23909 with Litton Bionetics,
Inc.

INTRODUCTION

Bacteriophage P1 encodes a site-specific recombination system that consists of a site (loxP) at which the
physical exchange of DNA sequences occurs and a gene
(cre) whose product is the only protein required for
recombination (1, 2). cre is located about 400-bp
to one side of loxP, and it encodes a protein of 35 kd.
The direction of cre transcription is away from loxP
(unpublished observations). In the P1 life cycle, this
recombination system has two known functions: the
cyclization of virion DNA after its injection into cells
(3, G. Cohen, unpublished observations) and the dissociation of plasmid prophage dimers (4). The latter function
has the effect of increasing the fidelity of prophage
segregation because it provides substrates for the segregation event. We discuss three aspects of the recombination system: the DNA sequence of the lox site, the
interaction of Cre with lox DNA, and the topological
features of lox recombination as determined in an in
vitro lox-Cre recombination system.

RESULTS AND DISCUSSION

The DNA Sequence of lox Sites.

Much has been learned about the sequence of lox sites
by two types of analyses: recombination between loxP and
a site in the bacterial chromosome, called loxB, which
generates two hybrid sites, loxL and loxR; and structure-
function studies using loxP deletions constructed in vitro.
loxP x loxB Recombination. loxP recombines with
loxP very efficiently. For example, in crosses between
two λ phages containing loxP, recombination frequencies
of >20% are normal (5). In contrast, recombination
between loxP and a site in the bacterial chromosome,
loxB, is very inefficient. Indeed in phage crosses
between a λloxP phage and a λloxB phage, recombination is not detectable over a backgound level of 0.02%
(2). The products of loxP x loxB recombination are two
hybrid sites, designated loxL and loxR. In vivo assays
indicate that loxL sites are as efficient in recombination
as loxP sites, whereas loxR sites are at least 20-fold
less efficient (2). The sequences of all four lox sites
have been determined (Figure 1). loxP contains a 34-bp

sequence consisting of a 13-bp inverted repeat, flanking
an 8-bp spacer. loxB contains a 25-bp sequence that
consists of a 8-bp inverted repeat flanking a 9-bp spacer.
The loxP and loxB inverted repeats are significantly
homologous, but the spacers are very different. Based on
the sequences of the loxL and loxR sites, the crossover
in loxP x loxB recombination can be localized to a TG-
sequence asymmetrically located in the spacer and common
to all four sites. The observed differences in the
recombination properties of the four lox sites may be
caused by differences in their inverted repeat sequences,
their spacer sequences or in the sequences that flank
the 34-bp site.

loxP Deletion Mutants. Deletions in loxP sequences
were generated in vitro using the enzyme Bal-31 starting
from unique restriction sites flanking those sequences
(6). The deletion end points were precisely determined
by DNA sequencing and the functional state of the site
was assessed by either of two assays: plasmid transduction or intraplasmid recombination. In the first assay,

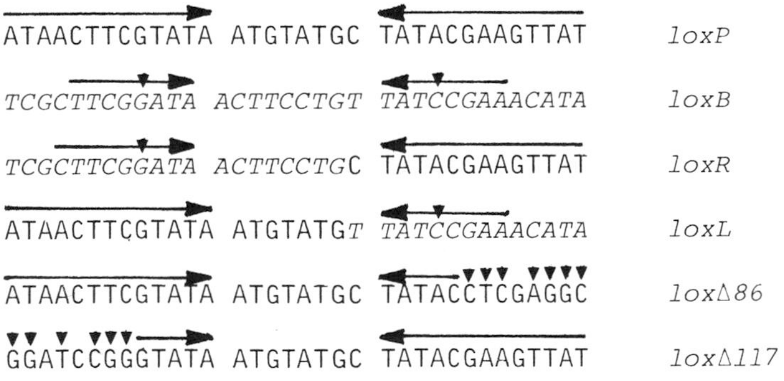

ATAACTTCGTATA	ATGTATGC	TATACGAAGTTAT	*loxP*
TCGCTTCGGATA	*ACTTCCTGT*	*TATCCGAAACATA*	*loxB*
TCGCTTCGGATA	*ACTTCCTGC*	TATACGAAGTTAT	*loxR*
ATAACTTCGTATA	ATGTATG*T*	*TATCCGAAACATA*	*loxL*
ATAACTTCGTATA	ATGTATGC	TATACCTCGAGGC	*lox△86*
GGATCCGGGTATA	ATGTATGC	TATACGAAGTTAT	*lox△117*

Figure 1. Sequences of loxP and various mutant
sites. Arrows indicate the inverted repeat structures
found in loxP and homologous ones found in the mutant
sites. Bacterial DNA in the loxB, loxR, and loxL sites
is shown in italics. Bases not homologous with those in
loxP are indicated with (▼).

a transduction test, plasmids with a functional loxP
site will recombine efficiently in cells with infecting
λ loxP$^+$cre$^+$ DNA to generate a λ plasmid cointegrate
that can be packaged into a phage virion. Since the new
recombinant phage DNA now contains the plasmid bla (ampR)
gene, its frequency in the phage population can be easily
determined by measuring the ability of the lysate to
transduce ampR. If the plasmid contains a functional
loxP site, the frequency of ampR transduction is 10^{-2}
per plaque-forming phage, and if it does not, the frequency is <10^{-6} per plaque-forming phage. The second
assay for loxP function uses a plasmid (pRH43) whose
structure is shown in Figure 2. This plasmid contains
two directly repeated loxP sites: one contained in an
80-bp fragment and the other within a 50-bp fragment.
The two sites flank segments of DNA that contain either
ampR and the pBR322 ori or a kanamycin resistance (kanR)
gene derived from the transposon Tn5. Intraplasmid
recombination between loxP sites should separate the two
DNA segments with only the ampR segment containing ori
capable of replication. Thus, if pRH43 DNA is transformed into a bacterial cell containing a single functional
copy of cre, fewer than 1% of the AmpR colonies will be
also KanR.

Results obtained from studies with deletion mutants
of loxP support the conclusion that the integrity of
most of the 34-bp inverted repeat sequence is important
for loxP function. Deletions that remove either the
left or the right repeat and part of the spacer are
defective. Interestingly, deletions that remove 7
bases from either the left or right repeat (Δ117
or Δ86, Figure 1; R. Hoess, unpublished observations)
are positive in the transduction tests. This finding
suggests that retention of the entire repeat is not
necessary for recombination. This conclusion, however,
must be tempered by the observation that when one of
the loxP sites in pRH43 is replaced by the Δ86 site,
intraplasmid recombination is largely defective; namely,
when that plasmid DNA is transfected into a cell containing Cre, >99% of the AmpR cells are also KanR. Thus,
with regard to this deletion, the two assays give different results. One possible explanation is the different
levels of Cre available under the two assay conditions.
In the transduction assay, the cell produces a large

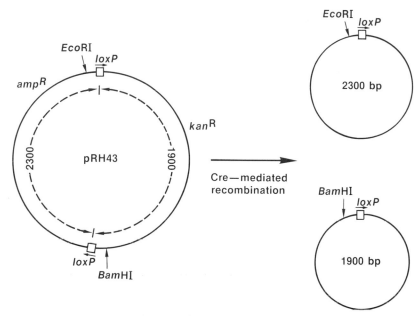

Figure 2. Genetic and physical map of pRH43 substrate DNA and the recombination products. The plasmid pRH43 contains two <u>loxP</u> sites in direct orientation and separated by 1900 bp. The DNA on one side of the <u>loxP</u> sites contains the <u>amp</u>R gene, an origin for replication in <u>Escherichia coli</u>, and a unique <u>Eco</u>RI site. The DNA on the other side contains the <u>kan</u>R gene and a unique <u>Bam</u>HI site. When this plasmid undergoes Cre-mediated recombination, two circular product molecules are produced. One of the products is 2300 bp long, contains the <u>amp</u>R gene, the origin, and an <u>Eco</u>RI site. The other product contains the <u>kan</u>R gene and the <u>Bam</u>HI site.

amount of Cre made from many copies of replicating λ<u>cre</u>[+] DNA. In the intraplasmid recombination assay, significantly less Cre is made from a single copy of a λ<u>cre</u>[+] prophage inserted in the chromosome of the cell. This latter result suggests that Δ86 is defective in its interaction with Cre and that this defect is overcome by a higher Cre concentration in the cell. Another conclusion we can draw from these deletion analyses is that two small <u>loxP</u> sites, such as found in pRH43, can recombine efficiently with each other.

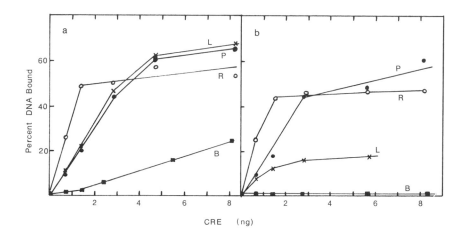

Figure 3. Binding of Cre to DNA containing lox sites. Purified Cre protein, in the amounts indicated, was added to 0.05 ml reaction mixtures containing 50 mM Tris-HCl, pH 7.5, 30 mM NaCl, 1 mM EDTA, 5% glycerol, 200 µg bovine serum albumin/ml, and 0.05 µg [^{3}H]thymidine-labeled lox site containing DNA. The reactions were incubated for 10 min at 30°C. Heparin was added to the samples in (b) at 5 µg/ml for 2 min. They were diluted with 1 ml of binding buffer (50 mM Tris-HCl, pH 7.5, 1 mM EDTA, 50 mM KCl and 10% glycerol), and filtered through a nitrocellulose filter. The filter was washed with 1 ml of binding buffer, dried, and the radioactivity was measured. The percentage of DNA bound was determined by the formula cpm of DNA bound to filter/cpm of total DNA X 100.

lox-Cre Binding and Footprinting Experiments.

We recently purified Cre to homogeneity (K. Abremski, unpublished observations) and used the purified protein to measure both its ability to bind to lox sites as well as its ability to protect lox sequences from nuclease degradation. As shown in Figure 3, we found that in the

absence of heparin, Cre binds efficiently to DNA containing loxP, loxL and loxR sites, but poorly to DNA containing loxB. In the presence of heparin, which should
compete for nonspecific binding of the protein to DNA,
the binding of Cre to loxP and loxR sites is unaffected.
The binding to loxL is reduced 3- to 4-fold, and the
binding to loxB is completely abolished. These results
contrast with other data showing that loxL is a much
more efficient recombination site than loxR and, therefore,
suggest that loxR is defective at some stage in the
recombination beyond the binding of Cre. We concluded
from footprinting experiments in which Cre was first
bound to an intact loxP site and the DNA then subjected
to limited degradation with DNase I that Cre protects a
segment of DNA that includes the 34-bp inverted repeat
and 2-3 flanking bases (Figure 4).

A Model For loxP-Cre Interaction During Recombination

    Important to the consideration of a model for loxP-
Cre interaction during recombination is the fact that
the lox site has directionality. That is, in pRH43
(Figure 2) both sites are oriented in the same direction
and, as a consequence of recombination at lox, the intervening DNA containing kanR is separated from the DNA
containing ampR and ori. When one of the sites is placed
in the opposite orientation, the DNAs between the loxP
sites are no longer separated from each other but are
simply inverted. Data from a combination of deletion
analysis and DNA footprinting experiments clearly show
that sequences outside of the 34-bp loxP site do
not play a role in recombination. This leaves only the
8-bp spacer region as the asymmetric portion of the
lox site, and hence we conclude that the spacer
gives the site its directionality.
    There are several possible ways that the spacer
might provide directionality to the recombination
reaction. One model attributes that directionality to
specific interactions between Cre and spacer sequences.
We propose that two molecules of Cre bind to a given lox
site, one to each inverted repeat, such that the molecules
are oriented head to head. Each of these molecules
would contact different bases in the spacer region, and
presumably these different contacts would affect the
cutting and ligation of lox sequences and would give the
site its directionality.

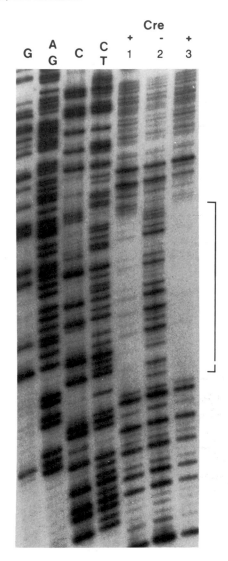

Figure 4.   Protection of the loxP site with purified Cre.

Gel shows an example of a footprinting experiment
using a 5' ^{32}P-end labeled restriction fragment containing loxP. Cre was bound under conditions described in
Figure 3 and then treated with DNaseI (200 ng/ml). The
DNA was then run on a sequencing gel adjacent to the
same fragment sequenced by the procedure of Maxam and
Gilbert (7). Lane 1 shows protection by Cre in the
presence of heparin; lane 2 serves as a control; no Cre
was added but the fragment was digested with DNaseI; and
lane 3 shows protection by Cre with no heparin added.
Protected sequences are outlined by the bracket. A
summary of protection experiments done on both strands
of loxP is shown diagrammatically below. Heavy lines
(▬) indicate bases protected by Cre against DNaseI
cleavage. Because DNaseI does not cleave at every
base, there is a 1-base uncertainty at each end noted by
(•) where the base may or may not be protected.

A slightly more complex model proposes that the spacer
region performs two functions. One, as already mentioned,
is to interact directly with Cre to initiate the recombination reaction. A second is to serve as a region of
homology for pairing between two lox sites. The necessity
for that pairing could account for the directionality
of loxP and could explain the properties of the various
lox sites. For instance, it could easily explain why
loxL is able to recombine efficiently with loxP, whereas
loxR and loxB cannot. The loxL spacer sequence is nearly
homologous (one base mismatch) to that of loxP, whereas
the loxR and loxB spacers are very different from the
loxP spacer. Thus, pairing could occur in the former
case, but not in the latter. Clearly, this explanation
makes a number of strong predictions regarding the
behavior of mutants in the spacer region. We hope soon
to test these predictions with the appropriately constructed mutants.

In Vitro lox-Cre Recombination

The purified Cre protein can carry out recombination
between loxP sites in vitro. The conditions of the in
vitro reaction are quite minimal (50 mM Tris-HCl pH 7.5,
30 mM NaCl, 10 mM MgCl$_2$) (8). This reaction does not

require an external energy source such as ATP. Under
conditions of protein excess (about 25 Cre molecules per
DNA molecule), about 70% of the starting DNA substrate
is recombined within 2-3 min of the addition of Cre at
25°C. The Cre protein does not act catalytically in
this system as evidenced by the observation that subop-
timal concentrations of Cre give correspondingly less
recombination even after prolonged incubation. Intramolecular recombination between two directly repeated
loxP sites occurs with equal efficiency whether the
initial substrate is linear, relaxed, or supercoiled.
Linear or supercoiled DNA with loxP sites in inverted
orientations are also efficiently recombined to invert
the sequences between the sites. Intermolecular recombination between a supercoiled circle and a linear molecule to generate a longer linear molecule also occurs but
somewhat less efficiently (~10-20% completion).

The most striking result obtained with the in vitro
system is the observation that, when directly repeated
loxP sites on a supercoiled circle recombine, about 50%
of the recombinant products are free supercoiled circles.
The remainder of the recombinants are catenated supercoiled circles. Thus, when pRH43 DNA is recombined in vitro,
two free supercoiled circles are produced that are easily
seen after gel electrophoresis (Figure 5). The catenated
circles migrate with unrecombined pRH43 molecules in
these gels. To determine the fraction of recombinants
that are free, the reaction products are treated with
either EcoRI or BamHI before electrophoresis. Each
enzyme cleaves once in pRH43 DNA in different segments
between loxP sites, resulting in the linearization of
one of the product circles, thereby dissociating the
catenated circles. The other circular product is untouch-
ed, and its relative intensity in gels before and after
restriction enzyme digestion is thus a reflection of the
fraction of recombinants that are free. To accurately
determine the fraction of free recombinant molecules,
[3H]thymidine-labelled pRH43 DNA was used and the recombinant bands after electrophoresis were cut out of the
gel and counted. The high yield of free circles is an
unexpected result because the initial starting substrate
is highly supercoiled and because the distance between
loxP sites is large, >2 kb. Thus, when loxP sites come
together before recombination, the DNA segments between
those sites are likely to be highly twisted around each
other. Subsequent recombination will produce multiply

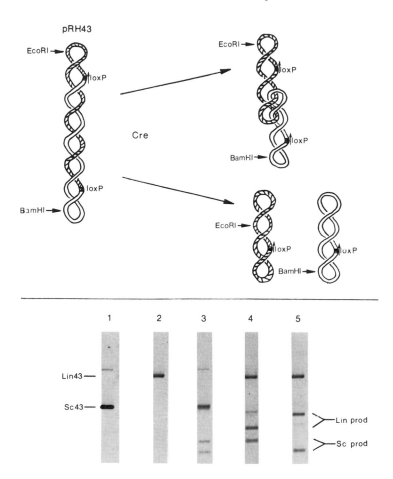

Figure 5. Production of free supercoiled molecules after in vitro site-specific recombination of pRH43. The upper part of the figure shows the types of products expected from recombination of supercoiled pRH43. The two small product molecules are supercoiled and could be topologically linked to one another, or they could be free circles.

The lower part of the figure shows the agarose gel patterns of the pRH43 DNA before and after treatment with Cre protein. The recombination reactions were carried out in a volume of 0.10 ml. The reactions contained 50 mM Tris-HCl, pH 7.5, 30 mM NaCl, 5 mM spermidine, 1 mM EDTA, 200 µg bovine serum albumin/ml, 0.6 µg of supercoiled [^{3}H]thymidine-labeled pRH43 DNA and ± 125 ng purified Cre protein. The samples were incubated at 37°C for 15 min, and the reactions were stopped by heating to 70°C for 10 min. After addition of 10 mM MgCl$_2$, some samples were treated with EcoRI or BamHI as indicated. Lane 1: pRH43 supercoiled DNA; lane 2: pRH43 DNA treated with EcoRI; lane 3: pRH43 supercoiled DNA + Cre; lane 4: pRH43 DNA + Cre, then digested with BamHI; lane 5: pRH43 DNA + Cre, then digested with EcoRI.

interlocked circles. In the case of two other in vitro recombination systems, λ and Tn3, recombination between sites on a similar supercoiled substrate produces <5% free circles. We conclude that the high yield of free supercoiled product in the lox-Cre system represents some special feature of that system.

Three models have been proposed to account for the above result: a topoisomerase model, a nucleosome model, and a tracking model (8). In the topoisomerase model, the actual product of recombination is predominantly catenated circles with Cre acting as a type II topoisomerase to separate those circles. This model predicts that the free circular recombinants would be largely relaxed since the effect of decatenation is to put positive turns in the DNA. This is not the result observed: the superhelical densities of the recombinant products do not differ significantly from that of the starting circles. Furthermore, all attempts to demonstrate Cre-mediated topoisomerase activity with the use of superhelical substrates with and without loxP have been negative. Therefore, we do not favor the topoisomerase model.

In the nucleosome model, Cre first binds to loxP and then wraps the adjacent DNA around itself producing a nucleosome-like structure. Binding of additional molecules of Cre to this initial complex would generate the processive formation of nucleosome structures along the length of the DNA and would have the effect of separating DNA domains between loxP sites. Recombination between those sites then would produce predominantly free circles. The difficulty with this model is that the formation of the final nucleosome structure must exactly relax the molecule. If the nucleosome wraps do

not take up all the negative turns in the original molecule, then the molecule will still have a net negative
twist and catenated products will be produced. If the
nucleosome wraps take up more negative twists than were
present in the original molecule, then the molecule will
have positive twists and catenated products will again
be produced by recombination. Because of these arguments,
this model seems very unlikely.

The model we favor is the tracking model. In this
model, Cre binds tightly to loxP and also nonspecifically
attaches to a second DNA chain segment. At this stage,
the DNA between the lox sites translocates past the
protein until the two sites on the molecule are brought
together. Recombination at this point would produce a
high yield of free circles since the DNA segments between
lox sites have now been separated into two domains.
Tracking models also have been proposed to account for
the binding of lac repressor to lac operator sequences
(9). The data presented here suggest that such tracking
models may also be applicable to P1 site-specific
recombination.

ACKNOWLEDGEMENTS

We thank Robert Grafstrom and Daniel Hamilton for
critical reading of this manuscript.

REFERENCES

1. Sternberg N, Hamilton D (1981). Bacteriophage P1
   site-specific recombination. I. Recombination
   between loxP sites. J Mol Biol 150:467.
2. Sternberg N, Hamilton D, Hoess R (1981). Bacteriophage P1 site-specific recombination. II. Recombination between loxP and the bacterial chromosome.
   J Mol Biol 150:487.
3. Sternberg N, Hamilton D, Austin S, Yarmolinsky M,
   Hoess R. (1981). Site-specific recombination
   and its role in the life cycle of bacteriophage
   P1. Cold Spring Harbor Symp Quant Biol 45:29.
4. Austin S, Ziese M, Sternberg N (1981). A novel
   role for site-specific recombination in maintenance
   of bacterial replicons. Cell 25:729.

5. Sternberg N (1978). Demonstration and analysis of P1 site-specific recombination using λ-P1 hybrid phages constructed in vitro. Cold Spring Harbor Symp Quant Biol 43:1143.
6. Hoess R, Ziese M, Sternberg N (1982). P1 site-specific recombination: nucleotide sequence of the recombining sites. Proc Natl Acad Sci USA 79:3398.
7. Maxam A, Gilbert W (1980). Sequencing end-labeled DNA with base-specific chemical cleavages. Methods Enzymol 65:499.
8. Abremski K, Hoess R, Sternberg N (1983). Studies on bacteriophage P1 site-specific recombination: Evidence for topologically unlinked products following recombination. Cell, 32:1301.
9. Berg OG, Winter RB, von Hippel, PH (1981). Diffusion-driven mechanisms of protein translocation on nucleic acids. I. Models and theory. Biochemistry 20:6929.

Mechanisms of DNA Replication and Recombination, pages 685–694
© 1983 Alan R. Liss, Inc., 150 Fifth Avenue, New York, NY 10011

ANALYSIS OF SITE-SPECIFIC RECOMBINATION ASSOCIATED
WITH THE YEAST PLASMID 2 MICRON CIRCLE*

M. Jayaram, Y.-Y. Li, M. McLeod and J.R. Broach

Department of Microbiology
State University of New York
Stony Brook, New York 11794

ABSTRACT The multicopy yeast plasmid, 2 micron circle,
encodes a specialized recombination system. The recombination is initiated at a specific site of less than
60 bp within inverted repeats in the plasmid and yields
intramolecular inversion.   The reaction requires the
protein product of only one 2 micron circle encoded
gene, FLP.   FLP is able to catalyze resolution and
dimerization, in addition to inversion. FLP mediated
intermolecular recombination can lead to instability of
chromosomes containing an integrated copy of 2 micron
circle. FLP recombination system can also act in yeast
on the inverted repeats of the bacterial transposon
Tn5.   These observations bear on the mechanism of specialized recombination in a eucaryote.

INTRODUCTION

Site specific recombination events causing development-
ally relevant genome rearrangements have been documented in
both procaryotic and eucaryotic cells. As an approach to
evaluating the mechanisms of such recombination events in a
eucaryotic cell, we have examined the recombination system
encoded by the yeast plasmid 2 micron circle.   The 2 micron
circle is a multicopy, 6318 bp, double stranded DNA plasmid
(1).   It contains two regions, each 599 bp in length, that
are precise inverted repeats of each other and that divide
the molecule into approximately equal halves (2; see Figure

*This work was supported by grants from NIH and NSF.
JRB is an Established Investigator of the Am Heart Assn

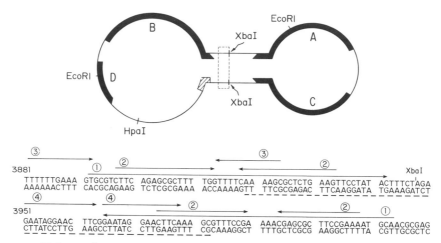

FIGURE 1. Structural features of the 2 micron circle. Upper: Schematic diagram of the 2 micron circle genome shows the large and small unique regions separated by the inverted repeats (parallel lines). Indicated are large open coding regions (heavy lines, taper at the 3' end), origin of replication (hatched box), and recombination initiation site (dashed box). Lower: Repeated segments and the recombination initiation site (dashed underline) are indicated on the nucleotide sequence from the middle of the inverted repeat.

1). In yeast recombination occurs readily between the inverted repeats, resulting in inversion of one unique region with respect to the other (3). This recombination requires the product of one and only one 2 micron circle gene, designated FLP (4-6). In addition, the recombination event is initiated near the center of the inverted repeat sequence at a site of less than sixty base pairs (6), the location and sequence of which are shown in Figure 1.

The role played by the FLP recombination system in the biology of the yeast plasmid is not yet known. Nonetheless, the system provides ready experimental access to the mechanism of a specialized eucayotic recombination reaction. In this communication we show that FLP protein catalyzes resolution and dimerization as well as inversion. Also we show that FLP mediated intermolecular recombination between sister chromatids containing insertions of the FLP site can result in the formation of dicentric chromosomes. Finally, we show that the inverted repeat of the bacterial transposon Tn5 contains a site that is recognized by the FLP recombination system.   Comparison of this site with that in the 2 micron circle inverted repeat provides a more precise definition of the sequences recognized by the FLP system.

MATERIALS AND METHODS

Yeast strains DC04 [cir[o]] and DC04 [cir[+]] are isogenic MATa leu2-04 ade1 strains, lacking or containing endogenous 2 micron circles, respectively. Construction of strain SB1-1D (MATα leu2-04 ade1 [cir[o]] and its isogenic derivative containing an integrated copy of plasmid CV7 (strain SB1-1D(CV7)) has been described (7).   Strain SB1-1D(Xho5) (strain SB1-1D containing an integrated copy of plasmid Xho5) was constructed in a manner analogous to that used to constructed SB1-1D(CV7). The structure of plasmid Xho5 is described in Results and its construction is described elsewhere (8). Strain S150-2B (MATa leu2-3,112 ura3-52 trp1-289 his3) was cured of 2 micron circles to yield strain MM1 (9). A derivative of strain MM1 containing a single copy of FLP integrated into the genome was isolated as follows.   The XbaI fragment containing the small unique region of 2 micron circle (see Figure 1) was inserted into an XhoI site lying within a cloned BamHl genomic restriction fragment spanning the yeast HIS3 gene.   The XhoI site lies external to the HIS3 gene.   The resulting BamHl fragment was used to transform strain MM1 to histidine prototrophy (10).   One His[+] transformant that was Flp[+] and possessed a single copy of FLP adjacent to HIS3 as judged by Southern analysis was retained and designated MM2. Additional procedures used in this study are described elsewhere (6,10).

RESULTS AND DISCUSSION

FLP Promotes Inversion, Resolution, and Dimerization.
Several different recombination reactions catalyzed by the FLP system are diagrammed in Figure 2.   The first reaction (Figure 2, top) — intramolecular recombination between the two inverted repeats — yields inversion of sequences bracketed by the repeats (3-6).   This FLP-mediated reaction is very efficent: we estimate that the frequency is at least 0.1 recombination event per cell per generation (1), although it could in fact be much higher.   In contrast, this reaction is not detectable in mitotically-growing yeast cells in the absence of FLP activity.
Although most plasmids in which the inversion reaction has been observed carry two complete 599 bp repeat sequen-

ces, the initiation site for FLP-mediated recombination consists only of 60 bp spanning the XbaI site in the repeat (see Figure 1). That is, only mutations within a 60 bp segment of the repeat abolish recombination (6). However, current in vivo assays for FLP-mediated recombination would not necessarily reveal reduction in the rate of recombination of an altered substrate. Thus although sequences outside the 60 bp initiation site are clearly not essential for recombination, we cannot preclude a role for these sequence in modulating efficiency of FLP-mediated recombination.

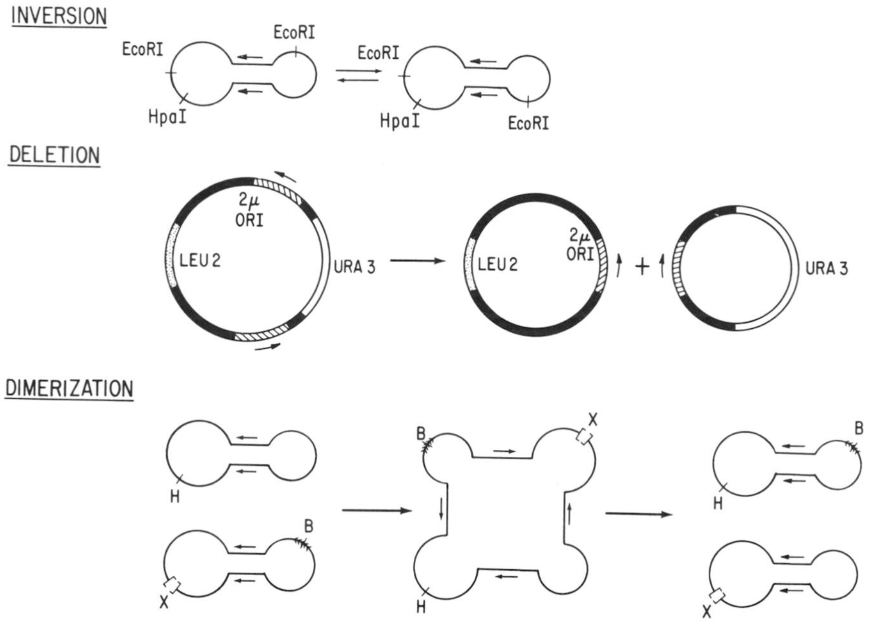

FIGURE 2. Reactions catalyzed by the FLP recombination system. Upper: Interconversion between the two forms of 2 micron circle is accomplished by recombination between inverted repeats (arrows). Middle: Plasmid pMMY (left), containing two copies of the 2 micron circle repeat (hatched lines) in direct orientation, and the two products that result from recombination between the repeats are shown. Lower: Intermolecular recombination between wild type 2 micron circle and plasmid Xho5 (lower left), which carries pBR322 sequences (feathered line) in the small unique region and a deletion plus XhoI linker insertion (bracket) in the large unique region, yields a dimer that can resolve into the monomers on the right.

FLP activity also promotes resolution, that is, intra-molecular recombination between repeats present on a plasmid in direct orientation. This has been demonstrated in vivo using plasmid pMMY as substrate (Figure 2, middle). Plasmid pMMY carries two copies of the 2 micron circle repeat sequence in a direct orientation. These sequences bracket the yeast URA3 gene in one half of the molecule and the LEU2 gene plus the 2 micron circle origin of replication in the other half. As outlined in Figure 2, recombination between the repeats yields two circular molecules, one containing the LEU2 gene and sequences that promote its propagation in yeast and one containing the URA3 gene but lacking sequences for propagation. We used plasmid pMMY to obtain Leu$^+$ transformants of strain MM1 (ura3 leu2 flp) and of strain MM2, an isogenic derivative of strain MM1 that contains a single copy of FLP integrated at the HIS3 locus (see Materials and Methods). We scored the Leu$^+$ transformants of the two strains for uracil prototrophy, that is for retention or loss of the URA3 gene initially present on plasmid pMMY. We found that greater than 99.9% of the Leu$^+$ transformants of strain MM1 were Ura$^+$, whereas only 0.01% of the Leu$^+$ transformants of strain MM2 were Ura$^-$. In addition plasmids resident in the Leu$^+$ Ura$^-$ transformants of strain MM2 were of a size and structure consistent with their having arisen by the recombination event shown in the middle of Figure 2.

The presence of a relatively high proportion (20%) of 2 micron circle DNA as multimeric plasmids in yeast strains provides circumstantial evidence that FLP can promote inter-molecular recombination, or dimerization (1). In addition, Falco et al (7) presented data demonstrating recombination between endogenous 2 micron circle plasmids and a chromosomally integrated copy of one of the 2 micron circle repeats. An additional confirmation of the ability of FLP to promote intermolecular recombination derives from the observation that the reaction outlined at the bottom of Figure 2 occurs in vivo, that is, that each half of a 2 micron circle plasmid can segregate reasonably independently in the cell.

We demonstrated the existence of the reaction outlined at the bottom of Figure 2 by monitering the segregation in a [cir$^+$] strain of a hybrid plasmid that carries the entire 2 micron circle genome with scorable markers included in each half of the plasmid. As shown in Figure 2 plasmid Xho5 consists of 2 micron circle DNA with pBR322 plus LEU2 cloned into the EcoRl site in the small unique region and a 200 bp deletion spanning the HpaI site in the large unique region (8). We isolated DNA from strain DC04 [cir$^+$] transformed with plasmid Xho5 or with plasmid pCV22, which is identical to Xho5 except that it carries a wild type large unique region. After digesting a sample of the DNA with BamH1 plus EcoR1 and fractionating it on an agarose gel, we probed it with labelled pBR322 DNA. If FLP promotes intermolecular recombination, then a reasonable proportion of plasmid Xho5 should exchange large unique regions with endogenous 2 micron circles during propagation in the [cir⁺] strain and thus be converted to a plasmid indistinguishable from plasmid pCV22. As demonstrated in Figure 3, this is in fact the case. It is impossible to determine from these data whether such a replacement event occurs by the two step reciprocal process outlined in Figure 2 or whether replacement is a result of FLP-mediated gene conversion. In either case, though, these results provide a clear confirmation that FLP will promote intermolecular recombination.

A second demonstration that FLP promotes intermolecular recombination emerges from our studies of the properties of chromosomally integrated 2 micron circle sequences (7). Chromosomal instability due to integrated 2 micron circle sequences can be explained by either of two models, both of which are diagrammed in Figure 4. One model (Figure 4A) postulates a FLP-induced double stranded break at the site of the integrated repeat. The loss of the acentric fragment and recombinational healing of the residual chromosome would account for the observed results. Alternatively, chromosomal instability could arise from unequal sister chromatid exchange between two contiguous 2 micron circle inverted repeats (Figure 4B). This would result in formation of a

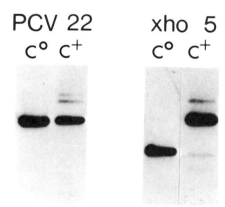

PCV 22

c⁰ c⁺

xho 5

c⁰ c⁺

FIGURE 3. Independent segregation of the two halves of 2 micron circle. Plasmid pCV22 and its deletion derivative Xho5 were transfromed into isogenic [cir⁰] and cir[⁺] strains. Samples of DNA from representative transformants were digested with BamH1 plus XhoI, fractionated, and probed with labeled pBR322 DNA. Digestion of purified Xho5 or pCV22 plasmid DNA yields the same pattern as that seen in the [cir⁰] host

dicentric chromosome as well as an acentric palindromic
fragment that would be lost during growth. The dicentric
chromosome can be lost during mitosis or can undergo breakage followed by recombinational healing of the broken arm.

To distinguish between the two models for FLP-induced
chromosomal instability, we constructed isogenic [cir⁰]
strains, 1D(CV7) and 1D(Xho5), that contain either one copy
of the 2 micron circle repeat or two copies in inverted
orientation integrated into the left arm of chromosome III.
These two constructions are represented by the upper lines
in Figure 4A and 4B, respectively. Each strain was crossed
to strain MM2, a [cir⁰] strain with a functional FLP gene
integrated in chromosome XV but lacking any copies of the 2
micron circle repeat. If the model outlined in Figure 4B is

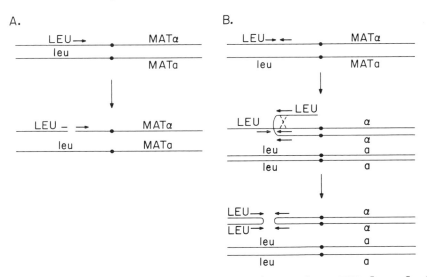

FIGURE 4. Possible explanations for FLP-dependent
chromosome instability. The chromosome III homologues from
diploids heterozygous for LEU2 and MAT and carrying either
one (A) or two (B) copies of the 2 micron circle inverted
repeat (arrows) are shown at the top (diploids 1D(CV7)/MM2
and 1D(Xho5)/MM2 discussed in the text have genomes similar
to those shown in A and B, respectively). Model A postulates that FLP protein makes a double-strand cut in the
repeat, resulting in loss of the LEU2 marker and destabilization of the remainder of the chromosome. Model B postulates that FLP promotes recombination between different
repeats on sister chromatids during G2. This produces an
acentric fragment and an unstable dicentric chromosome.

correct, then only the diploid formed from 1D(Xho5) should display instability of chromosome III. On the other hand, if the model in Figure 4A is correct, then both diploids should display instability. We found that >99.9% of the diploids formed from strain 1D(CV7) were Leu$^+$ non-maters. In contrast, approximately 90% of the diploids formed from strain 1D(Xho5) were Leu and of these approximately half were capable of mating as a cells. Thus chromosome III is stable if one copy of the repeat is integrated but unstable if two copies are present. That the instability of chromosome III in 1D(Xho5)/MM2 diploids is a result of FLP mediated recombination was confirmed by the fact that >99.9% of the diploids formed between strains 1D(Xho5) and MM1, the flp parent of strain MM2, were Leu$^+$ nonmaters. Thus, FLP-induced chromosomal instability at the site of integrated 2 micron circle sequences only obtains when at least two copies of the repeat are present. Therefore, we conclude that such chromosomal instability results from dicentric chromosome formation as outlined in Figure 4A. In addition, these results confirm the capacity of the FLP system to promote intermolecular recombination.

FLP mediated recombination can occur within the inverted repeats of the bacterial transposon Tn5.
     We have found that the FLP recombination system will promote recombination in yeast between the inverted repeats of the bacterial transposon Tn5. This is illustrated in Figure 5. Plasmid CV21(Tn5)#85, whose structure is shown on the left in Figure 5, was transformed into strain DC04 [cir$^+$]. DNA was isolated from a transformant, digested with the indicated restriction enzymes, fractionated and probed with labeled Tn5 DNA. In addition to fragments derived from the parent plasmid and that arising from FLP recombination between the 2 micron circle inverted repeats (fragments 1-3, Figure 4B track 3), there are three fragments that contain Tn5 sequences. From extensive restriction analysis, we have determined that these fragments correspond to plasmids derived from the CV21(Tn5)#85 by inversional recombination between the Tn5 repeats (11). That this recombination is dependent on the FLP gene product is documented by the absence of fragment 4 in digests of DNA from a [cir^0] strain harboring plasmid CV21(Tn5)#85 (Figure 5, lane 2).
     The search for homologies between the 2 micron circle and Tn5 repeats has revealed a region of the Tn5 repeat with an exceptionally striking resemblence to certain structural features within the 2 micron circle recombination site (Fig

A)

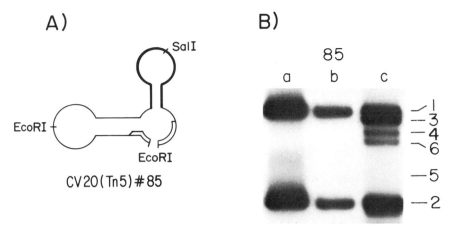

CV20(Tn5)#85

B)

FIGURE 5. FLP-mediated inversion of Tn5 in yeast. A. Plasmid CV20(Tn5)#85 contains Tn5 (heavy line) inserted in the C coding region and pBR322 plus LEU2 (not shown) inserted at the EcoRl site in the small unique region (indicated by the gap). B. CV20(Tn5)#85 DNA (a) or DNA from DC04 [cir⁰] (b) or DC04 [cir⁺] (c) transformed with CV20(Tn5)#85 was digested with SalI plus EcoRl, fractionated, and probed with labeled Tn5 DNA.

ure 6). The FLP recombination site resides within an imperfect dyad symmetry (ABC-C'B'A') consisting of a 49 bp inverted repetition separated by 22 bp. Within each 49 bp segment is a 16 bp dyad symmetry. At position 701 of the Tn5 repeat there is a 12 bp perfect match to a portion of

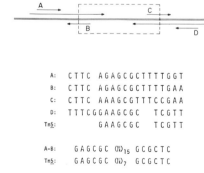

A: CTTC AGAGCGCTTTTGGT
B: CTTC AGAGCGCTTTTGAA
C: CTTC AAAGCGTTTCCGAA
D: TTTCGGAAGCGC TCGTT
Tn5:   GAAGCGC TCGTT

A-B: GAGCGC (N)₁₅ GCGCTC
Tn5: GAGCGC (N)₇ GCGCTC

FIGURE 6. Structural similarities between the Tn5 and 2 micron circle inverted repeats. The site at which FLP recombination occurs and the symmetries around this site are indicated on a diagram of a portion of the 2 micron circle repeat. The sequence of the symmetries indicated as well as that of Tn5 between positions 723 and 701 (12) are shown.

one of these smaller dyad symmetries (D). In addition, 6 bp of this homology are duplicated in inverted orientation after a 7 bp gap and thus closely resembles the A-B dyad symmetry in the 2 micron circle repeat. These similarities together with the lack of any significant homologies between the rest of the two repeats suggest that these small dyad symmetries play a significant role in the recombinational event.

REFERENCES

1. Broach JR (1981) The yeast plasmid 2 micron circle. In Strathern JN, Jones EW, Broach JR (eds): "The Molecular Biology of the Yeast Saccharomyces," New York: Cold Spring Harbor Laboratory, p 445.
2. Hartley JL, Donelson JE (1980) Nucleotide Sequence of the yeast plasmid. Nature 286:860.
3. Beggs JD (1978) Transformation of yeast by a replicating hybrid plasmid. Nature 275:104.
4. Gerbaud C, Fournier P, Blanc H, Aigle M, Heslot H, Guerineau M (1979) High frequency of yeast transformation by plasmids carrying part or entire 2 micron yeast plasmid. Gene 5:233.
5. Broach JR, Hicks JB (1980) Replication and recombination functions associated with the yeast plasmid 2 micron circle. Cell 21:501.
6. Broach JR, Guarascio VR, Jayaram M (1982) Recombination within the yeast plasmid 2 micron circle is site specific. Cell 29:227.
7. Falco SC, Li Y-Y, Broach JR, Botstein D (1982) Genetic properties of chromosomally integrated 2 micron plasmid DNA in yeast. Cell 29:573.
8. Jayaram M, Li Y-Y, Broach JR (1983) The yeast plasmid 2 micron circle encodes components required for its high copy propagation. Cell in press.
9. Broach JR (1983) Construction of high copy vectors using 2 micron circle sequences. Methods in Enzymol 101:in press.
10. Rothstein R (1983) Transformation of yeast using linear DNA. Methods in Enzymol 101:in press.
11. Jayaram M, Broach JR (1983) The yeast plasmid 2 micron circle promotes recombination within the bacterial transposon Tn5. PNAS submitted
12. Auerswald EA, Ludwig G, Schaller H. (1981) Structural Analysis of Tn5. Cold Spring Harbor Symp Quant Biol 45:107.

VII. GENERAL RECOMBINATION

Mechanisms of DNA Replication and Recombination, pages 697–707
© **1983 Alan R. Liss, Inc., 150 Fifth Avenue, New York, NY 10011**

HETERODUPLEX DNA WITH LARGE UNPAIRED REGIONS
MADE BY recA PROTEIN

Marco E. Bianchi and Charles M. Radding

Departments of Human Genetics and
Molecular Biophysics and Biochemistry,
Yale University School of Medicine,
New Haven, CT 06510

ABSTRACT E.coli recA protein promotes the pairing of
circular single strands with linear duplex DNA and the
subsequent formation of large heteroduplex joints. From
fd and M13 DNA, heteroduplex molecules are made that
incorporate every kind of single-base mismatch. In the
presence of E.coli single-strand binding protein (SSB)
and ATP regeneration, recA protein can incorporate
into heteroduplex DNA insertions that are hundreds of
base pairs long whether the extra DNA is located
initially in either the single-stranded or the double-
stranded substrate. The ability of recA protein to
span large insertions in the duplex DNA suggests that
recA protein unwinds a sizeable number of turns in
advance of the growing heteroduplex joint and therefore
possesses a true helicase activity. Given the importance
of recA protein for the major pathway of recombination
in E.coli, the observed facility with which it makes
extensively mismatched joints in vitro supports the
prevailing view that the formation of heteroduplex DNA
is a frequent source of conversion-like events.

This work was supported by Grant No. NP-90I from the
American Cancer Society.

INTRODUCTION

The recA protein from E. coli, acting in vitro without the aid of any other protein, promotes two processes that are central to general genetic recombination, namely synapsis, the conjunction and homologous pairing of DNA molecules, and strand exchange, which creates long regions of heteroduplex DNA (1-6). Because recA protein is essential for the major pathway of recombination in E. coli, the mechanism by which it makes heteroduplex DNA in vitro is particularly interesting with regard to inferences that have been drawn from genetic data about the role of hetero- duplex DNA in recombination. A particularly useful pair of substrates is provided by the combination of circular single-stranded DNA and homologous linear duplex DNA (See Figure 2, and references 1 and 4-8). The DNA from the small DNA phages provides a rich source of natural sequence variations and in addition can be manipulated readily to produce desired changes in sequence. Using this system, we have reported that recA protein can make heteroduplex DNA containing many single base pair mismatches (9). In other experiments we found that superhelicity favors the formation of more highly mismatched joints (10). The experiments described in this paper have revealed that recA protein can readily make heteroduplex joints that contain insertions and deletions hundreds of nucleotides long as well as extensively mismatched joints.

METHODS

Enzymes and DNA.

RecA protein and DNA were purified as described earlier (11, 12). E. coli exonuclease I was purified from the hyper- producing strain SK1447 (13) by the procedure of Lehman and Nussbaum (14). E. coli SSB protein was a gift from J. Chase.

Origin of the phages of the M13mp series and construction of their derivatives.

Phages M13mp7, M13mp8, M13mp9 are described by Messing and Vieira (15). From these we derived by standard cloning techniques the phages M13MEB42, M13MEB140 and M13MEB486. Phage M13MEB42 is a deletion derivative of phage M13mp7 from which 42 base pairs have been excised. The sequence of this phage corresponds to a deletion of 33 bases with respect to M13mp8 and M13mp9. Phage M13MEB140 is a derivative of M13mp9, into which a 140 bp fragment of E. coli has been cloned at the EcoRI site. Phage M13MEB486 is a derivative of M13mp8, into which a 486 bp fragment of human DNA (provided by Dr. Sherman Weissman) has been cloned at the HincII site.

Strand exchange reactions.

Standard recA reaction mixtures contained, unless otherwise indicated, 31 mM Tris-HCl (pH 7.5), 12.5 mM MgCl$_2$, 0.4 mM dithiothreitol, 6 mM phosphocreatine, 10 units/ml phosphocreatine kinase (Sigma), 4% glycerol, 4 μM circular single strands, 6 μM duplex DNA and 4 μM recA protein. The reaction mixture was preincubated at 37°C for 2 minutes and the reaction was started by the simultaneous addition of 1.5 mM ATP and 0.4 μM SSB protein.

Quantitation of the end-products of the strand exchange reaction by means of E. coli exonuclease I.

Strand exchange reactions were performed as described above in a total volume of 15 μL. After 60 minutes, 12 μL of 28 mM ADP were added to dissociate recA protein from DNA. Incubation at 37°C was continued for 15 minutes to allow intermediate joint molecules to dissociate by spontaneous branch migration (See Figure 2). Then, 3 μL of a solution containing 670 mM glycine buffer (pH 9.5) and 10 mM dithiothreitol, and 0.45 units of exonuclease I (1 μL) were added. After 20 minutes more of incubation, carrier DNA and cold 8% TCA were added and the amount of acid-soluble labeled DNA was determined. The amount of DNA degraded is expressed as a fraction of the theoretical maximum, viz one half of the input labeled duplex DNA.

RESULTS

Formation of heteroduplex DNA containing nearly 200 mismatches

Previous experiments have shown that in the presence
of ATP, recA protein promotes the transfer of a strand
from a duplex DNA more than 3000 base pairs long to homologous circular single strands (1). The rate and yield of
this exchange are improved by E. coli single-strand binding protein which therefore facilitates the formation of
nicked circular duplex DNA from circular single strands
and full length linear duplex DNA (4). The use of an ATP
regeneration system maintains a constant pool of ATP,
whose continued hydrolysis is required for the strand
exchange reaction, and forestalls the accumulation of ADP,
a potent inhibitor of a presynaptic phase in the pairing
reaction (8 and Kahn et al., manuscript in preparation).
RecA protein, in the presence of SSB and an ATP
regenerating system, promoted the complete transfer of

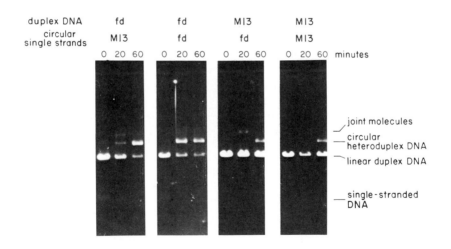

FIGURE 1. Formation of mismatched heteroduplex DNA
from M13 and fd DNAs. Reactions were performed as described
in Methods; at the indicated times 40 µL of the reaction
mixtures were withdrawn and quenched with 4 µL of 10% sodium
dodecylsulphate. The samples were loaded onto a 0.8% agarose
gel and electrophoresed at 1.5 V/cm for 8 hours.

a strand from a full length linear duplex DNA to a circular single strand even when M13 and fd DNA were paired.
Mixed pairs of fd and M13 DNA formed full length heteroduplex DNA about as well as the homologous combinations
(Figure 1). In the cases of the mixed pairs, however,
intermediates that migrated more slowly than form II DNA
were more prominent.

In the nucleotide sequence of M13 and fd there are
193 differences, or about 3% of their genome length (16).
These consist of 179 changes of a single base, 5 changes
of two consecutive bases, one change of 3 consecutive
bases, and one extra base in fd. Consequently, the mismatched fd/M13 heteroduplexes contain every possible kind
of single base-pair mismatch as well as the other limited
kinds of mismatch.

The sequence differences between fd and M13 phages
are also reflected in slight differences of their restriction maps: in particular, M13 contains one BamH1 site,
and fd contains two. Heteroduplex M13/fd DNA made by recA
protein had only one restriction site for BamH1 at position 5645. At position 2220, where fd and M13 sequences
are different, the heteroduplex fd/M13 DNA should have
contained a mismatched base pair that rendered the site
unavailable for restriction, and indeed the site was not
cleaved by BamH1.

Heteroduplex DNA with insertions and deletions hundreds
of bases long

In order to examine the ability of recA protein to
form heteroduplex DNA with large unpaired regions, we
constructed the series of derivatives of the phages of
the M13mp series as described in Methods. These were then
used in pairwise combinations as substrates for heteroduplex DNA formation.

Gel electrophoresis was a poor method for examining
the products made from substrates with the largest insertions, in part because yields were lower, and in part
because the products migrated differently than form II
DNA. The following assay, however, overcame that problem
and provided a rapid way to study the behavior of insertions of any size. Since recA protein promotes strand
exchange in the 5' to 3' direction (5, 6, 17), the linear
strand displaced from duplex DNA does not become sensitive
to E. coli exonuclease I until complete strand separation

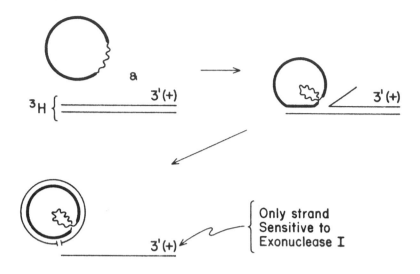

FIGURE 2. Diagram of the substrates used to study the formation of heteroduplex DNA. Because of the polarity of strand exchange promoted by recA protein, the production of single-stranded DNA that is sensitive to digestion by exonuclease I signals completion of strand exchange.

has occurred (See Figure 2 and Methods). Exonuclease I, under standard conditions, digests 75 to 80% of linear single-stranded DNA.

We incubated recA protein with ^{3}H-labeled linear duplex DNA and insertion or deletion-containing circular single strands under standard conditions for 60 minutes and stopped the reaction by adding excess ADP, which causes recA protein to dissociate (S. S. Tsang, personal communication). After 15 minutes of incubation we added exonuclease I, kept the reaction mixture at 37°C for 20 more minutes and measured the release of acid-soluble ^{3}H counts. When homologous duplex and single stranded DNA were reacted together, exonuclease I released the maximum number of counts expected if the (+) strand were completely displaced from every duplex molecule. The blank for a duplex molecule that had heterologous DNA at its ends, and therefore did not initiate the strand transfer reaction, was about 4%.

Using the assay described above, we screened insertions varying in length from 33 to 1308 nucleotide residues and located either in the single-stranded or double-
stranded DNA (Table 1). None of the insertions in the
single-stranded DNA prevented the formation of heteroduplex DNA: 20 to 30% of the labeled (+) strands were
digested by exonuclease I. A similar displacement of (+)
strands was observed when the duplex DNA contained insertions of 33 or 140 base pairs. With an insertion in the
duplex DNA of 486 base pairs, strand displacement was
still measurable, but it fell to background levels with
an insertion of 1308 base pairs.

To confirm the structure of the heteroduplex DNAs
with large mismatches, we separated them from the unreacted substrates and linear single-stranded DNA by means
of electrophoresis in a low-melting agarose gel. The DNA
recovered from the gel was filtered on nitrocellulose
filters in 1.5 M NaCl, 0.15 M Na citrate. The heteroduplex DNAs with large unpaired sequences were retained
quantitatively by the filters, while unreacted linear

TABLE 1. Efficiency of recA-promoted formation of full-
length heteroduplex DNA as a function of the size of the
insertion or deletion to be overcome during strand exchange.

duplex DNA	circular single strands	size of the insertion	amount of (+)strand DNA degraded by exoI
INSERTION IN THE SINGLE-STRANDED DNA			
M13MEB42	M13mp9	33 nucleotides	27%
M13MEB42	M13MEB140	173 nucleotides	24%
M13MEB42	M13MEB486	519 nucleotides	25%
M13	M13MEB486	1308 nucleotides	20%
INSERTION IN THE DOUBLE-STRANDED DNA			
M13mp9	M13MEB42	33 base pairs	50%
M13MEB140	M13mp9	140 base pairs	25%
M13MEB486	M13mp8	486 base pairs	12%
M13MEB486	M13	1308 base pairs	4%
HOMOLOGOUS CONTROLS			
M13MEB42	M13MEB42		61%
M13MEB140	M13MEB140		65%
M13MEB486	M13MEB486		64%
M13	M13		78%

TABLE 2. Single-Stranded Character of Heteroduplex DNA Containing Large Unpaired Sequences

| | Retention by nitrocellulose (%) | | | |
| | linear DNA | | heteroduplex DNA | |
Treatment by S_1 nuclease:	−	+	−	+
140 base pair insertion in dsDNA	17	2	108	2
homologous control	16	1	19	2
173 nucleotides insertion in ssDNA	7	1	101	1
homologous control	18	3	28	3

duplex DNAs and control homologous heteroduplex DNAs showed only background binding (Table 2). Treatment with saturating amounts of nuclease S_1 prior to filtration abolished the difference between mismatched and control DNAs: therefore, the unpaired sequences in heteroduplex DNAs were in a single-stranded form.

DISCUSSION

The experiments described here appear to provide an important clue to the mechanism by which recA protein promotes strand exchange. When the insertion is present in the single-stranded DNA (Figure 3A), strand exchange can proceed past the sequence non-homology simply by folding the extra single-stranded DNA out of the way. When a large insertion is present in the duplex DNA (Figure 3B), however, the extra DNA cannot be spanned, unless that whole DNA segment is unwound. As indicated in the figure, the matching of complementary sequences requires that a strand loop out of the duplex DNA in a way that is reminiscent of the helicase activity of the E. coli recBC enzyme (18).

Our data reveal that as insertions increase in length, recA protein spans them less readily in duplex DNA than in the single-stranded DNA. However, the ability of recA protein to span sizeable insertions in the duplex DNA indicates that recA protein possesses a true helicase activity that enables it to unwind a number of turns of duplex DNA in advance of the region of heteroduplex DNA. A corollary of this conclusion is that the unwinding of as many as

50 turns by recA protein should be independent of sequence homology.

These observations show further that in E. coli the enzymatic machinery exists for incorporating into heteroduplex DNA mismatches that are hundreds of nucleotides long. The mechanism resides in a single protein, which is recA protein. Because of the key role that recA protein plays in recombination in E. coli and the facility with which it makes extensively mismatched heteroduplex DNA, these data support the prevailing view that at least in prokaryotes the formation of mismatched heteroduplex DNA is a frequent source of conversion-like events such as localized negative interference.

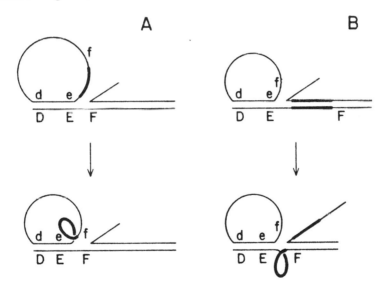

FIGURE 3. Physical dissimilarity of the mechanisms by which insertions in the single-stranded or the double-stranded DNA are included into heteroduplex DNA. When a large insertion is present in the duplex DNA, the non-homologous sequence cannot be spanned unless that whole DNA segment is unwound, presumably in advance of the point of strand exchange.

ACKNOWLEDGEMENTS

The authors gratefully acknowledge the technical assistance of Lynn Osber and Maureen Leahy. M.E.B. is also grateful to M. R. Cattadori for her patience and support.

REFERENCES

1. DasGupta C, Shibata T, Cunningham RP, Radding CM
   (1980). The topology of homologous pairing promoted
   by RecA protein. Cell 22:437.
2. Radding CM, Shibata T, DasGupta C, Cunningham RP,
   Osber L (1981). Kinetics and topology of homologous
   pairing promoted by Escherichia coli RecA protein.
   Cold Spring Harb. Symp. Quant. Biol. 45:385.
3. Cunningham RP, Wu AM, Shibata T, DasGupta C, Radding
   CM (1981). Homologous pairing and topological linkage
   of DNA molecules by combined action of E. coli RecA
   protein and topoisomerase I. Cell 24:213.
4. Cox MM, Lehman IR (1981). RecA protein of Escherichia
   coli promotes branch migration, a kinetically distinct
   phase of DNA strand exchange. Proc Natl Acad Sci USA
   78:3433.
5. Cox MM, Lehman IR (1981). Directionality and polarity
   in recA protein-promoted branch migration. Proc Natl
   Acad Sci USA 78:6018.
6. Kahn R, Cunningham RP, DasGupta C, Radding CM (1981).
   Polarity of heteroduplex formation promoted by
   Escherichia coli RecA protein. Proc Natl Acad Sci USA
   78:4786.
7. Cox MM, Lehman IR (1982). RecA protein promoted strand
   exchange: Stable complexes of recA protein and
   single-stranded DNA formed in the presence of ATP
   and single-stranded DNA binding protein. J Biol Chem
   257:8523.
8. Wu AM, Kahn R, DasGupta C, Radding CM (1982). Formation of nascent heteroduplex structures by RecA protein and DNA. Cell 30:37.
9. DasGupta C, Radding CM (1982). Polar branch migration
   promoted by recA protein: Effect of mismatched base
   pairs. Proc Natl Acad Sci USA 79:762.
10. DasGupta C, Radding CM (1982). Lower fidelity of
    RecA protein catalysed homologous pairing with a
    superhelical substrate. Nature 295:71.
11. Shibata T, Cunningham RP, Radding CM (1981). Homolo-
    gous pairing in genetic recombination. Purification
    and characterization of Escherichia coli RecA protein.
    J Biol Chem 256:7557.
12. Cunningham RP, DasGupta C, Shibata T, Radding CM
    (1980). Homologous pairing in genetic recombination:
    RecA protein makes joint molecules of gapped circular
    DNA and closed circular DNA. Cell 20:223.

13. Vapnek D, Alton NK, Bassett CL, Kushner SR (1976). Amplification in Escherichia coli of enzymes involved in genetic recombination: Construction of hybrid ColE1 plasmids carrying the structural gene for exonuclease I. Proc Natl Acad Sci USA 73:3492.

14. Lehman IR, Nussbaum AL (1964). The deoxyribonucleases of Escherichia coli. V. On the specificity of exonuclease I (Phosphodiesterase). J Biol Chem 239:2628.

15. Messing J, Vieira J (1982). A new pair of M13 vectors for selecting either DNA strand of double-digest restriction fragments. Gene 19:269.

16. van Wezenbeck PMGF, Hulsebos TJM, Schoenmakers JGG (1980). Nucleotide sequence of the filamentous bacteriophage M13 DNA genome: Comparison with phage fd. Gene 11:129.

17. West SC, Cassuto E, Howard-Flanders P (1981). Heteroduplex formation by recA protein: Polarity of strand exchanges. Proc Natl Acad Sci USA 78:6149.

18. Taylor A, Smith GR (1980). Unwinding and rewinding of DNA by the RecBC enzyme. Cell 22:447.

Mechanisms of DNA Replication and Recombination, pages 709–721
© 1983 Alan R. Liss, Inc., 150 Fifth Avenue, New York, NY 10011

A STABLE recA PROTEIN-SINGLE-STRANDED DNA COMPLEX:
AN INTERMEDIATE IN DNA STRAND EXCHANGE[1]

Daniel A. Soltis, Michael M. Cox,[2]
Zvi Livneh and I. R. Lehman

Department of Biochemistry, Stanford University
School of Medicine, Stanford, California 94305

ABSTRACT   The recA protein of Escherichia coli promotes the exchange of strands between linear duplex
and homologous, circular, single-stranded DNAs to
generate a nicked circular DNA duplex (RFII) and a
displaced linear single strand. At an early stage in
the reaction, recA protein exists in a stable, kinetically significant complex with the circular single-
stranded DNA, whose formation requires single-stranded
DNA binding protein (SSB) and ATP. After completion
of strand exchange, the recA protein is bound to the
RFII and the SSB is associated with the displaced
single strand. The recA protein-RFII complex can be
dissociated upon addition of ADP, permitting the recA
protein to participate in another cycle of strand
exchange.

INTRODUCTION

Mutations at a number of loci in the E. coli chromosome are known to affect general recombination; however,
strains bearing mutations in the recA gene are particularly defective in this process (1,2). Studies of recA

[1]This work was supported by grants from the National
Institutes of Health (GM06196) and the National Science
Foundation (PCM79-04638).
[2]Present address: Department of Biochemistry, University of of Wisconsin, Madison, WI 53706

mutants indicate that the <u>recA</u> gene product is essential during the early stages of homologous recombination, possibly in the transfer of strands between duplex DNA molecules (3-6). Models for general recombination propose that the formation of heteroduplex DNA is preceded by the pairing of strands from different DNA molecules to produce a heteroduplex joint (7,8), which is then extended by branch migration.

The product of the <u>recA</u> gene has been isolated (9-12) and shown to promote reactions <u>in vitro</u> that are relevant to its proposed role in homologous recombination <u>in vivo</u>. These reactions include the renaturation of complementary single-stranded (SS) DNAs (9) and the pairing of SS DNA molecules with homologous regions of duplex DNA (12,13). This latter reaction, termed strand assimilation or D-loop formation, is analogous to the pairing reaction believed to be an early step in homologous recombination.

Our studies have been aimed at defining the role of recA protein in homologous recombination. In particular, we have focused on a DNA strand exchange reaction in which a single strand is transferred from a linear duplex to a circular SS DNA, yielding a nicked circular duplex (RFII) and a linear SS DNA (14-16). This strand exchange reaction has been particularly useful in examining the role of recA protein in homologous recombination for several reasons. First, it is analogous to several of the steps in which the recA protein very likely participates during homologous recombination <u>in vivo</u>. Second, the substrates and products of the strand exchange reaction are well characterized and easily separated. Finally, there are a number of assays to follow this reaction including one which measures heteroduplex formation directly (14).

The exchange of strands is coupled to the hydrolysis of ATP and is stimulated by the single-stranded DNA binding protein of <u>E</u>. <u>coli</u> (SSB) (14,17,18). In fact, there is evidence which suggests that SSB stimulates homologous recombination <u>in vivo</u> (19,20). In this report, we describe a stable complex between recA protein and circular SS DNA that is formed in the presence of SSB and ATP and which is required for efficient DNA strand exchange.

RESULTS

Complex formation. The general strategy was to examine the strand exchange reaction at limiting recA protein
concentrations. At various times, the reaction was chal-
lenged with a second set of DNA substrates to determine
whether the recA protein was available for reaction with
the challenging DNAs.

Typically, recA protein was preincubated with both the
circular SS DNA and linear duplex DNA. After starting
strand exchange with ATP and SSB, the reaction was chal-
lenged with another equivalent of SS and duplex DNAs.
Heteroduplex formation occurring with the substrates in the
preincubation is referred to as reaction 1; heteroduplex
formation involving the challenging DNAs is termed reaction
2. By using [^{3}H]-labeled SS DNA in one or the other set of
substrates, reactions 1 and 2 could be followed independently. The amount of recA protein present was sufficient
to promote only half of the heteroduplex formation possible
in either reaction 1 or reaction 2 alone. Sufficient SSB
was added to provide optimal conditions for heteroduplex
formation in both reactions. An ATP regenerating system
was also included in each experiment to ensure that ATP was
not limiting.

When the reaction 1 and 2 substrates were preincubated
with recA protein and the reaction initiated with ATP and
SSB, approximately 25% of both SS DNA substrates were
incorporated into heteroduplex DNA indicating that they
could compete successfully for the recA protein (Table 1).
When recA protein was preincubated with reaction 1 substrates and the ATP and SSB added with the reaction 2 substrates, a competition for recA protein again occurred.
However, in this case, reaction 2 proceeded more efficiently than reaction 1 (Table 1, experiments 1A and 1B).
The lower extent of reaction 1 relative to reaction 2 in
this experiment was probably due to the formation of unproductive intermediates between the reaction components
during the preincubation in the absence of ATP. In contrast, when recA protein was preincubated with the substrates of reaction 1, and ATP and SSB were added to initiate the reaction 20 min prior to the addition of the reaction 2 substrates, reaction 1 was approximately 12-fold
more efficient than reaction 2 (experiments 2A and 2B).
Taken together, these results indicate that early in strand

TABLE 1
REQUIREMENTS FOR COMPLEX FORMATION

Experi-ment	Time of addition (min)				Extent of heteroduplex formation (%)
	-40	-30	-20	0	
1A	[³H]SS DNA DS DNA	recA		SS DNA DS DNA ATP/2xSSB	23.5
B	SS DNA DS DNA	recA		[³H]SS DNA DS DNA ATP/2xSSB	45.7
2A	[³H]SS DNA DS DNA	recA	ATP SSB	SS DNA DS DNA SSB	58.8
B	SS DNA DS DNA	recA	ATP SSB	[³H]SS DNA DS DNA SSB	4.7
3A	[³H]SS DNA DS DNA	recA	ATP	SS DNA DS DNA 2x SSB	26.2
B	SS DNA DS DNA	recA	ATP	[³H]SS DNA DS DNA 2x SSB	38.4

TABLE 1. Reactions were performed as described (18). The extent of heteroduplex formation is the average of three measurements taken at 40, 50 and 60 minutes after t=0. The concentrations resulting from the indicated additions are: unlabeled circular SS DNA (SS DNA), 3.3 µM; [³H]-labeled circular SS DNA ([³H]SS DNA), 3.3 µM; linear duplex DNA (DS DNA), 5.6 µM; recA protein, 0.9 µM; SSB, 0.15 µM; ATP, 1.2 mM. An ATP regenerating system was also included in each experiment.

exchange, the recA protein becomes sequestered in a complex with the substrate DNAs which is largely irreversible when both ATP and SSB are present.

Requirements for complex formation. When reaction 1
was initiated without SSB, both reactions 1 and 2 proceeded
to similar extents (experiments 3A and 3B); thus SSB is
necessary for the formation of the stable recA protein-SS
DNA complex. Omission of duplex DNA from the preincubation
gave results similar to those in experiments 2A and 2B,
demonstrating that the duplex DNA is not required for
complex formation. However, when reaction 1 was initiated
without ATP, and ATP was added with the reaction 2 substrates, the extent of reaction 1 decreased and the extent
of reaction 2 increased relative to experiments 2A and 2B.
These results indicate that recA protein had not been
sequestered within the complex and was able to interact
with both sets of substrates. Similarly, when SS DNA was
omitted from the preincubation and added with the reaction
2 substrates, recA protein was not sequestered as in experiments 3A and 3B. Thus, SS DNA, SSB and ATP are required
for the formation of the stable recA protein-SS DNA complex.

Interaction of recA protein with SS DNA. A complex
formed between recA protein and circular SS DNA in the
presence of SSB and ATP can be observed by sedimentation in
sucrose density gradients if the complex is stabilized with
the nonhydrolyzable ATP analog, ATPγS, prior to sedimentation (Fig. 1). This complex corresponds to the one inferred from the kinetic analyses described above and shown
to be an early intermediate in the strand exchange reaction
(18). The sedimentation coefficient of the recA protein-SS
DNA complex was approximately 78S. Addition of 150 mM NaCl
to the sucrose gradient did not affect this value, indicating that it is an accurate measure of the density of the
recA protein-SS DNA complex.

The kinetic analyses summarized in Table 1 showed that
SSB is required for the formation of a stable complex
between recA protein and SS DNA and suggested that SSB
might function in complex formation by interacting directly
with the recA protein. Thus, it might be expected that SSB
would be found associated with the recA protein-SS DNA
complex. As shown in Fig. 1, however, SSB sedimented near
the top of the gradient and not with the recA protein in
the 78S peak. Attempts to crosslink SSB chemically to the
recA protein-SS DNA complex using formaldehyde and/or
glutaraldehyde were unsuccessful. These results suggest
that SSB does not interact stably with the recA protein-SS
DNA complex, however, an alternative possibility is that

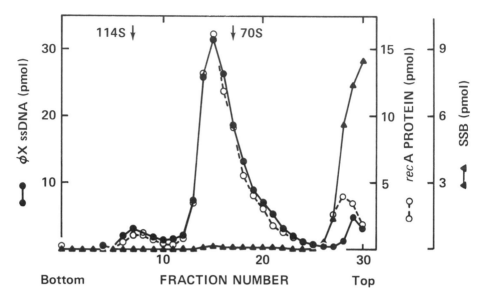

FIGURE 1.   RecA protein-SS DNA complex formation in the presence of SSB.   The reaction and sedimentation were performed as described (21).   The migration of normal (114S) and eclipsed (70S) forms of bacteriophage φX174 are indicated in this and succeeding figures.

SSB cannot be crosslinked to the recA protein-SS DNA complex by the methods used.

An analysis by sucrose gradient sedimentation of the interaction of recA protein with SS DNA in the absence of SSB is shown in Fig. 2.   Under these conditions, approximately 80% of the recA protein sedimented near the top of the gradient and was not bound to the SS DNA, whereas in the presence of SSB, essentially all of the recA protein was found associated with the SS DNA (Fig. 1).   Since ATPγS was added to stabilize any interaction between the recA protein and SS DNA, the recA protein found in the 78S peak would appear to represent a steady state level of recA protein bound to SS DNA in the absence of SSB.   Thus, in the absence of SSB, a stable complex does not form between recA protein and SS DNA, a result which is in agreement with our kinetic analysis.

RecA protein-RFII complexes.   DNA challenge experiments similar to those described above indicated that recA

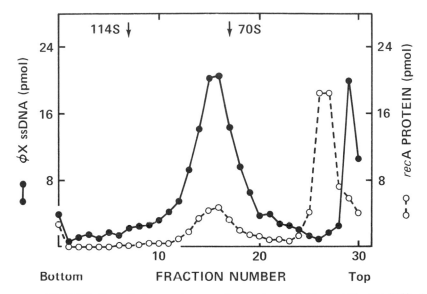

FIGURE 2. Interaction of recA protein with SS DNA in the absence of SSB. The reaction and sedimentation were performed as described (21).

protein was unavailable for further reaction after the completion of strand exchange when the ATP/ADP ratio was maintained at sufficiently high levels (18). The sucrose density gradient analysis of the products of a strand exchange reaction performed with labeled circular SS DNA, in the presence of SSB is shown in Fig. 3. Once again, ATPγS was added to stabilize any recA protein-DNA interactions. At the end of the reaction, most of the recA protein cosedimented with the labeled, circular SS DNA which had been incorporated into the RFII product. The sequestering of recA protein after completion of strand exchange was therefore the result of the formation of a complex between recA protein and the product RFII.

The use of labeled linear duplex DNA in a strand exchange reaction with unlabeled circular SS DNA performed in the presence of SSB provided additional evidence for the association of recA protein with the RFII product. As shown in Fig. 4, there were two major peaks of labeled DNA. The more rapidly sedimenting peak comigrated with the recA protein and sedimented to the same position in the gradient

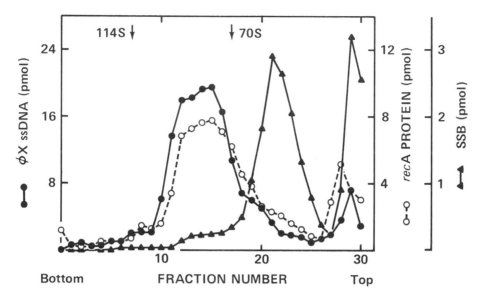

FIGURE 3. Association of recA protein with the RFII product of DNA strand exchange. Strand exchange was performed in the presence of SSB under standard conditions and the products sedimented as described (21).

as the recA protein-RFII complex shown in Fig. 3. The slower sedimenting peak comigrated with SSB (see Fig. 3). When analyzed by agarose gel electrophoresis, the DNA in the more rapidly sedimenting peak containing the recA protein was in the form of RFII, and the slower sedimenting SSB-DNA peak contained linear SS DNA. Thus, one strand of the linear duplex DNA is incorporated into the RFII product which is complexed with the recA protein and the other strand is associated with SSB.

The sedimentation coefficient of the recA protein-RFII complex shown in Figures 3 and 4 is approximately 85S. This value was not affected by the addition of 150 mM NaCl to the sucrose gradient indicating that the S value is again a reasonable measure of the size and density of the complex. However, when the products of the strand exchange reaction were incubated with 4 mM ADP prior to sedimentation, the recA protein and labeled DNA sedimented near the top of the gradient indicating that the recA protein-RFII complex had dissociated (Fig. 5). The dissociation of the

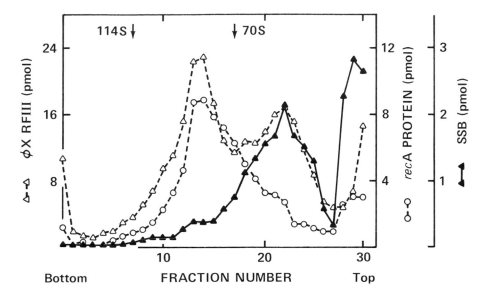

FIGURE 4. RecA protein-RFII complex formation during DNA strand exchange. The reaction was performed in the presence of SSB and the products sedimented as described (21).

recA protein-RFII complex by ADP had been inferred previously from kinetic analyses of the strand exchange reaction (17).

DISCUSSION

We have presented here a summary of our recent studies on the mechanism of recA protein-promoted DNA strand exchange. A plausible interpretation of the results of our DNA challenge experiments is that a specific recA protein-DNA complex is formed during strand exchange. The existence of this complex was confirmed by sedimenting the reaction intermediates and products in sucrose density gradients. Although SSB is necessary for the formation of the stable complex, we have been unable to demonstrate that SSB interacts directly with either the recA protein or SS DNA present in the complex as predicted by the model (18). It is possible, however, that the techniques used to exam-

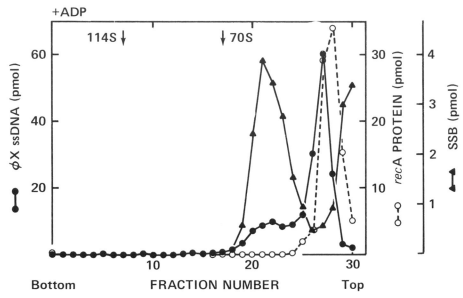

FIGURE 5. ADP mediated dissociation of recA protein from the RFII product of DNA strand exchange. Strand exchange in the presence of SSB was performed under standard conditions followed by incubation of the reaction products with ADP. Sedimentation was performed as described (21).

ine these complexes disrupted the interactions between the SSB and the recA protein-SS DNA complex. Flory and Radding (22) have reported differences in the structure of recA protein-SS DNA complexes formed in the presence and absence of SSB but they did not show directly that SSB was present in the complexes.

In the absence of SSB, the stable recA protein-SS DNA complex is not formed. Instead, recA protein equilibrates rapidly between free and bound forms. The inability of recA protein to form a stable complex with SS DNA in the absence of SSB causes the strand exchange reaction to proceed at a lower rate and to a lesser extent than when SSB is present (17,18). Thus, the formation of a stable complex between recA protein and SS DNA is necessary for efficient DNA strand exchange.

After completion of strand exchange in the presence of SSB, most of the recA protein is found in association with

the RFII product and the SSB is bound to the displaced
linear SS DNA. The recA protein remains associated with the
RFII as long as the level of ATP is maintained at a sufficiently high level. The accumulation of ADP causes the
disruption of the recA protein-RFII complex and enables the
released recA protein to interact with additional DNA
substrates once the excess ADP has been converted to ATP.
These results account for the previously observed inacces-
sibility of recA protein after the completion of strand
exchange and are in good agreement with the model for the
reaction proposed earlier (18).

ACKNOWLEDGMENTS

M.M.C. was supported by an A. P. Gianinni-Bank of
America Fellowship. Z.L. was supported by a Dr. Chaim
Weizmann Fellowship.

REFERENCES

1.   Clark AJ, Margulies AD (1965). Isolation and charac-
     terization of recombination-deficient mutants of
     Escherichia coli K12. Proc Natl Acad Sci USA 53:451.
2.   Clark AJ (1973). Recombination deficient mutants of
     E. coli and other bacteria. Ann Rev Genet 7:67.
3.   Birge EA, Low KB (1974). Detection of transcribable
     recombination products following conjugation in rec⁺,
     recB⁻ and recC⁻ strains of Escherichia coli K12. J
     Mol Biol 83:447.
4.   Benbow RM, Zuccarelli AJ, Sinsheimer RL (1975).
     Recombinant DNA molecules of bacteriophage φX174.
     Proc Natl Acad Sci USA 72:235.
5.   Potter H, Dressler D (1976). On the mechanism of
     genetic recombination: Electron microscopic observa-
     tion of recombination intermediates. Proc Natl Acad
     Sci USA 73.3000.
6.   Holloman WK, Radding CM (1976). Recombination pro-
     moted by superhelical DNA and the recA gene of
     Escherichia coli. Proc Natl Acad Sci USA 73:3910.
7.   Holliday R (1964). A mechanism for gene conversion in
     fungi. Genet Res 5:282.
8.   Whitehouse HLK (1973). "Towards an Understanding of
     the Mechanism of Heredity." New York: St. Martin's.

9. Weinstock GM, McEntee K, Lehman IR (1979). ATP-dependent renaturation of DNA catalyzed by the recA protein of E. coli. Proc Natl Acad Sci USA 76:126.

10. Ogawa T, Wabiko H, Tsurimoto T, Horii T, Masukata H, Ogawa H (1979). Characteristics of purified recA protein and the regulation of its synthesis in vivo. Cold Spring Harbor Symp Quant Biol 43:909.

11. Roberts JW, Roberts CW, Craig NL, Phizicky E (1979). Activity of the E. coli recA gene product. Cold Spring Harbor Symp Quant Biol 43:917.

12. Shibata T, DasGupta C, Cunningham RP, Radding CM (1979). Purified E. coli recA protein catalyzes homologous pairing of superhelical DNA and single-stranded fragments. Proc Natl Acad Sci USA 76:1638.

13. McEntee K, Weinstock GM, Lehman IR (1980). RecA protein-catalyzed strand assimilation: Stimulation by E. coli single-stranded DNA binding protein. Proc Natl Acad Sci USA 77:857.

14. Cox MM, Lehman IR (1981). RecA protein of E. coli promotes branch migration, a kinetically distinct phase of DNA strand exchange. Proc Natl Acad Sci USA 78:3433.

15. DasGupta C, Wu AM, Kahn R, Cunningham RP, Radding CM (1981). Concerted strand exchange and formation of Holliday structures by E. coli recA protein. Cell 25:507.

16. West SC, Cassuto E, Howard-Flanders P (1981). Heteroduplex formation by recA protein: Polarity of strand exchanges. Proc Natl Acad Sci USA 78:6149.

17. Cox MM, Soltis DA, Lehman IR, De Brosse C, Benkovic SJ (1983). ADP-mediated dissociation of stable complexes of recA protein and single-stranded DNA. J Biol Chem 258:2586.

18. Cox MM, Lehman IR (1982). RecA protein-promoted DNA strand exchange: Stable complexes of recA protein and single-stranded DNA formed in the presence of ATP and single-stranded DNA binding protein. J Biol Chem 257:8523.

19. Glassberg J, Meyer RR, Kornberg A (1979). Mutant single-strand binding protein of E. coli: Genetic and physiological characterization. J Bact 140:14.

20. Witkin E, personal communication.

21. Soltis DA, Lehman IR (1983). RecA protein-promoted DNA strand exchange. Isolation of recA protein-DNA complexes formed in the presence of single-stranded DNA binding protein. In press.

22. Flory J, Radding CM (1982). Visualization of recA protein and its association with DNA: A priming effect of single-strand DNA binding protein. Cell 28:747.

Mechanisms of DNA Replication and Recombination, pages 723–729
© **1983 Alan R. Liss, Inc., 150 Fifth Avenue, New York, NY 10011**

COMPLEXES OF recA PROTEIN WITH SINGLE STRANDED DNA

Th. Koller, E. Di Capua and A. Stasiak

Institute for Cell Biology,
Swiss Federal Institute of Technology,
ETH-Hönggerberg, 8093 Zurich, Switzerland

ABSTRACT   RecA protein cooperatively forms complexes
with single-stranded DNA even in the absence of ATP.
When complexes are made in the presence of ATP, the
average axial distance between successive bases is
increased to 3.4 $\overset{o}{A}$. The structural organization of
these complexes could not be deduced by conventional
electron microscopy. Since the value of 3.4 $\overset{o}{A}$
corresponds to the distance between base pairs in
double-stranded DNA in the B-conformation, we suggest
that recA in the presence of ATP forces single-
stranded DNA into a structure which can interact with
double-stranded DNA at the level of the nucleotide
sequence to initiate the search for homology and the
strand exchange reaction. When recA and single-
stranded DNA are first reacted in the presence of ATP
and then the non-hydrolyzable ATP analogue ATPγS is
added, helical complexes are formed which are similar
to those of recA complexes with double-stranded DNA
in the presence of ATPγS described earlier (6). The
average axial distance between successive bases in
these complexes is close to 5.1 $\overset{o}{A}$ and one helical
turn contains about 18.5 nucleotides. We suggest that
at some step of the recombination event a switch,
induced in vitro with ATPγS, occurs from a compact
recA helix   with pitch of about 34 $\overset{o}{A}$ to a stretched
helix with a pitch of about 95 $\overset{o}{A}$.

INTRODUCTION

While the conceptual model of Holliday and of Meselson & Radding for homologous recombination is generally accepted, no structural models exist so far for the molecular basis of homologous strand transfer, the reaction mediated by recA protein (for review see 1). This reaction has been shown in vitro to require ATP hydrolysis and stoichiometric amounts of recA protein (e.g. 2). In the presence of the non-hydrolyzable analog ATPγS, double-stranded DNA is found covered with recA in stable complexes (3,4,5), very likely at a stoichiometry of one recA per 3 base pairs (6). These filamentous recA-DNA complexes as seen by electron microscopy are helical, suggesting that recA polymerizes in a helical structure in order to force the DNA into a conformation useful for some step of the recA-mediated reaction, maybe the search for homology between DNA strands that are to be exchanged. We have shown that in recA complexes stabilized with ATPγS, double-stranded DNA is unwound from 10.5 base pairs per turn (B-DNA) into 18.5 base pairs per turn, following the protein helix (7). The DNA is stretched by a factor 1.5 (4). These data suggest an intercalative type of interaction. Our aim is to describe the conformations of DNA involved in the homologous strand exchange reactions mediated by recA protein. With this aim in mind we studied complexes between recA protein and single-stranded DNA in the presence of ATP and ATPγS.

METHODS

RecA protein and DNA (from phage ØX174, phage M13 or from a derivative of plasmid pBR 322 containing 4961 base pairs) were reacted at a ratio (w/w) of 100/1 (2 recA per nucleotide) for 10 to 30 min. at 37°C in 25 mM triethanolamine chloride (pH 7.6), 1 mM magnesium acetate and 4 mM ATP (Boehringer). Complexes were then fixed for 15 min at 37°C by adding glutaraldehyde to 0.2%. In some experiments ATP was omitted and in some experiments, prior to fixation, ATPγS (Boehringer) was added to a concentration of 0.5 mM and the incubation was continued for another 15 min. Complexes were purified by gel filtration on Sepharose 2 B (Pharmacia) and were eluted with 5 mM magnesium acetate. The further processing for electron microscopy was as described earlier (6).

Instead of calibrating the magnification of electron micrographs we mixed the samples with a size marker, namely complexes of recA with double-stranded M 13 or plasmid (see above) DNA formed in the presence of ATPγS and fixed with glutaraldehyde. In these reference molecules, the value of 18.5 base pairs per helical repeat and the average axial distance between successive base pairs of 5.1 Å given by DiCapua et al. (6) are taken as standard. In some experiments latex spheres were used as reference.

RESULTS and DISCUSSION

Since for the in vitro strand exchange reaction between homologous molecules the presence of single-strands appears to be essential (1), we studied the interaction of recA with single-stranded (ss) ØX 174 DNA. Even in the absence of ATPγS recA and ssDNA cooperatively form complexes with a smooth appearance (Fig. 1 and 4).

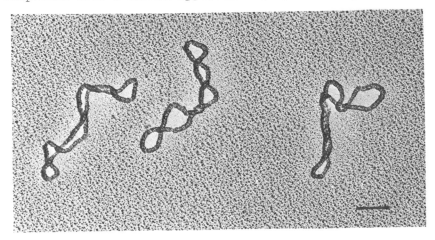

FIGURE 1. Complexes of recA with ssØX174 DNA in the absence of ATP. Bar represents 0.1 μm. Contour length of shown complexes is 1.14±0.02 μm (n = 14).

When ATP was present in the reaction mixture (see Methods) complexes of reproducible length were obtained. Their appearance was still smooth, and by shadowing with platinum or by negative staining we have not been able to resolve a substructure so far (Fig. 2 and 4). We call them compact

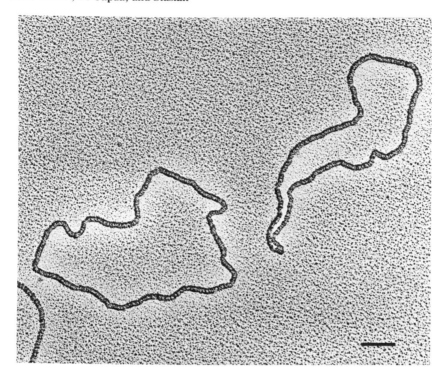

FIGURE 2. Complexes of recA with ssDNA in the
presence of ATP. Bar represents 0.1 μm.

complexes. Their contour length of 1.84+0.02 μm (n = 37)
corresponds to an average axial distance between successive
bases of close to 3.4 Å. Since this value is the same as
the distance between base pairs in double-stranded DNA in
the B-conformation (8) we speculate that by the complex
formation with recA and ATP, ssDNA adopts a structure which
is able to interact with double-stranded DNA at the level
of the nucleotide sequence. This interaction is thought to
be one of the first steps in the recognition reaction
between homologous recombining DNA strands.

When ssDNA was reacted with recA in the presence of
ATP and then, in a second step, ATPγS was added prior to
fixation with glutaraldehyde, clearly longer complexes were
obtained (Fig. 3). Many of them formed side-by-side
associates as described earlier for complexes with double-
stranded DNA (6). The average contour length of these

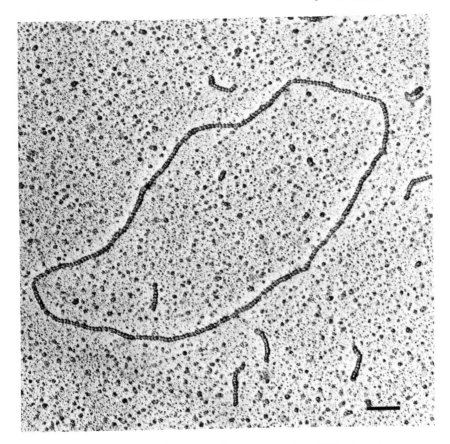

FIGURE 3. Complex of recA with ssDNA in the presence of ATP followed by treatment with ATPγS. Bar represents 0.1 μm. (Sample not purified by gel filtration.)

complexes of 2.65 ± 0.1 μm (n=18) corresponds to an average axial distance between successive bases of close to 5.1 Å. This value is the same as that obtained earlier (6) for the average distance between successive base pairs in recA complexes with double-stranded DNA in the presence of ATPγS. In addition, the detailed appearance of these stretched complexes with cross-striations on shadowed specimens (Fig. 3) and with a zigzag appearance in negative staining (not shown) is similar to complexes with double-stranded DNA, and even the periodicity fits to the 18.5 nucleotides per helical repeat.

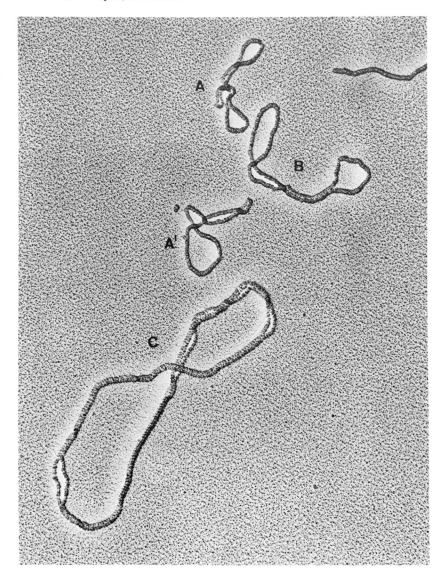

FIGURE 4. For comparison the 3 complexes of recA with ssDNA were coprepared on the same grid. (A,A'): without ATP, (B) in the presence of ATP, (C) in the presence of ATP and ATPγS.

Since the complexes of recA and double-stranded DNA in the presence of ATPγS are clearly helical (6), we assume that the complexes with ssDNA in the presence of ATPγS, and also those in the presence of ATP with no visible substructure, are helical. If this assumption is correct then the results reported here suggest that the recA helix containing ssDNA undergoes, upon reaction with ATPγS, a switch from a helix with a pitch of about 34 Å to a helix with a pitch of about 95 Å. RecA reacted with double-stranded DNA in the presence of ATP forms a compact helix similar to the complex with ssDNA (not shown). Therefore, the compact and the stretched helices may well represent two different functional states of recA in the early steps of the recombination event. The stretched helix is likely to be more open than the compact one and by this the bases might become accessible in the recognition and strand exchange reactions.

ACKNOWLEDGMENTS

This work was supported by Schweizerischer National-fonds zur Förderung der wissenschaftlichen Forschung (grant to Th.K.).

REFERENCES

1. Dressler D, Potter H (1982). Ann Rev Biochem 51: 727.
2. Cox MM, Lehman I R (1981). Proc Natl Acad Sci USA 78: 3433-3437.
3. West SC, Cassuto E, Mursalim J, Howard-Flanders P (1980). Proc Natl Sci USA 77: 2569-2573.
4. Stasiak A, DiCapua E, Koller Th (1981). 151: 557-564.
5. Flory J, Radding CM (1982) Cell 28: 747-756.
6. DiCapua E, Engel A, Stasiak A, Koller Th (1982). J Mol Biol 157: 87-103.
7. Stasiak A, Di Capua E (1982). Nature 299: 185-186.
8. Dickerson R E, Horace R D, Conner B N, Wing R M, Fratini A V, Kopka M L (1982). Science 216: 475-485.

RecA PROTEIN UNWINDS DUPLEX DNA BY 180 DEGREES FOR EVERY
17 BASE PAIRS IN THE FIBER FORMED WITH ATPɣS

James C. Register III, Joan M. Sperrazza,
and Jack Griffith

Department of Microbiology and Immunology
Cancer Research Center
University of North Carolina,
Chapel Hill, North Carolina 27514

ABSTRACT. When RecA protein binds to duplex DNA in the
presence of the ATP analog, adenosine
5'-O-(3'thiotriphosphate), it binds in a highly cooperative
manner as seen by electron microscopy, producing a stiff
nucleoprotein fiber with a distinct 8 nm axial repeat. The
fiber length is 1.6 times that of the protein-free DNA. For
the DNA to be extended by this amount, it must also be
partially unwound. To measure the degree of unwinding,
conditions were established under which we could synchronize
the progressive polymerization of the RecA protein tract
along a small supertwisted DNA beginning at a single, random
locus on the DNA. We observed that as the tract length
increased, the number of supertwists in the remaining
protein-free DNA diminished to zero and then increased
again. By measuring the length of the RecA protein tract on
SV40 DNA in molecules in which the protein-free DNA is fully
relaxed, we could estimate the angular unwinding due to RecA
protein binding. This value was refined by analysis of
molecules with 1 to 12 twists remaining, and determination
of the exact number of supertwists in the SV40 DNA under the
ionic conditions used in the preparation of the samples for
visualization by electron microscopy. From this, we show
that RecA protein unwinds duplex DNA by 180 degrees for
every 17-18 base pairs covered, or 10.5 degrees per base
pair.

1. This work was supported by grants from the NIH
(GM31819), the NSF (PCM77-15653), and from the NCI
(CA16086).

INTRODUCTION

The RecA protein of E. coli catalyzes the displacement and exchange of homologous DNA strands in vitro. These reactions are thought to be central to the mechanism of homologous recombination in vivo (1). Much is known about the biochemistry of RecA protein and the requirements for these reactions, and a variety of models have been proposed to explain their basis (2-6). The detailed mechanisms by which RecA mediates recombination however, remain unresolved. These events may occur on proteinaceous scaffolds formed from the several different RecA protein-DNA filaments which have been identified by electron microscopy (7-9). Because the duplex DNA partners must unwind during recombination, it is of central importance to examine how the RecA protein filaments promote this unwinding.

Recently we and others observed that when RecA protein binds to duplex DNA in the presence of the ATP analog, adenosine 5' -O-(thiotriphosphate) (ATPɣS), the DNA becomes encased in a very stiff nucleoprotein filament whose length is about 1.6 times that of the DNA alone; an 8 nm repeat is easily visualized along the axis of the fiber (7,9). From this 1.6 fold extension we surmised that the DNA in the filaments must also be partially unwound (7). In contrast, in the RecA protein-DNA filament formed when RecA protein binds to duplex DNA in the presence of ATP, a more compact fiber is observed whose length is exactly that of the protein free DNA (7). This led us to suggest that these filaments could undergo a cyclic extension and contraction concomitant with the hydrolysis of ATP (7). During strand exchange, this might aid in the unwinding of the DNA partners and movement of the heteroduplex joint. We assume that in the extended fiber produced with ATPɣS, the non-hydrolyzable ATP analog traps the complex at a crucial step in this cycle. Thus, a study of the structure of these fibers and a precise measurement of the unwinding of the DNA in the extended structure should contribute to our understanding of the mechanism of strand exchange.

RESULTS AND DISCUSSION

Our previous EM observations showed that negatively supertwisted DNA is partially unwound when RecA protein binds in the presence of Mg and ATPγS (7). Several features of this binding facilitated our measurement of the angular unwinding per RecA monomer. The binding is highly cooperative, and we established conditions under which polymerization of RecA proceeded rapidly, resulting in the growth of a single protein tract along the length of a small supertwisted DNA. The RecA covered portion is easily distinguished by EM and can be measured reproducibly. To determine the length of DNA which must be covered by RecA protein to relieve one supertwist, we determined the length of the protein-covered DNA required to remove all of the supertwists from covalently closed, supertwisted SV40 DNA. SV40 DNA purified from SV40 minichromosomes was used because the number of twists is known to be close to 26 (10).

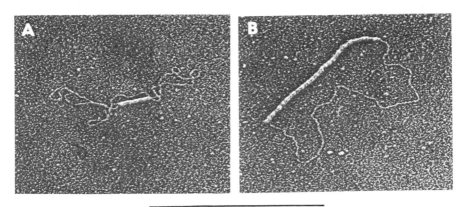

Figure 1. Unwinding of supertwisted SV40 DNA due to the polymerization of RecA protein along the DNA. RecA protein (10 μg) was incubated with supertwisted SV40 DNA (2 μg) in 10 mM Tris-HCl (pH 7.5) buffer containing 2 mM MgCl$_2$ and 0.4 mM ATPγS at 37' for 30 sec. (A) and 1 min. (B). The reaction was stopped by the addition of 5 mM EDTA, followed by filtration through Sepharose 4B to remove free protein. Complexes were fixed and mounted for EM as described (11). Bar equals 0.5 μm.

In this study, (conditions described in Fig. 1) we found that incubation of the DNA with RecA protein for 15 sec. to 4 min. led to the initiation of a single short tract of RecA protein on the still highly supertwisted SV40 DNA circle (0-30 sec.) (Fig. 1A). As the tract length increased, the supertwisting of the remaining protein-free DNA decreased until at times of about 60 sec, many of the molecules had a single, long protein tract and the remaining protein-free DNA was fully relaxed (Fig. 1B). Measurement of the length of the RecA protein tract in these molecules yielded a mean length of 0.45 ±0.05 μm. With increasing times of incubation, the tract length continued to increase and the free DNA portion became more and more twisted.

No evidence for a preferential site of initiation of the RecA protein tracts on the SV40 DNA circles under these conditions was found. Cleavage of molecules having a very short protein tract with a single cut restriction enzyme followed by examination of molecules by EM indicated a random site of initiation (data not shown).

Relating the length of an SV40 DNA circle covered by RecA protein (at the point of relaxation of the remaining DNA) to the number of supertwists which have been lost required knowing the relation between the twisting as seen by EM and that determined by solution methods. We have previously shown (12) that covalently closed circular molecules containing zero to 3 twists show a good correlation between the number of cross-overs in the DNA as seen by EM and the number of twists measured by agarose gel electrophoresis. We will show elsewhere that this correlation between EM and electrophoretic methods holds for DNA molecules with at least 15 twists (13).

From topological considerations of closed circular DNA molecules, we know that the addition or removal of one supertwist results in the gain or loss of one cross-over (defined by the duplex crossing itself in a two dimensional projection) as visualized by EM. Often it was possible to both measure the length of the RecA-covered segment, and to count apparent cross-overs on the same DNA molecule. Both parameters were measured for 200 molecules and the results are plotted in Figure 2. Although there was little evidence of nuclease contamination, no data is presented for the zero cross-over value to avoid including molecules which might have been nicked during preparation. In this plot molecules having RecA covered portions greater than 0.45 microns were assigned positive cross-over values (we will show elsewhere that this approach does not alter the y-intercept) (14). Here, the y-intercept, or length of the RecA covered tract at the point of relaxation is 0.44 μm for the best fit line.

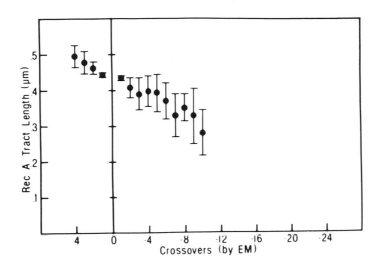

Figure 2. Correlation between the length of a RecA-covered tract of SV40 DNA and cross-overs as seen by EM. Supertwisted SV40 DNA circles containing 29±3 supertwists were incubated with RecA protein for 15 sec. to 4 min. and prepared for EM as described in Fig. 1. The length of each RecA-covered tract was measured and the apparent number of cross-overs counted on each of 200 randomly selected examples (micrographs taken at zero tilt). Tract lengths greater than 450 nm were assigned positive values (see text). The tract lengths were averaged for each cross-over value and these points plotted (bars indicate +/- one standard deviation). The best fit line obtained by linear least squares analysis gives a y-intercept of 0.440 μm at the point of relaxation. The correlation coefficient is -0.98.

In order to measure the number of supertwists in SV40 DNA removed by the binding of 0.45 μm of RecA protein, we had to determine the number of supertwists present in the DNA under the same conditions under which the samples were prepared for EM. Band counting agarose gel electrophoresis (14,15) was applied as described in Figure 3 to measure the number of supertwists in SV40 DNA (in the presence of 1 mM MgCl$_2$ and 0.6 mM ATPγS). A value of 29±3 supertwists was obtained by these measurements.

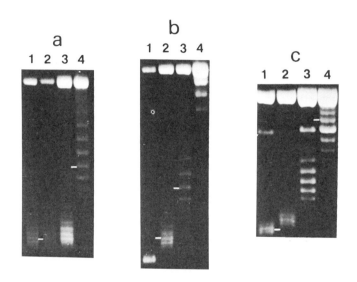

Figure 3. Determination of the number of supertwists in SV40 DNA under the conditions used for EM analysis. Band counting gel electrophoresis was performed by the method of Keller (14) using chloroquine in the gels and running buffer as described by Shure et al. (15). Gel (a) was run in the presence of 2.5 μg/ml chloroquine, gel (b) was without chloroquine, and gel c was run in the presence of 5 μg/ml chloroquine. Lanes a-1, a-2, and b-1 contain native SV40 DNA. Lanes c-1, and c-2 contain DNA relaxed in the presence of RecA buffer (see Figure 1) at room temperature and 37', respectively. The remaining lanes contain DNA fully relaxed by Drosophila topoisomerase 1 in the presence of increasing concentrations of ethidium bromide. The number of supertwists in SV40 DNA were counted as follows: The white bar in lanes a-1, and a-2 represents the center of mass in native DNA. The white bars in lanes a-4, and b-2 indicate

identical bands as do the white bars in lanes b–3 and c–4. The white bar in lane c–1 represents fully relaxed DNA (at room temperature at which the DNA was mounted for EM).

In summary, we have found that when a 0.44 μm tract of RecA protein forms along an SV40 DNA in the presence of Mg and ATPγS, all 29±3 supertwists are removed. Since the binding of RecA protein under these conditions extends the duplex DNA, this 0.44 μm length must be reduced appropriately to calculate the number of base pairs contained in it. Our previously reported value for the extension factor was 1.5 (7). We have recently revised this number to 1.6 as the result of additional measurements. Reduction of the 0.44 μm length 1.6 fold to 0.275 μm corresponds to 17.5% of the 1.57 μm length of protein free SV40 DNA or 917 base pairs of the 5243 base pair molecule (17). Thus for every 32 base pairs bound by RecA protein in the presence of Mg and ATPγS, one supertwist is removed.

Using electron microscopy, we and others have shown that these fibers exhibit a distinct axial repeat of 8 nm or 17–18 base pairs (7,8). Since 32 base pairs contains 2 repeat units of 17 base pairs, we conclude that the DNA is unwound by 180 degrees in each repeat unit. This corresponds to 10.5– 11.0 degrees of unwinding per base pair. Recently another approach to this measurement was reported by DiCapua and Stasiak (18) also utilizing EM. They examined the class of molecules in which the RecA protein tract had reached its maximal length on a covalently closed DNA circle, limited by the extreme torsional stress on the remaining protein free DNA. They reported 15 degrees of unwinding per base pair. The difference will be discussed in detail elsewhere (13).

We have reported here a characterization of the unwinding of duplex DNA in one of the several fibers that RecA protein forms with duplex DNA. RecA protein also forms complexes with single stranded DNA (7,9); it will be important in the future to define the ultrastructure of each RecA-DNA complex and to relate each to the others as a basis for understanding the molecular mechanics of genetic recombination.

REFERENCES.

1. Radding, C.M., (1982). Ann Rev Genet 16: 405–437.
2. Weinstock, G.M. (1982). Biochimie 64: 611–616.
3. Cunningham, R.P., Das Gupta, C., Shibata, T., and Radding, C.M. (1980). Cell 20: 223–235.
4. Shibata, T., Das Gupta, C., Cunningham, R.P., Williams, J.G.K., Osber, L., and Radding, C.M. (1981). J Biol Chem 256: 7565–7572.
5. West, S.C., Cassuto, E., Howard Flanders, P. (1981). Proc Natl Acad Sci USA 78: 2100–2104.
6. Cox, M.M., and Lehman, I.R. (1982). J Biol Chem 257: 8523–8532.
7. Dunn, K., Chrysogelos, S., and Griffith, J. (1982). Cell 28: 757–765.
8. Stasiak, A., DiCapua, E., and Koller, Th. (1981). J Mol Biol 151: 557–564.
9. Flory, J., and Radding, C.M. (1982). Cell 28: 747–756.
10. Griffith, J.D. (1978). Science 201: 525–527.
11. Griffith, J.D., and Christiansen, G. (1978). Ann Rev Biophys Bioeng 7: 19–35.
12. Griffith, J.D. (1978) Biopolymers 17: 237–241.
13. Sperrazza, J.M., Register, J.C., and Griffith, J (manuscript in preparation).
14. Chrysogelos, S., Register, J.C., and Griffith, J. (1983) submitted
15. Keller, W. (1975). Proc Natl Acad Sci USA 72: 4876–4880.
16. Shure, M., Pulleyplank, D.E., Vinograd, J. (1977). Nucleic Acids Res 4: 1183–1205.
17. Tooze, J. (1980) DNA tumor Viruses, Second Edition, Cold Spring Harbour Laboratory.
18. Stasiak, A., and DiCapua, E. (1982) Nature 299: 185–186.

Mechanisms of DNA Replication and Recombination, pages 739–751
© 1983 Alan R. Liss, Inc., 150 Fifth Avenue, New York, NY 10011

FORMATION AND RESOLUTION OF FIGURE-EIGHT
RECOMBINATION INTERMEDIATES

Stephen C. West and Paul Howard-Flanders

Departments of Therapeutic Radiology,
and Molecular Biophysics and Biochemistry
Yale University, New Haven CT 06511

ABSTRACT   We describe the formation and purification
of biparental figure-eight DNA molecules from two
partially homologous plasmids, using a purified enzyme
system. The figure-eights constructed in vitro contain
Holliday junctions and have all the basic properties
of similar structures isolated from whole cells. When
purified figure-eights were transfected into recA⁻
E. coli cells, they were resolved to produce monomeric
or dimeric plasmid progeny, apparently by the cutting
and joining of the single-strand crossovers of the
Holliday junction.

INTRODUCTION

    The RecA protein of E. coli plays an essential role in
general genetic recombination, catalyzing both the pairing
and exchange of strands between homologous DNA molecules to
form regions of heteroduplex DNA. However, while RecA
protein is necessary for the formation of recombination
intermediates in which two DNA molecules are covalently
linked, it is the subsequent process of resolution which
determines whether the flanking arms of DNA emerge in their
original form, or in a new recombinant linkage.
    Holliday (1) first proposed the idea that recombining
DNA molecules might be linked by single strand crossovers,
and evidence for such structures was provided by the
electron microscopic observation of figure-eight DNA
molecules containing Holliday crossovers, in extracts from

This work was supported by grants from the United States
Public Health service, CA06519, GM11014 and AM09397.

recombination proficient cells carrying the small circular DNA phages S13 and øX174 (2, 3), or plasmid DNA (4). Although the RecA protein-mediated reactions that lead to the formation of recombination intermediates have been studied in depth, little is known of the later steps in which intermediates are resolved into mature recombinant chromosomes. The aim of our current work is to identify the activity that resolves Holliday crossovers and to characterize its mechanism of action. Our approach in this study is to use RecA protein in vitro to catalyze the formation of figure-eight molecules from the DNA of two homologous plasmids, and to use the figure-eight DNA in assays directed at identifying the resolving activity.

In this paper we describe the formation of figure-eight recombination intermediates and present the results of experiments that suggest that figure-eights are resolved into mature recombinant DNA molecules upon transfection of recA⁻ E. coli cells.

RECIPROCAL STRAND EXCHANGE BY RecA PROTEIN

An important in vivo function of RecA protein is to catalyze strand exchanges between homologous duplex DNA molecules to form regions of heteroduplex DNA. This strand exchange activity can be demonstrated in vitro by incubating purified RecA protein with ^{3}H-labeled gapped circular and ^{32}P-labeled linear duplex DNA (5). As shown in Fig. 1A, 30 minutes after the start of the reaction, much of the ^{32}P-label can be observed in the position of form II (nicked duplex) DNA. This form II duplex now contains the intact strand of the gapped molecule base paired with the complementary strand of the ^{32}P-labeled linear. The other product of the reciprocal exchange reaction, a heteroduplex gapped linear molecule, cannot be seen on this gel since it comigrates with the unreacted linear duplexes.

The data accumulated over the past 2 years have led to the following conclusions concerning the mechanism of strand exchange (Fig. 2):

1) Pairing requires the presence of a single stranded region in one of the DNA molecules (6-8). RecA protein has not been shown to pair any combinations of intact or nicked duplex DNA.

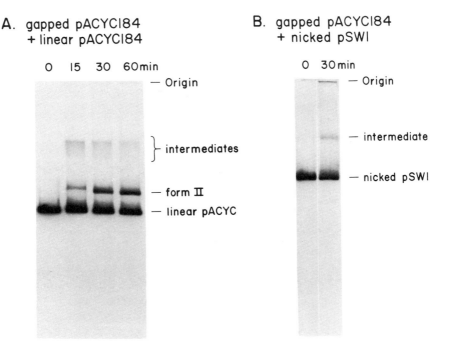

A. gapped pACYCI84 + linear pACYCI84

0 15 30 60min

— Origin

} intermediates

— form II
— linear pACYC

B. gapped pACYCI84 + nicked pSWI

0 30min

— Origin

— intermediate

— nicked pSWI

FIGURE 1. *Reciprocal strand exchanges by RecA protein.*
(A). Gapped ^{3}H-labeled plasmid DNA was reacted with homologous EcoR1-linearized ^{32}P-labeled DNA in the presence of RecA protein. At the times indicated, samples were taken, deproteinized and assayed by electrophoresis through a 1 % agarose gel (5). The ^{32}P-label was detected by autoradiography.
(B). Gapped ^{3}H-labeled plasmid DNA was reacted with EcoR1/ethidium bromide nicked ^{32}P-labeled pSW1 DNA in the presence of RecA protein, and assayed as above (23).
The gapped plasmid DNA was prepared by EcoR1/EtBr nicking followed by limited digestion with exonuclease III.

2) Pairing appears to be initiated at the site of the gap (Fig. 2b). RecA protein initially binds stoichiometrically to the ssDNA of the gap, perhaps using this site for the formation of a protein–DNA filament (6, 9–11). This structure is presumably required for RecA to bind to and unwind duplex DNA (12–14).

3) Synapsis, which results in the alignment of complementary base sequences, requires the presence of ATP but is not dependent upon its hydrolysis (5, 12, 15).

4) Although there is no requirement for a free end for synapsis and the formation of the initial heteroduplex joint (16), the extension of heteroduplex DNA by strand exchange (Fig. 2b) only occurs when there is a free end complementary to the single stranded region in the gapped duplex (5, 17).

5) Once initiated, strand exchange is driven by RecA protein in one direction. As indicated in Fig. 2c, the (-) strands of each molecule are tranferred with a 3' to 5' polarity (18-20), and the two molecules are linked by single strand crossovers characteristic of a Holliday junction (17, 21).

6) Strand exchange requires the continued hydrolysis of ATP (15), and proceeds at a rate of approximately 4 base pairs/sec (5, 15). At the conclusion of strand exchange the products of the reaction are two heteroduplex DNA molecules which separate (Fig. 2d).

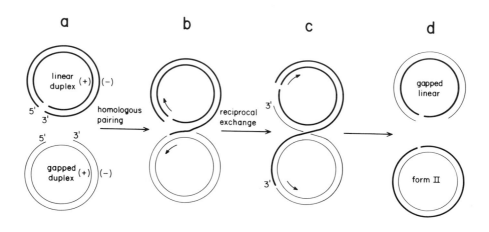

FIGURE 2. A schematic representation of RecA protein-mediated reciprocal strand exchanges between gapped circular and linear duplex DNA to form two heteroduplexes (5).

EXPERIMENTAL PROTOCOL

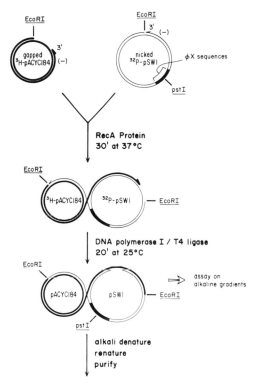

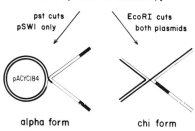

FIGURE 3. Experimental design to prepare, purify and assay figure-eight DNA.
The plasmids pACYC184 and pSW1 are homologous except for 872 bp of øX DNA in pSW1. Therefore, both plasmids contain a single EcoR1 site, and in addition, pSW1 contains a single Pst1 site within the øX sequences. 3'-termini are indicated with a half arrow.

FORMATION OF FIGURE-EIGHT MOLECULES
IN VITRO

   The known activities of three proteins were utilized
to prepare figure-eight DNA molecules in vitro by the
experimental scheme shown in Fig. 3. First, RecA protein
was used to catalyze reciprocal strand exchanges between
the partially homologous DNA of the plasmids pACYC184 (22)
and pSW1 (18). Second, since strand exchanges were limited
in length by a region of heterology, addition of DNA
polymerase I and DNA ligase converted the intermediates
formed by RecA protein (Fig. 1B) into covalently closed
figure-eight structures. Finally, the figure-eights were
purified by gel electrophoresis and assayed by alkaline
sucrose sedimentation, gel electrophoresis and electron
microscopy (23). An example of a typical electron
micrograph of a figure-eight constructed in vitro is shown
in Fig. 4.

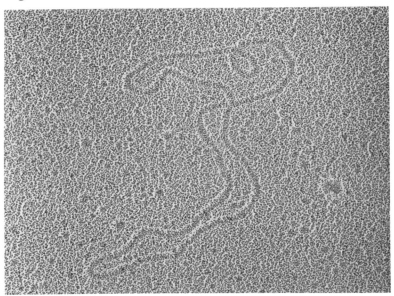

*FIGURE 4. Electron micrograph of a figure-eight
constructed in vitro.*

The figure-eights constructed with the purified enzyme system are similar to those recovered from intact cells that carry viral or plasmid DNA (2-4). A detailed description of the preparation and characterization of figure-eights is described elsewhere (23). In brief, we found that:

1) Approximately 60 % of the figure-eights were covalently closed, as determined by alkaline sucrose sedimentation.

2) The figure-eights were biparental, consisting of one monomer of each parental plasmid, and could be cleaved efficiently with restriction endonucleases Pst1 or EcoR1 to form alpha or chi shaped structures respectively (see Figure 3).

3) When spread for electron microscopy at high formamide concentrations, the two single strand connections in the crossover junctions were often visible and took the form of open junctions similar to those proposed by Holliday (1).

TRANSFECTION WITH FIGURE-EIGHT DNA

To investigate how figure-eight DNA molecules are matured into new genetic products within the cell, we wished to transfect genetically marked figure-eights into recA⁻ E. coli. To do this, we constructed the two plasmids shown in Fig. 5 from the DNA of pACYC184. pSW31 contains a short (approx 4 bp) deletion within the tetracycline resistance (tet) gene, and pSW44 has a long (430 bp) deletion within the chloramphenicol (cat) resistance gene. The plasmids are therefore homologous, yet contain different antibiotic resistance markers, and are also of different physical sizes.

Since RecA protein can drive reciprocal exchanges through short mismatched regions (24), but not through long regions of heterology (18), the figure-eights constructed from these plasmids have the two possible forms shown in Figure 5. Structure (i) contains only short lengths of heteroduplex DNA that involve neither gene for drug resistance, whereas structure (ii) has hybrid DNA that extends through the tet but not the cat gene. Both types of figure-eights contain Holliday crossovers and are presumed to be substrates for enzymes that resolve them, and in addition, those that contain heteroduplex tet genes

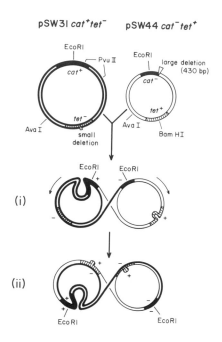

FIGURE 5. Structure of figure-eight DNA molecules used in the transfection experiments.
Since RecA-driven strand exchanges can extend hybrid DNA through short mismatched regions, but not through regions of gross heterology, the figure-eight molecules constructed from pSW31 and pSW44 can be either (i) homozygous (+/+ or -/-) or (ii) heterozygous (+/-) at the tetracycline genes. The tetracycline (tet) and chloramphenicol (cat) resistance genes are represented by the crosshatched and filled in areas, respectively, and the differences between the two plasmids (at the deletions) are represented in the figure-eights by the looped structures (23).

might be acted upon by mismatch repair enzymes.

The preparation of figure-eight DNA was found to be biologically active and tranfected cells to resistance for both antibiotic markers (23). To see the DNA produced by each figure-eight infected cell, we picked 68 colonies from plates that contained tetracycline and chloramphenicol, and prepared cleared lysates from cultures grown from each colony.

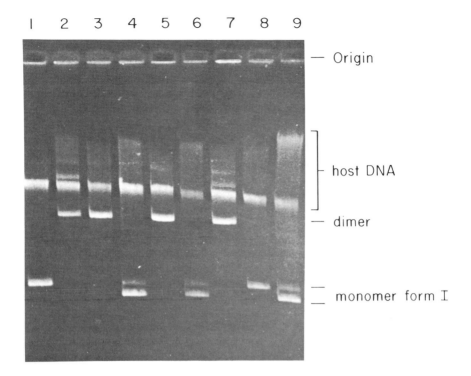

FIGURE 6. Gel analysis of cleared lysates prepared from recA⁻ host cells transfected with figure-eight DNA.

The results of a gel analysis of a random sample of nine such cleared lysates are shown in Fig. 6. Clones 2, 3, 5 and 7 contained supercoiled dimeric circles. Clones 4, 6 and 9 contained two monomers with sizes identical to the parental plasmids, and clones 1 and 8 contained a single monomeric plasmid. In a control experiment, cells of the same strain were cotransfected with pSW31 and pSW44 plasmids. However, in this situation we found no clones that contained dimeric progeny, the DNA being only of the two input types (23). These results indicate that upon transfection of a recA⁻ host each infecting figure-eight molecule is resolved in one of two ways to give rise to either dimeric or monomeric plasmid progeny.

RESOLUTION OF HOLLIDAY CROSSOVERS

Because of the inherent symmetry of a Holliday crossover, it is thought that a figure-eight may be resolved in either of two related ways to produce a dimer (cut at R) or two monomers (cut at P), as shown in Figure 7. Also shown in this figure are the expected products of resolution, repair and replication of figure-eights with heteroduplex tet genes. Resolution by cutting at R would give dimeric forms 1-4, whereas cutting at P would give monomeric forms 5-8.

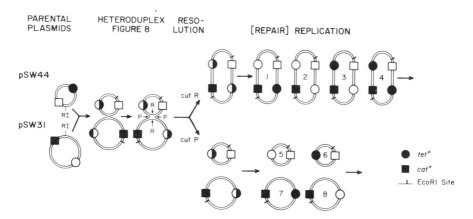

FIGURE 7. *Expected resolution products of a heterozygous figure-eight.*
Resolution can occur by cutting at R (for recombinant) to produce a dimer, or at P (for parental) to produce two monomers. As a result of the formation of hybrid DNA, the dimers or monomers can be heterozygous with respect to the tet *genes. Such structures may, or may not, be acted upon by mismatch repair enzymes to produce the different types of progeny indicated (23).*

When we analyzed the plasmid DNA produced by figure-eight infected cells, by restriction endonuclease analysis, we made the following observations (23):
1) The dimeric progeny were recombinant and contained DNA of both parental plasmids. Approximately half of the dimers appeared to derive from figure-eights that were heteroduplex at tet, and many showed evidence of mismatch

repair. 40 % of the dimeric clones tested were found to contain a heterogeneous population of dimers (types 1 and 4, Fig. 7), indicating that replication occurred before completion of mismatch repair.

2) Clones containing two monomeric species were found to contain the DNA of both parental plasmids (types 6 and 8, Fig 7).

3) The single monomeric species, found in clones 1 and 8 of Fig. 6, were recombinant plasmids that contained the tetracycline and chloramphenicol resistance genes from each parent (type 7, Fig 7).

4) Because of our selection for $Cm^R Tc^R$ we would not expect to detect dimeric type 2 or monomeric type 5. We detected all other types except dimeric type 3, which may be infrequent because of the need for mismatch repair in opposing directions at the two heteroduplex tet genes.

The results we obtained were consistent with the idea that each infecting figure-eight was resolved by cutting of the Holliday crossover in one of the two orientations shown in Fig. 7, and the products of resolution were of either recombinant or parental types. The frequency of resolution to the dimeric form was 25-45 %.

CONCLUSIONS

A preparation of DNA consisting mainly of biparental figure-eight molecules, made with purified enzymes, successfully transfected a recA⁻ E. coli host to produce drug resistant colonies containing viable plasmids. The progeny of each transfection were plasmids of either recombinant or parental types. Restriction analysis of the progeny indicated that mismatch correction could occur and that the Holliday crossover in each figure-eight was resolved in either of two symmetrical ways to produce monomeric or dimeric plasmids. It therefore appears likely that E. coli possesses an enzyme similar to the T4 gene 49 protein, endonuclease VII (25), which has been shown to cleave Holliday crossovers in vitro (26). In our future studies we will attempt to identify the enzyme(s) responsible for resolution and characterize the mechanism of maturation.

ACKNOWLEDGEMENTS

We would like to thank Jill Countryman for the electron micrograph of Figure 4.

REFERENCES

1. Holliday R (1964). Genet Res Camb 5:282.
2. Thompson BJ, Escarmis C, Parker B, Slater WC, Doniger J, Tessman I and Warner RC (1975). J Mol Biol 91:409.
3. Benbow RM, Zuccarelli AJ and Sinsheimer RL (1975). Proc Natl Acad Sci USA 72:235.
4. Potter H and Dressler D (1976). Proc Natl Acad Sci USA 73:3600.
5. West SC, Cassuto E and Howard-Flanders P (1982). Molec Gen Genet 187:209.
6. West SC, Cassuto E, Mursalim J and Howard-Flanders P (1980). Proc Natl Acad Sci USA 77:2569
7. Cassuto E, West SC, Mursalim J, Conlon S and Howard-Flanders P (1980). Proc Natl Acad Sci USA 77:3962.
8. Cunningham RP, DasGupta C, Shibata T and Radding CM (1980). Cell 20:223.
9. McEntee K, Weinstock GM and Lehman IR (1981). J Biol Chem 256:8835.
10. Flory J and Radding CM (1982). Cell 28:747.
11. Dunn K, Chrysogelos S and Griffith J (1982). Cell 28:757.
12. Cunningham RP, Shibata T, DasGupta C and Radding CM (1979). Nature 281:191.
13. Stasiak A, DiCapua E and Koller Th (1981). J Mol Biol 151:557.
14. DiCapua E, Engel A, Stasiak A and Koller Th (1981). J Mol Biol 157:87.
15. Cox MM and Lehman IR (1981). Proc Natl Acad Sci USA 78:3433.
16. Cunningham RP, Wu AM, Shibata T, DasGupta C and Radding CM (1981). Cell 24:213.
17. West SC, Cassuto E and Howard-Flanders P (1981). Proc Natl Acad Sci USA 78:2100.
18. West SC, Cassuto E and Howard-Flanders P (1981). Proc Natl Acad Sci USA 78:6149.
19. Kahn R, Cunningham RP, DasGupta C and Radding CM (1981). Proc Natl Acad Sci USA 78:4786.
20. Cox MM and Lehman IR (1981). Proc Natl Acad Sci USA

78:6018.
21. DasGupta C, Wu AM, Kahn R, Cunningham RP and Radding CM (1981). Cell 25:507.
22. Chang ACY and Cohen SN (1978). J Bacteriol 134:1141.
23. West SC, Countryman JK and Howard-Flanders P (1983). Cell 32:817.
24. DasGupta C and Radding CM (1982). Proc Natl Acad Sci USA 79:762.
25. Kemper B, Garabett M and Courage U (1981). Eur J Biochem 115:133.
26. Mizuuchi K, Kemper B, Hays J and Weisberg RA (1982). Cell 29:357.

Mechanisms of DNA Replication and Recombination, pages 753–759
© 1983 Alan R. Liss, Inc., 150 Fifth Avenue, New York, NY 10011

INDUCTION OF REC1 PROTEIN IN IRRADIATED
CELLS OF USTILAGO[1]

E. Kmiec, M. Brougham, M. Yarnall, P. Kroeger,
and W. Holloman

Department of Immunology and Medical Microbiology,
University of Florida College of Medicine,
Box J-266 JHMHC, Gainesville, Florida 32610

ABSTRACT   The level of Ustilago rec1 protein, a
protein with recombination activity in vitro,
increased significantly in cells irradiated with
ultraviolet light.

INTRODUCTION

A considerable body of evidence from prokaryotes
shows that DNA strand breakage stimulates recombination
(1-4).   Studies in vitro with purified E. coli recA
protein illustrate the governing role of single stranded
DNA in homologous pairing of DNA molecules and are in
accord with the in vivo studies indicating the importance
of strand interruptions in stimulating recombination
(5,6).   However, induction of recombination in vivo is not
brought about merely by introducing strand breaks in DNA.
Rather, a complex series of metabolic reactions is
triggered in response to DNA damage in which a number of
cellular functions including recombination are coordi-
nately expressed (7).   The recA gene of E. coli which is
required for homologous recombination (8) is also necessary for regulating its own induction as part of the
so-called SOS response (9).   In proficient cells the level
of recA protein is increased 20- to 55-fold after
ultraviolet irradiation (10).

[1]This work was supported by Grant GM27103 from the
National Institutes of Health.

In eukaryotes it is less clear whether or not there is an SOS response akin to that in bacteria (11,12). However, introduction of strand breaks in fungi is known to enhance strongly the rate of mitotic recombination (13-15). Futhermore, induction of mitotic recombination after irradiation can be prevented by blocking protein synthesis (16). In Ustilago the recl mutant, which is defective in repair of damage by ultraviolet light, is altered in a number of cellular functions including recombination (17). After irradiation there is little increase in mitotic recombination. These observations have been interpreted to mean the recl gene product regulates mitotic recombination and other cellular processes such as cell division and mutation perhaps as part of an SOS response.

We have purified a protein from Ustilago that promotes pairing of homologous DNA molecules (18). Like recA protein, the purified Ustilago protein is a potent ATPase dependent on single stranded DNA for activity. The protein is not detectable in the recl mutant, an observation strengthening the view that the recl lesion is in the structural gene for the protein.

In this study we were interested in the idea that the recl gene product is part of an SOS response system in Ustilago. To provide evidence in support of this notion we asked whether or not the Ustilago homologous DNA pairing protein, which we have named recl protein, could be induced in cells irradiated with ultraviolet light.

METHODS

Strain.

Ustilago maydis M133 (17) is a diploid phenotypically recombination proficient and heteroallelic at the narl, inosl, and nicl loci. We monitored mitotic recombination by selecting for nar+ (nitrate reductase proficient) recombinants on nitrate minimal medium supplemented with inositol and nicotinic acid (13). We measured cell survival by plating cells on rich medium composed of 1% yeast extract, 2% peptone and 2% sucrose (YEPS).

Cell Growth and Irradiation.

Cells were grown in 16 liters of YEPS at 25°C in a
VirTis fermentor under vigorous aeration. When the
density reached 2.8 x 10^8 cells/ml as determined by
direct counting using a Neubauer counting chamber, we
collected the cells by centrifugation and resuspended them
in 16 liters of water. We irradiated 8 liters of cell
suspension for 3 minutes using a submersible ultraviolet
lamp and withheld the remainder for control. Each suspension was split into 1 liter aliquots in flasks, mixed with
one tenth volume of 10X concentrated YEPS, and individual
flasks incubated in a gyratory water bath for various
times.

Partial Purification of Rec1 Protein.

Assays for rec1 protein are not reliable if the
protein fraction is impure. Therefore, we partially
purified the protein from individual 1 liter cultures
using an abbreviated version of the procedure described
before (18). Cells from 1 liter of culture were harvested
by centrifugation and resuspended 20 ml buffer A (50 mM
Tris-HCl, pH 7.5, 1 mM EDTA, 2 mM 2-mercaptoethanol, 10%
glycerol). Cells were crushed at 20,000 psi in a French
pressure cell and centrifuged at 15,000 x g for 75
minutes. The supernatant was discarded and the pellet was
resuspended in 20 ml buffer A containing 1 M KCl. After
gentle stirring for 20 hours at 4°C the suspension was
clarified by centrifugation and the supernatant brought to
7% polyethylene glycol 67000. After 90 minutes the
precipitate was removed and the supernatant dialyzed
against two 2 liter changes of buffer A. A small column
of DEAE-cellulose (0.6 x 6 cm) pre-equilibrated with
buffer A was loaded with the dialyzed polyethylene glycol
fraction. The column was washed with 5 column volumes of
buffer A containing 0.1 M NaCl and then 5 column volumes
of buffer A containing 0.5 M NaCl. The 0.5 M NaCl wash
was dialyzed against buffer A and applied to a column of
phosphocellulose (0.6 x 6 cm) equilibrated with buffer A.
This column was washed with buffer A containing 0.1 M and
0.5 M NaCl as above. The 0.5 M NaCl wash from phosphocellulose was the fraction used to determine rec1
activity.

Assays.

Recl protein was assayed by measuring DNA dependent ATPase as described before (18). Protein was measured according to Lowry et al. (19).

RESULTS

We followed the kinetics of induction of recl protein using a recl+ diploid heteroallelic at the nar locus. This strain was selected because we wished to correlate mitotic recombination with the level of recl activity. Scoring is convenient because as a result of intrageneic recombination the heteroallelic diploid will revert to nar+ and utilize nitrate as the sole source of nitrogen. Ultraviolet (UV) irradiation greatly enhances the frequency of this event, which is due to gene conversion rather than reciprocal exchange (20). For the optimal dose of ultraviolet light to use in our induction experiment we empirically determined the level of radiation which yielded a high level of gene conversion with minimal

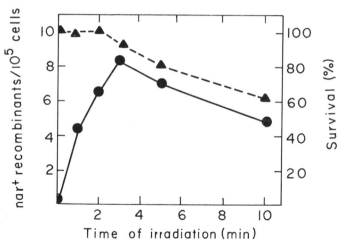

FIGURE 1. Induction of recombinants by ultraviolet light. Cells resuspended in 8 liters of water at 2.3 x 10⁸ cells/ml were irradiated with a submersible ultraviolet lamp. Aliquots were removed at the indicated times and disluted samples (0.1 ml) were spread on nitrate minimal medium to determine recombination or on YEPS to determine survival. Recombinants (●); viability (▲).

cell killing. Using a submersible UV lamp to irradiate 8
liters of cells at high density, we found that irradiation
for 3 minutes induced a high level of recombinants with
little cell killing (Fig. 1).

Having optimized the dose of UV light light on a
large scale to induce recombination, we irradiated a
comparable batch of cells and monitored the level of rec1
activity. Lacking a reliable assay for rec1 activity in
crude extracts, we partially purified the protein to a
degree where interferring activities were removed. This
involved selective extraction in high salt, phase partition with polyethylene glycol and step wise chromatography
on DEAE- and phosphocellulose. We observed (Fig. 2) a
substantial increase in the level of DNA-dependent ATPase
activity in cells irradiated with UV light. The level

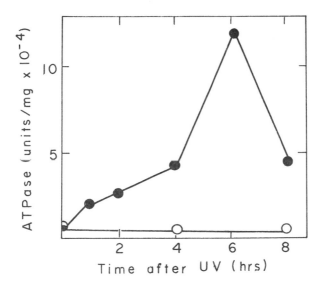

FIGURE 2. Induction of ATPase activity in cells
irradiation with UV light. Cells were grown and
irradiated as described in METHODS. At time zero cells
were irradiated for 3 minutes. Cultures were continued
for the indicated periods of time before harvesting for
determination of rec1 activity. Irradiated (●);
unirradiated (○).

rose progressively with time reaching a peak 6 hours after the cells had been irradiated. The level achieved was greater than 20-fold higher than the level in uninduced cells.

DISCUSSION

After UV irradiation the frequency of mitotic recombination in Ustilago is greatly increased. Holliday found that the increased level could be blocked by mutation at the recl locus (17) or by addition of the protein synthesis inhibitor cycloheximide. Stimulation of recombination was effectively blocked by cycloheximide (16) provided that the drug was added within 2 hours after irradiation. Later than this time, addition of the drug had less effect. From this and other studies done with gamma irradiated cells (21), Holliday concluded that recombination is complete sometime between 2 and 4 hours after irradiation and that enzymes for recombination are induced, if indeed they are induced, by 2 hours. Our results, although not exactly in accord with the time frame observed by Holliday, do suggest a specific enzyme capable of recombination activity in vitro is induced in response to irradiation. It is clear that more studies with the recl protein may well illuminate the process of SOS repair in eukaryotes.

REFERENCES

1. Benbow RM, Zuccarelli AJ, Sinsheimer RL (1974). A role for single-strand breaks in bacteriophage ϕX174 genetic recombination. J Mol Biol 88:629.
2. Howard-Flanders P (1975). Repair by genetic recombination in bacteria: overview. In Hanawalt PC, Setlow RB (eds): "Molecular Mechanisms for Repair of DNA," New York: Plenum, p 265.
3. Miller RC (1975). Replication and molecular recombination of T-phage. Ann Rev Microbiol 29:355.
4. Konrad EB (1977). Method for the isolation of Escherichia coli mutants with enhanced recombination between chromosomal duplications. J Bact 130:167.
5. Radding CM (1981). Recombination activities of E. coli recA protein. Cell 25:3.

6. Radding CM (1982). Homologous pairing and strand exchange in genetic recombination. Ann Rev Genet 16:405.

7. Gottesman S (1981). Genetic control of the SOS system in E. coli. Cell 23:1.

8. Clark AJ (1973). Recombination deficient mutants of E. coli and other bacteria. Ann Rev Genet 7:67.

9. Witkin E (1976). Ultraviolet mutagenesis and inducible DNA repair in Escherichia coli. Bacteriological Rev 40:869.

10. Salles B, Paoletti C (1983). Control of UV induction of recA protein. Proc Natl Acad Sci USA 80:65.

11. Radman M (1980). Is there SOS induction in mammalian cells? Photochemistry and Photobiology 32:823.

12. Nomura S, Oishi M (1983). Indirect induction of erythroid differentiation in mouse Friend cells: evidence for two intracellular reactions involved in the differentiation. Proc Natl Acad Sci USA 80:210.

13. Holliday R (1964). The induction of mitotic recombination by mitomycin C in Ustilago and Saccharomyces. Genetics 50:323.

14. Fabre F, Roman H (1977). Genetic evidence for inducibility of recombination competence in yeast. Proc Natl Acad Sci USA 74:1667.

15. Game JC, Johnston LH, von Borstel RC (1979). Enhanced mitotic recombination in a ligase-defective mutant of the yeast Saccharomyces cerevisiae. Proc Natl Acad Sci USA 76:4589.

16. Holliday R (1975). Further evidence for an inducible recombination repair system in Ustilago maydis. Mutation Res 29:149.

17. Holliday R, Halliwell RE, Evans MW, Rowell V (1976). Genetic characterization of rec1, a mutant of Ustilago maydis defective in repair and recombination. Genet Res Camb 27:413.

18. Kmiec E, Holloman WK (1982). Homologous pairing of DNA molecules promoted by a protein from Ustilago. Cell 29:367.

19. Lowry OH, Rosebrough NJ, Farr AL, Randall RJ (1951). Protein measurement with the Folin phenol reagent. J Biol Chem 193:265.

20. Holliday R (1966). Studies on mitotic gene conversion in Ustilgao. Genet Res 8:323.

21. Holliday R (1971). Biochemical measure of the time and frequency of radiation-induced allelic recombination in Ustilago. Nature New Biol 232:233.

Mechanisms of DNA Replication and Recombination, pages 761–772
© **1983 Alan R. Liss, Inc., 150 Fifth Avenue, New York, NY 10011**

GENETIC RECOMBINATION OF PLASMIDS
IN <u>ESCHERICHIA</u> <u>COLI</u>[1]

Anthony A. James and Richard Kolodner

Dana-Farber Cancer Institute and
Department of Biological Chemistry,
Harvard Medical School, Boston, MA. 02115

ABSTRACT The recombination of plasmids in <u>E</u>. <u>coli</u>
has been examined by looking at the interconversion
of monomers and oligomers in strains carrying mutations known to affect conjugation-mediated recombination. Similarly, intermolecular recombination
between two compatible plasmids carrying different
mutations of a homologous gene was examined. The
results indicated that plasmids can recombine via
a pathway that required the wild-type function of
the gene products of at least two loci, <u>recA</u> and
<u>recF</u>. In addition, recombination via this pathway
required the presence of a recombinogenic element
in cis or in trans. Plasmids were also able to
recombine via an <u>sbcA</u>-inducible pathway that did
not require recombinogenic elements.

INTRODUCTION

Genetic recombination results in the rearrangement of
genetic information and provides for genetic diversity on
both cellular and organismic levels. Evidence for recombination events can easily be obtained from experiments
designed to detect changes in the linkage arrangement of a

[1]This work was supported by grants from the American
Cancer Society and Boston Medical Foundation, Inc. to
AAJ, and NIH grant GM26017, ACS grant MV-125, and an ACS
Junior Faculty Research Award, JFRA-35 to RK.

gene or part of a gene. As in most metabolic processes, recombination most likely results from the interaction of enzymes with a substrate molecule, in this case DNA. However, the enzymatic mechanisms involved in recombination and the control of these mechanisms are generally unknown. Our approach to understanding the mechanisms of recombination involves characterization of the effects on plasmid recombination of various recombination-deficient mutations of E. coli, characterization of the proteins involved in recombination, and an examination of substrate DNA molecules that have special recombinogenic properties.

The work of Clark and his co-workers (for reviews see references 1-3) led to the discovery of many mutations that decrease the frequency of conjugation-mediated recombination in E. coli. The effects of various single mutations and combinations of different mutations on recombination led Clark (1) to propose the existence of at least two recombination pathways, the RecA RecBC pathway and the RecA RecF pathway. The RecA RecBC pathway appears to be the major pathway for conjugation-mediated recombination in wild type E. coli.

Further information on the mechanism of recombination has come from the study of recombinogenic elements. In this case recombination is dependent on or stimulated by specific regions of chromosomes, and hence specific DNA sequences, that elevate the frequency of recombination at or near the regions containing the recombinogenic element. The best characterized examples of DNA sequences that stimulate genetic recombination are the Chi sites that stimulate phage lambda recombination via the E. coli RecA RecBC recombination pathway but not via the RecA RecF pathway (4). These DNA sequences are eight base-pairs long (5) and were isolated because they stimulated the growth of λ red$^-$ gam$^-$ by increasing via recombination the formation of circular dimers which could be packaged into viable phages (6). Chi sites could be detected in mutant λ phages and wild type E. coli chromosomes and were active only when present in one of the two possible orientations.

Recently we isolated DNA segments that stimulated the recombination-mediated formation and interconversion of circular oligomers of certain plasmids in wild type E. coli (7). The observed stimulation of recombination appears to occur by the E. coli RecA RecF recombination pathway and occurs regardless of the orientation of the DNA segment inserted in the plasmid (8). These and other properties distinguish these DNA segments from Chi sites and indicate

that they are a new class of recombinogenic elements.
Some of the properties of these elements and the effects
of certain recombination-deficient mutations are discussed
below.

RESULTS

Initial studies of plasmid recombination focused on
the recombination-mediated interconversion of different
oligomeric species (9-15). The critical observation was
that some monomeric plasmids used to transform wild type
E. coli strains were often recovered from those strains as
a mixture of monomers and different oligomeric species.
The predominant oligomeric species were shown to be
covalently-closed circular molecules composed of an integral number of tandemly repeated monomeric units. In
addition, catenated and figure-8 molecules were also
identified. However, in strains carrying a recA mutation,
very few oligomeric molecules were observed. Thus it was
concluded that recombination played a role in the formation
and/or maintenance of the circular oligomeric molecules.

In addition to the formation of oligomers by intermolecular recombination of monomers, the breakdown of
oligomers to monomers and different oligomeric species by
intramolecular recombination, was observed (10-15). We
used various plasmid tetramers to examine intramolecular
recombination and found that tetramers were converted to a
mixture of monomers and other oligomers in wild type
strains while remaining predominantly tetramers in a recA
strain (8). Other mutations that we have tested that
affect plasmid recombination are listed in Table 1.

Not all plasmids had the ability to form circular
oligomers in wild type strains. We reasoned that such
plasmids would be able to do so if provided an appropriate
DNA sequence. In a series of experiments using the plasmid
pVH51 (16), a plasmid that normally does not form
oligomers in wild type strains, we cloned random EcoRl
fragments into the unique EcoRl-site of pVH51 in order to
search for fragments of DNA that would provide a recA$^+$ -
mediated ability to oligomerize. Several elements were
found and some of their properties characterized.

In addition to the plasmids we constructed, plasmids
contructed by other workers for other purposes were also
determined to contain recombinogenic elements. Most
notable among these plasmids were pBR322 and pACYC184.

TABLE 1

MUTATIONS THAT AFFECT PLASMID RECOMBINATION

Mutation	Principle Effect[a]	References
recA	A,B,C	8, 18
recB recC	A,C	8, 18
recE	C (A,B)[b]	Joseph and Kolodner, unpublished
recF	A,B,C	8, 18
ssb	C (A,B)[b]	Kolodner, unpublished
lexA71[c]	D	Kolodner, unpublished
sbcA	D	8, 18
recJ	C (A,B)[b]	A.J. Clark pers.comm.
topA	A,B,C	Fishel, Kolodner, and Wang, unpublished

[a]Letters refer to: A, decreases the formation of oligomers from monomers; B, decreases the formation monomers and different oligomeric species from tetramers; C, decreased the recovery of tetracycline-resistant colonies in a genetic test (8,18); D, increased the frequency of recovery of recombinant plasmid molecules.
[b]Parentheses indicate not measured.
[c]Results in the constitutive expression of SOS response (19).

The plasmid pACYC184 was constructed using the origin of replication of a plasmid, p15A, the tetracycline-resistance gene of pSC101, and the chloramphenicol-resistance gene of pKT002 (17). Figure 1 illustrates the p15A lacks the ability to oligomerize in wild type strains. However, in the construction of pACYC184 some sequences were introduced that promote the recA$^+$-mediated interconversion of monomers and tetramers.

The majority of the recombinogenic elements recovered were cis-acting, only one trans-acting element was found, and all required the wild type function of the recA, recB, recC, and recF genes in order to convert monomers into oligomers (7,8). Intramolecular recombination of tetrameric plasmids to form lower-order oligomers and monomers also required the wild type functions of the recA and recF genes but did not require the wild type function at the recB and recC genes. In all cases where recA, recB, recC and recF

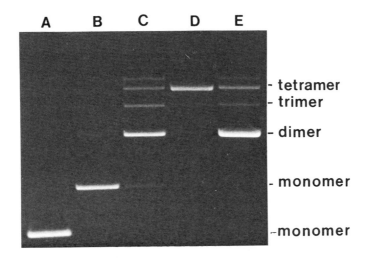

FIGURE 1. The $\underline{recA^+}$-mediated interconversion of
pACYC184 monomers and tetramers. DNA prepared and run on
.8% agarose gel as described in reference 8. Lane A, p15A
in a wild type strain; Lane C, pACYC184 monomer transformed
into a wild type strain (AB1157); Lane B, pACYC184 monomer
transformed into a $\underline{recA13}$ strain (JC2926); Lane E, pACYC184
monomer transformed into wild type; Lane D, pACYC184 tetramer transformed into a $\underline{recA13}$ strain.

blocked the formation of circular oligomers, these effects
could be reversed by the presence of an $\underline{sbcA}$ mutation.

A possible explanation for the activity of the recombinogenic elements is that instead of promoting the formation of oligomers they somehow prevent the rapid breakdown
of oligomers that could occur in the recombination-
deficient strains. This seemed initially unlikely because
the recombinogenic elements promoted the interconversion of
both tetramers and monomers. However, we reasoned that if
the elements were stabilizing oligomeric forms then an
oligomer lacking an element should be unstable. Tetramers
of pVH51, a plasmid lacking a recombinogenic element, were
transformed into various recombination-deficient strains
and examined by gel electrophoresis (Figure 2). Tetramers
were stable indicating that the effect of the elements is
most likely to stimulate via recombination the interconversion of oligomers.

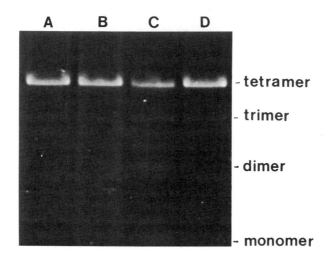

FIGURE 2. Stability of pVH51 tetramers in wild type and different mutant strains. DNA prepared and run on .5% agarose gel as described in reference 8. Lane A, pVH51 tetramer in wild type strain (AB1157); Lane B, pVH51 tetramer in a recA13 strain (JC2926); Lane C, pVH51 tetramer in a recB21 recC22 strain (JC5519); Lane D, pVH51 tetramer in a recF143 strain (JC9239).

Further studies on plasmid recombination were carried out using a genetic assay. Tetracycline-sensitive derivatives of pBR322 and pACYC184 were made by mutating restriction endonuclease sites in the tetracycline-resistance genes of both plasmids. Using a pBR322 derivative and pACYC184 derivative each containing a different mutation in the tetracycline-resistance genes, wild type E. coli strains and various mutant strains were co-transformed to chloramphenicol and ampicillin resistance. Recombination was monitored by quantitating the formation of tetracycline-resistant progeny during growth of these transformants. The results for wild type, recA, recF, and strains containing an sbcA mutation reflected the data seen for oligomerization (8). Recombination to produce tetracycline resistance was high in wild type and strains containing sbcA (.02 to .1%) and low in recA and recF strains (.0003 and .002%, respectively). Interestingly recombination in recB recC strains was quite high (.16%) indicating that in recB recC strains recombination does take place between monomers.

To further study a cis-acting recombinogenic element, we have carried out a number of studies with pBR322 and pACYC184, two plasmids that contain cis-acting recombinogenic elements. In one set of experiments a series of deletions in pACYC184 were constructed using recombinant DNA techniques and these plasmids were tested for their ability to recombine to form circular oligomers (James and Kolodner, in preparation). In this way we identified a region between nucleotide positions 375 and 650 within the tetracycline-resistance gene that was required for recombination to form high levels of circular oligomers. In another series of experiments a number of genetic crosses between different pairs of tet⁻ mutations in pBR322 and pACYC184 were carried out as described above. The crosses were performed in wild type E. coli strains where recombinogenic elements are required for the interconversion of oligomers and, as a control, in a recA recB recC sbcA strain in which the recombinogenic elements are not required for oligomer interconversion (18). The results of these crosses were used to construct genetic maps of the tetracycline-resistant gene. An example of these genetic maps along with the physical map of the tetracycline-resistance gene are presented in Figure 3. These maps illustrate that with recombination in an recA recB recC sbcA strain, the genetic map corresponds very well with the physical map. However, in wild type strains there appears to be 100-200 fold more recombination in the amino terminus of the gene compared to the carboxy terminus of the gene resulting in a large distortion of the genetic map compared to the physical map. This suggests that in wild type strains, the recombinogenic elements we have identified have a large effect on genetic recombination.

DISCUSSION

Plasmid recombination in wild type E. coli strains has been shown to convert monomeric or tetrameric plasmids into a mixture of different oligomeric forms. Recombination of the genetically marked plasmid recombination substrates discussed above to produce tetracycline-resistant progeny during the growth of transformants occurred at frequencies of 10^{-3} to 10^{-4} in wild type E. coli strains. Among the mutations tested, mutations in the recA, recB, recC, recF, recJ, ssb and topA gene have been shown to

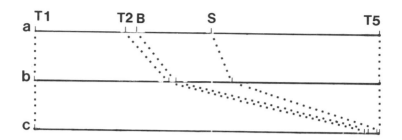

FIGURE 3. Genetic and physical maps of a portion of
the tetracycline-resistance genes in pACYC184 and pBR322.
a) Physical map of tetracycline-resistance gene and several
restriction-endonuclease sites based on Sutcliffe (22), b)
genetic map generated in a recA56 recB21 recC22 sbcA23
strain (JC9604), c) genetic map generated in a wild type
strain (AB1157). Abbreviations; Tl, Taql site or Clal site
at position 24; T2, Taql site at position 339; B, BamHl
site at position 375; S, Sall site at position 650; T5;
Taql site at position 1267.

affect plasmid recombination. The large affect of the
recA, recF and ssb mutations on all of the recombination
reactions tested suggests that the RecA protein, single
stranded DNA binding protein and the RecF protein all play
important role in plasmid recombination. The recB and
recC mutations do not decrease the frequency of plasmid
recombination but do affect the types of products formed.
This affect has been interpreted to suggest that the product of the recB and recC genes, exonuclease V, is involved
in the resolution of some type of plasmid recombination
intermediate (8).
        Conjugation-mediated recombination in wild type E.
coli strains is inhibited by recA recB and recC mutations
but is not inhibited by recF or ssb mutations. These
mutations define the host RecA RecBC pathway for conjugation mediated recombination which is the major recombination pathway in E. coli. The fact that plasmid recombination is strongly inhibited by recF and ssb mutations and is
not inhibited by recB and recC mutations suggests that
plasmid recombination does not occur by the RecA RecBC
pathway. Instead, this genetic analysis suggests that
plasmid recombination occurs by the RecF pathway that was
generally thought to be substantially inactive in wild

type E. coli strains.

SbcA mutations map in a cryptic λ-like prophage in E. coli and were originally identified because they suppressed the effect of recB and recC mutations on conjugation-mediated recombination. Furthermore, conjugation-mediated recombination in strains containing sbcA mutations is known to be inhibited by recA, recE, recF and recJ mutations. We have extensively analyzed plasmid recombination in recB recC sbcA strains. Our results, show that plasmid recombination in recB recC sbcA strains is 10-20 times as frequent as in wild type E. coli strains and is not inhibited by either recA or recF mutations (8). More recently we have shown that recB recC sbcA recE strains have only 10-15% of the plasmid recombination that recB recC sbcA strains have indicating that the sbcA-induced pathway for plasmid recombination requires the recE gene product, exonuclease VIII (Joseph and Kolodner, unpublished results). Our results indicate that the sbcA-induced pathway for plasmid recombination is different from the sbcA-induced pathway for congugation-mediated recombination. This pathway for plasmid recombination may be related to the bacteriophage λ red pathway because sbcA induced functions can replace the λ red gene products during λ recombination (2). Further studies of this pathway should help us understand how recombination can occur effeciently in the absence of RecA protein.

The specific role of recombinogenic elements in plasmid recombination is unclear. Although they appear to be relatively abundant in genomic E. coli DNA (7), the fact that some are cis-acting and one trans-acting, suggests that all the elements are not identical. In fact, a Southern hybridization (20) analysis of the EcoRl fragment containing the trans-acting elements clearly demonstrated that it is present in the E. coli: genome in one copy (Lee, James and Kolodner, unpublished). Whether or not the cis-acting elements are identical is not known. The deletion analysis of pACYC184 located the recombinogenic element to the tetracycline-resistance gene. Since the tetracycline resistance genes of pACYC184 and pBR322 both came from pSC101 (17,21) it is reasonable to assume that the recombinogenic elements of these two plasmids are identical. We are in the process of checking other cis-acting elements not in tetracycline-resistance genes for their sequence homology.

Although we have demonstrated a clear relationship between recombinogenic elements and the interconversion of oligomers, the relationship between these elements, oligomer interconversion, and the production of tetracycline-resistant cells due to intragenic recombination is unclear. Several pieces of evidence indicate that oligomer interconversion and the events that produce a wild type tetracycline gene share some related dependence on recombinogenic elements. Certain deletions in the tetracycline-resistance gene that decreases the ability of pACYC184 to oligomerize and the results from intragenic fine structure mapping suggest a relationship between the two effects (James and Kolodner, in preparation). In addition, the deletions that affect oligomer interconversion also decrease the frequency of recovery of wild type recombinants during genetic crosses. We are currently exploring the relationship of these observations.

The specificity of the recombinogenic elements for the RecA RecF pathway suggests that either the RecA, RecF, or some other proteins in the pathway may recognize these elements. Whether this recognition involves an initiation or termination event is unknown. Plasmid recombination in sbcA strains appears not to require a recombinogenic element (8,18), but plasmid recombination via this alternate pathway may be so high that the effects of recombinogenic elements are masked.

ACKNOWLEDGMENTS

The authors would like to thank their colleagues in the Laboratory of Molecular Genetics for constructive criticism and helpful discussions and Sandra Phillips for typing the manuscript.

REFERENCES

1. Clark, AJ (1980). A view of the RecBC and RecF pathways of E. coli recombination. In: Alberts, B, Fox, CF (eds), Mechanistic Studies of DNA Replication and Genetic Recombination. ICN-UCLA Symposia on Molecular and Cellular Biology Volume 19", New York: Academic Press, 891:899.
2. Gillen, JR, Clark, AJ (1974). The recE pathway of bacterial recombination. In: Grell, RF (ed), "Mechanisms in Recombination", New York: Plenum, 123:126.
3. Clark, AJ (1973). Recombination-deficient mutations of E. coli and other bacteria. Ann Rev Genet 7, 67:86.
4. Stahl, FW (1979). Special sites in generalized

recombination. Ann Rev Genet 13, 7:24.

5. Smith, GR, Comb, M, Schultz, DW, Daniels, DL, Blattner, FR (1981). Nucleotide sequence of the Chi recombinational hotspot X$^+$D in bacteriophage lambda. J Virol 37, 336:342.

6. Malone, RE, Chattoraj, DK, Faulds, DH, Stahl, MM, Stahl, FW (1978). Hotspots for generalized recombination in the Escherichia coli chromosome. J Mol Biol, 121, 473:491.

7. James, AA, Morrison, PT, Kolodner, R (1983). Isolation of genetic elements that increase frequencies of plasmid recombinants. Nature, in press.

8. James, AA, Morrison, PT, Kolodner, R (1982). Genetic recombination of bacterial plasmid DNA: Analysis of the effect of recombination-deficient mutations on plasmid recombination. J Mol Biol 160, 411:430.

9. Hobom, G and Hogness, D (1974). The role of recombination in the formation of circular oligomers of the λ dvl plasmid. J Mol Biol 88, 65:47.

:combination

le formation of

mechanism of
pic observation
tl Acad Sci

e mechanism of
recombination
4, 4168:4172.
combination:
ng Harbor

of bacterial
s of in vitro
cad Sci USA

, Doherty, MJ
plasmid DNAs
x, CF (eds),
nd Genetic
ecular and
ademic Press,

ki, DR (1976).
J Bacteriol

17. Chang, ACY, Cohen, SN (1978). Construction and Characterization of amplifiable multicopy DNA cloning vehicles derived from the P15A cryptic mimiplasmid, J Bacteriol 134, 1141:1156.
18. Fishel, RA, James, AA, Kolodner, R (1981). recA-independent general genetic recombination of plasmids. Nature 294, 184:186.
19. Krueger, JH, Elledge, SJ, Walker, GC (1983). Isolation and characterization of Tn5 insertion mutations in the lexA gene of Escherichia coli. J Bacteriol 153, 1368:1378.
20. Southern, EM (1975). Detection of specific sequences among DNA fragments separated by gel electrophoresis. J Mol Biol 98, 503:517.
21. Bolivar, F, Rodriques, RL, Greene, PJ, Betlach, MC, Heyneker, HL, Boyer, HW, Crosa, JH, Falkow, S (1977). Construction and characterization of new cloning vehicles. II. A multipurpose cloning system. Gene 2, 95:113.
22. Sutcliffe, JG (1978). Complete nucleotide sequence of the Escherichia coli plasmid pBR322. Cold Spring Harbor Symp Quant Biol 43, 77:90.

Mechanisms of DNA Replication and Recombination, pages 773–783
© 1983 Alan R. Liss, Inc., 150 Fifth Avenue, New York, NY 10011

CHI IS ACTIVATED BY A VARIETY OF ROUTES[1]

Franklin W. Stahl, Ichizo Kobayashi, and Mary M. Stahl

Institute of Molecular Biology, University of Oregon
Eugene, Oregon 97403

ABSTRACT Chi, a recombinator in E. coli, is activated
by cos, the packaging origin of phage λ. It can be
activated also by EcoRI endonuclease acting at a site
far from Chi. These results and others suggest that a
recombination machine enters DNA at a double chain cut
and travels inertly until it encounters a Chi properly
oriented with respect to its own orientation. The
machine is activated as it passes Chi so that it can
then effect some step in recombination. A Chi in the
absence of an activating cos is partially active in the
presence of DNA replication. Perhaps the recombination
machine can enter DNA also at replication forks.

INTRODUCTION

Since the 1960's it has been evident that the
probability of meiotic recombination in fungi is conditioned
by the proximity of special sites (recombinators) sprinkled
along the chromosomes (1, 2). Indeed, it is current wisdom
that most eukaryotic recombination depends on such sites
(3), and most recombination models feature them (3-5). The
role played by the recombinator differs from one model to
another and is a prominent, distinguishing feature of
each model. The discovery, in phage λ, of a recombinator,
Chi, for the E. coli recombination system encouraged our
hopes of understanding how one such system works.

[1]This work was supported by grants from the National
Science Foundation and the National Institutes of Health to
FWS.

Chi was recognized in λ following its accidental introduction via two distinct routes (6). In one route, Chi was introduced into λ along with a piece of bacterial DNA during the in vivo formation of plaque-forming bio-transducing strains. Many of these λ pbio are Red⁻ Gam⁻ as a result of the substitution. Red⁻ Gam⁻ λ, infecting rec⁺ E. coli are dependent on recA·recBC-mediated crossing over to achieve the concatemeric state, which is prerequisite to in vivo DNA packaging into phage particles. The λ pbio phage make nice plaques, indicating that they use the bacterial recA·recBC recombination pathway effectively. In fact, they can be seen to have a high rate of exchange in and near the bio substitution (7). In the other route, Chi was introduced into λ by spontaneous mutation (single-base change). Pure deletions of red and gam cause λ to make tiny plaques on rec⁺ bacteria. Upon repeated cycles of such growth, Red⁻ Gam⁻ λ sports particles that form nice plaques. The large-plaque mutants arise at several well-spaced sites in λ. Each confers upon its chromosome a high rate of exchange in it own neighborhood (7-9).

Chi brought into λ on bacterial DNA and Chi arising by spontaneous mutation have apparently identical properties, and both have been shown to be the octomeric sequence (5' GCTGGTGG) (10, 11). The following properties of Chi were established:

1. Chi increases exchange in its own neighborhood with the increase being detectable as far away as 10-20 kb (8). The magnitude of the recombination stimulation implies that Chi is about 10^5 times as recombinogenic as a nonChi octomer (12).

2. Chi functions only in the simultaneous presence of recA and recBC gene functions; i.e., it is specific to the RecBC pathway (13, 14). This observation suggests that either recA or recBC protein (ExoV) recognizes Chi.

3. Chi-stimulated recombination proceeds in the absence of chromosome replication. The chromosome segments donated by the two parents are "spliced" (3) together. The splices are revealed by heterozygosity for a marker located near the site of the exchange (8).

4. χ^+, the active Chi allele, is dominant to $\chi°$, the inactive one from which it was derived by single base change. Only one of the two recombining DNA molecules need be χ^+ in order for the major stimulation to be observed (8). Holliday's (4) model demanding simultaneous cuts at homologous recombinators is evidently inapplicable.

5. Chi functions even when one parent is heterologously

substituted for several kb across the region in which its partner carries the χ^+ sequence (15). Thus a Chi site cannot function by being a special point at which "synapsis" or any other interaction requiring sequence similarity is initiated.

6. A given Chi has "bias". It stimulates exchange primarily to one side of itself (15-17). All Chi examined in λ showed leftward bias. This unexpected result was clarified by the observations below.

7. When Chi is inverted in λ, its activity as a recombinator (and hence as a plaque-size enhancer) is so reduced as frequently to escape detection (18). This result says that Chi is an asymmetric sequence (in harmony with the sequencing data) and that its ability to act is conditioned by its orientation with respect to some feature(s) of the λ chromosome. The result explains why all Chi's examined had shown leftward bias in their action. Chi must have an intrinsic "directionality", but when Chi is oriented so as to act rightward it fails to act (well).

8. The orientation-dependence of Chi activity (item 7) is the same throughout λ (19). Thus, the active orientation of Chi is set by a global feature, not by local features, of λ.

9. The orientation of cos, λ's packaging origin, determines the active orientation of Chi (20). When cos is oriented normally (rightward), only leftward Chi can act well. When cos is inverted (leftward), only rightward (i.e., inverted) Chi can act well.

10. Item 9 suggests that the RecBC pathway enzymes involved with Chi interact somehow with either the process of DNA packaging or that of DNA injection, which, we presume, is likewise polarized by cos. However, the cos-Chi interaction is demonstrably separable from packaging (21), and high Chi activity does not require injection (22). Furthermore, the cos-Chi interaction is demonstrably separable from injection (unpublished).

The failure to relate the cos-Chi interaction to the known activities of cos forced us to consider a radical alternative. Although cut cos has never been physically detected among unpackaged chromosomes (eg., ref. 23), we supposed that such cutting does occur but is rapidly reversed. During the brief open phase, the recombination machine (recBC protein?) enters the λ duplex. The entry is one way (into λ's right end, going leftward from cos) because terminase, the enzyme that cuts cos, is bound to the right of the cut (the gene A side) (24, 25). The traveling machine then moves leftward through the DNA in an inert state. When it passes Chi, the machine is activated, perhaps by picking up a subunit. It continues to travel, still going leftward, but is now either doing something or is able to do something when circumstances permit.

Support for the idea that reversible cos-cutting activates Chi was sought by asking whether some other double-chain cut could activate Chi. The extensive work done to make λ into a cloning vehicle for EcoRI restriction fragments (26-28) provided material suitable for the test.

We designed crosses in which a $\chi°$ parent was uncuttable by EcoRI due to RI modification. The other parent, which was alternately χ^+ or $\chi°$, had a pair of intact EcoRI sites distant from the Chi site. The Chi was oriented rightward and had no detectable activity in these replication-blocked crosses. The χ^+ and $\chi°$ crosses were conducted with and without cutting. The cutting was in vivo thanks to an EcoRI plasmid carried by the host cell. No-cutting was achieved either by having both parents RI-modified or by conducting the crosses in a host without the plasmid. Chi was seen to be activated whenever the Chi-carrying parent was subject to cutting (29).

The activation of Chi by a distant EcoRI cut lends strong support to the idea of a recombination machine that enters DNA at an end. ExoV, the recBC gene product, has been shown to enter DNA at an end and to travel through it driven by ATP hydrolysis (30, 31). Since ExoV is required for Chi activity (point 2, above), it is likely that ExoV is (part of) the recombination machine.

We have previously (17) made the suggestion that the Chi-recognizing enzyme acts destructively as it moves away from Chi. Generation of a single-chain gap seemed a reasonable possibility. The notion came from observations on replication-blocked crosses in which Chi was in one parent only. In these crosses ($a^+\chi^+b^-$ x $a^-\chi°b^+$) complementary Chi-stimulated recombinants were not recovered equally among mature progeny. It was the Chi-containing recombinant ($a^-\chi^+b^-$) that was the less well recovered one, suggesting that Chi action is destructive. Recently we have shown this interpretation to be wrong. When coses not involved in Chi activation are used to package the recombinants, the complements are equally well recovered (unpublished). Perhaps packaging tends to use the very cos that activated a Chi. That would lead to the preferential packaging of one recombinant from the dimeric product of exchange.

The re-interpretation of these "nonreciprocality" data
allows us to entertain the possibility that Chi is, in fact,
highly reciprocal in its action. Such reciprocality would
be a reasonable expectation based on the reported
reciprocality of the E. coli system as measured in λ single
bursts (32). The possibility is supported by our
observation (33) that Chi stimulates the incorporation of
the circular plasmid λdv into λ by homologous exchange.
Such incorporation of circle-into-circle is one criterion
for reciprocality.

The λ x λdv crosses cited above show that Chi
stimulates "patch", as well as splice, formation. These
patches are revealed as the pick-up of markers from λdv
without simultaneous incorporation of the plasmid into λ.
The patches can be understood as one of two possible
outcomes of resolving a Holliday junction. To test further
the reciprocality of Chi-induced recombination, we have
asked whether a Chi-bearing chromosome donates and receives
patches equally. Our experiments (Stahl and Lieb,
unpublished) demonstrate equality and are, therefore, compatible with reciprocality of Chi-stimulated recombination.

The apparent reciprocality of Chi-induced recombination
puts few constraints on models for what the recombination
machine does after it is activated by Chi. One plausible
possibility is that the activated machine resolves Holliday
junctions, which, since presumed to form independently of
Chi, need show no asymmetry in χ^+ x $\chi°$ crosses. The
unfortunate feature of this view is that it avoids the
interesting question of how Holliday junctions are formed.

RESULTS

In vegetative λ crosses, Chi activation is mostly due
to cos, presumably via transient, terminase-induced,
openings. In E. coli there is one Chi per 5 kb (18, 34),
but there is unlikely to be any active cos, and we may
wonder what features of the E. coli chromosome serve to
activate Chi during bacterial recombination. In
transduction and in conjunction, recombination is frequent,
and Chi is demonstrably active (35). In both of these
processes, ends and/or gaps are present, and these ends may
allow the recombination machine to enter and find Chi. Our
experiments with EcoRI-activation suggest another source of
Chi-activation. We noted that a rightward Chi-rightward cos
combination had detectable activity when replication was

allowed but none at all when replication was blocked (Figure 1 and Table 1).

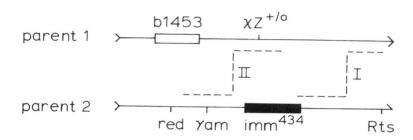

FIGURE 1. The activity of the rightward Chi (χZ^+) is measured as recombinants I/II for the χ^+ cross divided by the same ratio for the $\chi°$ cross. The dependence on replication of the activity of this rightward Chi-rightward cos combination is shown in Table 1. Parent 1 incidentally carries the nin5 deletion in the imm-R interval. For further technical details see ref. 29.

TABLE 1

REPLICATION ACTIVATION OF RIGHTWARD CHI

| | Chi Activity [a] | |
	A	B
With Replication[b]	1.8	1.7
Without Replication[c]	.98	1.1

[a]Defined in Figure 1. In column B all phages were EcoRI-modified. In column A only parent 2 was modified. The modification was incidental to these experiments.

[b]In the Su⁻ rec⁺ host 594 at 39°.

[c]In the Su⁻ rec⁺ dnaB host FA77 at 39°. Data from ref. 29.

Extension of that observation to a leftward Chi-leftward <u>cos</u> combination (Figure 2 and Table 2) revealed its generality. It seems possible from these preliminary experiments that replication forks can activate Chi by serving as entries for the recombination machine. That possibility is being tested.

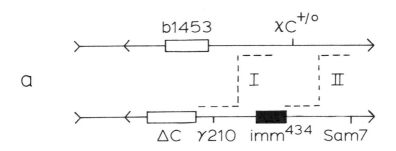

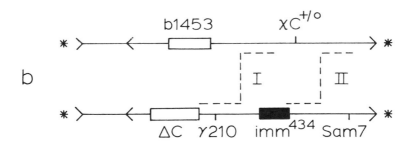

FIGURE 2. The activity of the leftward Chi (χc^+) is measured as recombinants I/II for the χ^+ cross divided by the same ratio for the χ^0 cross. The phages in (a) have a standard <u>cos</u> (>) as well as an additional, cloned <u>cos</u> in inverse orientation (<). In (b), the standard <u>coses</u> of the two phages have been inactivated by the deletion <u>cos2</u> (*> and >*). The dependence on replication of the Chi activities is shown in Table 2. For further technical details, see ref. 21.

TABLE 2

REPLICATION ACTIVATION OF LEFTWARD CHI

	Chi Activity[a]	
	A	B
With Replication[b]	14.8±4.0	3.8±1.2
Without Replication[c]	10.9	.88

[a]Defined in Figure 2. Columns A and B correspond to crosses (a) and (b) of that Figure, respectively.

[b]In the Su⁻ rec⁺ host 594 at 37°. Data from ref. 21.

[c]In the Su⁻ rec⁺ dnaB host FA77 at 39°.

ACKNOWLEDGEMENTS

Mary Gilland made the figures and Nancy Hirata typed the paper.

REFERENCES

1.   Murray NE (1963). Polarized recombination and fine structure within the me-2 gene of Neurospora crassa. Genetics 48:1163.
2.   Lissouba P, Mousseau J, Rizet G, Rossignol J-L (1962). Fine structure of genes in the Ascomycete Ascobolus immersus. Adv Genet 11:343.
3.   Stahl FW (1979). "Genetic Recombination. Thinking about it in Phage and Fungi." San Francisco: Freeman.
4.   Holliday R (1968). Genetic recombination in fungi. In Peacock WJ, Brock RD (eds): "Replication and Recombination of Genetic Material," Canberra: Australian Academy of Science, p157.
5.   Meselson MS, Radding CM (1975). A general model for genetic recombination. Proc Natl Acad Sci USA 72:358.
6.   Henderson D, Weil J (1975). Recombination-deficient deletions in bacteriophage λ and their interaction with

chi mutations. Genetics 79:143.

7.   McMilin KD, Stahl MM, Stahl FW (1974). Rec-mediated recombinational hot spot activity in bacteriophage lambda. I. Hot spot activity associated with Spi⁻ deletions and bio subsitutions. Genetics 77:409.

8.   Lam ST, Stahl MM, McMilin KD, Stahl FW (1974). Rec-mediatd recombinational hot spot activity in bacteriophage lambda. II. A mutation which causes hot spot activity. Genetics 77:425.

9.   Stahl FW, Crasemann JM, Stahl MM (1975). Rec-mediated recombinational hot spot activity in bacteriophage lambda. III. Chi mutations are site mutations stimulating Rec-mediated recombination. J Mol Biol 94:203.

10. Smith GR, Kunes SM, Schultz DW, Taylor A, Triman, KL (1981). Structure of Chi hot-spots of generalized recombination. Cell 24:429.

11. Triman KL, Chattoraj DK, Smith GR (1982). Identity of a Chi site of Escherichia coli and Chi recombination hotspots of bacteriophage λ. J Mol Biol 154:393.

12. Malone RE, Chattoraj DK (1975). The role of Chi mutations in the Spi⁻ phenotype of phage λ: lack of evidence for a gene delta. Mol Gen Genet 143:35.

13. Gillen J, Clark AJ (1974). The RecE pathway of bacterial recombination. In Grell RF (ed): "Mechanisms in Recombination," New York: Plenum, p421.

14. Stahl FW, Stahl MM (1977). Recombination pathway specificity of Chi. Genetics 86:715.

15. Stahl FW, Stahl MM (1975). Rec-mediated recombinational hot spot activity in bacteriophage λ. IV. Effect of heterology on Chi-stimulated crossing over. Mol Gen Genet 140:29.

16. Chattoraj DK, Craseman JM, Dower N, Faulds D, Faulds P, Malone RE, Stahl FW, Stahl MM (1979). Chi. Cold Spring Harbor Symp Quant Biol 43:1063.

17. Stahl FW, Stahl MM, Malone RE, Crasemann JM (1980). Directionality and nonreciprocality of Chi-stimulated recombination in phage λ. Genetics 94:235.

18. Faulds D, Dower N, Stahl MM, Stahl FW (1979). Orientation-dependent recombination hot-spot activity in bacteriophage λ. J Mol Biol 131:681.

19. Yagil E, Dower NA, Chattoraj D, Stahl M, Pierson C, Stahl FW (1980). Chi mutation in a transposon and the orientation-dependence of Chi phenotype. Genetics 96:43.

20. Kobayashi I, Murialdo H, Crasemann JM, Stahl MM, Stahl

FW (1982). Orientation of cohesive end site cos determines the active orientation of χ sequence in stimulating recA•recBC-mediated recombination in phage λ lytic infections. Proc Natl Acad Sci USA 79:5981.

21. Kobayashi I, Stahl MM, Leach D, Stahl FW (1983). The interaction of cos with Chi is separable from DNA packaging. Submitted.

22. Stahl FW, Stahl MM, Young L, Kobayashi I (1983). Injection is not essential for high level of Chi activity during recombination between λ vegetative phages. John Innes Symposium (1982). Norwich: John Innes Horticultural Institute. In press.

23. Laski F, Jackson EN (1982). Maturation cleavage of bacteriophage P22 DNA in the absence of DNA packaging. J Mol Biol 154:565.

24. Feis M, Widner W (1982). Bacteriophage λ DNA packaging: scanning for the terminal cohesive end site during packaging. Proc Natl Acad Sci USA 79:3498.

25. Feiss M, Kobayashi I, Widner W (1983). Separate sites for binding and nicking of bacteriophage lambda DNA by terminase. Proc Natl Acad Sci USA 80:955.

26. Murray NE, Murray K (1974). Manipulation op restriction targets in phage λ to form receptor chromosomes for DNA fragments. Nature 251:476.

27. Rambach A, Tiollais P (1974). Bacteriophage λ having EcoRI endonuclease sites only in the nonessential region of the genome. Proc Natl Acad Sci USA 71:3927.

28. Thomas M, Cameron JR, Davis RW (1974). Viable molecular hybrids of bacteriophage lambda and eukaryotic DNA. Proc Natl Acad Sci USA 71:4579.

29. Stahl MM, Kobayashi I, Stahl FW, Huntington SK (1983). Activation of Chi, a recombinator, by the action of an endonuclease at a distant site. Proc Natl Acad Sci USA, In press.

30. Rosamond J, Telander KM, Linn S (1979). Modulation of the action of the recBC enzyme of Escherichia coli K12 by Ca^{++}. J Biol Chem 254:8646.

31. Taylor A, Smith GR (1980). Unwinding and rewinding of DNA by the RecBC enzyme. Cell 22:447.

32. Sarthy PV, Meselson M (1976). Single burst study of rec- and red-mediated recombination in bacteriophage lambda. Proc Natl Acad Sci USA 73:4613.

33. Stahl FW, Stahl MM, Young L, Kobayashi I (1982). Chi-stimulated recombination between phage λ and the plasmid λdv. Genetics 102:599.

34. Malone RE, Chattoraj DK, Faulds DH, Stahl MM, Stahl FW

(1978). Hotspots for generalized recombination in the Escherichia coli chromosome. J Mol Biol 121:473.

35. Dower NA, Stahl FW (1981). χ activity during transduction-associated recombination. Proc Natl Acad Sci USA 78:7033.

Mechanisms of DNA Replication and Recombination, pages 785–795
© 1983 Alan R. Liss, Inc., 150 Fifth Avenue, New York, NY 10011

A SITE-SPECIFIC ENDONUCLEASE FROM SACCHAROMYCES CEREVISIAE WHICH PLAYS AN ESSENTIAL ROLE IN MATING TYPE INTERCONVERSION[1]

Richard Kostriken, Jeffrey Strathern,
Amar J.S. Klar, James B. Hicks, Carolyn Moomaw
and Fred Heffron[2]

Cold Spring Harbor Laboratory
Cold Spring Harbor, NY   11724

ABSTRACT   We have detected a site-specific endonuclease in some strains of Saccharomyces cerevisiae. This endonuclease, which we call YZ endo, is present in yeast strains which are undergoing mating type interconversion.   YZ endonuclease cleaves the yeast mating type locus (MAT). The location of the double strand break generated in vitro corresponds to the in vivo double stand break observed in switching yeast cells by Strathern et al. (1). Furthermore, several mutations of the MAT locus, known to impair mating type interconversion are also shown to be resistant to cleavage by YZ endonuclease in vitro.   YZ endo makes a site-specific double-strand break at the sequence

              ↓
C G C A A C A G T A A A

G C G T T G T C A T T T
  ↑

[1] This work supported by NSF PCM8140063, PCM 8217002 and the Robertson Fund
[2] Present address: Scripps Clinic and Research Foundation, La Jolla, California

to generate a four base 3' extension terminating in a 3' hydroxyl in the Z region of the MAT locus. The molecular genetics of mating type interconversion is discussed in the context of this endonucleolytic activity.

INTRODUCTION

Haploid yeast can be grouped into two types a and α, defined by the ability of one type to mate with the other type forming non-mating diploids (2). The mating type of a cell is determined by a single genetic locus called the mating type locus (MAT) (3). A haploid yeast will exhibit the a or α mating type depending on whether the a or α allele is present at the MAT locus (4). MATa and MATα either directly or indirectly regulate a number of mating type specific genes and responses such as those involved in pheromone production, degradation and response. In addition to specifying mating type, MATa and MATα also play an important role in determining whether or not a diploid cell may undergo meiosis and sporulation (5).

There are three copies of mating type information all present on chromosome three in budding yeast (6). In addition to the expressed copy at MAT, unexpressed α information is located at HML and unexpressed a information at HMR. These three loci, HMLα, HMRa and MAT share regions of homology called 'X' and 'Z' which flank the allele specific region 'Y' (Figure 1) (7). Regulatory genes, SIR1, MAR1 (SIR2), SIR3 and SIR4 have been identified as being necessary for the repression of HML and HMR (8, 9, 10).

Haploid yeast can switch mating types (11). Mating type interconversion entails replacement of the allele resident at MAT with a copy of the alternate allele from one of the two silent loci (12). The frequency at which interconversion occurs varies from 10^{-6} to 1 switch per

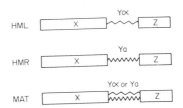

FIGURE 1. A comparison of <u>HML</u>α and <u>MAT</u>(<u>a</u> or α) is shown above illustrating the relative locations of the homologous sequences. This figure is not to scale.

generation depending on the genetic background of the strain. For interconversion to occur frequently the haploid cell must be wild-type for <u>RAD51</u>, <u>RAD52</u> (11) and <u>RAD54</u> (three genes required for meiotic recombination and mating type interconversion) as well as <u>HO</u> (12) and <u>SWI1</u> (15) (genes apparently affecting only mating type interconversion).

Recent work by Strathern et al. (1) has demonstrated a correlation between interconversion and the presence of a double-strand break at the <u>MAT</u> locus. They have proposed a model for interconversion based on the recognized recombinogenic propensity of double-strand breaks (16).

To determine the mechanism by which this double-strand break was generated <u>in vivo</u>, extracts of yeast were assayed for endonuclease activity capable of acting on <u>MAT</u> sequences <u>in vitro</u>. We have discovered such an activity which we have named YZ endo. We show that it is present in strains able to switch and not present in <u>ho</u> or <u>swi</u> haploids or <u>a</u>/α diploids. The exact structure of the double-strand break that is generated has been determined as well as preliminary characterization of the sequence required for endonuclease recognition. Furthermore, mutant <u>MAT</u> loci known to be poor substrates for in vivo mating type interconversion are shown to be poor substrates for YZ endo in vitro. We take this as evidence for the involvement of YZ endo in the process of yeast mating type switching.

RESULTS

I. Detection of the endonuclease. Detection of small amounts of site specific endonuclease activity was accomplished with the use of a sensitive endonuclease assay. The assay consists of incubating end labeled substrate DNA with extracts made from interconverting and non-interconverting yeast. After recovery from the extract, the DNA is electrophoresed on agarose gels and autoradiographed. In this way, conversion of less than 0.1% of the substrate to a uniquely cleaved product can be detected.

Some of the strains from which extracts were made yielded an endonucleolytic activity capable of cleaving the cloned MATa locus in the vicinity of the YZ junction (Figure 2). We have termed this activity YZ endo. Only those strains which are HO, SWI1 and phenotypically haploid yield YZ endo activity (Figure 2).

Since the YZ endonuclease cleaves very near the junction of Y and Z we expected to detect a difference between the ability of MATa and MATα to act as substrates for cleavage if the recognition sequence of the enzyme included part of the Y region. However, our results show that the YZ endo derived from cells of the a mating type is as effective in cutting MATα sequence as it is in cutting the MATa sequence (Figure 2). This suggests that the sequence differences between Ya and Yα regions are not part of the YZ endo recognition site. Silent cassette DNA is also cut by YZ endo (Figure 2).

The endonuclease does not appear to be covalently bound to digested DNA. Although we normally terminate our reactions with pronase to facilitate purification of the digested DNA, omission of this deproteination step does not alter the characteristic migration of the products on acrylamide gels nor does the ratio of product to uncleaved substrate change.

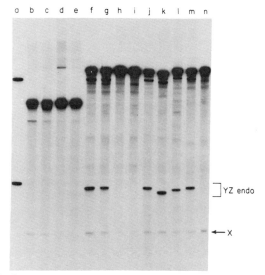

FIGURE 2.

The extracts from four strains of haploid yeast were tested against several substrates. Yeast DNA's were <u>Eco</u>RI <u>Hind</u>III fragments cloned into pBR322. All the substrates were end-labeled at their unique <u>Bam</u>HI site. All extracts examined possessed an endonucleolytic acitvity which cleaves pBR322 (near the <u>Ava</u>I site) generating the fragment labeled X. Cleavage of the substrate molecule by YZ endo produces a fragment in the region of the gel indicated by the bracket.

a)   molecular weight markers
b)   <u>MATa</u> <u>HO</u> <u>SWI</u>/pBR322
c)   <u>MATα</u> <u>HO</u> <u>SWI</u>/pBR322
d)   <u>MATa</u>⁻ <u>ho</u> <u>SWI</u>/pBR322
e)   <u>MATα</u> <u>HO</u> <u>swi</u>⁻/pBR322
f)   <u>MATa</u> <u>HO</u> <u>SWI</u>/<u>MATa</u> clone
g)   <u>MATα</u> <u>HO</u> <u>SWI</u>/<u>MATa</u> clone
h)   <u>MATa</u>⁻ <u>ho</u> <u>SWI</u>/<u>MATa</u> clone
i)   <u>MATα</u> <u>HO</u> <u>swi</u>⁻/<u>MATa</u> clone
j)   <u>MATa</u> <u>HO</u> <u>SWI</u>/<u>MATα</u>clone
k)   <u>MATa</u> <u>HO</u> <u>SWI</u>/<u>HML</u>α clone
l)   <u>MATa</u> <u>HO</u> <u>SWI</u>/<u>HMR</u>a clone
m)   <u>MATa</u> <u>HO</u> <u>SWI</u>/<u>MAT</u>Δ143 clone
n)   <u>MATa</u> <u>HO</u> <u>SWI</u>/<u>MAT</u>Δ147 clone

II. <u>Sequence specificity of the YZ endonuclease; cutability is correlated with switchability</u>. Deletion mutants of <u>MATa</u> (1) shown in Figure 2 exhibit a correlation between the ability to be interconverted <u>in vivo</u> and the ability to be cut <u>in vitro</u>. The deletion end point of <u>MATa</u>Δ143 falls 25 bp short of the YZ junction. This deletion is cut <u>in vitro</u> (Figure 3) and is interconverted <u>in vivo</u> (1), whereas the deletion in <u>MATa</u>Δ147 which extends 9 bp into the Z region is not cut <u>in vitro</u> and is not interconverted <u>in vivo</u> (1).

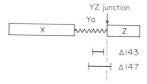

FIGURE 3.

Two deletion mutants of <u>MATa</u> are shown relative to the YZ junction. The bars indicated the region deleted. This figure is not to scale.

Another mutant <u>MAT</u> locus characterized by its failure to be interconverted efficiently, <u>MATα</u>inc3-7 (17), was tested as a substrate for YZ endo. The <u>in vitro</u> cutting rate of <u>MATα</u>inc3-7 was approximately 20-fold less than that of wild-type <u>MATα</u> (data not shown).
Studies of the sequence specificity of YZ endo are summarized in Figure 4. It appears that the recognition site is confined to a specific sequence at the YZ junction (CGCAACAGTAAA). A detailed description of the recognition site awaits further study.

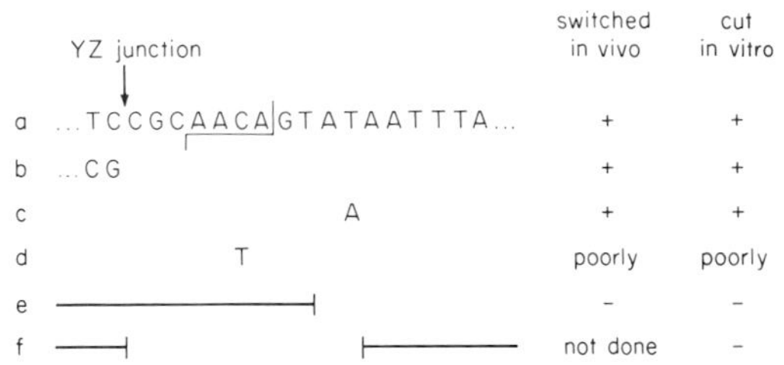

		switched in vivo	cut in vitro
YZ junction			
a	...TCCGC AACA GTATAATTTA...	+	+
b	...CG	+	+
c	A	+	+
d	T	poorly	poorly
e		–	–
f		not done	–

FIGURE 4.

The correlation between in vivo interconversion and in vitro endonculeolytic cleavage is shown.

a) MATa, b) MATα, c) MAT polymorphism, d) MATα inc3-7, e) MATΔ147, f) synthetic 11 base pair sequence flanked by non-homologous vector sequence.

III. Nature of the double strand break made by YZ endo. The precise site of endonucleolytic cleavage was determined by incubating an end-labeled substrate with partly purified YZ endo followed by gel electrophoresis of the products alongside the products of base specific chemical cleavage reactions derived from the end-labeled substrate. Thus, enzymatic cleavage of the substrate generated a band on the gel whose position relative to the sequencing ladder revealed the exact location at which the enzymatic cleavage took place (data not shown). The site of cleavage by YZ endo lies within the sequence defined by the deletion studies (Figure 4). The enzyme generates a 4 base pair 3' extension. Double strand cutting appears to be concerted; we have not detected any specific nicking at this sequence. End labeled substrate was incubated with YZ endo so as to give incomplete digestion of the substrate. The products were run out on an agarose gel and the band corresponding to nicked and/or uncut substrate was recovered and reelectrophoresed on a denaturing agarose gel. No label was observed at the positions corresponding to a single strand nick within the YZ endo site (data not shown).

The endonuclease appears to leave ends with a 3' hydroxyl group and a 5' phosphate. Careful comparison of the electrophoretic mobilities of the 5' end labeled fragment generated by YZ endo and of those produced by the chemical cleavage reactions reveal a slight difference in mobility. Base specific chemical cleavage reactions produce fragments terminating in 3' phosphates which, by virtue of their additional net negative charge, migrate faster than their counterparts having 3' hydroxyls. Thus we can attribute this difference in mobility to the presence of a free hydroxyl on the 3' end of the fragment generated by YZ endo.

IV. No covalent in vivo modification of the YZ endo site is present at HML and HMR. The YZ endo recognition sequence occurs at least three times in the yeast genome, at MAT HML and HMR yet, during mating type interconversion, the in vivo double strand break is observed only at MAT and not at either HML or HMR. The double strand break present at MAT in switching cells is therefore correlated with its ability to act as a recipient in the transfer of mating type information. In mar⁻ and sir⁻ cells, HML and HMR serve as recipients, as well as incur in vivo double strand breaks (1). We were therefore interested in testing whether the YZ endo site present at HML and HMR in wild type yeast (MAR⁺, SIR⁺) is resistant to YZ endonuclease because it had been covalently modified, e.g., by methylation, as a consequence of MAR-SIR regulation. To test for the presence of covalently modified DNA we treated purified plasmid borne HMR DNA, isolated from MAR⁺ and mar⁻, yeast with YZ endo. We found no evidence of covalent DNA modification; both DNA's were equally good substrates for the enzyme (data not shown).

DISCUSSION

Our evidence for the in vivo role of YZ endonuclease in the mating type interconversion process is that
- All frequently interconverting strains examined have detectable YZ endonuclease activity.
- The only strains which interconvert at a low frequency yet have detectable YZ endonuclease activity are those which possess mutant MAT loci.
- Mutant MAT loci which are refractory to YZ endo in vitro do not participate in high frequency mating type interconversion in vivo.

It is not possible to conclude, however, that the double-strand endonucleolytic activity is the sole in vivo activity of this enzyme necessary for mating type interconversion. For example, the resolvase of transposon Tn3 has been observed to cleave double strand DNA in vitro yet functions as a site-specific recombinase in vivo (18). Questions about other potential activities of YZ endonuclease will have to await its further purification.

There are two classes of MAT mutations which prevent high frequency mating type interconversion. The MATinc (inconvertible) class switches very infrequently. However, once switched the MAT locus regains its normal phenotype. In constrast MATstk (stuck) mutations are not healed upon mating type interconversion and remain unable to participate in high frequency interconversion. The lesions producing the inc phenotype have been shown to occur at the YZ junction (17, and A.J.S. Klar, personal communication) and, as we have shown, produce an altered YZ endo recognition site which is refractory to endonucleolytic cleavage. Since mutations of this class lie close to the allele specific Y region they are replaced, along with the Y region, with wild-type sequences from either <u>HML</u> or <u>HMR</u> as a consequence of mating type interconversion.

Stuck mutations have not been as well characterized. We speculate the stuck phenotype could be the result of mutations outside <u>MAT</u>, hence not healed upon interconversion, which effect the accessibility of the YZ junction to YZ endo possibly in a manner analogous to the way the silent cassettes are made refractory to YZ endo.

There appear to be two pathways for mating type interconversion. The efficient YZ endonuclease catalysed pathway requires in addition to YZ endonuclease, a <u>MAT</u> locus which has an intact accessible YZ endonuclease recognition site. A deficiency in one or both of these produces a strain having the low frequency interconversion phenotype in which, presumably, interconversion events are the result of relatively rare non-specific mitotic gene conversions.

The ability of YZ endo to act on <u>MATa</u> and <u>MAT</u>α with equivalent efficiency has an important biological consequence. The rate of <u>a</u>-->α interconversion will equal the α--><u>a</u>. Hence in a population of interconverting haploid cells the ratio of <u>a</u> and α cells should equal unity. This favors the formation of a diploid population which has recently been shown to have an adaptive advantage over the corresponding haploid population (19).

The gene encoding the endonuclease has not been identified. Of the genes identified as being required for mating type interconversion the <u>HO</u> gene appears to be the best candidate for YZ endonuclease (Stern, M., Jensen, R., and Herskowitz, I., personal communication).

REFERENCES

1.   Strathern, J.N., Klar, J.J.S., Hicks, J.B., Abraham, J.A., Ivy, J.M., Nasmyth, K.A., and McGill, C. (1982) Cell <u>27</u>, 15-23.
2.   Lindegren, C.C., and Lindegren, G. (1943) Proc. Natl. Acad. Sci. USA <u>29</u>, 306-308.

3. Astell,C.R., Ahlstrom-Jonasson, L., Smith, M., Tatchell, K., Nasmyth, K.A., and Hall, B.D. (1981) Cell 27, 15–23.
4. Strathern, J.N., Hicks, J.B., and Herskowitz, I. (1981) J. Mol. Biol. 147, 357–372.
5. Roth, R., and Lusnak, K. (1970) Science 168, 493–494.
6. Harashima, S., and Oshima, Y. (1976) Genetics 84, 437–451.
7. Klar, A.J.S., Fogel, S., and MacLeod, K. (1979) Genetics 93, 37–50.
8. Haber, J.E., and George, J.P. (1979) Genetics 93, 13–35.
    Klar, A.J.S., Fogel, S., and MacLeod, K. (1979) Genetics 93, 37–50.
    Rine, J., Strathern, J., Hicks, J.B., and Herskowitz, I. (1981) Genetics 93, 877–901.
9. Winge, O., Roberts, C. (1949) Compt. Rend. Trav. Lab. Carlsberg Ser. Physiol. 24, 341–346.
10. Hicks, J.B., Strathern, J.N., and Herskowitz, I. (1977) In DNA Insertion Elements, Plasmids, and Episomes, Bukhari, A.I., Shapiro, J.H., and Adhya, S.L. (eds.), Cold Spring Harbor, New York pp 457–462.
11. Malone, R.E., and Esposito, R.E. (1980) Proc. Natl. Acad. Sci. USA 77, 503–507.
12. Winge, O., Roberts,C. (1949) Compt. Rend. Trav. Lab. Carlsberg Ser. Physiol. 24.
13. Haber, J.E., and Garvik, B. (1977) Genetics 87, 33–50.
14. Orr-Weaver, T.L., Szostak, J.W., and Rothstein, R. (1981) Proc. Natl. Acad. Sci. (USA) 78, 6354–6358.
15. Weiffenbach, B., Rogers, D.T., Haber, J.E., Zoller, M., Russell, D.W., and Smith, M. Proc. Natl. Acad. Sci. (USA), in press.
16. Krasnow, M.A., and Cozzarelli, N.R. (1983) Cell, 32, 1313–1324.
17. Paquin, C., and Adams, J. (1983) Nature 302, 495–500.

Mechanisms of DNA Replication and Recombination, pages 797–817
© 1983 Alan R. Liss, Inc., 150 Fifth Avenue, New York, NY 10011

DISSECTION OF TRANSPOSITION IMMUNITY:
THE TERMINAL 38 BASE PAIRS OF Tn3 ARE REQUIRED FOR TRANSPOSITION IMMUNITY

Chao-Hung Lee, Ashok Bhagwat, Mary McCormick[2]
Eiichi Ohtsubo[3] and Fred Heffron[4]

Cold Spring Harbor Laboratory
Cold Spring Harbor, NY   11724

ABSTRACT   Transposons related to Tn3 exhibit a phenomenon called transposition immunity which was first described by Robinson et al. (1977).   Immunity works in cis.   A plasmid containing Tn3 is immune to further insertions of Tn3 but other plasmids in the same cell are not immune.   Using a cointegration assay to detect transposition (Ohtsubo et al., 1979) we have determined that (1) tnpR is not required for immunity (2) only the terminal 38 bp of Tn3 need be present to confer immunity (3) other parts of Tn3 do not confer immunity.   However, some internal deletions of Tn3 are non-immune in one plasmid but immune in another plasmid.   Our results limit the sort of models that can be used to explain transposition immunity.

[1]   This work supported by NSF PCM8140063 and PCM 8217002 and Damon Runyon-Walter Winchell Cancer Fund.
[2]   Present address: NIH, Bethesda, Maryland
[3]   Present address: University of Tokyo, Toyko, Japan
[4]   Present address: Scripps Clinic and Research Foundation, La Jolla, California

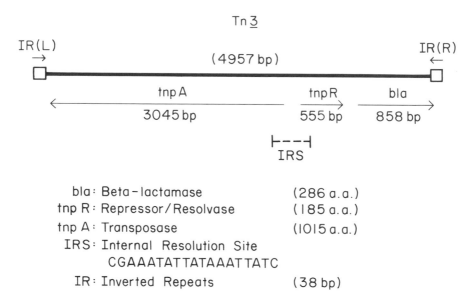

Tn<u>3</u>

bla: Beta-lactamase (286 a.a.)
tnp R: Repressor/Resolvase (185 a.a.)
tnp A: Transposase (1015 a.a.)
IRS: Internal Resolution Site
      CGAAATATTATAAATTATC
IR: Inverted Repeats (38 bp)

FIGURE 1. General structure of Tn<u>3</u>

Tn<u>3</u> is a 4957 bp transposon which encodes three genes. The <u>tnpA</u>, <u>tnpR</u> and <u>bla</u> genes encode the transposase, repressor/resolvase and β-lactamase respectively. Tn<u>3</u> is terminated by 38 bp inverted repeats required for its transposition. A site required for resolution of cointegrates is encoded in the intercistronic region between <u>tnpA</u> and <u>tnpR</u>.

INTRODUCTION

Tn<u>3</u> is a 5 kb transposon which encodes at least three genes and is terminated by 38 base pair inverted repeats. The structure of Tn<u>3</u> is shown in Figure 1. There are a number of transposons which are related to Tn<u>3</u>. Members of this family share several common structural features. They are flanked by a short terminal repeat (about 40 bp). Terminal repeats of different members of this group share sequence homology. They transpose by a two step mechanism and encode related genes which are essential for transposition. In the first step of transposition a cointegrate between the donor and the recipient molecule is formed. In the second step a site-specific recombination process resolves the cointegrate. Many members of the Tn3 family of transposons show transposition immunity. In this and several other respects they are different than the composite transposons such as Tn5 and Tn10 (see Heffron, 1983 for a review).

A plasmid containing Tn3 is immune to further insertions of Tn3 (Robinson et al., 1977; Figure 2). However, a plasmid which contains one

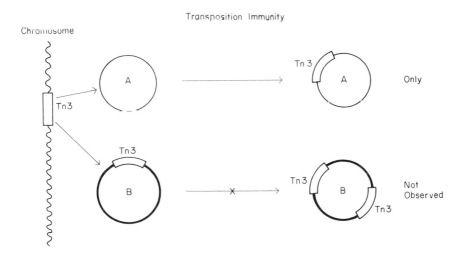

Transposition Immunity

FIGURE 2. Transposition immunity

Transposons related to Tn3 show transposition immunity. A Tn3 residing in the chromosome transposes from the chromosome to a plasmid at normal frequencies. However, transposition of that same Tn3 to another plasmid in the same cell which already contains Tn3 occurs at a very low frequency.

transposon of the Tn3 group need not be immune to other members. For example, plasmids containing Tn501 are not immune to Tn3 insertion (Grinsted et al., 1978; Stanisich et al. 1977). Tn501, Tn1721, and Tn21 form a group of Tn3 like transposons more closely related to each other than to Tn3 or γδ. Their transposition functions are interchangeable with each other but not with Tn3's. Furthermore, they show cross immunity to each other but not to Tn3 (Schmitt et al., 1981). This suggests that immunity is highly specific to the transposon and that the functions involved in transposition may also be involved in immunity.

Immunity to Tn3 insertion is not always correlated with the presence of Tn3 on the recipient molecule. Tn3 transposes much less efficiently to some plasmids than to others. Transposition to R100-1 and to R391 occures very infrequently (Bennett and Richmond, 1976). Transposition to the E. coli chromosome is also inefficient (Kretschmer and Cohen, 1977). It is not clear whether this natural immunity comes from the same phenomenon as immunity conferred by Tn3 itself.

RESULTS

We employ a cointegration assay to detect transposition (Ohtsubo et al.,1979 and McCormick et al., 1981). The strategy is shown in Figure 3. A ts donor plasmid containing Tn3 is introduced into the same cell with a recipient plasmid. We use a tnpR⁻ derivative of Tn3 and thus cointegrates formed in transposition are not resolved. Selection for the antibiotic resistance carried by the ts plasmid at 42°C kills all cells except those containing cointegrates between the donor and recipient plasmids. We use the Luria-Delbruck fluctuation test to determine the cointegration frequencies (Luria and Delbruck, 1943). Cointegrates between the donor [pHS1::Tn3(linker mutant #5, tnpR⁻, hereafter referred to simply as #5)] and a recipient plasmid can arise at frequencies which range from 10^{-2} to 1 depending on the recipient

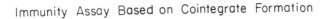

Immunity Assay Based on Cointegrate Formation

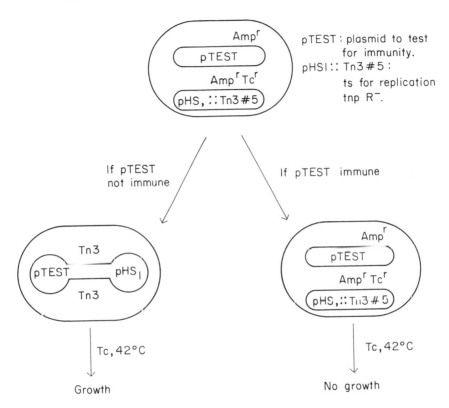

FIGURE 3. Immunity assay based on cointegrate formation

A plasmid to be tested for immunity (pTEST) is introduced into the same cell with a donor plamid, pHS1:Tn3 #5. The latter is a ts plasmid containing a tnpR⁻ derivative of Tn3. Transformants are grown to stationary phase at 30°C, diluted 10^6-fold to 10^7-fold and grown up to late log phase again in L-broth containing 5 µg/ml tetracycline at 30°C. The cointegrates are scored by plating out the cells on L-agar plates containing 10 µg/ml tetracycline and incubating them at 42°C. The transposition frequencies are determined using the Luria-Delbruck fluctuation test (Luria and Delbruck, 1943).

plasmid. For a recipient plasmid containing Tn$\underline{3}$ (tnpR$^-$) and therefore immune to Tn$\underline{3}$ transposition the frequency with which we observe cointegrates is at least 10^2 (and as much as 10^6) fold lower than the parent plasmid not containing any Tn$\underline{3}$ sequence.

We first wanted to determine whether only a part of Tn$\underline{3}$ need be present on a plasmid to confer immunity. To answer this we employed two different strategies as shown in Figures 4 and 5. pACYC184 is not immune to Tn$\underline{3}$ transposition. In the first strategy we transposed a copy of Tn$\underline{3}$ #5, to pACYC184 and then deleted various restriction fragments from it. The three independent Tn$\underline{3}$ insertions used in this study cover a 400 bp region of pACYC184 as shown in Figure 4. As expected, these and other insertions make pACYC184 immune to further Tn$\underline{3}$ insertions. Deletions were constructed by deleting restriction fragments from a site within Tn$\underline{3}$ to a site within pACYC184. All such deletions were found to be immune to Tn$\underline{3}$ suggesting that either end of Tn$\underline{3}$ is sufficient to make pACYC184 immune. Plasmids carrying both ends of Tn$\underline{3}$ confer approximately the same amount of immunity as one end. Furthermore, the Tn$\underline{3}$ sequences are able to confer immunity in either orientation with respect to pACYC184 (Figure 4).

To further localize the immunity sequences, various restriction fragments of Tn$\underline{3}$ were cloned into pACYC184 and these derivatives tested for immunity. These constructions cover the entire Tn$\underline{3}$ except for a small fragment between nucleotides 84 and 284 from the left end (figure 5). We find that restriction fragments containing a terminal repeat are immune whereas none of the internal fragments are immune. Furthermore, we can detect no significant difference in the level immunity conferred by an 84 base pair fragment of Tn$\underline{3}$ containing the left terminal repeat and the level conferred by the complete transposon.

From these two studies it appears that a small sequence from either end of Tn$\underline{3}$ can make pACYC184 immune. While either end is proficient in making pACYC184 immune, presence of both the ends on the plasmid does not make it

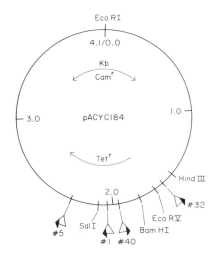

Plasmid	F	Ratio
pACYC184	0.0012	1
#1	7.3×10^{-6}	6.1×10^{-3}
#5	$\leq 7.9 \times 10^{-6}$	$\leq 6.6 \times 10^{-3}$
#32	$\leq 1.5 \times 10^{-5}$	$\leq 1.3 \times 10^{-2}$
	$\leq 6.2 \times 10^{-5}$	$\leq 5.2 \times 10^{-2}$

FIGURE 4a. Tn3 insertions in pACYC184

Cointegrates between pHS1::Tn3 #5 and pACYC184 were selected for by plating cells containing these two plasmids on ampicillin plates at 42°C. Several thousand colonies from these plates were pooled together, grown up and plasmid DNA was extracted. This DNA was used to transform KL16, an E. coli strain that carries a copy of γΔ in the chromosome. Transformants were selected for chloramphenicol-resistance and ampicillin-resistance at 42°C. Colonies were screened for tetracycline-resistance and the Tn3 insertions within several tetracycline-sensitive derivatives were mapped. Positions and orientations of four such insertions (#1, #5, #32 and #40) are shown in the figure. The arrows, representing Tn3 insertions, point towards the end containing the bla gene. These insertions were tested for transposition immunity as described in Figure 3. F is the frequency of cointegrate formation.

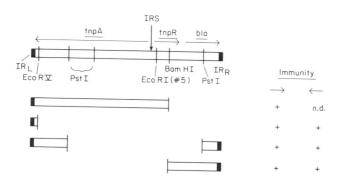

FIGURE 4b. Deletion derivatives of pACYC184::Tn3 #5

Deletions were generated by digesting the appropriate Tn3 insertion with restriction enzyme BamHI, EcoRV or PstI, and self-ligating the fragments. By selecting two Tn3 insertions in opposite orientations and on the two sides of a restriction site (#1 and #32 for Bam HI) deletions were generated that left behind the same Tn3 sequences in opposite orientations. All the constructions were tested for immunity by the cointegrate assay.

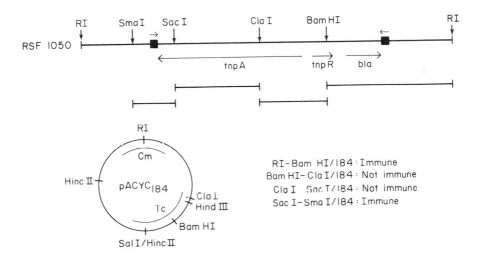

FIGURE 5a. Cloning of Tn<u>3</u> sequences in pACYC184

A Tn<u>3</u> containing plasmid, RSF1050, was digested with <u>Eco</u>RI, <u>Bam</u>HI, <u>Sac</u>I and <u>Sma</u>I restriction enzymes. Restriction fragments were isolated and ligated to <u>Hin</u>dIII linkers and cloned into pACYC184 at the <u>Hin</u>dIII site. The resulting recombinant plamids containing portions of Tn<u>3</u> sequences were then assayed for immunity. Two fragments, <u>Eco</u>RI-<u>Bam</u>HI and <u>Sac</u>I-<u>Sma</u>I when cloned into pACYC184 conferred immunity.

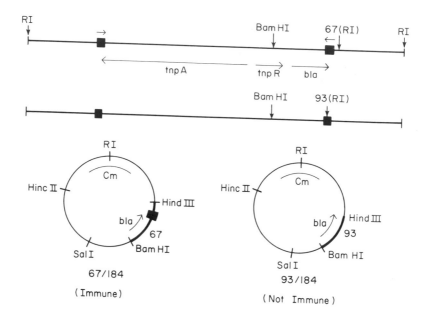

FIGURE 5b. Finer mapping of Tn<u>3</u> immunity sequence

Two linker mutants of RSF1050 (designated as #67 and #93) had <u>Eco</u>RI linkers inserted just inside and just outside the right-hand end of Tn<u>3</u>. These two plasmids were first digested with <u>Eco</u>RI and <u>Bam</u>HI and then cloned into pACYC184 between the <u>Hind</u>III and <u>Bam</u>HI restriction sites. The resulting plasmid, 67/184, confers immunity whereas 93/184, which is missing only the 39 bp at the right hand end of Tn<u>3</u> including the IR, is no longer immune.

significantly more immune. These sequences are able to confer immunity in either orientation with respect to pACYC184. Fragments not containing at least one end of Tn3 are incapable of conferring immunity.

These rules hold true for other plasmids besides pACYC184. As shown in Table 1 we have examined pMB8 (one parent of pBR322; Bolivar et al. 1977), pBR322 (which contains the left hand end of Tn3; Sutcliff, 1978; Heffron et al. 1979), and several derivatives of pBR322 for immunity. Like pACYC184, pMB8 is not immune to transposition whereas pBR322 and all pBR322 derivatives that we have constructed are immune (see below). There is one surprise from this study, however. The frequency of cointegration for the two parent plasmids, pMB8 and pACYC184, are different. The frequency of cointegration for pMB8 is approximately 1 whereas the frequency for pACYC184 is at least $10-2$ fold lower. These results are summarized in Table 1. The method of construction of pACYC184 (Chang and Cohen, 1978) makes it unlikely that it contains any Tn3 sequences.

TABLE I. Transposition immunity for different plasmids*

Plasmid	Frequency of Cointegrates**
pMB8	$>10^{-2}$
pACYC184	10^{-3}
pBR322	$<10^{-6}$
pMB8::Tn3-#5	$<10^{-6}$
pACYC184::Tn3-#5	$<10^{-5}$

*Transposition immunity assay was carried out as escribed in Figure 3 using pHS1::Tn3-#5 as the donor lasmid.

** Frequency per generation per cell.

To test if the terminal repeat by itself can confer immunity we have carried out several experiments. We have constructed a series of deletions within a fragment containing the right terminal repeat. Bal31 deletions which come very close to, but do not remove the 38 bp repeat, are still immune (data not shown). On the other hand, a deletion which just removes the terminal repeat is no longer immune. This deletion was constructed from an EcoRI linker insertion in the right terminal repeat of Tn3 (Heffron et al., 1979).

We have synthesized the 38 base pairs that constitute the terminal repeat of Tn3 by filling in two overlapping complementary single-stranded fragments. The overlapping fragments (30 and 15 nucleotides long) were synthesized using phosphate triester chemistry on an automated DNA synthesis machine by Biosearch. The synthetic terminal repeat was cloned in a pACYC184 derivative as shown in Figure 6. Sequence of several such inserts was determined by the Maxam and Gilbert (1977) protocol. Several clones were found to contain complete copies of the terminal repeat. The results of the immunity assay for two plasmids containing one copy of the 38-mer in either orientation is shown in Table II. Both these plasmids are immune and the level of immunity is similar to that observed for the complete Tn3. These results clearly show that the terminal 38 base pairs of Tn3 are sufficient for transposition immunity.

Table II. Immunity of pIR38*

Plasmid	Frequency of Cointegrates
pCHL884	10^{-4}
pIR38-6**	$<10^{-6}$
pIR38-8**	$<10^{-6}$

*pIR38 was constructed as described in Figure 6 and the immunity assay was performed as described in Figure 3.

**pIR38-6 and pIR38-8 are two clones containing the 38 bp IR located at the SmaI site of pCHL884 in opposite orientations.

Some deletion derivatives of RSF1050 (pMB8::Tn3) carrying both the Tn3 terminal repeats are not immune to Tn3 insertion. The deletions we have tested are shown in Figure 7. As expected, deletions which remove the left hand end of Tn3 are still found to be immune but surprisingly deletions which remove the intercistronic region between tnp A and tnp R are not immune. This result is not consistent with the data from Tn3 deletions in pACYC184 described before (Figures 4 and 5). For example, the PstI deletion in Tn3 which also removes the intercistronic region, is immune (Figure 4).

We have used the fragment of Tn3 already present in pBR322 to reconstruct an intact copy of the transposon (pCL1, Figure 8). Both pBR322 and pCL1 are immune to Tn3. We have constructed a series of deletions in the Tn3 part of pCL1 and find that in accord with earlier results deletions which leave one end of Tn3 intact continue to be immune. However, in contrast with the results with RSF1050, deletions that remove the intercistronic region between tnpR and tnpA are immune in pCL1 (Figure 8). This is particularly surprising because, as noted before, both RSF1050 and pBR322 are derived from the same parent – pMB8. Thus internal deletions such as Δ596 have different phenotypes on different plasmids. To ascertain that the Tn3 sequences within the two plasmids are identical we removed a restriction fragment containing the 596 deletion from RSF1050 and exchanged this with the analogous fragment from pCL1 and from the intact RSF1050 (see Figure 8). We found that once again this deletion made RSF1050 non-immune while leaving the phenotype of pCL1 unchanged. The reason for this discrepancy is not yet clear.

We have mapped a number of independent insertions of Tn3 into an 'immune' plasmid. The distribution of these insertions and the distribution of insertions into its non-immune parent are shown in Figure 9. The distribution is complex and is changed by the presence of the terminal repeat. In the non-immune plasmid insertions are clustered at the amino terminal end of the tetracycline resistance gene and at the carboxyl

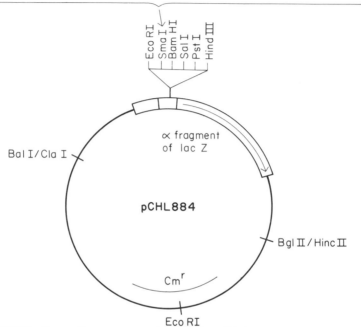

5'...C T T A A C G T G A G T T T T C G T T C C A C T G A G C G T...3'
                                    3'...A C T C G C A G T C T G G G G... 5'

5'...C T T A A C G T G A G T T T T C G T T C C A C T G A G C G T **C A G A C C C C**
    G A A T T G C A C T C A A A A G C A A G G T G A C T C G C A G T C T G G G G... 5'

EcoRI
SmaI
BamHI
SalI
PstI
HindIII

α fragment
of lac Z

BalI/ClaI

pCHL884

BglII/HincII

Cm^r

EcoRI

FIGURE 6.   Construction of pIR38

Two complementary DNA fragments were synthesized, one of 30 nucleotides and the other 15 nucleotides using the phosphite triester chemistry on an automated machine by Biosearch. The 30-mer was a generous gift from Biosearch. These oligomers are shown in bold print in the figure. These fragments were annealed and the single-stranded region filled in by the PolI Klenow fragment in the presence of 200 μM dNTPs. The blunt-ended 38 bp molecule was then ligated to SmaI digested pCHL884 (a pACYC184 derivative containing the chloramphenicol resistance gene from pACYC184 and a fragment of lacZ gene from M13 mp8). These plasmids, containing the terminal repeat of Tn3, are immune as shown in Table II.

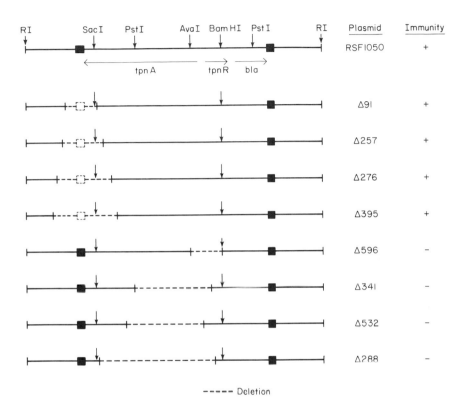

FIGURE 7. Immunity assay on deletion mutants

Several deletion mutants were tested for transposition immunity. All deletion mutants were either tnpR⁻ (Δ596, Λ341, Δ532) or were made BamHI resistant and therefore TnpR⁻ prior to the assay (Δ91, Δ276, Δ295, Δ395). Mutants, Δ91 Δ276, Δ295 and Δ395 had deletions at left hand end of Tn3 sequence. Whereas Δ596, Δ341 and Δ532 removed the intercistronic region between tnpR and tnpA. Their phenotypes in immunity are as indicated.

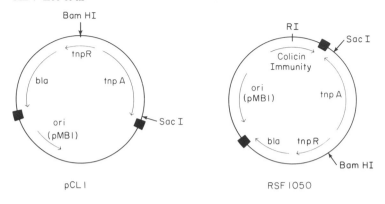

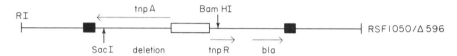

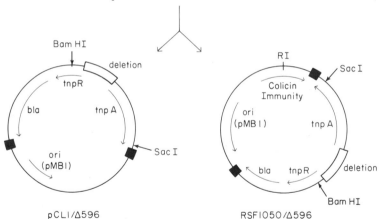

FIGURE 8. Phenotype of a deletion of the intercistronic region. pCL1, and RSF1050 are pMB8 derivatives containing an intact copy of Tn3. RSF1050/Δ596 contains a Tn3 with an internal deletion indicated as boxed the area. Replacement of the BamHI/SacI fragment from RSF1050 and pCL1, with BamHI/SacI fragment from RSF1050/Δ596 results in two identical Tn3 deletion mutants.

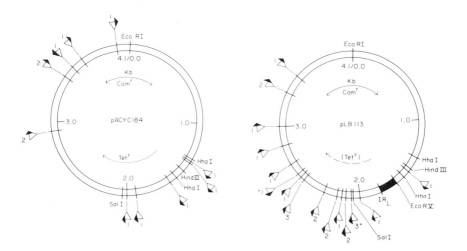

FIGURE 9. Tn3 insertions that overcome immunity

pLB113 is an EcoRV deletion derivative of pACYC184::Tn3 #5 described in Fig. 4. The residual Tn3 sequence within pLB113 is shaded in black. Cells containing pHS1::Tn3 #5 were transformed with pACYC184 or pLB113. A transformant from each was grown in LB +20µg/ml chloramphenicol +20µg/ml ampicillin +5µg/ml tetracycline at 30°C overnight. The cultures were diluted 2×10^8-fold in the same medium, divided in 0.2 ml aliqoutes and incubated at 30°C for 36 hrs. Typically about half the cultures showed growth. 0.5µl (for pACYC184) or 10µl (for pLB113) of each culture was plated on LB +50µg/ml chloramphenicol +50µg/ml ampicillin +10µg/ml tetracycline and incubated at 42°C. One colony from each plate was picked and grown up at 37°C or above. Plasmid DNA from several such colonies was prepared and the position and orientation of Tn3 insertions determined by restriction mapping. The arrows representing Tn3 insertions point from the tnpA to the bla genes. Numbers next to the insertions are the number of independent insertions isolated at those positions.

end of the chloramphenicol resistance gene. In the immune derivative, the insertions are clustered on one side of the terminal repeat, within few hundred base pairs. We also note that most insertions into pACYC184 and its immune derivatives take place in only one orientation.

DISCUSSION

We have found that only a very small fragment of Tn3, namely the terminal 38 bp, need be present on a plasmid to confer transposition immunity to the whole plasmid. In most constructions, no other part of Tn3 need be present. The enigmatic exceptions to this rule are some internal deletions in Tn3 which remove the intercistronic region between tnpA and tnpR. These Tn3 derivatives confer immunity to most plasmids but not to some pMB8 insertions. We have no satisfactory explanation for these exceptions. We are carrying out resection experiments as well as in vitro mutagenesis on the synthetic terminal repeat to identify the precise nucleotides essential for immunity and to determine whether these same nucleotides are essential for transposition.

Although a plasmid containing the terminal repeat is said to be immune to insertion, we are able to select insertions into an immune plasmid. These insertions are not likely to be due to a mutation in the terminal repeat which renders the recipient non-immune. Plasmids containing two copies of the terminal repeat are no more immune than plasmids with one. The distribution of insertion sites in an immune plasmid and its non-immune parent is compared in Figure 9. We find that most insertions in either plasmid take place in only one orientation. This may suggest that the two ends of Tn3 are non-identical. Thus an asymmetry in the plasmid structure such as in unidirectional DNA replication may lead to a preference of one orientation over the other. We are currently testing this hypothesis. We do find that insertions are directed into a particular region of the plasmid by the presence of an terminal repeat. Most insertions occur into a region on one side near the terminal repeat. It is interesting to note that this distribution is unaltered if the orientation of the terminal repeat is reversed (data not shown) and that the insertions are not clustered diametrically away from the terminal repeat.

Immunity is not likely to be due to a _cis_- acting product. The size of the terminal repeat makes it highly unlikely that it encodes a _cis_- acting function. Furthermore, all known transposition functions of Tn_3_ work well _in trans_. As all the Tn_3_ constructions used in these experiments are tnpR⁻, the product of _tnpR_ can not be required for immunity. Transposition immunity must act at some point during the transposition process itself and cannot result from possible instability of a plasmid carrying two copies of Tn_3_. It is possible to construct plasmids containing two copies of Tn_3_ and these are stable provided that the copies are in inverted orientation (Robinson et al., 1978). Plasmids which are immune to one transposon need not show immunity to another related transposon. Plasmids containing Tn_501_ are immune to Tn_501_ but not to Tn_3_ insertion (Schmitt et al., 1981). These transposons share a common mechanism of transposition, a similar structure including the terminal repeats and similar but non-complementing transposition functions. Therefore, a change in the overall plasmid structure due to the terminal repeat, such as supercoiling, is unlikely to be a sufficient explanation for transposition immunity. Together, these observations suggest that immunity is a result of a small _cis_-acting sequence within the terminal repeat which acts only on closely related transposons and probably acts during the transposition process itself.

One possible model for the mechanism of transposition immunity is that an immune plasmid is covalently modified due to the presence of a terminal repeat. That is, the terminal repeat provides an entrance site for a modification enzyme(s) which acts in cis. This modification makes the plasmid a poor recipient for Tn_3_. The distribution of insertions we observe into an immune plasmid does not fit the simplest forms of this model. If the modification enzyme binds the terminal repeat and slides along the plasmid a gradient of modification would be expected from this point. Depending on whether the sliding is unidirectional or bidirectional the gradient will be symmetrical or asymmetrical with respect to the terminal repeat. If the gradient is symmetric one would expect most insertions to be clustered diametrically across from the terminal repeat. If the gradient is asymmetric one would expect that reversing the orientation of the terminal repeat should reverse the distribution of insertions. Our data does not agree with either of these predictions.

During transposition the integration event in the recipient molecule may be slow or inefficient. As a consequence an integration complex composed of the transposase joined to the ends of Tn3 may have a long half life. If the integration complex tracks along a recipient molecule it would thus have a high probability of scanning the entire recipient molecule before integration. Such a model is supported by transposition specificity studies (Grinsted, 1978) and suggests a mechanism for transposition immunity. Tn3 transposition shows regional specificity, that is, insertions tend to cluster in some regions of the plasmid and are infrequent in other regions of the plasmid. For example, Tn501 is found to be a hot spot for Tn3 insertion although the presence of Tn501 does not appear to change the frequency of Tn3 insertion (Grinsted et al., 1978). A model for immunity could be that the integration complex dissociates itself when it encounters an immunity sequence on the recipient molecule. Such a model does not require a cis-acting function and would be specific for each transposon. This model assumes several details about the early events in Tn3 transposition that are poorly worked out at this time. If the model is essentially accurate its specific predictions can help dissect the transposition process.

REFERENCES

1. Bennett, P.M. and Richmond, M.H. (1976) J. Bacteriol. 126, 1-6.
2. Chang, A.C.Y. and Cohen, S.N. (1978) J. Bacteriol. 134, 1141-1156.
3. Grinsted, J., Bennett, P.M., Higginson, S. and Richmond, M.H. (1978) Mol. Gen. Genet. 166, 313-320.
4. Heffron, F., McCarthy, B.J., Ohtsubo, H. and Ohtsubo, E. (1979) Cell 18, 1153-1164.
5. Heffron, F. (1983) Transposable Elements. In Mobile Genetic Elements (ed. Shapiro, J.) Academic Press, New York.

6. Kretschmer, P.J. and Cohen, S.N. (1977) J. Bacteriol. 139, 515-519.
7. Luria, S.E. and Delbrück, M. (1943) Genetics 28, 491-511.
8. Maxam, A. and Gilberg, W. (1977) Proc. Natl. Acad. Sci. USA 74, 560-564.
9. McCormick, M., Wishart, W., Ohtsubo, H., Heffron, F. and Ohtsubo, E. (1981) Gene 15, 103-118.
10. Ohtsubo, S., Ohmori, H., and Ohtsubo, E. (1979) Cold Spring Harbor Symp. Quant. Biol. 45, 1269-1278.
11. Robinson, M.K., Bennett, P.M., Grinsted, J. and Richmond, M.H. (1978) Mol. Gen. Genet. 160, 339-346.
12. Robinson, M.K., Bennett, P.M. and Richmond, M.H. (1977) J. Bacteriol. 129, 407-414.
13. Schmitt, R., Altenbuchner, J., Wiebauer, K., Arnold, W., Puhler, A., and Schoffl, F. (1981) Cold Spring Harbor Symp. Quant. Biol. 45, 59-65.
14. Stanisich, V.A., Bennett, P.M., and Richmond, M.H. (1977) J. Bacteriol. 129, 1221-1223.
15. Sutcliff, J.G. (1978) Cold Spring Harbor Symp. Quant. Biol. 43, 77-90.

Mechanisms of DNA Replication and Recombination, pages 819–848
© 1983 Alan R. Liss, Inc., 150 Fifth Avenue, New York, NY 10011

INITIATION OF DNA REPLICATION IN VITRO PROMOTED
BY THE BACTERIOPHAGE λ O AND P REPLICATION PROTEINS[1]

Roger McMacken, Marc S. Wold, Jonathan H. LeBowitz,
John D. Roberts, Joanne Bednarz Mallory, Jo Anne K.
Wilkinson, and Christine Loehrlein

Department of Biochemistry
The Johns Hopkins University
Baltimore, Maryland 21205

ABSTRACT   Initiation of bacteriophage λ DNA replication requires the products of λ genes O and P as well
as numerous Escherichia coli replication proteins.
Metabolic poisons, such as cyanide and arsenate, block
the rapid intracellular degradation of the λ O protein
that normally occurs after λ infection. This finding
suggests that ATP-dependent proteolysis is responsible
for the extreme instability of the λ O initiator in
vivo.
     Both of the λ initiators have been purified to
homogeneity, in a functionally active form, from cells
containing amplified levels of the O or P proteins.
The isolation protocol developed for O protein yields
several mg of the initiator per gram of cell paste.
Purified O protein binds specifically to the λ replication origin and promotes the in vitro replication of
λdv plasmids. The properties of the λ P protein suggest that it is a phage analogue of the host dnaC
protein. P protein forms a stable complex with the E.
coli dnaB protein, displacing bound dnaC protein and
significantly altering the replicative, priming, and
ATPase activities of dnaB protein.
     A soluble enzyme system, containing the λ O and P
proteins, that specifically replicates λdv plasmid
molecules to supercoiled daughter molecules has been
developed. Extensive characterization of the system in

[1]This work was supported by grants from the American
Cancer Society and the National Institutes of Health.

regard to protein requirements, template specificity, and replication intermediates demonstrated that the features of λdv replication in vitro were identical to the known properties of the circle-to-circle phase of λ DNA replication in vivo.

We have discovered that the λ O and P proteins, together with six E. coli replication proteins, promote the rifampicin-resistant conversion of single-stranded M13 viral DNA to the duplex replicative form. In the rate-limiting step of this strand initiation reaction, which is non-specific with respect to DNA sequence, the λ O and P proteins participate in the ATP-dependent transfer of dnaB protein onto the template DNA strand. Further characterization of this system may help define the mechanisms by which the λ initiators direct the bacterial replication apparatus to act at the viral replication origin.

INTRODUCTION

A detailed picture of the genetics and properties of bacteriophage λ DNA replication in vivo has emerged from more than two decades of intensive research on this complex process (see references 1 and 2 for recent reviews). Shortly after infection, the linear chromosome of coliphage λ is converted to a supercoiled form and is replicated via a circular mode to generate daughter supercoiled monomers (3). Lambda DNA replication is initiated at a unique origin (oriλ) situated within the λ O gene (Fig. 1) from which replication forks are propagated in a bidirectional fashion (4). Classical theta structures (cairns forms) and, subsequently, catenated dimer circles are generated as replication intermediates during the production of supercoiled daughter molecules (3-5).

Replication of λ DNA is accomplished by the cooperative action of both phage and host replication proteins (1,2). Just two viral proteins, the products of genes O and P, directly participate in λ DNA replication. In contrast, a major portion of the E. coli replication machinery, consisting of at least 20 bacterial gene products, is required for replication of the phage chromosome. The necessary E. coli proteins include DNA polymerase III holoenzyme, primase (dnaG protein), single-stranded DNA binding protein (SSB), dnaB protein, dnaJ protein, dnaK protein, RNA polymerase, and DNA ligase. However, certain host proteins that func-

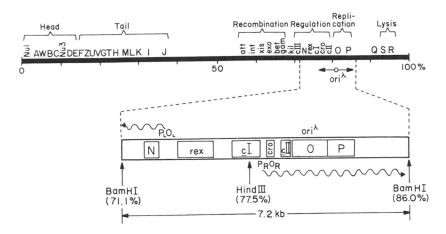

FIGURE 1. Genetic map of bacteriophage λ. The genes and sites discussed in this paper are depicted in the expanded region.

tion in the initiation of E. coli DNA replication, such as the dnaA and dnaC proteins, are not required for λ DNA replication. Genetic studies indicate that the specificity for initiation of λ DNA replication resides solely in the λ O protein, for it alone, among the numerous viral and host proteins required for λ DNA replication, is replicon specific (6).

Initiation of λ DNA replication is under the negative control of the phage cI repressor, which directly blocks λ DNA replication, even in the presence of all necessary proteins and DNA sequences (7,8). Genetic studies of this regulation by Dove and colleagues revealed that transcription of the λ replication origin region by E. coli RNA polymerase is required for initiation of DNA replication (8,9). Presumably, the λcI repressor regulates λ DNA replication by blocking formation of the RNA polymerase-mRNA complex responsible for "transcriptional activation" of the origin.

Soluble enzyme systems that support the specific initiation of λ DNA replication have recently been developed (10-12). Small supercoiled λdv plasmid molecules containing the λ replication origin served as effective templates in these *in vitro* systems. The phage-specific O and P initiator proteins were added to the reaction mixtures either as purified proteins (11,12) or as crude extracts containing amplified levels of the λ replication proteins (10). Only with the advent of recombinant DNA technology for protein amplification was it possible to develop these soluble _in vitro_ systems for λ DNA replication. Heretofore, the extreme proteolytic sensitivity of the λ O protein _in vivo_ ($t_{1/2}$ = 1.5 min) (13,14) precluded isolation of sufficient quantities of this essential initiator protein.

A primary goal of our research is to assemble and characterize a soluble enzyme system, composed of purified λ and E. _coli_ replication proteins, that is capable of specifically initiating authentic, bidirectional λ DNA replication. This article outlines our initial studies toward defining the molecular events which occur in this process.

RESULTS AND DISCUSSION

Chemical Stability of the λ O Polypeptide _In Vivo_

Our initial efforts toward establishing a system capable of replicating phage λ DNA _in vitro_ focused on the isolation of the λ O and P replication proteins. Prior to attempting to amplify and purify the phage-specific initiators, however, we decided to try to identify the factors which affect the intracellular stability of the λ O polypeptide. This decision was prompted by reports that the λ O protein is highly susceptible to proteolytic inactivation in λ-infected E. _coli_ (13,14). The chemical stability of the λ O polypeptide can be determined by densitometric monitoring of its disappearance on polyacrylamide gels following pulse-chase labeling (14). We have confirmed the marked instability of the λ O protein _in vivo_ (Fig. 2). The 34,000 dalton O polypeptide, identified by its absence in cells infected with λOam8 phage (Fig. 2, lane C), decays with a half-life of approximately 1.4 min.

We examined the use of protease mutants, protease inhibitors, or metabolic poisons as possible approaches for blocking the turnover of the λO protein _in vivo_. In none of the known protease mutants of E. _coli_ (15), including strains deficient in the _lon_ protease or proteases I, II or III, was the stability of the λ O polypeptide significantly altered[2]. Likewise, most of the protease inhibitors known to decrease protein catabolism in growing E. _coli_ (16) had no appreciable effect on the stability of this phage initia

[2]J. Wilkinson and R. McMacken, unpublished data

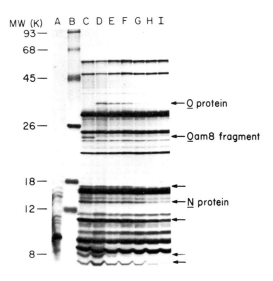

FIGURE 2. Pulse-chase labeling of the λ O protein and
other early λ proteins in UV-irradiated cells. The methods
used for analysis of the chemical stabilities of λ early
proteins by pulse-chase labeling have been described (14).
E. coli strain 159 (uvrA⁻) was grown in RM medium (14),
irradiated with ultraviolet light (600-900 J/m²), infected
with λcI857Sam7 and labeled with [³⁵S]methionine for one
minute starting 5 min after infection. At 6 min after
infection (zero time) the incorporated label was chased by
addition of a large excess of unlabeled methionine. At
various times during the chase the cells were lysed at 100°C
in SDS-mercaptoethanol. Labeled proteins were separated by
SDS-polyacrylamide gel electrophoresis in a 14-20% linear
gradient gel and visualized by autoradiography. Lane A,
uninfected cells; lane B, ¹⁴C-labeled molecular weight
markers; lane C, λcI857cl7Oam8-infected cells, no chase;
lanes D-I, λcI857Sam7-infected cells chased for 0, 1, 2, 4,
6, and 8 min, respectively. The unlabeled arrows locate the
positions of other unstable early λ proteins.

tor². Approximately a two-fold increase in the half-life of
the O polypeptide was observed, however, when λ-infected
cells were grown in the presence of 0.5 mM tosyl phenylala-
nine chloromethyl ketone, an inhibitor of chymotrypsin.
    Finally, we investigated the possibility that metabolic
poisons would impair O protein breakdown, since agents which

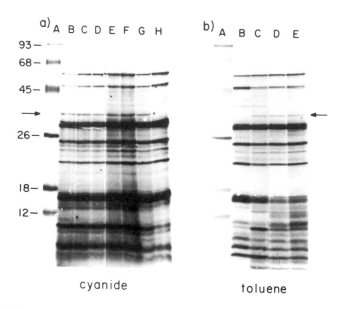

cyanide   toluene

FIGURE 3. Effect of metabolic poisons on the chemical stability of the λ O polypeptide. Fig. 3a). λ-infected cells were pulse-labeled with [^{35}S]methionine as described in the legend to Fig. 2, except that the RM media contained 20 mM α-methylglucoside (lanes B-F) or 0.4% maltose (lanes G and H). Fifteen seconds after the addition of excess unlabeled methionine, KCN was added to a final concentration of 5 mM (zero time). Lane A, ^{14}C-labeled molecular weight standards; lanes B-F, λ-infected cells chased for 0, 4, 8, 16 and 64 min, respectively, in the presence of cyanide and α-methylglucoside; lanes G and H, λ-infected cells chased for 0 and 64 min, respectively, in the presence of cyanide and maltose. Fig. 3b). λ-infected cells were treated as previously described (Fig. 2) except that 0.05 volumes of toluene were added 15 sec after starting the chase. Lane A, ^{14}C-labeled molecular weight markers; lane B, λ-infected cells chased for 64 min in the absence of toluene; lanes C-E, λ-infected cells chased for 0, 8 and 64 min, respectively, in the presence of toluene. The arrows identify the positions of the λ O polypeptide.

interfere with energy production inhibit intracellular protein degradation in E. coli (17). Addition of potassium cyanide together with an inhibitor of glycolysis (α-methylglucoside) to λ-infected cells caused a nearly instantaneous stabilization of the λ O polypeptide (Fig. 3a). Densitometric analysis indicated that the intracellular O protein was chemically stable for at least one hour following addition of cyanide (Table 1).

TABLE 1
EFFECT OF METABOLIC INHIBITORS ON THE STABILITY
OF THE λ O PROTEIN[a]

Treatment	Surviving fraction of O protein at:					
	2	4	8	16	32	64(min)
None	0.33	0.11	0.02	0	0	0
Cyanide	-	1.09	1.03	1.14	1.08	1.00
Arsenate	-	0.75	0.88	0.90	0.69	0.65
Azide	-	0.48	0.52	0.56	0.48	0.48
Toluene	0.76	-	0.80	-	-	0.63
Ice	-	0.30	0.27	0.25	0.22	0.21

[a]The quantity of pulse-labeled O polypeptide surviving relative to that present at zero time was determined by densitometry of SDS-polyarylamide gel autoradiograms as described (14) (Fig. 2 legend). The amount of O polypeptide present in each sample was normalized to the amount of the stable, 22,000 dalton early λ protein contained in the sample (14). The metabolic inhibitors were added to pulse-labeled λ-infected cells (zero time = 15 sec after addition of unlabeled methionine) to the following final concentrations: KCN, 5 mM; Na_2HAsO_4, 100 mM; NaN_3, 20 mM. Arsenate-treated cells were grown in low phosphate (*ca*. 0.2 mM) media and azide-treated cells were grown at pH 5. Cyanide- and toluene-treated cells were handled as described in the legend to Fig. 3.

If cyanide causes stabilization of the λ O polypeptide as a consequence of its capacity to impair oxidative phosphorylation, thereby reducing intracellular concentrations of ATP and other high energy phosphates, then other agents which lower intracellular concentrations of high energy compounds should also stabilize this λ initiation protein. Indeed, not only toluene, which makes E. coli permeable to small molecules, but also arsenate and azide, which inhibit aerobic phosphorylation, blocked degradation of the λ O polypeptide (Fig. 3b, Table 1). These results suggest that ATP-dependent proteolysis is responsible for the rapid decay of the O protein in vivo. The requirement for ATP to fuel O protein turnover is also suggested by the fact that the presence of sugar (maltose or glucose) in the growth medium obviates the effect of cyanide on O protein stability (Fig. 3a, lanes G and H). Apparently, the ATP generated by the substrate-level phosphorylations of glycolysis is sufficient to energize proteolysis of the O protein. In this regard, our preliminary results indicate that those metabolic poisons which effect the quickest lowering of intracellular ATP levels cause the most rapid stabilization of the λ O polypeptide[2]. The O protein appeared to be almost completely stable once ATP levels had dropped below a critical level (Table 1). It would be interesting to determine if regulation of the activity of specific chromosomal initiation proteins by means of ATP-dependent proteolysis is a general mechanism of controlling initition of DNA replication in prokaryotic organisms.

The addition of the metabolic poisons to λ-infected cells increased the stability of most of the unstable λ early proteins (e.g., compare lanes B and E of Fig. 3b), including the λ N protein and an 8000 dalton polypeptide that may be the bacteriophage λ Xis protein, whose functional activity is known to be partially stabilized by cyanide (18).

Overproduction and Purification of the Bacteriophage λ O Replication Protein

Before attempting to purify the λ O protein from λ-infected cells that had been treated with extremely toxic poisons to inhibit ATP-dependent proteolysis, we decided to try to amplify intracellular levels of the O protein by cloning the phage O gene downstream from strong promoters situated on multicopy plasmids. The discovery of small quantities of λ O protein in partially purified preparations of the λ P protein (derived from cells containing cloned λ O and P genes--see below) (Mallory and McMacken, unpublished data) convinced us that the problem of O protein instability could be overcome by massively overproducing this phage initiator.

Plasmid pKC30, a protein amplification vector (19), is a derivative of pBR322 that carries the powerful P_L promoter of phage λ. We isolated the λ O gene on a 1.06 kb HindII fragment and inserted it in both orientations into the pKC30 chromosome downstream from the plasmid P_L promoter (20). Plasmid pRLM71 contains the O gene in the proper orientation for expression directed by the P_L promoter, whereas the O gene is positioned in the opposite orientation in plasmid pRLM72. In each case transcription of the cloned O gene is controlled by regulating the activity of the P_L promoter using a thermosensitive λ cI repressor produced by a defective λ prophage that is also present in the cells. Thermal inactivation of the cI repressor by cell growth at 42°C led

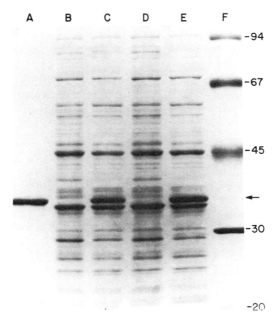

FIGURE 4. Thermal induction of λ O protein synthesis. E. coli N99(λ bio10cI857Pam3) cells carrying plasmids pRLM65 (22) (lane B), pRLM71 (lane C), pRLM72 (lane D) and pRLM73 (lane E) were grown at 30°C to 2 x 10^8 cells/ml and then aerated at 42°C for 100 min. Cell lysates (25 μg total protein) were subjected to SDS-polyacrylamide gel electrophoresis (20). Lane A, 3.2 μg of purified O protein; lane F, protein molecular weight standards.

to the accumulation of a 35,000 dalton polypeptide, approximately the predicted size of the λ O polypeptide, in cells containing pRLM71 (lane C, Fig. 4). And, as expected, pKC30 plasmid derivatives not containing the O gene or containing the O gene inserted in the wrong orientation did not produce the 35,000 dalton polypeptide (lanes B and D, Fig. 4). Surprisingly, however, the largest amplification of the putative O polypeptide (lane E, Fig. 4) was produced by plasmid pRLM73, which contains a 0.92 kb HindII fragment from the λ nin region located promoter-distal to the properly oriented O gene. We estimated from densitometric measurements that the amplified O protein constituted at least 20% of the total cellular protein, approximately 3×10^5 molecules per cell, in thermally induced cells carrying pRLM73 (20).

This huge amplification of the λ O replication protein and its concomitant stabilization have enabled us to develop a simple and rapid purification protocol that yields several mg of the homogeneous λ initiator from each gram of cell paste (20). Gently lysed cells were centrifuged at high speed and λ O protein was purified from the soluble fraction by ammonium sulfate fractionation and successive chromatography on phosphocellulose and hydroxyapatite (Fig. 5).

Characterization of the λ O Protein

Amino-terminal sequence analysis of the isolated protein (20) confirmed its identity as the product of the λ O gene and demonstrated that translation in vivo initiates only at the first of the two possible initiation codons predicted from the DNA sequence of the gene (21). The purified O protein was screened for the presence of likely enzymatic functions and found to be devoid of any exonuclease, endonuclease, or ATPase ativities.

Tsurimoto and Matsubara have demonstrated that a partially purified O protein fraction specifically bound to a quartet of repeating, tandem 19 bp sequences present within the λ replication origin (23). Using an agarose gel electrophoretic assay to detect protein-DNA interactions, we determined whether our highly purified O protein preparation retained a capacity to specifically bind to oriλ. The λdv-like plasmids pRLM4 and pRLM5 (11), which contain the replication origins of related phages λ and 82, respectively, were each cleaved into several fragments by digestion with restriction endonucleases. Each mixture of fragments was

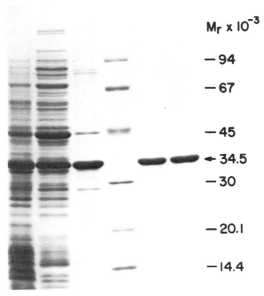

FIGURE 5. Purification of the bacteriophage λ O replication protein. Fractions generated during the purification of the O protein were analyzed by SDS-polyacrylamide gel electrophoresis. Samples were (from left to right): induced N99(λbio10cI857Pam3)/pRLM73 (whole cells, 200 µg total protein); soluble cell lysate (fraction I, 200 µg); ammonium sulfate fraction (fraction II, 52 µg); molecular weight standards; phosphocellulose fraction (fraction III, 10.5 µg); and hydroxyapatite fraction (fraction IV, 10.7 µg).

incubated for 30 min at 37°C with purified λ O protein and the protein-DNA mixtures were immediately subjected to agarose gel electrophoresis (Fig. 6) (20). In the presence of sufficient O protein, DNA fragments containing oriλ were specifically converted to O protein-DNA complexes with reduced electrophoretic mobilities. The specificity of the λ O protein-oriλ interaction is demonstrated by the inability of O protein to form a detectable complex with a fragment containing the structurally related replication origin of lambdoid phage 82 (Fig. 6). Three, and in some instances four, discrete O protein-oriλ DNA complexes were detected by the agarose gel electrophoretic analysis. We presume, but have not proven, that the four more slowly moving bands

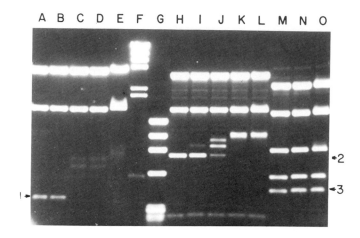

FIGURE 6. Analysis of specific binding of λ O protein to _ori_λ by agarose gel electrophoresis. Plasmid pRLM4 DNA (11) was digested to completion with either _Bgl_ II + _Eco_RI (lanes A-E) or _Bgl_ II + _Hpa_ I (lanes H-L). Plasmid pRLM5 DNA (11) was digested to completion with _Bgl_ II + _Eco_RI + _Bam_HI (lanes M-O). Various amounts of purified λ O protein (Fraction IV) were incubated with 315 ng of digested plasmid DNA, and the mixtures were subjected to electrophoresis through agarose. The levels of O protein used were: none, lanes A, H, and M; 11 ng, lanes B and I; 22 ng, lanes C, J, and N; 43 ng, lanes D and K; 86 ng, lanes E, L, and O. Fragment size standards were: lane F, _Hin_dIII digest of λ_cI_857_Sam_7 DNA (23.1, 9.4, 6.6, 4.4, 2.3, 2.0 and 0.56 kb); lane G, _Hae_ III digest of φX174 replicative form DNA (1.35, 1.08, 0.87, 0.60, 0.31, 0.28, 0.23, and 0.19 kb). The arrows indicate the positions of the protein-free fragments containing either _ori_λ (arrows 1 and 2, lanes A-E and H-L, respectively) or the replication origin of phage 82 (arrow 3, lanes M-O).

contain 1, 2, 3, and 4 moles of native O protein per mole of _ori_λ DNA fragment. In addition to the capacity of the purified O protein to bind to _ori_λ, the physiological activity of the O protein preparation is perhaps best demonstrated by its proficiency in stimulating the initiation of λ_dv_ replication _in vitro_ (11) (see below).

Overproduction and Purification of the Bacteriophage λ P
Replication Protein

Data garnered from both genetic and biochemical approaches suggest that the bacteriophage λ P replication
protein functions to deliver elements of the bacterial replication apparatus to the λ chromosomal origin[3] (6, 24-28).
It is notable in this regard that the λ P protein has been
isolated from λ-infected cells as a complex with the E. coli
dnaB protein (27) and that the P protein appears to specifically interact with the host dnaK protein (28). Since the
physical and functional properties of the λ P protein had
not been characterized, we sought to isolate sufficient
quantities of P protein to pursue such studies.
We constructed a series of hybrid plasmids that were
designed to amplify intracellular levels of the λ P protein[3]. The concentration of soluble P protein produced by
each plasmid was measured, using as an assay the capacity of
the P protein to inhibit the in vitro conversion of φX174
viral single-stranded DNA to the duplex replicative form
(26). High level expression of the λ P protein inhibits E.
coli DNA replication (29) and cell growth[3]. Therefore,
expression of the cloned P gene needed to be strictly regulated. Plasmids which yielded the highest levels of soluble
P protein contained a 7.2 kb BamHI DNA fragment (Fig. 1)
(from λcI857cro27 phage) that carries the λ genes involved
in viral replication and in transcription control. Thermal
inactivation of the temperaturesensitive cI repressor encoded by the recombinant plasmids led to the rapid accumulation of approximately 1000 molecules of P protein per cell
before cell growth ceased[3]. Plasmids which contain the P
gene inserted downstream from the strong λ P_L promoter
yielded substantially more P protein, but, as reported by
others (30), the additional P protein is tightly bound to
membranous components and cannot be released except by
treatment with protein denaturing agents[3].
The soluble portion of amplified P protein was purified
to apparent homogeneity from gentle cell lysates[3] (Fig. 7).
From 714 g of wet cell paste, 4.5 mg of purified λ P protein
was obtained. The hydrophobic P protein was found to exist
as a monomer in solution (M_r = 26,500) and its N-terminal
and C-terminal sequences were precisely those predicted from
the sequence of the P gene (31). The purified P protein
neither bound to single- or double-stranded DNA nor did it

[3]J. B. Mallory and R. McMacken, manuscript in preparation

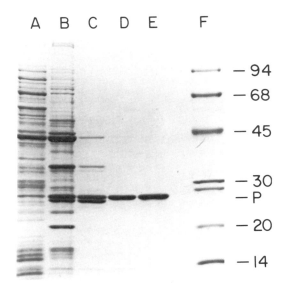

FIGURE 7. Purification of the λ P protein. Fractions generated during the isolation of the P protein were analyzed by SDS-polyacrylamide gel electrophoresis. Lane A, fraction I (cell lysate, 80 μg); lane B, fraction II (ammonium sulfate, 72 μg); lane C, fraction III (DEAE-cellulose, 21 μg); lane D, fraction IV (Blue Dextran-Sepharose, 7 μg); lane E, fraction V (Bio-Rex 70, 7 μg); lane F, molecular weight standards.

have associated endonuclease, exonuclease or ATPase activities. However, as demonstrated previously (11), the functionality of the isolated P protein is clearly evident from its capacity to promote the initiation of λdv DNA replication in vitro.

Functional Properties of the λ P Protein: Interaction with the E. coli dnaB Protein

The dnaB protein of E. coli plays a pivotal role in the replication of the bacterial chromosome and of the chromosomes of phages φX174 and λ. It is apparently transferred onto E. coli and φX174 DNA through specific interactions with the E. coli dnaC protein, whereupon the dnaB protein migrates on the lagging strand of a replication fork as part of the primosome and participates in the priming of Okazaki fragments (32,33). Since the E. coli dnaC protein is not essential for λ DNA replication and since the λ P protein interacts with the host dnaB protein, it is reasonable to suggest that the λ P protein functions as a viral-encoded analogue of the bacterial dnaC protein. To characterize the P protein-dnaB protein interaction more thoroughly, we have examined the effect of added P protein on the multiple enzymatic activities of the host replication protein.

Purified λ P protein was found to be a potent inhibitor of the in vitro replication of φX174 viral DNA and of plasmids containing the E. coli chromosomal replication origin[3]. Replication of both of these chromosomes depends on the formation of a dnaB protein-dnaC protein complex (34). The DNA-dependent ATPase activity of dnaB protein was also completely supressed by the addition of P protein[3]. Approximately 2.5 moles of P protein monomer per mole of native hexameric dnaB protein (M_r = 312,000) were required to cause a 50% reduction in ATPase activity. Significantly more dnaC protein, 11.6 moles per mole of dnaB protein, was necessary to cause an equivalent reduction in dnaB protein ATPase activity.

We determined the effect of both the λ P protein and the E. coli dnaC protein on the general priming reaction of dnaB protein[3]. In this non-physiological reaction, dnaB protein and E. coli primase (dnaG protein) act to synthesize short RNA transcripts in a nonspecific fashion on naked single-stranded DNA (35). P protein strongly inhibited this aspect of dnaB protein function also (Fig. 8a). On the other hand, dnaC protein markedly stimulated general priming (Fig. 8b). Because of the contrasting effects of P protein and dnaC protein on general priming, it was possible to determine which protein predominated when a mixture of the two proteins was used. The P protein effectively supressed general priming even when a preformed dnaB-dnaC protein complex was employed (Fig. 8a). In the reverse experiment, a nearly five-fold excess of dnaC protein (M_r = 29,000) over P protein only partially relieved the P protein-mediated inhibition (Fig. 8b). The predominance of the λ P protein over the bacterial dnaC protein observed here is consistent with the notion that the P protein produced by an infecting λ chromosome can outcompete the host dnaC protein for the E. coli dnaB protein and direct the latter to act specifically in λ DNA replication. The reversal by excess dnaC protein of the P protein-mediated inhibition of general priming indicates that P protein suppresses, but does not destroy,

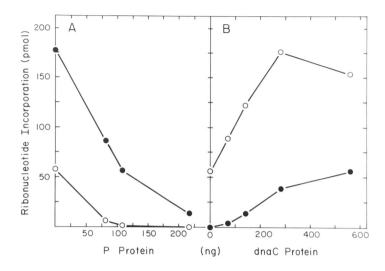

FIGURE 8. Effect of λ P protein and E. coli dnaC protein on general priming. RNA priming reactions were carried out as previously described (35) using 280 ng dnaB protein, 180 ng primase, and 500 pmol φX174 SS DNA as template. A). E. coli dnaB protein was preincubated for 10 min at 30°C in the presence of ATP either with (-●-) or without (-O-) dnaC protein (280 ng). P protein was subsequently added, as indicated, prior to the start of RNA synthesis. B). E. coli dnaB protein was preincubated for 10 min at 30°C in the presence of ATP either with (-●-) or without (-O-) λ P protein (108 ng). E. coli dnaC protein was subsequently added, as indicated, prior to the start of RNA synthesis.

the functional activity of dnaB protein.

The data presented thus far suggest that the purified λ P protein is capable of forming a complex with the E. coli dnaB protein, even if the dnaB protein is already bound to the bacterial dnaC protein. Because of the large difference in the sizes of the native P and dnaB proteins, such complex formation can be readily detected by monitoring the conversion of P protein activity into a rapidly sedimenting species. In the absence of dnaB protein, P protein sedimented as a monomer slightly behind carbonic anhydrase (M_r = 30,000). In the presence of dnaB protein, a significant portion of the P protein activity, as measured by its capacity to promote λdv DNA replication in vitro (11), sedi-

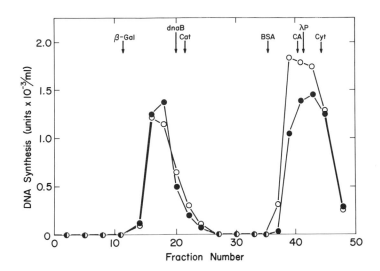

FIGURE 9. Glycerol gradient sedimentation of a P protein-dnaB protein complex formed in the presence or absence of dnaC protein. λ P protein (20 µg) was mixed in the presence of ATP either with dnaB protein (20 µg) (-●-) or with a dnaB-dnaC protein complex (20 µg dnaB protein, 8.5 µg dnaC protein) (-O-). Samples were sedimented in 15-35% glycerol gradients. P protein activity was detected by its capacity to promote λ<u>dv</u> replication <u>in vitro</u> (11). Arrows mark the positions of marker proteins sedimented in a parallel gradient.

mented more rapidly in a glycerol gradient than native dnaB protein[3] (Fig. 9). Moreover, P protein was also converted to this rapidly sedimenting form when it was incubated with a preformed dnaB-dnaC protein complex (Fig. 9). Densitometric analysis of the polypeptides in these glycerol gradient fractions, separated by electrophoresis in SDS-polyacrylamide gels, indicated (i) that each fast sedimenting P protein peak contained approximately 3 moles of P protein per mole of hexameric dnaB protein, and (ii) that more than 80% of the dnaC protein was displaced from its complex with dnaB protein upon the addition of the λ P protein. We conclude that the λ P protein is a phage coded analogue of the bacterial dnaC protein and we suggest that the P protein plays an essential role in the transfer of the dnaB protein onto the λ chromosome during the initiation of viral DNA replication.

Initiation of λdv DNA Replication in Vitro

With the availability of purified preparations of the λ O and P replication proteins, we succeeded in establishing a soluble in vitro system that specifically replicates λdv plasmid DNA (11). The system is a modified version of that developed by Fuller et al. (36) for replicating plasmids bearing the origin of the E. coli chromosome. A partially purified, but still relatively crude, protein fraction is used to supply the necessary E. coli replication proteins (11). The λdv-like plasmid pRLM4 (11), used as a template in these studies, contains the λ replication origin and regulatory signals on a segment that includes the region of the λ chromosome from the rightward promoter-operator (P_R O_R, Fig. 1) through gene P. Extensive characterization of the properties of λdv replication in the in vitro system indicates that the plasmid DNA is replicated via mechanisms that closely resemble those described for replication of the phage λ chromosome in vivo[4] (11).

Initiation of λdv DNA replication in vitro requires both the λ O and P initiator proteins, as well as an E. coli enzyme fraction (Table 2). Initiation of DNA replication in the soluble enzyme system containing the phage initiators occurs specifically on plasmid chromosomes containing the λ replication origin (Table 2). Initiation of DNA replication from the E. coli chromosomal origin is partially suppressed by the presence of the λ P protein (Table 2).

Because of the crude nature of the E. coli enzyme fraction used in the in vitro system, it was necessary to use specific inhibitors to establish the identity of some of the bacterial replication proteins that participate in the enzymatic replication of λdv DNA. Such studies suggested the involvement of dnaB protein, primase (dnaG protein), single-stranded DNA binding protein, DNA gyrase, and RNA polymerase in λdv replication and ruled out a role for the host dnaC protein (11). Recent studies indicate that the E. coli dnaJ and dnaK proteins are also required for initiation of λdv DNA replication in vitro[5]. These results are completely consistent with the requirements found for λ DNA replication in vivo (1,2). Although initial experiments (11) suggested a role in λdv replication for E. coli pre-priming proteins n and n' (37), the lack of an apparent

[4]M. S. Wold and R. McMacken, unpublished data
[5]J. LeBowitz, M. Zylicz, C. Georgopoulos, and R. McMacken, unpublished data

TABLE 2

REQUIREMENTS FOR λdv REPLICATION IN VITRO
AND TEMPLATE SPECIFICITY OF IN VITRO REPLICATION[a]

Template	O protein	P protein	DNA synthesis pmol
None	+	+	9
pRLM4 (oriλ)	-	-	4
pRLM4	-	+	4
pRLM4	+	-	5
pRLM4	+	+	157
pRLM4	+	+	3[b]
pRLM5 (ori82)	+	+	7
pKN402 (oriR1)	+	+	7
M13oriC26 RF	-	-	161
M13oriC26 RF	-	+	44

[a]Replication of supercoiled plasmid DNA templates (650 pmol) in vitro was performed as previously described (11).
[b]E. coli enzyme fraction was omitted.

requirement for these proteins in the O and P protein-dependent single-strand replication system described below raises questions about the validity of this assessment.

The requirement for RNA polymerase for λdv replication in vitro (10-12) was investigated further[4]. If the in vivo phenomenon of transcriptional activation of λ DNA replication (8,9) is operative in vitro as well, then addition of purified λ cI repressor to the soluble enzyme system should block the replication of those λdv-like plasmids, such as pRLM4 (11), that contain a wild type λ right operator. Physiological concentrations of the λ cI repressor (10^{-7} M) can indeed specifically inhibit replication of pRLM4 DNA in vitro, as illustrated in Fig. 10. Replication of oriC-containing plasmids, on the other hand, remains unaffected even at high concentrations of repressor. As a further test of the specificity of the cI repressor-mediated block of λdv replication, we constructed a plasmid carying oriλ that should be capable of replicating in the presence of repressor and examined its behavior in the soluble enzyme system[4].

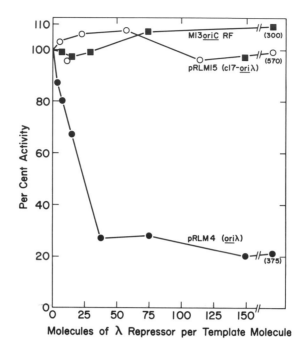

FIGURE 10.   Effect of λ cI repressor on the initiation
of λdv replication in vitro.   Purified λ repressor was
incubated for 10 min at 30°C with superhelical plasmid DNA
(650 pmol).   This mixture was subsequently added to the
soluble λdv replication system and DNA synthesis measured as
previously described (11).   The number of repressor monomers
added is indicated, although only dimeric repressor binds to
operator DNA.

This plasmid, pRLM15 (Fig. 11), was derived from λc17, a
phage carrying a spontaneous mutation which generates a
repressor-insensitive   promoter positioned upstream from
oriλ.   λc17 phage can replicate in a λ lysogen, since its
replication origin can still be "transcriptionally activated" in the presence of λ repressor (8).   Thus, it is
comforting that plasmid pRLM15, which bears the c17 promoter, is also readily replicated in the in vitro system in
the presence of the cI repressor (Fig. 10).
We have previously reported that the soluble enzyme
system replicates exogeneously added λdv supercoils through
a single round to produce catenated, covalently closed,

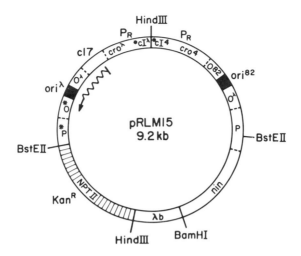

FIGURE 11. Genetic and restriction map of plasmid pRLM15. Plasmid pRLM15 was constructed by standard recombinant DNA techniques from plasmid pRLM5 (11) and phage λc17Oam8Pam3.

dimeric daughter molecules (11). More extensive electron microscopic and electrophoretic analyses of the λdv replication intermediates indicates that initiation occurs at or near oriλ in a bidirectional fashion. Replication proceeds through cairns (theta) forms to produce what are apparently highly intertwined catenated dimers that are ultimately segregated to monomeric daughter circles, much as has been described for SV40 replication intermediates[4] (38).

The Bacteriophage λ Initiators Promote the Replication of Single-Stranded DNA

The *E. coli* dnaB protein apparently migrates along the lagging strand of a replication fork as part of the primosome protein complex that functions in the rifampicin resistant priming of nascent DNA chains (32,33). Lagging strand initiation determinants have been identified in the chromosomes of φX174, ColE1, and pBR322 (39-41). These determinants, which are believed to be specific DNA sequences that direct the assembly of the multi-protein primosome, have been located at significant distances (up to 2 kb) from the primary leading strand origin. The involvement of the bacterial dnaB protein, the prototypical primosomal protein, in λ DNA replication and the proposal (42) that the λ ice site (located within the λ cII gene, Fig. 1) is a lagging strand replication origin prompted us to search for primosome assembly sites in the λ chromosome that are active in the soluble enzyme system (11) that replicates λdv DNA.

Adopting an approach similar to that pioneered by Ray and colleages (43), we inserted fragments from the λ origin region into the single-stranded cloning vector M13mp7. The recombinant chromosomes were subsequently examined for the presence of rifampicin-resistant single-strand (SS) initiation determinants by using the hybrid DNAs as templates in the in vitro λdv replication system[6]. Phage M13mp7 DNA is employed as the cloning vector in this type of experiment, since the initiation of M13 replication is strictly dependent on priming by E. coli RNA polymerase, a process that is readily blocked by rifampicin.

When hybrid M13 chromosomes bearing both the λori and ice sites from either the λ L or R strands were used as templates in the λ replication system, both recombinant DNAs were extensively replicated to the duplex form in a rifampicin resistant reaction[6] (Table 3). Initiation of complementary strand DNA synthesis required the presence of both the λ O and P replication proteins. Surprisingly, however, the O and P protein-dependent replication of single-stranded DNA was independent of λ DNA sequences as is shown by the vigorous replication of M13mp7 DNA in the complete system (Table 3, line 5). Moreover, the λ initiators were found to stimulate the replication of φX174 viral DNA in vitro[6].

In spite of the fact that this strand initiation reaction apparently occurred non-specifically in regard to DNA sequence, we decided to further characterize its properties, anticipating that the SS reaction might be used to clarify the physiological roles of the λ O and P initiators. We have found[5], using specific inhibitors, that the SS reaction requires many of the proteins identified as participating in λdv replication in vitro and λ DNA replication in vivo (see previous section). In addition to the viral O and P proteins, the proteins required for DNA strand initiation include six E. coli replication proteins: SSB, primase, DNA polymerase III holoenzyme, and the dnaB, dnaJ and dnaK proteins[5,6] (28). As expected, neither DNA gyrase nor RNA polymerase participate in the SS reaction.

[6]J. LeBowitz and R. McMacken, manuscript in preparation

TABLE 3
REPLICATION OF SINGLE-STRANDED DNA IN VITRO[a]

Template	Omission	DNA synthesis pmol
1. M13mp7:oriλL	None	170
2. M13mp7:oriλL	O & P proteins	6
3. M13mp7:oriλR	None	191
4. M13mp7:oriλR	O & P proteins	4
5. M13mp7	None	178
6. M13mp7	O protein	6
7. M13mp7	P protein	5

[a]M13mp7:oriλ DNA contains a 1.50 kb Alu I fragment of λ DNA (positions 38453-39956 on the λ chromosome) inserted at the HincII site of M13mp7. The reaction conditions were those previously described for λdv DNA replication in vitro (11), except that rifampicin (17 μg/ml) was present.

The products of the SS reaction consist of a mixture of two species, a supercoiled RF I form (75%) and a nicked circular RF II form (25%) containing a full length complementary strand[6] (Fig. 12, lane A). We presume that the nick or small gap present in the complementary strand of the RF II form product is located at the origin-terminus of complementary strand synthesis. By mapping this interruption relative to unique restriction endonuclease cleavage sites, it is possible to determine whether the strand initiations occur at unique or multiple sites. In the absence of rifampicin and the λ initiators, M13mp7 complementary strand synthesis in the soluble enzyme system was initiated at a discrete site located at or near the physiological origin (43). Digestion of labeled RF II product at its solitary Bgl II site, followed by electrophoretic separation of the DNA chains in alkaline agarose, revealed the presence of two fragments of the size expected if initiation occurred at the physiological complementary strand origin (Fig. 12, lanes J and N). A similar characterization of the rifampicin resistant, O and P protein-dependent, SS reaction product yielded a broad distribution of fragment sizes (Fig. 12, lanes E and L). We conclude that initiation of DNA

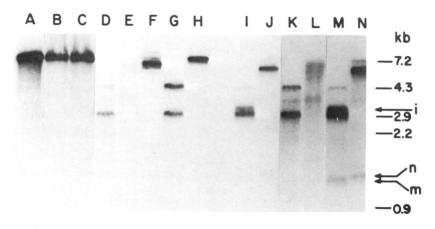

FIGURE 12. Alkaline agarose gel electrophoretic analysis of the products produced by replication of M13mp7 DNA in vitro. M13mp7 DNA was replicated in the soluble λ replication system containing λ O and P proteins and ^{32}P-labeled dNTPs as previously described for λdv DNA (11), except that the reaction mixture also contained rifampicin (17 μg/ml) and nicotinamide mononucleotide (NMN) (1 mg/ml). In control reactions, M13mp7 DNA was replicated in the absence of O and P proteins and rifampicin. RF II reaction products were isolated from ethidium bromide/CsCl gradients and digested with restriction endonucleases as indicated below. Labeled DNA fragments were then separated by electrophoresis in a 1% agarose gel at pH 12.5. The complementary strand origin of M13mp7 DNA (7238 bases) is located at position 5740; the single Bgl II site is at position 6923; the two Cla I sites are at positions 2527 and 6870. Lane A, RF II from O and P protein-dependent reaction; lanes B and C, RF II from the control reaction carried out in the absence or presence of NMN, respectively; lanes D and E, RF II from the O and P protein-dependent reaction digested with Cla I or Bgl II, respectively; lanes F, G, and H, supercoiled M13mp7 RF I, undigested, and digested with Cla I or Bgl II, respectively; lanes I and J, RF II from the control reaction digested with Cla I or Bgl II, respectively; lanes K-N, longer autoradiographic exposure of lanes D, E, I, and J, respectively.

strand synthesis promoted by the λ O and P proteins occurs at multiple, perhaps random, sites.

The complementary strand of φX174 is also initiated at multiple sites, as a direct consequence of the mobility of the multi-protein primosome (32,33). In the case of φX174, the rate limiting step of complementary strand synthesis is the assembly of the preprimosome (37) on the φX174 viral strand, after which the activated template is rapidly primed and replicated (44,45). Since dnaB protein also participates in the O and P protein-dependent SS reaction, it seemed plausible that a primosome-like complex was responsible for the multiple starts observed. To examine this possibility, we determined if the M13mp7 template DNA could be converted to an activated form when it is preincubated with λ and E. coli replication proteins prior to initiating DNA synthesis. Indeed, the initial rate of M13mp7 complementary strand DNA synthesis was elevated more than 20-fold if the template was first preincubated with the required proteins (Fig. 13a). The kinetics of template activation promoted by this λ replication system (Fig. 13b) are nearly

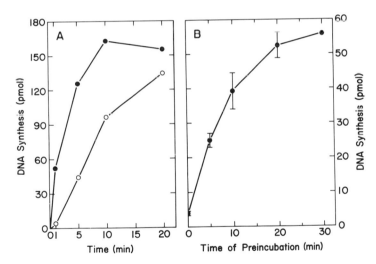

FIGURE 13. Lambda O and P protein-dependent formation of activated M13mp7 DNA. A). Time course of DNA synthesis in the O and P protein-dependent replication of M13mp7 DNA. Open circles, normal time course. Closed circles, time course using template DNA that had been preincubated for 20 min at 30°C with λ and E. coli replication proteins. B). Time course of activation of M13mp7 DNA during its preincubation with required proteins. DNA synthesis in the first minute of replication is measured.

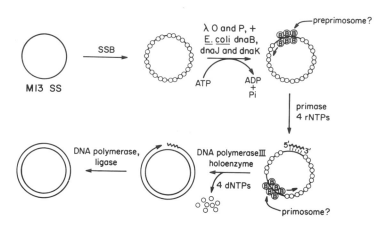

FIGURE 14. Hypothetical scheme for the λ O and P protein-dependent replication of single-stranded, circular DNA. See text for details.

identical to those reported for the assembly of the preprimosome on φX174 DNA (45).

Recent investigations indicate that formation of activated M13mp7 DNA is an ATP-dependent process that requires the λ O and P proteins and four E. coli proteins, namely SSB, dnaB protein, dnaJ protein and dnaK protein[5,6]. Furthermore, the activated template, isolated free of unassociated protein, was found to contain bound dnaB protein[6]. We presume, then, that the λ O and P proteins function, at least in part, to transfer the E. coli dnaB protein onto SSB-coated single-stranded DNA (Fig. 14). The nature and stoichiometry of the polypeptides present on the activated template remain to be determined. It is also uncertain at this time if the dnaB protein present on the activated template is capable of movement along the coated strand. In any event, primase interacts with the bound dnaB protein and catalyzes the synthesis of RNA transcripts that are utilized as primers for chain elongation by DNA polymerase III holoenzyme (Fig. 14).

We believe that the λ O and P protein-dependent SS replication reaction reported here, although nonspecific in regard to DNA sequence, is physiologically relevant to the protein transfer reactions that occur in the initiation of bidirectional λ DNA replication. It is even conceivable that similar protein transfer reactions are also required for each of the multiple initiation events that occur on the lagging strand at a λ replication fork. We hope to exploit this relatively simple single-stranded system to probe the biochemical mechanisms used by the λ O and P initiators to direct elements of the host replication apparatus to act at the λ chromosomal origin.

REFERENCES

1. Skalka AM (1977). DNA replication: Bacteriophage lambda. Curr Top Microbiol Immunol 78:201.
2. Furth ME, Wickner SH (1983). Lambda DNA replication. In Hendrix R, Roberts J, Stahl F, Weisberg R (eds): "The Bacteriophage Lambda", Second Edition, Cold Spring Harbor: Cold Spring Harbor Laboratory, in press.
3. Sogo J, Greenstein M, Skalka A (1976). The circle mode of replication of bacteriophage lambda: the role of covalently closed templates and the formation of mixed catenated dimers. J Mol Biol 103:537.
4. Schnos M, Inman R (1970). Position of branch points in replicating lambda DNA. J Mol Biol 51:61.
5. Ogawa T, Tomizawa J, Fuke M (1968). Replication of bacteriophage DNA. II. Structure of replicating DNA of phage lambda. Proc Natl Acad Sci USA 60:861.
6. Furth ME, McLeester C, Dove WF (1978). Specificity determinants for bacteriophage lambda DNA replication. I. A chain of interactions that controls the initiation of replication. J Mol Biol 126:195.
7. Thomas R, Bertani LE (1964). On the control of the replication of temperate bacteriophage superinfecting immune hosts. Virology 24:241.
8. Dove WF, Inokuchi H, Stevens WF (1971). Replication control in phage lambda. In Hershey AD (ed): "The Bacteriophage Lambda", Cold Spring Harbor: Cold Spring Harbor Laboratory, p. 747.
9. Furth ME, Dove WF, Meyer BJ (1982). Specificity determinants for bacteriophage λ DNA replication. III. Activation of replication in λ*ri*[c] mutants by transcription outside of *ori*. J Mol Biol 154:65.
10. Anderl A, Klein A (1982). Replication of λ*dv* DNA *in vitro*. Nucleic Acids Res 10:1733.
11. Wold MS, Mallory JB, Roberts JD, LeBowitz JH, McMacken R (1982). Initiation of bacteriophage λ DNA replication *in vitro* with purified λ replication proteins. Proc Natl Acad Sci USA 79:6176.

Sci USA 77:799.

40. Nomura N, Low RL, Ray DS (1982). Identification of ColEl DNA sequences that direct single strand-to-double strand conversion by a φX174 type mechanism. Proc Natl Acad Sci USA 79:3153.

41. Marians KJ, Soeller W, Zipursky SL (1982). Maximal limits of the Escherichia coli replication factor Y effector site sequences in pBR322 DNA. J Biol Chem 257:5656.

42. Hobom G, Kroger M, Rak B (1981). Primary and secondary replication signals in bacteriophage λ and IS5 insertion element initiation systems. In Ray DS (ed): "The Initiation of DNA Replication", New York: Academic Press, p. 245.

43. Ray DS, Hines JC, Kim MH, Imber R, Nomura N (1982). M13 vectors for selective cloning of sequences specifying initiation of DNA synthesis on single-stranded templates. Gene 18:231.

44. Wickner S, Hurwitz J (1974). Conversion of φX174 viral DNA to double-stranded form by purified Escherichia coli proteins. Proc Natl Acad Sci USA 71:4120.

45. Weiner JH, McMacken R, Kornberg A (1976). Isolation of an intermediate which precedes dnaG RNA polymerase participation in enzymatic replication of bacteriophage φX174 DNA. Proc Natl Acad Sci USA 73:752.

Mechanisms of DNA Replication and Recombination, pages 849–861

Domains of simple sequences or alternating purines and pyrimidines
are sites of divergences in a complex satellite DNA

D. M. Skinner
R. F. Fowler
V. Bonnewell

Biology Division, Oak Ridge National Laboratory
and University of Tennessee-Oak Ridge Graduate School
of Biomedical Sciences
Oak Ridge, Tennessee 37830

ABSTRACT

Three cloned variants of a very complex, G+C-rich
(63%), satellite DNA of the Bermuda land crab have some
highly conserved domains (from 84 to 96% homology) interspersed with other domains of marked sequence divergence.
The conserved domains can be as long as 594 bp; they contain
"garden variety" DNA. The divergent domains contain
repeated sequences, such as homopolymers [n = 7-23] of (C)n
or (G)n but not (A)n or (T)n; or homocopolymers [n = 4-17]
of (CCT)n, (CA)n, (AGGG)n, or (AG)n but not (AT)n. Other
regions adjacent to major sequence variations are characterized by long stretches of alternating purines and pyrimidines (pu/py). That the latter adopt a Z conformation
appears likely, since at least one stretch of pu/py of complex sequences which should not permit strand slippage,
exhibits S_1 sensitivity.

INTRODUCTION

A complex G+C-rich (63%) satellite DNA (8) with an
average repeat unit of 2.07 ± 0.1 kb (9) accounts for 3 percent of the total DNA of the Bermuda land crab, Gecarcinus
lateralis (13). Three distinct variants of the satellite
have been recovered from insertion of an Eco RI digest of
cellular satellite into the Eco RI site of pBR322 (14).

One of these three variants (RU) has a repeat unit of 2.089 kb similar to the sizes of the average repeat units recovered from Eco RI digests of cellular satellite (9) and to 13 other satellite variants cloned into the Eco RI site of pBR322. Another satellite variant (truncated, TRU) has a repeat unit of 1.674 kb. It is truncated by the presence of an Eco RI site 360 $\pm$ 4 bp upstream from the Eco RI site common to all three variants. TRU also has several smaller internal deletions.

The third variant (extended, EXT) contains a fivefold tandem amplification of a purine-rich 142 bp segment; a close homologue (83% homology) of that segment occurs in a similar position in RU and TRU (2). An inverted tetranucleotide borders the ends of the amplified region in EXT. The downstream arm of the inverted tetranucleotide is present in all three variants, while the upstream arm of the inverted tetranucleotide is missing from RU and TRU. In addition to the amplification, EXT also suffers several deletions and insertions, resulting in a 2.639 kb repeat unit.

The sequences of these three satellite variants are characterized by domains of varying levels of complexity. Changes in the sequences of the variants occur at some, but not all, domains of simple, repeated sequences.

MATERIALS AND METHODS

Satellite DNA--cellular and cloned: The G+C-rich satellite (ρ=1.721 g/cm) was purified by three sequential centrifugations in neutral CsCl gradients (1). Eco RI (9) or Pst I digests of purified satellite were inserted into the respective sites of pBR322 (6) and amplified in E. coli HB101.

Analyses of cloned satellite variants: Restriction enzyme digests were carried out according to the manufacturer's specifications. Sequencing was by chemical degradation (10) of Hinf I and Hpa II (or Msp I) fragments.

RESULTS AND DISCUSSION

General characteristics: Similarities and differences among cloned repeat units of the satellite are obvious in restriction maps prepared from digests with only a few restriction enzymes. The maps in Figure 1 contain data assembled on RU, EXT, and TRU (Part A), the three of sixteen Eco RI variants that have been most thoroughly characterized, and data on six variants recovered when fragments from a Pst I digest of the satellite were inserted into the Pst I site of pBR322 (Part B); the latter have been less well characterized. A map of cellular satellite is included for comparison (Part C).

Is every copy of the satellite different? In Figure 1, the sites of restriction enzymes in the nine individual satellite repeats and in cellular satellite have been aligned to maximize homologies. Although the variants share a similar pattern of Hinf I, Pst I, and Hind III restriction sites, each of the nine differs from the others by one or more base changes as detected by digestion with restriction enzymes. To date, no two variants of the satellite are identical; if this holds for all 160 cloned satellite fragments on hand, it will lend strong statistical support to the possibility that all 16,000 copies/genome of the satellite are different.

An average length for the most common variants is indicated by the line at the bottom of Figure 1. The sizes of homologous restriction fragments produced by digestion of RU, EXT, and TRU with many restriction enzymes differ significantly. Fragments produced by digestion with Hinf I and Hpa II are shown in Fig. 2. The sizes of Hinf I fragments are shown in Table 1. In Hinf I digests, the major differences among the variants lie within domains of variance located in fragments between sites 1 and 2, 3 and 4, and 5 and 6. Variations in the sequences of RU, EXT, and TRU are obvious in Figure 1, Part A. These include several point mutations, notably those leading to the absence of a Hinf I site (number 4) from pG1E15 (RU), as well as to the presence of an extra Hinf I site (number 8) in pG1E2 (TRU). Deletions of 1 to 24 bp have occurred in highly conserved

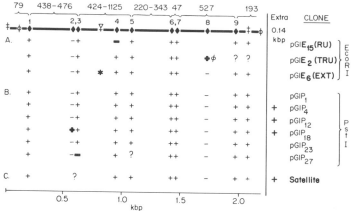

Figure 1. <u>Positions of Hinf I sites in nine satellite variants</u>. Eco RI or Pst I digests of purified satellite were inserted into the Eco RI (A) or Pst I (B) site of pBR322. Clones in the Eco RI site are designated: pGlEn = 1-16 (Gl for <u>G. lateralis</u>, E for Eco RI); those in the Pst I site are designated: pGlPn = 1-144. Each of the seven Hinf I fragments of the RU from pGlE15 is bracketed; the range of sizes of the Hinf I fragments in RU, EXT, and TRU is indicated at the top of the figure. All Hinf I sites (♦) found in any variant are indicated; they are numbered 1 to 9 from the 5' end. The presence (+) or absence (-) of any site is known from sequencing data of RU, TRU, and EXT or has been calculated from the sizes of Hinf I restriction fragments displayed on acrylamide gels (16). The Hinf I restriction pattern of cellular satellite (Part C) is shown for comparison. One fragment of 140 to 150 bp present in the cellular satellite and Pst I clones pGlP4, pGlP12, and pGlP18, has not been accounted for (column entitled EXTRA, Parts B and C). Exceptional sites (i.e. those found in only one variant) are indicated by enlarged symbols (+ or -). Uncertain sites are indicated by a question mark; two are included in the TRU variant because the 370 bp Eco RI fragments cut from copies of what we assume to be that or a related variant in cellular satellite DNA have not been characterized by digestion with restriction enzymes. The amplified DNA that accounts for the extra DNA in the EXT variant is indicated by an asterisk. Its origin and characterization have been described (See Text and ref. 2). The differences in the patterns of the satellite variants inserted into the Pst I site (Part B) can be accounted for by differences in the lengths of Hinf I fragments 2, 3, and 4. In addition, pGlP18 has an extra Hinf I site while pGlP27 is missing one and possibly two Hinf I sites.

Table I

Sequences of Divergent or Amplified Domains in Hinf I Fragments 2,3,4, and 6 of RU, EXT, and TRU

	Hinf 2 Fragment Length bp	Hinf 2 Lengths of Divergent Domains bp	Hinf 2 Sequences	Hinf 3 Fragment Length bp	Hinf 3 Length of Amplified Domain bp	Hinf 3 Sequences	Hinf 4 Fragment Length bp	Hinf 4 Length of Divergent Domain bp	Hinf 4 Sequences	Hinf 6 Lengths of Divergent Domains bp	Hinf 6 Sequences
RU	476	a.126	$G_4("AG_3")_{13}A^CAG_{20}N_{47}$	424	127	$[("AAC/G'")_5A_3G'"AAC/G'")_{18} - N_{43}CTCC_3TCTCC]$	343	189	$("CCT")_{27}CC(CCT)_{15}CTTAAC_3TC_{22} - (CGCAC)_5CGAAC$	a. 61	$C_4AC_3G(AC)_3GC_2TCGTC_5TC_3 - TCAC_{13}TC_4TCCTC$
		b. 48	$GTC_5TT(AC)_2AA(AC)_4AAC_{23}$							b. 25	$C_3TCTTCGC_{10}ACCTTC$
										c. 53	$Z_6N_{39}Z_4Z_5NZ_{15}NZ_5Z_4$
EXT	444	a. 120	$C_4AC_3AG_{18}(AC)_{17}X_{36}G_{23}$	989 + 136	$(142)_6$	$[("AAC/G'")_7A_3G'("AAC/G'")_{21} - N_{43}CTCC_3TCTCC]_{1+5}$	220	66	$("CCT")_9ACA/ /C_{21}(CGCAC)_3$	a. 72	$C_4AC_3G(AC)_3GC_4TCGTC_5TC_3 - TCAC_{13}TCTC_3(TC_4)_3$
		b. 22	$GTC/ /(A_3C)_2/ /C(AC_4)_2$							b. 25	As RU.
										c. 41	$Z_3N_2/ /NZ_4Z_5NZ_4NZ_5Z_4$
TRU	438	a. 101	$G_4(AC_3)_{10}AC_{17}N_{39}$	292 + 136	130	$[("AAC/G'")_6A_3G'("AAC/G'")_{18} - N_{43}CTCC_3TCTCC]$	326	173	$("CCT")_{10}C_4("CCT")_{17}CC(CCT)_4 - CTT(CCT)_3ACTC_{19}(CGCAG)(CGCAC) - (CGCAG)(CGCAC)_4CGAAC$	a. 62	$C_4AC_3G(AC)_3GC_2TCGTC_5TC_3 - TAAC_{11}AC_2(TC_4)_2$
		b. 33	$CAC_5AC/ /AAC_{17}TCTCC$							b. 24	$C_3TCATCG(C_4A)_2C_3TC$
										c. *	

Fragments 1, 5, and 7 contain few variations. Fragment length: length of fragment including conserved and divergent regions. a,b,c: Divergent regions, interspersed with conserved regions. N: Part of divergent domain but not homoco- or homopolymer. X: As N, but not conserved. (" "): variations on sequence listed. Z_n: unbroken stretches of pu/py. / /: deletion. *: truncated end of TRU. Hinf I site that yields fragments 3 and 3a (136 bp) in EXT and TRU missing from RU; amplified domain is in fragment 3. Fragments 3a are homologous to that segment in RU. TRU: extra Hinf I site that yields Hinf 6a (66 bp); diverged sequences are in 290 bp fragment.

Figure 2: <u>Homologous</u>
<u>restriction fragments of RU, EXT,</u>
<u>and TRU have very different sizes</u>:
Acrylamide gel of restriction
digests of EXT (Lanes 1, 7), RU
(Lanes 2, 6) and TRU (Lanes 3, 5).
Size marker: Hinf I digest of
ØX174 (Lane 4). Hinf I digests:
Lanes 1-3; Hpa II digests:
Lanes 5-7. Fragments are
numbered from their 5' to 3' ends
of the satellite inserts, (Hinf I
digests: fragments 1-7; Hpa II
digests: fragments 1-14).
Homologous fragments in each
variant have the same numbers.
Fragments 3 and 3a (Lanes 1
and 3) reflect the extra Hinf I
site in EXT and TRU (See below).
A description of similarities
and differences in the seven
Hinf I fragments follows.
Specific changes that lead to
the very different sized Hpa II
fragments (Lanes 5-7) are not
detailed here.

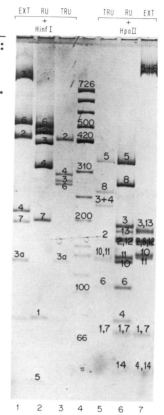

H1: 92 % homology between RU, EXT, and TRU.
H2: Deletions shorten fragment to 444 bp in EXT, 438
     in TRU.
H3: A. Transition (G→A) produces Hinf I site in EXT
         and TRU that is missing from RU. This accounts
         for 136 bp fragment (3a) in both EXT and TRU.
     B. Amplified DNA accounts for 989 bp fragment in EXT.
H4: 30 bp deletion in TRU, deletions totaling 120 bp
     in EXT.
H5: 98 % homology between RU, EXT, and TRU
H6: A. 96 % homology between RU and TRU. 1 to 12 bp
         deletions or insertions in EXT, resulting in net
         increase of 5 bp compared to RU.
     B. Compared to RU, TRU has a C→G transversion;
         compared to EXT, an A→G transition that yield
         extra Eco RI site with loss from TRU of 3' end
         of H6 and all of H7.
H7: 95 percent homology between RU and EXT.

regions of RU with homologies of 84 to 96% to EXT and TRU.
From the sequences of RU, EXT, and TRU, we realize that the
variations among the pG1E clones are far more striking than
was apparent in restriction maps. For example, there are
other deletions of as many as 108 bp and/or amplifications
and insertions in the three variants. Some of these are
described in Table I and in the following text.

The major truncation in TRU occurs at a second Eco RI
site 364 bp and 357 bp upstream from the Eco RI site at the
3' end of RU and EXT, respectively. When TRU is compared to
RU and EXT, different base changes at a common site have
occurred to produce the additional Eco RI site in TRU.
Specifically, there has been a C→G transversion and an A→G
transition from the two variants. These changes in the same
base have "corrected" the first base of the Eco RI site,
leading to a perfect second site in TRU which defines its 3'
end. In digests of cellular satellite, fragments of 1.69 kb
(9) and 0.32 and 0.37 kb have been seen. Thus we conclude
that the truncation is not due to cloning accidents.

The differences in the patterns of the inserts cloned
into the Pst I site (Fig. 1, Part B) appear as variations in
lengths of fragments produced by digestion with Hinf I, by
the presence of an additional Hinf I site (number 2) in
pG1P18, and by the absence of a Hinf I site (number 3) from
pG1P27. They are also due to the presence of the extra 140
bp fragment in pG1P4, P12, and P18 described in the legend
to Figure 1.

General characteristics of conserved regions: Copies
of cellular satellite are arranged in tandem (15). Therefore, the largest block of highly conserved sequences encom-
passes the single Eco RI site found in most repeat units of
the satellite (Fig. 3). This conserved region includes
parts of four Hinf I fragments (fragments 1 and 7, the first
231 bp of fragment 2, and the last 91 bp of fragment 6).
There is 96% homology between RU and EXT over the 594 bp
domain as well as between TRU and RU and EXT over Hinf I
fragment 1 and the first 231 bp of fragment 2. Other
domains approximately 60 to 590 bp of high homology
(>84%) are interspersed among the divergent regions. Such
levels of homology are considerably higher than those
described as "conserved" that occur upstream from the sites
of initiation of transcription of a series of immunoglobulin
V_H genes (4).

General characteristics of variable regions: Regions of marked sequence divergence are common to RU, EXT, and TRU (Table I and Fig. 3). Three of these divergent regions are characterized by simple sequences in all three of the satellite variants. Each region has a specific repeating oligonucleotide theme which has degenerated to varying degrees in the three variants. The characteristics of the sequences of RU, EXT, and TRU are described with reference to one strand from their 5' to 3' ends as depicted in the Figures.

In fragment 2 produced by digestion with Hinf I, one variable sequence domain is a purine-rich tract, while in Hinf I fragment 4, the region of divergence is pyrimidine-rich. The last third of Hinf I fragment 6 is exceptionally rich in pu/py.

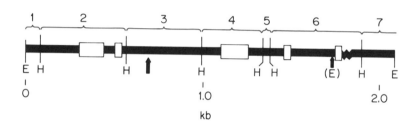

Figure 3. Diagram of RU: Regions conserved in TRU and EXT: ■■■; regions that vary in TRU and EXT: ▭; alternating purines and pyrimidines: W . Hinf I sites indicated (H and vertical lines); Hinf I fragments are numbered from 1 to 7 from Eco RI site at 5' end. Segment that undergoes fivefold amplification in EXT begins at arrow in Hinf I fragment 3 and continues downstream for 127 bp in RU and 130 bp in TRU. Second Eco RI site in TRU indicated by arrow and (E) in Hinf fragment 6. Details of simple sequences and variations present in each diverged region are given in the text and Tables I and II.

Most, but not all, simple sequence regions in RU, EXT and TRU -- even homopolymer tracts -- are sites of sequence divergence (Table II). One short C tract, C8 in Hinf I fragment 5, is not only conserved in all three variants, it does not abut any changes in sequence. Another C7 tract in

Hinf I fragment 4 is conserved in RU and TRU but has undergone a single C→T transition in EXT. However, some of the longer C (or G) tracts are associated with greater divergences (Tables I and II).

Comparison of regions of divergence in Hinf I fragment 2 of RU, EXT, and TRU: The repeating oligonucleotide theme in this domain is the tetramer AGGG (a. in Table I). In all three variants, the theme is preceded by GGGG. TRU has 10 tandem copies of AGGG while RU has seven invariant copies scattered among six others that have single base changes. The purine-rich domain ends in G tracts of 20 and 17 residues in RU and TRU respectively. In EXT, following the GGGG and one AGGGA, the rest of the purine-rich region is replaced by a tract of G18, a homocopolymer of (AG)17, a 36 bp purine-rich sequence, and a tract of G23. Following the purine-rich region of high divergence, there is a region of high homology. The homologous region in RU is interrupted by a pyrimidine-rich region based on a repeating (CA) theme and terminated by a C23 tract (b. in Table I). Despite deletions of 26 and 14 bp in EXT and TRU respectively, all three variants maintain cytosine-rich sequences in this region.

TABLE II

Divergences in Tracts of C Residues

RU	EXT	TRU
C_7	CTC_5	+
C_8	+	+
C_{10}	+	$(C_4A)_2$
C_{13}	$\overset{*}{+}$	$C_{11}ACC$
C_{22}	C_{21}	C_{19}
C_{23}	$C(AC_4)_2$	$C_{17}TCTCC$

Compared to RU:
+ = no divergence within tract.
* = divergence adjacent to tract.

<u>Comparison of Hinf I fragment 3 of RU, EXT, and TRU</u>:
Although the sequence is conserved in all three variants,
divergence in Hinf I fragment 3 takes the form of an amplification. The repeating oligonucleotide theme in this fragment is AAG/C (Table I). The sequence is highly conserved
between RU and TRU (95% homology) but less so between EXT
and either of the other two variants (83%). It is in this
region that EXT contains 6 tandem copies of a 0.142 kb segment of DNA due to a fivefold amplification. Just upstream
from the amplified region, EXT has suffered a deletion of 24
bp that are present in both RU and TRU. The homology
between RU, EXT, and TRU is greater than 90% for approximately 100 bases upstream and several hundred bases downstream from the amplified sequence. The amplified region is
bordered by a short inverted repeat as are some hotspots for
mutations in other DNAs (3, 5). The upstream arm of the
inverted repeat is missing from RU and TRU (2).

<u>Comparison of regions of divergence in Hinf I fragment
4 of RU, EXT, and TRU</u>: The theme in this domain is the trimer (CCT) (Table I). RU has 15 and TRU four tandem copies
of the trimer, while EXT suffers a long deletion (108 bp)
which leaves only three CCT's scattered in a pyrimidine-rich
region. Similarly, a tract of nine CCT's that is present in
the flanking region upstream from two cloned immunoglobulin
V_H genes, is deleted from three others (Cohen et al., 1982).

Upstream from the perfect CCT's in RU there is a
pyrimidine-rich region of 83 bp based on the CCT theme. All
three variants align at a C tract: RU = C22; EXT = C21; TRU
= C19. Following the C tract, there is a series of tandemly
repeated pentamers (CGCAC). In RU, five perfect tandem
CGCAC's are followed by four pentamers that contain one or
two changes. EXT and TRU have sequences that are highly
homologous to the RU pentamer region except that only three
tandem copies of the pentamer are present in EXT and two
additional copies with a single base change are present in
TRU.

This CCT-rich domain is contained in a 464 bp Taq I
fragment in RU; it has been subcloned into the Cla I site of
pBR322. Of six transformants recovered, only one hybrid
plasmid (pGlT464,1) had the fragment inserted in one direction; five others (pGlT464,2-6) had it inserted in the
other. Although hybrid plasmids with the fragment inserted
in either orientation have an S1 sensitive site in the same

position in the insert, the topological properties of
pG1T464,1 differ from those of the plasmids with the opposite orientation. Native pG1T464,1 plasmids as isolated
from E. coli HB101 or following relaxation by topoisomerase
I assume an unusually wide range of negative superhelical
densities. Digestions with restriction enzymes that nibble
away sequences flanking the divergent domain, ligation to
pBR322, and subsequent cloning of those plasmids would verify or disprove that the topological determinants are within
the divergent region.

Comparison of regions of divergence in Hinf I fragment
6 of RU, EXT, and TRU: The homology among the three variants is closer here than in the other regions that foster
sequence divergence (Table I). RU contains a C13 tract that
is associated with sequence divergences in TRU and EXT.
Specifically, the C13 tract in RU lies in a pyrimidine-rich
domain that is longer by an additional 11 pyrimidines in EXT
and a single inserted A residue in TRU. The pyrimidine
tract displays marked S1 nuclease sensitivity in supercoiled
hybrid plasmids with all three variants.

The extra Eco RI site in TRU has removed the terminal
residues from that variant (360 ± 4 bp; 167 ± 4 bp from Hinf
6 and all of Hinf 7). In RU and EXT, this region is unusually rich in alternating purines and pyrimidines. A tract
of 15 residues in perfect alternating register sits in a bed
of other pu/py where the alternation is interrupted by an
occasional base change. This pu/py-rich region extends for
a total of 53 bp in RU. In EXT, the region hosts a 12 bp
deletion. Preliminary data on negatively supercoiled
plasmids as isolated from E. coli HB101 indicate S1 sensitivity near the 15 bp region of uninterrupted pu/py; we suggest that this domain has adopted a Z conformation, leading
to single-stranded junctions between it and B-form DNA
(11,12).

SUMMARY

In RU, EXT, and TRU, there are simple repeated
sequences corresponding to sites of sequence divergence.
Also implicated in the pattern of divergence among the three
variants are exceptionally long tracts of C or G residues at
which homologies may be aligned. In general, divergences

take the form of base changes within repeating oligonucleotides, or insertions or deletions of a given repeating oligonucleotide unit. In addition, one variant (EXT) has
undergone a fivefold amplification of an internal 0.142 bp
sequence. The amplified sequence is present once in the
other two variants. Finally, in a region enriched in pu/py,
there is a 12 bp deletion in EXT relative to RU; the entire
region is lost from TRU due to the truncation. S_1 sensitivity in this region portends the presence of B-Z junctions.

Digestion of cellular satellite and six other cloned
variants with restriction enzymes also indicates many variations. Despite all the variations detailed here, there has
been evolutionary conservation of the satellite, or related
sequences, over time (7).

ACKNOWLEDGMENTS

We thank M. Spann for technical assistance and D. Million for computer advice. Supported by NSF grant PCM 78-
23373 (to D.M.S.), by seed money from Oak Ridge National
Laboratory, and by the Office of Health and Environmental
Research, U.S. Department of Energy under contract W-7405-
eng-26 with Union Carbide Corporation.

REFERENCES

1. Beattie, W. G. and Skinner, D. M. (1972) Biochim.
Biophys. Acta 281, 169
2. Bonnewell, V., Fowler, R. F. and Skinner, D. M. (1983)
Science In press.
3. Calos, M. P., Galas, D. and Miller, J. H. (1978) J.
Molec. Biol. 126, 865.
4. Cohen, J. B., Effron, K., Rechavi, G., Ben-Neriah, Y.,
Zakut, R., and Givol, D. (1982) Nucleic Acids
Res. 10, 3353.
5. Edlund, R. and Normark, S. (1981) Nature 292, 269.
6. Fowler, R. F. and Skinner, D. M. In preparation.
7. Graham, D. E. and Skinner, D. M. (1973) Chromosoma 40,
135.
8. Gray, D. M. and Skinner, D. M. (1974) Biopolymers 13,
843.

9. LaMarca, M.E., Allison, D. A., and Skinner, D. M. (1981) J. Biol. Chem. 256, 6475.
10. Maxam, A. and Gilbert, W. (1980) Methods Enzymol. 65, 499.
11. Singleton, C. K., Klysik, J., Stirdivant, S. M., and Wells, R. D. (1982) Nature 299, 312.
12. Singleton, C. K., Klysik, J. and Wells, R. D. (1983) Proc. Natl. Acad. Sci. 80, 2247.
13. Skinner, D. M. (1967) Proc. Natl. Acad. Sci. 58, 103.
14. Skinner, D. M., Bonnewell, V. and Fowler, R. F. (1982) Cold Spring Harbor Symp. Quant. Biol. 47, 1151.
15. Skinner, D. M. and Kerr, M. S. (1971) Biochemistry 10, 1864.
16. Smith, H. O. and Birnstiel, M. (1976) Nucleic Acids Res. 3, 2387.

Index